Fundamentals of Stream Processing

Stream processing is a distributed computing paradigm that supports the gathering, processing, and analysis of high-volume, heterogeneous, continuous data streams to extract insights and actionable results in real time. This comprehensive, hands-on guide, combining the fundamental building blocks and emerging research in stream processing is ideal for application designers, system builders, analytic developers, as well as for students and researchers in the field. This book introduces the key components of the stream processing computing paradigm, including the distributed system infrastructure, the programming model, design patterns, and streaming analytics. The explanation of the underlying theoretical principles, illustrative examples, and implementations using the IBM InfoSphere Streams SPL language and real-world case studies provide students and practitioners with a comprehensive understanding of stream processing applications and the middleware that supports them.

Henrique C. M. Andrade is a vice president at JP Morgan and an adjunct associate professor in the Electrical Engineering Department at Columbia University. Along with Dr. Gedik, he is the co-inventor of the SPADE and the SPL stream processing languages. He has published over 50 peer-reviewed articles and is the co-recipient of the ACM SoftVis 2009, IEEE DSN 2011, and ACM DEBS 2011 best paper awards.

Buğra Gedik is in the faculty of the Computer Engineering Department, Bilkent University, Turkey. He is the co-inventor of the SPADE and the SPL stream processing languages. He has published over 50 peer-reviewed articles and is the co-recipient of the IEEE ICDCS 2003, IEEE DSN 2011, ACM DEBS 2011 and 2012, and IEEE ICWS 2013 best paper awards. He has been an Associate Editor for the *IEEE Transactions on Services Computing*. He has filed over 30 patents. He was named an IBM Master Inventor and is the recipient of an IBM Corporate Award.

Deepak S. Turaga is the manager of the Exploratory Stream Analytics department at the IBM T. J. Watson Research Center in Yorktown Heights and an adjunct associate professor in the Electrical Engineering Department at Columbia University. He has published over 75 peer reviewed articles, and has received the 2006 IEEE TCSVT best paper, and 2008 IEEE ICASSP best student paper awards. He has been an Associate Editor for the *IEEE Transactions CSVT* as well as *IEEE Transactions Multimedia*.

"This is the first comprehensive text on stream processing, covering details of stream analytic algorithms, programming language and application design, and finally systems issues. The use of several illustrative examples and real-world scenarios, coupled with advanced research topics, makes it very well suited for undergraduate and graduate classes."

Shih-Fu Chang, Columbia University

"In a world flooded with information, yet hungry for wisdom, you would find this refreshing and thorough treatment of stream computing an excellent resource for building systems that need to analyze live data to derive actionable insights."

Hans-Arno Jacobsen, University of Toronto

"A comprehensive guide to the field of stream processing covering a wide spectrum of analytical patterns against a specialized architecture for continuous processing. This reference will prove invaluable to those engaging in the fascinating field of continuous analysis. I wish it had been written when I started in this field!"

George Long, Senior System Architect

"This book is an excellent guide for anyone involved with stream processing or data-in-motion analytics in general and is a must-read for those using the InfoSphere Streams platform."

Jim Sharpe, President of Sharpe Engineering Inc.

"This book provides a very timely introduction to stream processing for engineers, students, and researchers. With the advent of Big Data, there is pressing need for real-time systems, algorithms, and languages for distributed streaming analysis. This book provides a comprehensive overview of the topic and is great for course work and also as a practitioner guide."

Mihaela van der Schaar, University of California, Los Angeles

"This is a first-of-its-kind book that takes a holistic approach to introduce stream processing – a technology that can help overcome the data deluge. The authors guide you through various system-level and analytical techniques for harnessing data-in-motion, using a clear exposition, supplemented with real-world scenarios. You will find this book an invaluable companion, whether you are an application developer, system builder, or an analytical expert."

Philip S. Yu, University of Illinois at Chicago

Fundamentals of Stream Processing

Application Design, Systems, and Analytics

HENRIQUE C. M. ANDRADE
JP Morgan, New York

BUĞRA GEDIK
Bilkent University, Turkey

DEEPAK S. TURAGA
IBM Thomas J. Watson Research Center, New York

CAMBRIDGE
UNIVERSITY PRESS

CAMBRIDGE
UNIVERSITY PRESS

University Printing House, Cambridge CB2 8BS, United Kingdom

Cambridge University Press is part of the University of Cambridge.

It furthers the University's mission by disseminating knowledge in the pursuit of
education, learning and research at the highest international levels of excellence.

www.cambridge.org
Information on this title: www.cambridge.org/9781107015548

© Cambridge University Press 2014

First published 2014

A catalogue record for this publication is available from the British Library

ISBN 978-1-107-01554-8 Hardback

Henrique dedicates this book to his parents, Gercil and Maria José, and to Kristen.

Buğra dedicates this book to his father Yusuf and to his mother Meral, and to all his teachers and mentors through life.

Deepak dedicates this book to his parents Sudha and Ravi, his wife Swapna, and daughter Mythri.

Contents

Preface

Stream processing is a paradigm built to support natural and intuitive ways of designing, expressing, and implementing *continuous* online high-speed data processing. If we look at systems that manage the critical infrastructure that makes modern life possible, each of their components must be able to *sense* what is happening externally, by processing continuous inputs, and to *respond* by continuously producing results and actions. This pattern is very intuitive and is not very dissimilar from how the human body works, constantly sensing and responding to external stimuli. For this reason, stream processing is a natural way to analyze information as well as to interconnect the different components that make such processing fast and scalable.

We wrote this book as a comprehensive reference for students, developers, and researchers to allow them to design and implement their applications using the stream processing paradigm. In many domains, employing this paradigm yields results that better match the needs of certain types of applications, primarily along three dimensions.

First, many applications naturally adhere to a sense-and-respond pattern. Hence, engineering these types of applications is simpler, as both the programming model and the supporting stream processing systems provide abstractions and constructs that match the needs associated with continuously sensing, processing, predicting, and reacting.

Second, the stream processing paradigm naturally supports extensibility and scalability requirements. This allows stream processing applications to better cope with high data volumes, handle fluctuations in the workload and resources, and also readjust to time-varying data and processing characteristics.

Third, stream processing supports the use of new algorithmic and analytical techniques for *online* mining of both structured data (such as relational database-style records) as well as unstructured data (such as audio, video, text, and image). This breaks the cycle of storing the incoming data first to analyze it later, and makes it possible to considerably shorten the lag between sensing and responding.

After more than a decade of research in this space, stream processing has had a prolific and successful history in academic and industrial settings. Several advances in data analysis and management, signal processing, data mining, optimization theory, as well as in distributed systems technology, have provided a strong foundation for the development of research and commercial stream processing systems. In essence, stream processing is no longer an emerging paradigm, it is now ready for prime time.

The stream processing paradigm can now be harnessed in at least two ways. First, it can be used to transition existing legacy applications into true streaming implementations, making them more flexible, scalable, and adaptive. Second, stream processing can also be used to implement new analytics-intensive, high-performance, and innovative applications that could not be practically engineered earlier. Indeed, as will be seen in this book, stream processing applications can now be elegantly designed to be adaptive and autonomic, as well as self-evolving and able to continuously make use of newly learned knowledge.

Considering all of these aspects, this book is designed to provide a comprehensive foundation on stream processing techniques and on the skills necessary to design and develop stream processing applications. The book is divided into five major parts.

In Part I, we start with a discussion on the trends that led to development of the stream processing paradigm, providing also an overview of the initial academic efforts on analytical techniques, and on the engineering of some of the early stream processing system prototypes.

In Part II, we focus on application development. We describe core concepts of stream processing application development and illustrate them using the SPL language. SPL is the dataflow programming language provided by InfoSphere Streams, a commercial distributed stream processing system.

In Part III, we shift our attention to the architecture of stream processing systems. We first describe a conceptual middleware software architecture and its required services to support efficient, scalable, and fault-tolerant stream processing applications. We then illustrate these concepts with the architectural organization of InfoSphere Streams, shedding light on its internal components, and on the application runtime environment it exposes to a developer.

In Part IV, we build on the foundation provided in the earlier two parts to discuss how to best structure and design a stream processing application. The focus in this part of the book is on design patterns and principles common to stream processing applications, as well as on the algorithms used to implement online analytics.

In Part V, we describe a few case studies, detailing the end-to-end process of designing, implementing, and refining stream processing applications. This part brings together all of the information distilled in the earlier parts of the book. The case studies include real-world applications and showcase typical design decisions that must be made by developers.

We have designed this book to be used for undergraduate- as well as graduate-level courses on stream processing. The book's content is also structured so that application developers and system analysts can quickly develop the skills needed to make use of the stream processing paradigm.

While there are many ways to structure a semester-long course on stream processing, we recommend the following breakdown:

- For an undergraduate course, we believe that substantial focus should be devoted to the algorithmic and application-building aspects of stream processing. We believe that the majority of the time should be spent in Part II, where the focus should be on teaching the SPL programming language. This training will provide a solid foundation for tackling Part IV, where the algorithmic and analytical techniques are discussed. This hands-on portion of the class should be concluded with the contents of Part V, to discuss case studies that show how a complete application might be designed. Information from Part III, particularly from Chapter 8, can be used as needed to provide basic system administration knowledge on managing the Streams system. We stress that a stream processing undergraduate-level course must be hands-on, so that students can pick up important technical skills along the way. At the end of some chapters, we suggest exercises that can be used to solidify a working knowledge of the SPL language and the Streams platform. Finally, we think the class should culminate with a medium size final project of the magnitude of the case studies described in Part V.

- For a graduate course, we believe that the emphasis should be on the theoretical foundations and research issues surrounding the algorithmic, analytical, software engineering, and distributed processing architectural foundations of stream processing. Nevertheless, a semester-long course should also provide a solid programming foundation and an understanding of the practical aspects of building stream processing applications. Hence, our suggestion is to follow a similar organization as for an undergraduate course on a compressed time scale, but augmented with selected readings from the bibliography included at the end of each chapter. For this extra reading, we suggest focusing primarily on a subset of the foundational papers listed in Chapters 1, 2, 10, and 11. Despite its maturity, stream processing is still a very fertile area of research. We offer suggestions of possible research topics and open problems in Section 13.2. These areas can be used for individual short-term research projects, as well as for more advanced studies leading to a thesis. One final suggestion we can offer is to make use of the case studies, discussed in Part V, as the motivation for the course's final research projects. These projects can tackle one or more of the supporting analytical- or distributed system-related parts of stream processing, particularly where the state-of-the-art can be further advanced.

We believe that this book is self-contained and no specific formal background is necessary. Yet a reader might benefit from prior knowledge of a modern programming language such as Java and C++, as well as experience with scripting languages such as Perl and Python. Likewise, previous experience with relational databases, data mining platforms, optimization theory, and signal processing can also be directly leveraged. Content that is complementary to this book covering several of the practical aspects of using Streams, including its set of documentation manuals and the Streams' IBM RedBook, are linked from www.thestreamprocessingbook.info/ibm/streams-infocenter and www.thestreamprocessingbook.info/ibm/streams-redbook, respectively.

We think that the best way to learn how to use a new technology is by trying it out and we include several code excerpts and examples of how to use InfoSphere Streams, to illustrate the fundamental concepts appropriately. For readers interested in obtaining an InfoSphere Streams license, commercial users can contact IBM directly as well as any of its authorized resellers. IBM also provides time-limited trial licenses for InfoSphere Streams. Additional information and specific conditions on the trial program can be found at www.thestreamprocessingbook.info/ibm/streams-main. IBM also maintains a program that enables academic users to obtain a license free of charge. This type of license can be used in a teaching environment. Additional information and specific conditions on the IBM academic program can be found at www.thestreamprocessingbook.info/ibm/academic-initiative.

As we mentioned, InfoSphere Streams and its SPL language are used throughout this book as examples of a stream processing system and programming language. In this way, we can provide a conceptual overview, coupled with practical foundations and code examples, application design challenges, and finally system administration issues. We believe that the abstractions, concepts, and examples included in the book are general enough to be used with a different stream processing system, both in a teaching or commercial setting.

As a commercial product, Streams is evolving and, periodically, new versions will become available. In this book, we have made an effort to provide working code and other usage examples consistent with version 3.0, the latest available version as of November 2012.

We hope that the reader will find this book as exciting to read as it was to write. We have attempted to balance the content such that it is useful both to readers who are attempting to familiarize themselves with the stream processing paradigm, and for advanced readers who intend to develop new stream processing applications, systems, and algorithms.

Finally, we welcome feedback on this book as well as accounts of experiences using stream processing in academic and industrial settings. The authors can be contacted through the website that accompanies this book at www.thestreamprocessingbook.info, where readers will also find a code repository with example applications and code excerpts (www.thestream processingbook.info/apps) as well as this book's errata (www.thestream processingbook.info/errata).

Foreword

Humans are deeply curious and expend boundless energy and thought in sensing and interpreting that which surrounds them. Over time, direct perception through the five physical senses was extended by the creation of ingenious instrumentation designed to magnify and amplify weak signals, bringing what was beyond the visible into focus. Telescopes and light microscopy revealed natural phenomena that enabled richer and more sophisticated theories and understanding.

In recent decades industrialization has filled the world with machines and complex systems that manufacture, transport, track and deliver, communicate and mediate financial and social transactions, entertain and educate, treat and repair, and perform thousands of other tasks. As was true with the natural world, human curiosity seeks to understand the operation of these machines and systems and their interactions, but now with the added urgency to understand how and how well systems are operating, and often why they are not working as intended. Direct perception is no longer effective, nor is observation through the mere amplification of our five senses. Specialized sensors capture phenomena such as vibration, frequency, complex movements, or human-generated data and messages produced by the machines and systems as a side-effect of their operation, and so perception must now be through computer-aided interpretation of the digital signals.

Up until recently, no systematic approach existed for the creation of digital signal interpretation required to engineer this new class of perception mechanisms. IBM has long recognized this, and early in 2004 initiated a research program to create such a new approach from the ground up. As the lead for this program, I assembled a multi-disciplinary team of experts in distributed systems, mathematics, programming languages, machine learning and data mining, and computer science theory, and, over a five-year period, we (~60 researchers and engineers) developed *System S* – the research precursor to the IBM InfoSphere Streams product described in this volume, to illustrate stream processing concepts and techniques. During this period, the stream processing model and its underlying system structures evolved through feedback and application of the technology in a wide variety of real-world contexts.

The authors of this book were central members of this research and development effort, and have been closely involved in all aspects of the program – from conceptualization of the objectives, to the design of the architecture and the programming model, to engineering and implementing the system, and finally designing analytic applications and deploying them in real-world settings. Each of the three authors focused their work on a core area: the weaving of analytics into sophisticated applications, the language in which applications are described, and the system and runtime that support execution of the applications. This provided the authors with a complete and unique insider's perspective on both the fundamentals of this new model and realizations of the model in practical implementations and deployments.

This magnificent volume has captured this knowledge for students, researchers, and practitioners. The book provides an in-depth introduction to the stream processing paradigm, its programming model, its distributed runtime, and its analytic applications that will enable readers to use these techniques effectively in various settings for complex environmental monitoring and control applications. The book also includes several sample implementations, algorithms, and design principles, along with real-world use cases to provide hands-on training to practitioners. Finally, the book provides advanced readers and researchers with the necessary fundamentals to enable the design and extension of the stream processing paradigm to solve future problems.

All of us involved with the creation of this technology are convinced that stream processing will become a permanent element in the quest to create ever more sophisticated instrumentation to better understand our world, the machines and systems that serve us, and the interaction among them. The impact will be felt in virtually all modern industrial sectors and its use will lead to a safer, more efficient, and more productive society.

Nagui Halim
IBM Fellow
IBM T. J. Watson Research Center
Yorktown Heights, NY
United States

Acknowledgements

Henrique, Buğra, and Deepak were colleagues at IBM Research for several years, and were part of the team that designed and implemented InfoSphere Streams. This team included researchers, architects, designers, project managers, programmers, and testers under the direction of Nagui Halim. The techniques and lessons described here are the product of numerous discussions and refinements with this much larger team. Therefore, we acknowledge the collaboration of colleagues from multiple IBM sites, including Rochester (MN), Toronto (Canada), Silicon Valley (CA), Raleigh (NC), the China Research Lab, and from the Thomas J. Watson Research Center (NY). This book is also a tribute to our talented IBM colleagues. Thank you!

It really takes a village to build a system such as InfoSphere Streams, as we ourselves learned in the process. So it came as no surprise that the same applies when writing a book. We owe a debt of gratitude to the expertise of our colleagues who also helped reviewing an early version of this manuscript. We are particularly thankful to these early "settlers" who braved the elements to provide us with many helpful suggestions and corrections: Tarık Arıcı (İstanbul Şehir Üniversitesi), John Cox (US Government), Renato A. Ferreira (Universidade Federal de Minas Gerais), Andy Frenkiel (IBM Research), Martin Hirzel (IBM Research), Gabriela Jacques da Silva (IBM Research), Paul Jones (HM Government), Rohit Khandekar (Knight Capital Group), Senthil Nathan (IBM Research), Scott Schneider (IBM Research), Robert Soulé (Cornell University), William Szewczyk (US Government), Rohit Wagle (IBM Research), Brian Williams (IBM Software Services Federal), and Kun-Lung Wu (IBM Research).

A special thanks is due to our colleague Wim De Pauw (IBM Research) who graciously produced the image we use on this book's cover, a depiction of a stream processing application.

Acronyms

A/D Analog-to-Digital
AAS Authentication and Authorization Service
ACID Atomicity, Consistency, Isolation, and Durability
ACL Access Control List
ADL Application Description Language
ADT Abstract Data Type
AES Advanced Encryption Standard
AIS Agrawal, Imielinski, Swami
AMS Alon, Matias, and Szegedy
ANSI American National Standards Institute
API Application Programming Interface
ARIMA Auto Regressive Integrated Moving Average
ASCII American Standard Code for Information Interchange
ATE Automated Test Equipment
BIH Beth Israel Hospital
BI Business Intelligence
BJKST Bar-Yossef, Jayram, Kumar, Sivakumar, Trevisan
CART Classification And Regression Tree
CC Command and Control
CDR Call Detail Record
CEP Complex Event Processing
CIS Clinical Information System
CKRM Class-based Kernel Resource Management
CORBA Common Object Request Broker Architecture
CPU Central Processing Unit
CQL Continuous Query Language
CQ Continuous Query
CSV Comma-Separated Value
CVFDT Concept-adapting Very Fast Decision Tree learner
DAG Directed Acyclic Graph
DBMS Data Base Management System
DBSCAN Density-Based Spatial Clustering of Applications with Noise
DCOM Distributed Component Object Model
DCT Discrete Cosine Transform

DDL Data Definition Language
DDoS Distributed Denial of Service
DFT Discrete Fourier Transform
DHT Distributed Hash Table
DMG Data Mining Group
DML Data Manipulation Language
DNS Domain Name System
DOM Document Object Model
DoS Denial of Service
DPI Deep Packet Inspection
DSL Domain-Specific Language
DSO Dynamically Shared Object
DSS Decision Support System
DTD Document Type Definition
ECA Event-Condition-Action
ECG Electrocardiogram
EDA Electronic Design Automation
EEG Electroencephalogram
EKG Elektrokardiogramm
EM Expectation Maximization
EMS Emergency Medical Services
EPL Event Processing Language
EPN Event Processing Network
EPFL École Polytechnique Fédérale de Lausanne
ER Entity Relationship
ESP Event Stream Processor
ETL Extract/Transform/Load
FDC Fault Detection and Classification
FFT Fast Fourier Transform
FPGA Field-Programmable Gate Array
FSF Free Software Foundation
FTP File Transfer Protocol
Gbps gigabits per second
GMM Gaussian Mixture Model
GPS Global Positioning System
GPU Graphics Processing Unit
GRAM Globus Resource Allocation Manager
GSM Global System for Mobile Communications
GSN Global Sensor Networks
GSQL Gigascope SQL
GUI Graphical User Interface
HA High Availability
HC Host Controller
HDFS Hadoop Distributed File System

HMM Hidden Markov Model
HTML HyperText Markup Language
HTTP HyperText Transfer Protocol
HTTPS HyperText Transfer Protocol Secure
Hz hertz
I/O Input/Output
ICU Intensive Care Unit
IDDQ Direct Drain Quiescent Current
IDE Integrated Development Environment
IDL Interface Definition Language
IDS Intrusion Detection System
IEEE Institute of Electrical and Electronics Engineers
IOR Interoperable Object Reference
IP Internet Protocol
IPC Inter-Process Communication
IT Information Technology
JDBC Java Data Base Connectivity
JMS Java Message Service
JNI Java Native Interface
JSON JavaScript Object Notation
JVM Java Virtual Machine
kbps kilobits per second
KHz kilohertz
KLT Karhunen–Loève Transform
KNN k-Nearest Neighbors
LAN Local Area Network
LBG Linde–Buzo–Gray
LDAP Lightweight Directory Access Protocol
LDA Linear Discriminant Analysis
LFUP Least Frequently Updated Partition
LLM Low Latency Messaging
LF Line Fit
LPC Linear Predictive Coding
LRUP Least Recently Updated Partition
MIMD Multiple Instruction Multiple Data
MIT Massachusetts Institute of Technology
MLA Manifold Learning Algorithm
MLE Maximum Likelihood Estimation
MLP Multi-Layer Perception
MPI Message Passing Interface
MPQ Moment Preserving Quantization
MP megapixel
ms millisecond
MTDF Mean Time to Detect Failures

MUSCLES MUlti-SequenCe LEast Squares
mV millivolt
NB Naïve Bayes
NFS Network File System
NICU Neonatal Intensive Care Unit
NN Nearest Neighbors
NOAA National Oceanic and Atmospheric Administration
NP Non-deterministic Polynomial time
NPMR Non-Parametric Multiplicative Regression
NS Name Service
NYSE New York Stock Exchange
ODBC Open Data Base Connectivity
ODBMS Object Data Base Management System
OLAP Online Analytical Processing
OLTP Online Transaction Processing
OMG Object Management Group
OM Operations Monitoring
OO Object-Oriented
OP Oldest Partition
OS Operating System
PAM Pluggable Authentication Module
PCA Principal Component Analysis
PCR Parent–Child Relationship
PDF Probability Density Function
PDMS Patient Data Management System
PEC Processing Element Container
PE Processing Element
PIPES Public Infrastructure for Processing and Exploring Streams
PLY Performance Limited Yield
PMF Probability Mass Function
PMML Predictive Model Markup Language
PM Patient Monitoring
POJO Plain Old Java Object
PSRO Performance Sort Ring Oscillator
PVM Parallel Virtual Machine
QoS Quality of Service
RAD Rapid Application Development
RBF Radial Basis Function
RDF Random Decision Forest
RDMA Remote Direct Memory Access
RFID Radio-Frequency IDentification
ROC Receiver Operating Characteristic
RPC Remote Procedure Call
RSS Really Simple Syndication

RTP Real-time Transport Protocol
SAM Streams Application Manager
SAN Storage Area Network
SAX Simple API for XML
SCADA Supervisory Control and Data Acquisition
SCH Scheduler
SDE Semi-Definite Embedding
SGD Stochastic Gradient Descent
SIMD Single Instruction Multiple Data
SLR Single Logistic Regression
SMS Short Message System
SMTP Simple Mail Transfer Protocol
SOAP Simple Object Access Protocol
SOM Self-Organizing Map
SPA Stream Processing Application
SPADE Stream Processing Application Declarative Engine
SPC Semiconductor Process Control
SPIRIT Streaming Pattern dIscoveRy in multIple Timeseries
SPS Stream Processing System
SQL Structured Query Language
SQuAl Stream Query Algebra
SRM Streams Resource Manager
SSH Secure SHell
SSL Secure Sockets Layer
SVD Singular Value Decomposition
SVM Support Vector Machine
SWS Streams Web Server
TB terabytes
TCP Transmission Control Protocol
TEDS Transducer Electrical Data Sheet
TelegraphCQ Telegraph Continuous Queries
UCLA University of California, Los Angeles
UDP User Datagram Protocol
UML Unified Modeling Language
URI Uniform Resource Identifier
UX User Experience
VFDT Very Fast Decision Tree
VFML Very Fast Machine Learning
VLSI Very Large Scale Integration
VoIP Voice over IP
VWAP Volume Weighted Average Price
WAN Wide Area Network
WSDL Web Services Description Language
WTTW Who is Talking To Whom

XHTML eXtensible HyperText Markup Language
XML eXtensible Markup Language
XPATH XML Path Language
XSD XML Schema Definition
XSLT eXtensible Stylesheet Language Transformations
ZIP Zone Improvement Plan

Part I

Fundamentals

1 What brought us here?

1.1 Overview

The world has become information-driven, with many facets of business and government being fully automated and their systems being instrumented and interconnected. On the one hand, private and public organizations have been investing heavily in deploying sensors and infrastructure to collect readings from these sensors, on a continuous basis. On the other hand, the need to monitor and act on information from the sensors in the field to drive rapid decisions, to tweak production processes, to tweak logistics choices, and, ultimately, to better monitor and manage physical systems, is now fundamental to many organizations.

The emergence of stream processing was driven by increasingly stringent data management, processing, and analysis needs from business and scientific applications, coupled with the confluence of two major technological and scientific shifts: first, the advances in software and hardware technologies for database, data management, and distributed systems, and, second, the advances in supporting techniques in signal processing, statistics, data mining, and in optimization theory.

In Section 1.2, we will look more deeply into the data processing requirements that led to the design of stream processing systems and applications. In Section 1.3, we will trace the roots of the theoretical and engineering underpinnings that enabled these applications, as well as the middleware supporting them. While providing this historical perspective, we will illustrate how stream processing uses and extends these fundamental building blocks. We will close this chapter with Section 1.4 where we describe how all these different technologies align to form the backbone of stream processing software.

1.2 Towards continuous data processing: the requirements

In 2010, Walmart, a retailer, was reported to handle more than 1 million customer transactions every hour [1]. According to OCC, an equity derivatives clearing organization, average daily volumes of options contracts[1] by October 2012 were around 14 million [2]. The New York Stock Exchange (NYSE), the world's largest stock exchange

---

[1] An *option* is a derivative financial instrument specifying a contract for a future transaction on an asset at a reference price, for instance, the right to buy a share of IBM stock for $300 on February 1, 2016.

by market capitalization of its listed companies, traded more than 800 million shares on a typical day in October 2012 [3]. In other business domains, IBM in a comprehensive report on data management and processing challenges [4] pointed out that: (1) by the end of 2011 there were about 30 billion Radio-Frequency IDentification (RFID) tags in circulation, each one a potential data source; (2) T-Mobile, a phone company, processed more than 17 billion events, including phone calls, text messages, and data traffic over its networks; and (3) the avionics of a modern commercial jet outputs 10 terabytes (TB) of data, per engine, during every half hour of operation. These examples are just a few illustrations of the data processing challenges organizations must face today, where all of this data must be ingested, processed, and analyzed quickly, potentially as it is being continuously produced.

Indeed, as the world gets more connected and instrumented, there is a deluge of digital data coming from various software and hardware sensors in the form of continuous streams of data [5]. Examples can be found in several domains ranging from financial markets [6], multi-modal surveillance and law enforcement [7, 8], manufacturing [6], healthcare [9, 10], interconnected city infrastructure and traffic systems [11, 12], large-scale infrastructure monitoring [8], to scientific and environmental domains such as radio astronomy [13, 14] and water management [15, 16].

In all of these domains there is an increasing need to gather, process, and analyze live streams of data to extract insights in real-time and to detect emerging patterns and outliers. While the requirements are very clear, there are several challenges that must be overcome for this to be possible. We will discuss some of these challenges in the rest of this section, and provide a description of the techniques and software development strategies that can be used to address them in the rest of this book.

Processing large quantities of distributed data

The fundamental goal of stream processing is to process live data in a fully integrated fashion, providing real-time information and results to consumers and end-users, while monitoring and aggregating new information for supporting medium- and long-term decision-making.

The high volume of streaming data flowing from distributed sources often makes it impossible to use the storage- and Input/Output (I/O)-centric relational database and data warehousing model, where *all* data is first stored on disk and later retrieved for processing and analysis. Additionally, continuous monitoring across *large*, potentially multi-site applications that connect remote sensing and processing, needs to rely on a distributed computing infrastructure. The real-time monitoring, analysis, and control nature of these applications also implies the need to adapt to dynamic characteristics of their workloads and to instantaneous resource availability of the systems.

Addressing stringent latency and bandwidth constraints

Stream Processing Applications (SPAs) need to satisfy several performance requirements in terms of latency and throughput. Specifically, the data processing must keep up with data ingest rates, while providing high quality analytical results as quickly as possible.

Consider, for example, certain types of finance engineering applications. Increasingly, large amounts of market data must be processed with very low latencies to determine trading opportunities [17]. These requirements translate into performance constraints in terms of data decoding, demultiplexing, and delivery between different components in an application, where peak rates of millions of messages per second and sub-millisecond end-to-end latency are very common.

Processing heterogeneous data

The raw data streams analyzed by SPAs are often heterogeneous in format, content, rates, and noise levels, making generic data processing and analysis difficult without appropriate pre-processing steps. Data streams may also consist of unstructured data types such as audio, video, image, and text that cannot easily be handled using a traditional data management infrastructure. The fundamental issues here range from ingesting the data from a disparate collection of sensors with different protocols and data formats, to cleaning, normalization, and alignment so the raw data is adequately prepared for in-depth processing.

Applying complex analytics to streaming data

The types of analysis required by typical SPAs range from simple monitoring and pattern detection, to more complex *data exploration* tasks, where new information must be distilled from raw data in an automated fashion. Complex analysis is usually needed in applications that are part of sophisticated Decision Support Systems (DSSs).

The set of relevant data sources for stream data exploration applications is time-varying, typically large, and potentially unbounded in contrast with the finite computational cycles available. Thus, decisions such as which sources to ingest data from and what fraction of data per source to consider must be autonomic and driven by findings in the data, which are periodically fed back to the earlier data processing stages of an application.

Hence, the analytics employed by a data exploration application must be adaptive, to handle the dynamic and potentially unbounded set of available data streams. There is also a need for both *supervised* and *unsupervised analysis*. The constraint on the available computational resources requires using *approximation algorithms* [18] that explicitly trade off accuracy for complexity, making the appropriate choices given the instantaneous conditions of processing resources and workload.

These applications also often need to support *hypothesis-based analysis* [19], where the processing is dynamically changed based on the results being produced. In other words, based on validated (or discarded) hypotheses, different sources, different processing, and different analytic algorithms may need to be switched on and off as the application runs. Finally, these applications are also mostly "best-effort" in nature, attempting to do as best as possible to identify useful information with the resources that are available.

Traditionally, these analytic tasks have been implemented with data exploration and modeling tools such as spreadsheets and mathematical workbenches, to analyze medium-size stored datasets. Stream processing should be used to supplement traditional tools by providing continuous and early access to relevant information, such that decisions can be made more expeditiously. Distributed stream processing also enables these applications to be scaled up, allowing more analysis to be performed at wire speeds.

Providing long-term high availability
The long-running and continuous nature of SPAs requires the construction and maintenance of state information that may include analytical models, operating parameters, data summaries, and performance metric snapshots. Thus, it is critical that applications have mechanisms to maintain this internal state appropriately without letting it grow unboundedly, which can lead to performance degradation and potential failures. Clearly, these applications are uniquely prone to suffering from these problems due the continuous nature of the processing.

Additionally, there is a need for tolerance to failures in application components or data sources, as well as failures in the processing hardware and middleware infrastructure. For these reasons, applications must be designed with *fault tolerance* and resilience requirements in mind.

An important aspect in terms of fault tolerance is that different segments of a SPA may require different levels of reliability. For instance, tolerance to sporadic data loss is, in many cases, acceptable for certain parts of an application, as long as the amount of error can be bounded and the accuracy of the results can be properly assessed. Other application segments, however, cannot tolerate any failures as they may contain a critical persistent state that must survive even catastrophic failures. Hence, the application design needs to consider the use of *partial fault tolerance* techniques [20, 21] when appropriate.

Finally, because of the additional overhead imposed by fault tolerance, taking into account *where* and *how* fault-tolerant capabilities should be deployed can make a substantial impact on minimizing computational and capital costs.

Distilling and presenting actionable information
Enabling faster information analysis and discovery is only one part of designing and deploying a new SPA. Deciding what data to keep for further analysis, what results to present, how to rank these results, and, finally, how to present them in a *consumable* form, on a continuous basis, are equally important.

Tightening the loop from data collection to result generation creates additional challenges in presenting freshly computed information. Such challenges also come with additional opportunities to increase the interactivity and relevance of the information being displayed [22].

1.3   Stream processing foundations

The development of Stream Processing Systems (SPSs) was the result of the constant evolution over the past 50 years in the technology used to store, organize, and analyze the increasingly larger amounts of information that organizations must handle.

These trends have also enabled the development of modern relational databases and, later, data warehousing technologies, as well as object-oriented databases and column-oriented databases.

More recently, Google and Yahoo have made the *MapReduce* framework popular for processing certain types of large-scale data aggregation/summarization tasks. This development builds upon several data processing techniques proposed by the research community [23, 24, 25, 26]. Additionally, multiple technologies have been developed to provide historical overviews and data summarization over long periods of time [25, 27, 28, 29]. Some of these technologies have been commercialized as part of data warehousing products as well as Business Intelligence (BI) platforms. We discuss all of these technologies in more detail in Section 1.3.1.

The need for tackling larger processing workloads has led to the development of *distributed systems* with the capability of dividing up a workload across processors. This area has been a particularly fertile ground for hardware and software experimentation, dating back to the beginning of the modern computer era in the early 1950s. The field has produced a large variety of platforms that can be found commercially today, ranging from simple multicore processors, to clusters, to specialized acceleration hardware, and supercomputers. Despite the astounding hardware improvements seen over the last few decades, substantial software engineering challenges still exist today in terms of how to effectively write software for these platforms and, more importantly, to ensure that the software can evolve and benefit from architectural improvements. We discuss these topics in more detail in Section 1.3.2.

The need for large-scale data analysis has led to several advances in *signal processing* and *statistics*. Several signal processing techniques for pre-processing, cleaning, denoising, tracking, and forecasting for streaming data have been developed over the last few decades. Statistical techniques for data summarization through *descriptive statistics*, *data fitting*, *hypothesis testing*, and *outlier detection* have also been extensively investigated with new methods frequently emerging from researchers in these fields.

More recently, researchers in statistics and computer science have coined the term *data mining* as a way to refer to techniques concerned primarily with extracting patterns from data in the form of models represented as *association rules*, *classifier models*, *regression models*, and *cluster models*. Of particular interest are the advancements that have been made on continuous (or online) analytical methods, which allow the incremental extraction of the data characteristics and incremental updates to these models. We discuss these topics in more detail in Section 1.3.3.

Finally, all of these developments have been supported by advances in *optimization theory*. These techniques make it possible to obtain optimal or *good enough* solutions in many areas of data analysis and resource management. While the roots of optimization theory can be traced as far back as Gauss' *steepest descent method* [30], modern optimization methods such as *linear programming* and *quadratic programming* techniques have only been developed over the last few decades. We discuss these topics in more detail in Section 1.3.4.

1.3.1 Data management technologies

The most prevalent data management technology in use nowadays is the *relational* database. Data management applications require transaction management, resiliency, and access controls. All these requirements and associated problems have been progressively studied by the research community, and solutions have been implemented by open-source and commercially available database and other data management systems. The long history behind data management is deeply intertwined with the development of computer science.

In the late 1960s and early 1970s, commercial data management relied on mainly on Data Base Management Systems (DBMSs) implementing either the *network model* [31] or the *hierarchical model* [32]. In this context, a *model* describes how the information in the database system is organized internally.

The network model, defined originally by the Codasyl Data Base Task Group [33], represented information in terms of a *graph* in which objects (e.g., an employee record) are *nodes* and relationships (e.g., an employee working in a department is represented by a relationship between two objects: a specific employee object and a specific department object) are *arcs*. The network model was tightly coupled with the Cobol language [34], which was used for writing applications that queried and manipulated the data stored in a database. GE's ID database system [35], relied on this model and was released in 1964.

The hierarchical model structured information using two basic concepts: records (e.g., a structure with attributes representing an employee and another representing a department) and parent–child relationships (PCR) (e.g., a department/employee PCR). Note that where the hierarchical model structured data as a tree of records, with each record having one parent record and potentially many children, the network model allowed each record to have multiple parent and child records, organized as a graph structure. Hierarchical model-based database management systems first became available with IBM's IMS/VS in the late 1960s. Many of these systems are still around and can be programmed using a plethora of host languages, from Cobol, to Fortran [36] and PL/1 [37].

In the early 1970s, the term *relational model* was coined by Codd [38]. The relational model sits atop a theoretical framework referred to as relational algebra, where the data is represented by *relations*. Some of the seminal relational data processing, including the original implementation of its SQL query language, came from IBM's System R implementation [39, 40] as well as from other efforts that were taking place in academia [41].

The theory and technological underpinnings behind relational databases ushered an enormous amount of evolution in data management, due to the simplicity, elegance, and expressiveness embodied in the relational model.

Relational model: fundamental concepts
In relational databases, the data is primarily organized in *relations* or *tables*, which comprise a set of *tuples* that have the same *schema*, or set of attributes (e.g., a transaction identifier, a client identifier, a ticker symbol, for a hypothetical relation representing a stock market transaction) [42].

From a mathematical standpoint, a *relation* is any subset of the Cartesian product of one or more domains. A *tuple* is simply a member of a relation. The collection of relation schemas used to represent information is called a relational *database schema*, and *relationships* can be established between the different relations.

From this brief description it can be seen that relational databases rely on three fundamental components. First, the Entity Relationship (ER) model, which depicts data management schemas in the form of a graph diagram. One such diagram represents *entity* sets (e.g., the set of `employees`), entity *attributes* (e.g., an employee's `name`), and *relationships* (e.g., an employee works for a `department`, i.e., another entity).

Second, the relational algebra, which defines operators for manipulating relations. Arguably, the most important breakthrough of the relational algebra has been the conceptual and intuitive nature of these operators: the *union* $R \cup S$ of relations R and S results in the set of tuples that are in either relation; the *set difference* $R - S$ results in the set of tuples in R and not in S; the *Cartesian product* $R \times S$ results in the set of every tuple in R combined with every tuple in S, each resulting tuple including the attributes from both original tuples; the *projection* $\pi_{a_1, a_2, \dots, a_n}(R)$ is the operation whereby tuples from a relation R are stripped of some of the attributes or have their attributes rearranged, resulting in attributes $a_1, a_2, \dots, a_n$ being output; and, finally, the *selection* $\sigma_F(R)$ operation employs a formula F, comprising operands such as constants or attribute names combined with arithmetic comparison operators as well as logical operators, and computes the set of tuples resulting from applying the function F to each tuple that belongs to relation R. Numerous other operations can also be formulated in terms of these basic operators.

Third, the Structured Query Language (SQL), formally specified by an ISO/IEC standard [43], which is a relational database management system's main user interface. The SQL concepts *table*, *row*, and *column* map directly to counterparts in relational algebra, *relation*, *tuple*, and *attribute*, respectively. Likewise, all the relational algebra operators can be translated directly into SQL constructs. SQL also includes constructs for data and schema definition (effectively working as both a Data Definition Language (DDL) and as a Data Manipulation Language (DML)), as well as constructs for querying, updating, deleting, and carrying out data aggregation tasks [44].

Data warehousing

A data warehouse is a large-scale data repository and the associated system supporting it [31]. A warehouse is different from a regular database in the sense that it is primarily intended as the processing engine behind decision-support applications, where transactional support (also referred to as Online Transaction Processing (OLTP)) and *fine-grain* query capabilities are of lesser importance.

Furthermore, a data warehouse typically has a data acquisition component responsible for fetching and pre-processing the input data that it will eventually store. This batch processing nature leads to hourly, daily, and sometimes longer update cycles, which require drawing data from different database sources and cleaning, normalizing, and reformatting it. This mode of operation contrasts with regular databases where data updates are incremental, resulting from the query workload imposed by an application.

The types of applications supported by warehouses, namely Online Analytical Processing (OLAP), DSS, and data mining, operate on aggregated data. Hence, efficient bulk data retrieval, processing, cleaning, and cross-database integration, coupled with extensive aggregation and data presentation capabilities, are the hallmark of data warehousing technologies.

Also in contrast with relational databases, typical operations include *roll-ups* (i.e., summarizing data with a progressive amount of generalization, for example, to produce weekly sales report from daily sales data), *drill-downs* (i.e., breaking down a summarization with a decreased amount of generalization), and *slicing* and *dicing* (i.e., projecting out different data dimensions from a multi-dimensional, aggregation data product such as a *data cube*^{2}) [45].

While the basic data operations possible within a data warehouse can provide multiple perspectives on large quantities of data, they often lack capabilities for automated knowledge discovery. Hence, a natural evolution in data warehousing technology involves integrating *data mining* capabilities as part of the data warehouse toolset. Not surprisingly, products such as IBM InfoSphere Data Warehouse [46], IBM SPSS Modeler [47], Oracle Data Mining [48], Teradata Warehouse Miner [49], among others, provide these capabilities. Data mining is further discussed in Section 1.3.3.

Business intelligence frameworks
Building on data warehousing technologies, several BI middleware technologies have sprung up. Their primary focus is on building visualization *dashboards* for depicting up-to-date data and results. As is the case with stream processing, these technologies aim at providing actionable intelligence to decision makers, as well as predictive capabilities based on trends observed in historical data. Nevertheless, the technological focus in this case is less on data management and more on assembling and post-processing results by relying on other systems as data sources.

In many ways, modern business intelligence frameworks such as those provided by IBM (IBM Cognos Now! [50]), Oracle (Oracle Hyperion [51]), MicroStrategy [52], draw on basic ideas from data warehousing. In many cases, they integrate with data warehouses themselves. Most of the added benefits include providing (1) additional operators to post-process the data coming from warehouses, from data mining middleware, as well as from regular DBMSs, (2) continuous processing and exception handling and notification for addressing abnormal conditions arising from newly arrived data, and (3) toolkits to produce interactive data visualization interfaces organized in executive dashboards.

Emerging data management technologies
While relational databases are currently dominant in the marketplace, several technologies have been developed over the last two decades to address certain limitations in relational DBMSs or, simply, to complement them.

---

^{2} A data cube is a data structure designed for speeding up the analysis of data using multiple aggregation perspectives. It is usually organized as a multi-dimensional matrix [45].

A first example is *object-oriented databases*. Object Data Base Management Systems (ODBMSs) were conceived to address issues that arise from how data is represented in relational databases. Particularly, ODBMSs allow developers to define the structure of complex *objects* and provide the data manipulation *methods* to operate on them [31], much like what object-oriented (OO) languages allow developers to do. Indeed, ODBMSs provide mechanisms for managing a *type hierarchy*, as well as for representing *inheritance*, *polymorphism*, and other traditional abstractions common in these languages [53]. Hence, applications written in OO languages can, in most cases, operate directly on data stored in an ODBMS.

While ODBMSs have found their niche in application domains where such finer data representation capabilities are needed, these capabilities may impact performance and scalability adversely. In certain situations, the more expressive data model implies more complex optimization and data storage methods, yielding inferior performance when compared to relational databases.

A second example is *in-memory* and *read-only databases* [54, 55, 56, 57, 58], which also aim to speed up certain types of data management workloads. The basic premise is to provide similar query capabilities, and some of the same assurances provided by regular relational databases, relying primarily on main memory as opposed to secondary storage.

Implementations differ on how fault tolerance and durability are provided (e.g., via replication or by backing up to a regular database) as well as in the types of operations that can be carried out. For instance, no update operations can be carried out in read-only databases. In many cases, the reliance on main memory, while being constrained in storage to the amount of server memory, results in performance benefits with additional gains when the data is partially or totally read-only.

A third example is *column-oriented* DBMSs [59]. The increasing reliance on large data warehouses, mostly triggered by BI applications, has led to performance concerns when employing regular relational databases for tackling these workloads. Relational DBMSs were originally conceived to process OLTP workloads. With the advent of multi-dimensional aggregations and related operations, it has become clear that specialized data storage structures are essential for these new workloads.

A column-oriented DBMS is a database management system, which stores the contents of relational tables by column rather than by row. This data organization may provide substantial advantages for data warehouses where multi-dimensional aggregates, commonly found in OLAP workloads, are computed over attributes stored in large columns of similar data items [60]. For instance, when computing an aggregate over a large number of rows, such as the average transaction value for the set of customers who visited a particular grocery store over a time horizon.

Finally, the intersection of emerging data management frameworks and newer distributed system paradigms (discussed in more detail in the next section) has been a fertile area since the late 1990s. The widespread availability of clusters of (cheaper) workstations, and the type of workloads that must be processed by many of the companies operating web-based businesses, has led to the development of specialized middleware [61, 62]. One of them, MapReduce [63], which was proposed by Google, has been particularly successful, finding rapid adoption in corporate and research environments fueled by Yahoo's Hadoop [64] open-source implementation.

MapReduce is a programming model for processing large data sets, inspired by constructs found in functional programming languages [65], and building on a number of earlier distributed processing ideas [23, 24, 26, 66, 67]. Its main strength lies in the simplicity of its programming model. Application developers must specify a *map* and a *reduce* function. The *map* function processes key/value pairs and generates intermediate results, also in the form of key/value pairs. The *reduce* function merges all of the values from the earlier step associated with the intermediate key. Data retrieval and partitioning require additional functions (for partitioning and comparing data chunks), as well as handlers for reading the data and writing results. Existing MapReduce implementations such as Hadoop can retrieve the input data from a distributed file system or a database.

While MapReduce provides a powerful and simple abstraction, the scalability of MapReduce applications hinges on the mapping step being carried out on different chunks of data, running independently and simultaneously on different processors. The simplicity of the framework is a by-product of MapReduce implementations providing middleware services that carry out workload distribution, monitoring, and a degree of fault tolerance, ensuring that failed subtasks are restarted and that straggler tasks are also handled.

Initially, MapReduce gained popularity due to its support for parallelizing the computation of the *page rank* for webpages [68], as part of the data- and computation-intensive process of indexing the web. Since then, several data mining algorithms have been implemented using MapReduce [63] or similar distributed processing frameworks [27, 28]. As will be discussed in Chapter 10, constructing offline models, by employing a MapReduce step, to be used in conjunction with SPAs, can be a fruitful approach.

Human–computer interaction

Beyond the engineering and data management aspects we discussed, depicting results effectively is central to the design of data analysis and decision support SPAs.

Designing human–computer interfaces for these tasks requires addressing softer and, sometimes, intangible requirements. First, *parsimony* in what to present is important and usually achieved by, for example, ranking the most *relevant* information first. Second, *accuracy* augmented with supporting evidence is important for situations where an analyst needs to drill down through the summarized results. Finally, *intuitiveness* is key and usually achieved by providing visualizations that do not require technical computing skills, and are *consumable* by analysts and by decision makers alike.

The foundations supporting these requirements have long been established in the *User Experience* (UX) literature [69], which from the 1980s onwards has been prolific in studying human cognitive processes [70], new hardware for human–computer interaction [71], new user interfaces and methods of interaction [72, 73], as well as algorithms for data summarization and presentation [74, 75].

As discussed earlier, presenting information in an effective form is fundamental to BI middleware, which generally provides programmable visual widgets that can be integrated with data sources. Standalone toolkits such as Adobe Flex [76], JFreeChart [77], SpotFire [78], among others, also provide the barebones UX constructs for building visualizations.

Data management trends and stream processing

Modern data management technologies, including large-scale data warehousing and online transaction processing have scaled up to handle vast amounts of data and large numbers of transactions per second [79]. Many of these advances originate from the use of distributed processing and parallel programming techniques [80, 81]. Furthermore, the use of BI dashboards and visualization tools have made available an unprecedented amount of *actionable* information to decision makers.

Nevertheless, not even in-memory or read-only databases have the capability to ingest and process high data rate streams with complex analytics [82]. This observation was partially responsible for spawning the development of stream processing middleware and streaming analytics. Nevertheless, given the prevalence of these traditional data management systems, it is often the case that SPAs both complement and interact, with them.

As will be described in more detail in Parts II and IV of this book, when a developer designs an application within a larger existing environment, an important part of the design effort lies in identifying the *streaming* parts of the solution, as well as the synergistic integration points with relational databases, data warehouses, and BI front-ends. These components can act as command centers, as sources of data and context, as well as destination repositories for SPAs.

Finally, several of the data management issues discussed in this brief overview will be revisited in the context of stream processing. This is because simple extensions of the earlier data models, data manipulation techniques, as well as visualization and system infrastructure are not sufficient to meet the needs of stream processing. The advances that were required in the data model, algebra, and language issues will be described in Chapters 2–4, while system integration issues will be covered in Chapter 5.

1.3.2 Parallel and distributed systems

Early on in the development of computing technology, parallel programming became important, as several of the problems in physics, in modeling, and in national security, required intensive numeric computation. Starting with discussions in the late 1950s when Cocke and Slotnik [83] observed that many of those problems could be broken into subproblems, and these subproblems could then be solved simultaneously. They also observed that in certain cases the same problem could be solved for different data, organized in grids, independently and simultaneously. The types of parallel computation paradigms they identified are now referred to as Single Instruction Multiple Data (SIMD) and Multiple Instruction Multiple Data (MIMD), respectively.

Early developments
In the early 1960s, Burroughs, Honeywell, and IBM started manufacturing commercial computers that provided different degrees of parallel processing. On the software front, parallel programming techniques proved effective in domains ranging from linear algebra [84] and dynamic programming [85], to n-body and Monte Carlo simulations [86].

Over the years, along with substantial progress in the general field of parallel programming, the subfield of *distributed computing* and systems emerged. A distributed system is simply a distributed computer system in which the processing elements are connected by a network. Its early roots can be traced to the development of the technologies that culminated with the Ethernet Local Area Network (LAN) [87] in the 1970s and Arpanet Wide Area Network (WAN) in the late 1960s. These building blocks provided the capability of integrating disparate computer systems under one distributed computer [88]. Further development in this subfield included the development of several distributed computing software architectures, including client-server, multi-tiered, tightly coupled, and, more recently, peer-to-peer systems [88].

Distributed computers can be highly scalable and, therefore, of interest to a large range of problems in scientific and business areas. Despite the substantial improvements in hardware and networking capabilities, programming distributed computing systems has always been more complex than sequential programming. Multiple parallel and distributed programming frameworks have been devised to mitigate this problem.

Parallel and distributed computing frameworks
Each distributed computing architecture relies on middleware software or on parallel programming languages to support the application development process. In general, to express parallelism, middleware and programming languages have either explicit constructs (typically, in the case of parallel middleware) or implicit constructs (typically, in the case of parallel languages) to help with operations such as spawning a new parallel task, carrying out inter-task communication, and achieving synchronization.

The parallel programming landscape includes technologies such as OpenMP [89], in which a programmer employs parallelization compiler directives in programs written in sequential languages such as Fortran and C. The landscape also includes fully implicit parallel programming languages, albeit with limited/niche use in the commercial world such as SISAL [90], Parallel Haskell [91, 92], and Mitrion-C [93], which is used for programming Field-Programmable Gate Arrays (FPGAs).

Considering that a substantial amount of existing high-performance code is written in Fortran, C, and C++, and the fact that these languages lack explicitly parallel constructs, parallel programming middleware has emerged. Examples include the Parallel Virtual Machine (PVM) [94] and the Message Passing Interface (MPI) [95]. While these frameworks present different abstractions, they share a common objective: to simplify the development of large-scale distributed applications. A common criticism to approaches like these, however, is that the application parallelism is *hidden* in the source code as one must use specific Application Programming Interfaces (APIs) for carrying out communication, synchronization, and task management. Another common criticism is that optimizing applications built with these frameworks is usually left to application developers. Both issues have been the subject of intense research with the development of numerous tools designed to lessen the burden on developers designing distributed applications [96, 97, 98, 99, 100].

More recently, we have witnessed the emergence of distributed computing middleware aimed at simplifying the complexities of writing multi-component and large-scale distributed applications. This includes the set of technologies grouped under the *component-based* category as well as under the MapReduce framework (discussed in Section 1.3.1).

Examples of component-based technologies include Sun's Remote Procedure Call (RPC) [101], Object Management Group (OMG)'s Common Object Request Broker Architecture (CORBA) [102] (which includes numerous implementations supporting a variety of languages [103, 104, 105]), Microsoft's Distributed Component Object Model (DCOM) architecture [106], and the *web services* infrastructure [107].

Applications relying on these frameworks are organized as a collection of cooperating components. In this context, a component is an independent object that offers a remote interface (or a service broker) through which service can be obtained by a client component. A component's service interface is described using an Interface Definition Language (IDL) [104, 105], to use the CORBA nomenclature. The IDL-based definitions are then used by a compiler to stub out client- and server-side source code for each component of a distributed application. The component implementation is done by a developer by plugging the service implementation into the stubbed out code. During runtime, a component either registers its interface using endpoints that are known a priori, or the service interface is located via a directory service lookup performed by a client component.

Distributed systems and stream processing

The widespread use of distributed systems technology is now routine in business and scientific computing environments. Such degree of system interconnection has fueled the need for better integrated and faster data analysis, thereby spurring the development of stream processing technologies. Indeed, distributed system technologies are the basis of stream processing middleware, allowing it to scale and connect with existing systems.

In other words, many existing distributed systems act as the *sources* (producers) for or the *sinks* (consumers) of data for SPAs. For example, in finance engineering, distributed Monte Carlo simulations form the basis of streaming price engines for market-making applications [6]. Similarly, results computed on streaming data may be used to provide initial conditions for offline parallel runs, e.g., to bootstrap simulation models to re-price volatile assets more frequently.

Stream processing as a distributed computing paradigm has also grappled with the issue of how to structure and develop code for applications that span a large number of computing nodes. As discussed in Chapters 2, 5, and 7, different stream processing middleware and stream processing programming languages have made distinct decisions regarding application decomposition, distribution, planning, and scheduling when employing a distributed computing environment. Along the way, substantial progress has been made in simplifying parallel programming for SPAs.

Hence, distributed system technologies are a building block for stream processing, but they also enable a synergistic integration with other technologies and systems that rely on them. Ultimately, this synergy contributes to the goal of processing information in a streaming, real-time manner.

1.3.3   Signal processing, statistics, and data mining

Signal processing is the field of study grouping techniques concerned with handling time series data, or an input signal, for the purposes of performing in-depth analysis of this signal. The roots of signal processing can be traced to the 1960s and 1970s when digital computers first became available [108]. Signal processing methods include techniques for *filtering*, *smoothing* as well as performing *correlation*, spectral analysis, and *feature/pattern extraction*.

Filtering methods are designed to eliminate unwanted signal components, *de-noising* the signal, enhancing certain signal components or simply to detect the presence of these components. *Smoothing* methods aim at removing short-term variations to reveal the underlying form of the data. These techniques are used, for example, to *de-blur* images. *Correlation* methods aim at identifying the time delay at which two signals "line up" or have the highest degree of similarity. Spectral analysis, using the Fourier as well as other transforms [109], can be used to characterize the magnitude and phase of a signal, to perform more involved computations such as *convolution* or other types of correlation [108].

Typically, all these basic techniques are employed for carrying out tasks such as pre-processing, cleaning, and *feature extraction* [110] in several different types of applications operating on time series data. Such data includes structured operating sensor readings, or numeric measurements, as well as unstructured data such as speech [111], image [112], and video. Signal processing techniques are used in applications for *pattern recognition* (e.g., detecting bursts [113]), *correlation analysis* (e.g., finding stock prices that move in tandem [114]), and *prediction* [115], where information of past behavior of a signal is used to summarize and extract trends.

Statistics is the branch of mathematics that provides methods for organizing and summarizing data (*descriptive statistics*) as well as for drawing conclusions from this data (*inferential statistics*) [116]. Fundamentally, these techniques are designed to provide quantitative descriptions of the data in a manageable form. For example, an application might have to process a continuous stream of sensor readings with environmental information such as temperature, pressure, and the amount of certain chemicals in the water. In most cases, the general characteristics of the data are more important than the individual data points. Descriptive statistics help to distill such large amounts of data in a domain-specific and application relevant manner, by either producing graphical representations summarizing the data or by computing one or more representative values (e.g., min, max, and mean) for a set of data samples.

Descriptive statistics methods range from representing the data using visual techniques (e.g., stem-and-leaf display, histograms, and boxplots) to quantitative methods ranging from obtaining measures of location (e.g., *mean* and *median*) to measures of variability (e.g., *standard deviation* and *variance*).

Once information is summarized using descriptive statistics, it can be used to draw conclusions about the data [19]. Statistical inference makes propositions about a *population* – the conceptual, complete set of values a random variable might take using data drawn from this population via random sampling.

The field of data mining, which is closely related to the field of statistics, is concerned with extracting information from raw data, learning from it, and turning it into insights [117, 118, 119]. While there is overlap across the fields of data mining and statistics, one of the early motivations behind data mining was the *size* of problems being tackled, at the time considerably beyond what statisticians would look at. Indeed, in the 1990s it became clear that the amount of information being accumulated by business and scientific institutions had already outpaced the ability to extract "knowledge" from it without automation.

Data mining includes the set of methods to provide the ability to "intelligently and automatically assist humans in analyzing the mountains of data for nuggets of useful knowledge" [120]. It focuses on two main goals *description* and *prediction*.

Discovering knowledge, in the form of rules and patterns in vast quantities of data has found applications in domains such as marketing [121], finance, manufacturing [122, 123], healthcare [31, 124], and several others.

The data mining community has developed a large number of algorithms to address different problems, as well as to handle functionally and structurally different types of data [125, 126]. These techniques may be categorized into *summarization, clustering, classification, regression, dependency modeling*, and finally *change detection*. We discuss many of them and their use in stream processing in Chapters 10 and 11.

With the increasing maturity of data mining technology, not only have many software platforms become available (as discussed in Section 1.3.1), but standards have also emerged to allow expressing data mining operations in structured ways. One such formalism is the eXtensible Markup Language (XML)-based Predictive Model Markup Language (PMML) [127, 128] established by the Data Mining Group (DMG) [129] so that models can be exchanged between different data mining platforms. PMML also allows the definition of data dictionaries, the specification of pre-processing data transformations, the definition of post-processing and output handling steps, as well as the definition of the actual data mining model.

Considerable effort in using signal processing, statistics, and data mining efficiently centers on identifying the appropriate techniques as well as adapting them to the problem at hand. The research literature as well as the existing commercial toolkits provide a large collection of methods to draw from. Each of these methods is paired with different modeling, validation, and evaluation strategies to determine the best approach to solve a particular problem.

As we will show throughout this book, designing and implementing a SPA involves tackling a combination of these tasks, as well as dealing with the engineering aspects related to the continuous nature of the analytic task.

Signal processing, statistics, data mining, and stream processing
These different knowledge discovery techniques are central to many SPAs. Specifically, signal processing techniques are employed to handle different types of structured and unstructured time series data [13, 130]. Descriptive statistics are commonly used in the ingest stage of SPAs to summarize and aggregate the large volume of incoming data. Several correlation techniques alongside with statistical inference and stream data mining techniques are used for in-depth analysis and decision-making.

In the stream processing setting mining tasks need to be interactive and time-adaptive [120] due to the continuous nature of the analytic tasks. Thus, while early work in data mining was centered on batch processing, recent efforts [131] have been devoted to *continuous* and *incremental* versions of mining algorithms and methods.

The progress in developing these techniques has enabled the implementation of advanced stream processing analysis applications that can address complex requirements associated with business, scientific, and other domains, as discussed in Chapters 10 to 12.

1.3.4   Optimization theory

Optimization theory is the study of the extreme values, specifically, the minima or maxima, of a function. The theory includes topics ranging from the *conditions* that define the existence of extreme values, to analytic and numeric *methods* for finding these values. The theory also includes techniques for computing the specific values of a function's independent variables for which the function attains its extreme values.

Optimization theory has been widely utilized in business and scientific applications. At its core, it is characterized by the use of mathematical models which provide the means to analysts so they can make *effective* decisions that positively affect a system (such as a merchandise distribution system, a manufacturing process, a scheduling application, or even a software system). In many application domains, the tools from optimization theory are also used to direct the effort towards seeking further information to improve a system, when the current knowledge is insufficient to reach a proper decision.

A brief history of optimization theory
The first documented optimization technique, referred to as the *least squares fit*,[3] is attributed to Johann Gauss, and is a combination of number theory and statistics dating back to the late 1700s [132]. Other fundamental techniques such as the *steepest descent* also have long histories that can be traced back to the 1800s [133]. During the past

---

[3] The least square fit is a method for finding the curve that best fits a given set of points by minimizing the sum of the squares of the residuals of the points from the curve. The curve can later be used to make predictions about other data points.

century substantial theoretical work was carried out in the optimization field. Indeed, in the last decades of the twentieth century, computational infrastructure has become available to tackle larger problems, widening the use of these methods across a variety of applications.

In 1947, George Dantzig, who was part of a research group funded by the US Air Force, developed the *simplex method* for solving the general *linear programming* problem, a maximization/minimization problem where the *objective function* and its *constraints* are expressed in terms of linear functions. Not only did the computational efficiency and robustness of the simplex method make its applicability vastly common but the availability of high-speed computer-based implementations made linear programming one of the most fundamental methods employed by business applications [134].

Fundamentals of optimization theory
Optimization theory comprises a set of mathematical and computational methods that span mathematics and computer science. Optimization problems may be of different types: *unconstrained* or *constrained optimization* as well as *stochastic/evolutionary search*.

Unconstrained optimization, in its simplest form, is the optimization problem intent on computing the minimum/maximum value that an objective function $f(x)$ can take, with no conditions restricting what values the independent variable x can assume. Techniques addressing this problem include several methods, such as gradient search [135], the Newton method and its variations [136], the conjugate gradient method [137], among others.

Constrained optimization problems seek the maximization/minimization of an objective function, subject to constraints on the possible values that can be assumed by the independent variables. Constraints may be defined as equalities, inequalities, or even in terms of specific values an independent variable might take (e.g., integer values only, positive values only, or binary values only). These constraints may be *hard constraints* that must be met, or *soft constraints* that translate into *preferences* for certain kinds of solutions. Soft constraints may be specified as *cost functions* [138], and one might prefer solutions that have a *high* or *low* cost over alternative solutions. Methods for solving constrained optimizations include include Lagrangian and augmented Lagrangian methods [136], linear programming [134], quadratic programming[4] [134, 136], and nonlinear programming methods[5] [136].

Additionally, there are also stochastic and evolutionary search techniques that use computational methods for *evolving* or *searching* for a solution in a *search space*. These methods typically require procedures where the selection of candidate (intermediate) solutions happens with some degree of tunable randomness. The use of these

---

[4] Quadratic programming is the optimization technique seeking the maximization/minimization of a quadratic objective function subjected to linear constraints.

[5] Nonlinear programming is the optimization technique seeking the maximization/minimization of an objective function with non-linearities in the form of either a non-linear objective function or non-linear constraints.

techniques has grown considerably over the past several years [139] and include genetic algorithms [140], evolutionary programming [141], evolution strategy [141], genetic programming [141], swarm intelligence [142], and simulated annealing [143].

Irrespective of the technique used to solve a particular optimization problem, it is often possible to combine an evolutionary/search technique with a search paradigm called *branch and bound* [85]. This is a divide-and-conquer strategy [85], where the search space is partitioned and partitions deemed *unproductive* (i.e., better solutions have already been seen in other partitions) are pruned as early as possible, removing them as well as any sub-partitions they would generate from the search space. The search in the remaining partitions can proceed recursively by further sub-dividing each partition, when possible, until a partition is exhaustively searched or pruned.

Dynamic programming [85, 134, 144] is an optimization approach that transforms a complex problem into a collection of simpler problems. Its essential characteristic is the multi-stage nature of the optimization procedure. More so than the optimization techniques described previously, dynamic programming provides a general framework for analyzing many types of problems, from computing the shortest path in a graph [85], to solving the traveling salesman problem [145], to a multitude of string processing algorithms [146] including genetic sequence alignment problems. Within this framework a variety of optimization techniques can be employed to solve the invidividual sub-problems embedded in the original formulation.

Finally, *sensitivity analysis* [134] is the study of how the variation (and/or uncertainty) in the *output* of a mathematical model can be apportioned, qualitatively or quantitatively, to different sources of variation in the *input* of the model. Hence, it includes techniques for systematically changing parameters in a model to determine the effects of such changes (e.g., how will fluctuations in the interest rate affect a particular bond price [147]?).

Modeling and optimization
When applying modeling and optimization theory techniques to solve a problem, the following steps usually take place: *model formulation, data gathering, optimal solution seeking, results and sensitivity analysis,* and *testing and implementation of the solution.*

The first step focuses on the main foundation of these techniques, model building, which embodies "an attempt to capture the most significant features of the decision under consideration by means of a mathematical abstraction" [134]. The aim in this step is to extract a simplified representation of the real-world problem in the form of variables and constraints. In this context, models have to be *simple* to understand and *easy* to use, as in being analytically or computationally *tractable*, but also *complete* and *realistic*, incorporating the elements required to characterize the essence of the problem.

The second step, data gathering, requires an in-depth knowledge of the physical/logical problem being modeled. An analyst must be able to extract the necessary coefficients for the optimization function and the constraints (if any) to be plugged into the model.

The third step, solution seeking, typically requires the utilization of optimization software [148, 149, 150]. This step begins with the optimization problem being translated into the representation required by the software. This can be time-consuming and may require multiple iterations to make further adjustments to the model. Sometimes, it may be necessary to gather additional data from the physical/logical system being modeled, and may require revisiting the first and second steps. Once the model is properly represented, the software can compute the solution, a process that might be computationally expensive and require considerable time.

The fourth step, results and sensitivity analysis, consists of vetting the results in practice and understanding how these results can be used, in particular, the effect of modifications to constraints as well as to the objective function, arising from changes in the real-world problem being modeled. This step ensures that the solutions are indeed optimal and that the model embodies a certain degree of flexibility, enabling it to be used even when facing changes in the real-world system.

Finally, in the fifth step, the model is used in practice along with additional mechanisms that ensure that the model pre-conditions are maintained and proper safeguards are put in place to, in some cases, re-generate the model, triggering a continuous optimization process as necessary.

Optimization theory and stream processing
Optimizations of different kinds are often employed within analytic tasks implemented by SPAs. In addition, SPSs often use optimization techniques internally for task scheduling and placement [151, 152, 153, 154, 155].

The continuous nature of SPAs introduces a unique challenge concerning the use of optimization techniques. Optimization models, like any other type of models, may require periodic tuning and updating to reflect changes in the underlying variables and constraints. This is simply a consequence of changes in the real-world system being modeled. Hence, the optimization problem *may*, in certain situations, have to be solved again, or incrementally, once the parameters of the model are shown to be outside of the bounds computed via sensitivity analysis. Different techniques can be used to ensure that, even under evolving conditions, up-to-date solutions are available. These techniques will be discussed in more detail in Chapter 11.

Finally, large distributed applications, in particular, the ones that demand high computational or I/O capabilities, must be carefully tuned and optimized. Such optimizations range from application decomposition (i.e., how to break a large application into smaller functional blocks [156]) and algorithm design (e.g., how to implement a certain piece of analytics trading off computational cost for accuracy [10]), to compile-time and runtime decisions (e.g., how to coalesce or break functional parts of an application and how to best place them on a set of computational nodes [154, 157]). While some of these decisions can be automatically made by an SPS' internal scheduler many of these problems lie at the intersection between the application and the middleware supporting it, requiring joint optimizations.

In summary, the advancement in optimization techniques has allowed these systems to scale up and manage a large number of distributed applications and computational resources.

1.4   Stream processing – tying it all together

SPAs are used in *sense-and-respond* systems [158] that *continuously* receive external signals in the form of one or more streams, from possibly multiple sources, and employ *analytics* aimed at detecting actionable information. In this chapter, we described the trends and data processing requirements that led to the development of the stream processing paradigm. We also surveyed many of the fundamental techniques and technologies that serve either as stepping stones leading to the stream processing framework, or that must work in concert with, and hence, be integrated with SPAs in a larger data processing ecosystem.

Early examples of SPSs include Supervisory Control and Data Acquisition (SCADA) [159] systems deployed for monitoring and controlling manufacturing, power distribution, and telecommunication networks, as well as environmental monitoring systems, and several modern business applications. All these systems share the need for calculating baselines for multiple samples of incoming signals (e.g., instantaneous electricity production levels, the fair price of a security, etc.) as well as computing the *correlation* of the signal with other signals (e.g. the instantaneous electricity consumption levels, the ask or offer price of a security, etc.).

Research on stream processing emerged in earnest in the late 1990s with the development of several early prototypes, including academic and commercial platforms. In academia, systems such as STREAM [160], Aurora [161], Borealis [151], Tele-graphCQ [162], have focused on providing stream processing middleware and, in some cases, declarative languages for writing applications. Specifically, on the programming language side, examples such as StreamIt [163] and the Aspen language [164] are representative of the effort to create Domain-Specific Languages (DSLs) for continuous processing. On the industry side, SPSs have been available since the early 2000s, including products from several companies, such as InfoSphere Streams [165] and StreamBase [166].

While SPAs are becoming central to a variety of emerging areas, designing and developing these applications is challenging. Not only is it essential to ensure that the application performs the required analytical processing correctly, but it is also important to ensure that it is efficient, high-performance, and adaptive. To meet these goals, application developers must carefully decompose the application, mapping it onto data flow graphs, design and implement individual components to perform the analytic tasks, distribute the application across the computational infrastructure, and, finally, continuously tune it. Additionally, developers also need to account for fault tolerance and dynamic adaptation in response to data and processing variations. In our experience, these characteristics require a shift in the developer's thought process to incorporate new analytic and engineering methods during application design, development, and ongoing evolution.

This book is designed to address all of these challenges and is broken into five main parts: Fundamentals, Application development, System architecture, Application design and analytics, Case studies, and Closing notes.

In the remaining portion of Part I, *Fundamentals*, we will explore the roots of SPSs by tracing recent developments in the research, commercial, and open-source communities. In Chapter 2 we will also discuss basic concepts, and describe application characteristics along with technological solutions for implementing and supporting these applications.

In Part II, *Application development*, we focus on the main software engineering challenges associated with building SPAs. We begin by presenting the basic building blocks of stream processing programming languages in Chapter 3, followed by discussions on data flow programming in Chapter 4, large-scale application development issues in Chapter 5, and application engineering challenges such as debugging and visualization in Chapter 6. In this part, we also introduce IBM's SPL programming language used by the IBM InfoSphere Streams SPS.

In Part III, *System architecture*, we shift the emphasis to the system runtime. In Chapter 7, we present the software architecture supporting SPAs by discussing the inner workings of stream processing middleware. This discussion focuses on the most important features provided by the runtime middleware, including job management, the communication substrate and execution environment, as well as scheduling and security. In Chapter 8, we illustrate each of these features using the IBM InfoSphere Streams internal architecture and components. We also provide a hands-on overview of the interfaces exposed by each of the Streams components and the tooling that allows developers and administrators to interact with the SPS.

In Part IV, *Application design and analytics*, we shift the discussion towards design principles as well as analytic techniques particularly suitable for SPAs. In Chapter 9, we describe a set of design principles and associated implementation patterns for edge adaptation, flow manipulation, application decomposition, parallelization, and fault tolerance.

In Chapter 10, we describe different types of analytics that are suitable for pre-processing and data transformations to allow the efficient and incremental extraction of information from continuous data. In Chapter 11, we describe data mining techniques for streaming data analysis. In both chapters, our goal is to provide a practical perspective on the techniques, as well as concrete implementation guidelines for some of the well-studied algorithms that are currently available.

In Part V, *Case studies*, we examine realistic case studies from three different domains (infrastructure monitoring in IT systems, patient monitoring in healthcare, and manufacturing process monitoring) that illustrate the multi-phase process required to design and develop the software infrastructure and analytics encapsulated by production-grade SPAs. We draw from the lessons in the preceding chapters to provide an end-to-end overview of the process of engineering a new application. Specifically, in Chapter 12, we start by identifying the requirements, along with the process of formalizing the problem and decomposing the application into its different elements. This is complemented by defining the analytics to be implemented, as well as by the necessary system support that makes the application scalable and reliable, briefly touching upon the fault tolerance and parallelization issues.

In Part VI, *Closing notes*, we summarize the key ideas in this book and discuss some of the open research questions and emerging topics that are shaping how stream processing research and technologies have been evolving more recently.

We have designed this book to address the needs of readers intent on learning the fundamentals of stream processing, but we also addressed many of the questions faced by practitioners working on designing and implementing SPAs. Considering these goals, we have chosen to center the practical examples in this book on IBM's InfoSphere Streams and its programming language SPL, while also surveying other existing languages and programming environments. We believe that the combination of technical discussions and hands-on experience, will bring concrete benefits to the reader, allowing those who wish to work on real applications to do so with practical knowledge of one of the programming platforms currently available in the market.

References

[1] Data, data everywhere; retrieved in October 2012. The Economist, February 25, 2010. http://www.economist.com/node/15557443.

[2] Daily Volume Statistics October 2012; retrieved in October 2012. http://www.theocc.com/webapps/daily-volume-statistics.

[3] NYSE Technologies Market Data; retrieved in October 2012. http://www.nyxdata.com/Data-Products/Product-Summaries.

[4] Zikopoulos PC, deRoos D, Parasuraman K, Deutsch T, Corrigan D, Giles J. Harness the Power of Big Data – The IBM Big Data Platform. McGraw Hill; 2013.

[5] Carney D, Çetintemel U, Cherniack M, Convey C, Lee S, Seidman G, *et al.* Monitoring streams – a new class of data management applications. In: Proceedings of the International Conference on Very Large Databases (VLDB). Hong Kong, China; 2002. pp. 215–226.

[6] Park Y, King R, Nathan S, Most W, Andrade H. Evaluation of a high-volume, low-latency market data processing sytem implemented with IBM middleware. Software: Practice & Experience. 2012;42(1):37–56.

[7] Verscheure O, Vlachos M, Anagnostopoulos A, Frossard P, Bouillet E, Yu PS. Finding "who is talking to whom" in VoIP networks via progressive stream clustering. In: Proceedings of the IEEE International Conference on Data Mining (ICDM). Hong Kong, China; 2006. pp. 667–677.

[8] Wu KL, Yu PS, Gedik B, Hildrum KW, Aggarwal CC, Bouillet E, *et al.* Challenges and experience in prototyping a multi-modal stream analytic and monitoring application on System S. In: Proceedings of the International Conference on Very Large Databases (VLDB). Vienna, Austria; 2007. pp. 1185–1196.

[9] Sow D, Biem A, Blount M, Ebling M, Verscheure O. Body sensor data processing using stream computing. In: Proceedings of the ACM International Conference on Multimedia Information Retrieval (MIR). Philadelphia, PA; 2010. pp. 449–458.

[10] Turaga D, Verscheure O, Sow D, Amini L. Adaptative signal sampling and sample quantization for resource-constrained stream processing. In: Proceedings of the International Conference on Biomedical Electronics and Devices (BIOSIGNALS). Funchal, Madeira, Portugal; 2008. pp. 96–103.

[11] Arasu A, Cherniak M, Galvez E, Maier D, Maskey A, Ryvkina E, *et al.* Linear Road: a stream data management benchmark. In: Proceedings of the International Conference on Very Large Databases (VLDB). Toronto, Canada; 2004. pp. 480–491.

[12] Biem A, Bouillet E, Feng H, Ranganathan A, Riabov A, Verscheure O, *et al.* IBM Info-Sphere Streams for scalable, real-time, intelligent transportation services. In: Proceedings of the ACM International Conference on Management of Data (SIGMOD). Indianapolis, IN; 2010. pp. 1093–1104.

[13] Biem A, Elmegreen B, Verscheure O, Turaga D, Andrade H, Cornwell T. A streaming approach to radio astronomy imaging. In: Proceedings of the International Conference on Acoustics, Speech, and Signal Processing (ICASSP). Dallas, TX; 2010. pp. 1654–1657.

[14] Schneider S, Andrade H, Gedik B, Biem A, Wu KL. Elastic scaling of data parallel operators in stream processing. In: Proceedings of the IEEE International Conference on Parallel and Distributed Processing Systems (IPDPS); 2009. pp. 1–12.

[15] Angevin-Castro Y. Water resources: quenching data thirst the first step to water security. CSIRO Solve. 2007;10(4).

[16] Water Resources Observation Network; retrieved in March 2011. http://wron.net.au/.

[17] Zhang X, Andrade H, Gedik B, King R, Morar J, Nathan S, *et al.* Implementing a high-volume, low-latency market data processing system on commodity hardware using IBM middleware. In: Proceedings of the Workshop on High Performance Computational Finance (WHPCF). Portland, OR; 2009. Article no. 7.

[18] Williamson D, Shmoys D. The Design of Approximation Algorithms. Cambridge University Press; 2011.

[19] Asadoorian M, Kantarelis D. Essentials of Inferential Statistics. 5th edn. University Press of America; 2008.

[20] Jacques-Silva G, Gedik B, Andrade H, Wu KL. Language-level checkpointing support for stream processing applications. In: Proceedings of the IEEE/IFIP International Conference on Dependable Systems and Networks (DSN). Lisbon, Portugal; 2009. pp. 145–154.

[21] Jacques-Silva G, Gedik B, Andrade H, Wu KL. Fault-injection based assessment of partial fault tolerance in stream processing applications. In: Proceedings of the ACM International Conference on Distributed Event Based Systems (DEBS). New York, NY; 2011. pp. 231–242.

[22] Bouillet E, Feblowitz M, Liu Z, Ranganathan A, Riabov A. A tag-based approach for the design and composition of information processing applications. In: Proceedings of the ACM International Conference on Object-Oriented Programming, Systems, Languages, and Applications (OOPSLA). Nashville, TN; 2008. pp. 585–602.

[23] Beynon M, Chang C, Çatalyürek Ü, Kurç T, Sussman A, Andrade H, *et al.* Processing large-scale multi-dimensional data in parallel and distributed environments. Parallel Computing (PARCO). 2002;28(5):827–859.

[24] Chang C, Kurç T, Sussman A, Çatalyürek Ü, Saltz J. A hypergraph-based workload partitioning strategy for parallel data aggregation. In: Proceedings of the SIAM Conference on Parallel Processing for Scientific Computing (PPSC). Portsmouth, VA; 2001.

[25] Li X, Jin R, Agrawal G. A compilation framework for distributed memory parallelization of data mining algorithms. In: Proceedings of the IEEE International Conference on Parallel and Distributed Processing Systems (IPDPS). Nice, France; 2003. p. 7.

[26] Riedel E, Faloutsos C, Gibson GA, Nagle D. Active disks for large-scale data processing. IEEE Computer. 2001;34(6):68–74.

[27] Glimcher L, Jin R, Agrawal G. FREERIDE-G: Supporting applications that mine remote. In: Proceedings of the International Conference on Parallel Processing (ICPP). Columbus, OH; 2006. pp. 109–118.

[28] Glimcher L, Jin R, Agrawal G. Middleware for data mining applications on clusters and grids. Journal of Parallel and Distributed Computing (JPDC). 2008;68(1):37–53.

[29] Jin R, Vaidyanathan K, Yang G, Agrawal G. Communication and memory optimal parallel data cube construction. IEEE Transactions on Parallel and Distributed Systems (TPDS). 2005;16(12):1105–1119.

[30] Snyman J. Practical Mathematical Optimization: An Introduction to Basic Optimization Theory and Classical and New Gradient-Based Algorithms. Springer; 2005.

[31] Elmasri R, Navathe S. Fundamentals of Database Systems. Addison Wesley; 2000.

[32] McGee W. The Information Management System IMS/VS, Part I: General structure and operation. IBM Systems Journal. 1977;16(2):84–95.

[33] Metaxides A, Helgeson WB, Seth RE, Bryson GC, Coane MA, Dodd GG, *et al.* Data Base Task Group Report to the CODASYL Programming Language Committee. Association for Computing Machinery (ACM); 1971.

[34] ISO. Information Technology – Programming Languages – Cobol. International Organization for Standardization (ISO); 2002. ISO/IEC 1989.

[35] Bachman C, Williams S. A general purpose programming system for random access memories. In: Proceedings of the American Federation of Information Processing Societies Conference (AFIPS). San Francisco, CA; 1964. pp. 411–422.

[36] ISO. Information Technology – Programming Languages – Fortran – Part 1: Base Language. International Organization for Standardization (ISO); 2010. ISO/IEC 1539-1.

[37] ISO. Information Technology – Programming Languages – PL/1. International Organization for Standardization (ISO); 1979. ISO/IEC 6160.

[38] Codd E. A relational model for large shared data banks. Communications of the ACM (CACM). 1970;13(6):377–387.

[39] Astrahan MM, Blasgen HW, Chamberlin DD, Eswaran KP, Gray JN, Griffiths PP, *et al.* System R: relational approach to data management. ACM Transactions on Data Base Systems (TODS). 1976;1(2):97–137.

[40] Astrahan MM, Blasgen MW, Chamberlin DD, Gray J, III WFK, Lindsay BG, *et al.* System R: a relational data base management system. IEEE Computer. 1979;12(5):43–48.

[41] Hellerstein J, Stonebraker M, editors. Readings in Database Systems. MIT Press; 2005.

[42] Ullman J. Principles of database and knowledge-base systems. Computer Science Press; 1988.

[43] ISO. Information Technology – Database Languages – SQL. International Organization for Standardization (ISO); 2011. ISO/IEC 9075.

[44] Beaulieu A. Learning SQL. 2nd edn. O'Reilly Media; 2009.

[45] Jarke M, Lenzerini M, Vassiliou Y, Vassiliadis P, editors. Fundamentals of Data Warehouses. 2nd edn. Springer; 2010.

[46] IBM InfoSphere Data Warehouse; retrieved in March 2011. http://www-01.ibm.com/software/data/infosphere/warehouse/.

[47] IBM SPSS Modeler; retrieved in March 2011. http://www.spss.com/software/modeler/.

[48] Oracle Data Mining; retrieved in March 2011. http://www.oracle.com/ technetwork/database/options/odm/.

[49] Teradata Warehouse Miner; retrieved in March 2011. http://www.teradata.com/ t/products-and-services/teradata-warehouse-miner/.

[50] IBM Cognos Now!; retrieved in September 2010. http://www-01.ibm.com/ software/data/cognos/products/now/.

[51] Oracle Hyperion; retrieved in September 2010. http://www.oracle.com/us/ solutions/ent-performance-bi/index.html.

[52] MicroStrategy; retrieved in September 2010. http://www.microstrategy.com/.

[53] Meyer B. Object-Oriented Software Construction. Prentice Hall; 1997.

[54] Garcia-Molina H, Salem K. Main memory database systems: an overview. IEEE Transactions on Data and Knowledge Engineering (TKDE). 1992;4(6): 509–516.

[55] Garcia-Molina H, Wiederhold G. Read-only transactions in a distributed database. ACM Transactions on Data Base Systems (TODS). 1982;7(2):209–234.

[56] Graves S. In-memory database systems. Linux Journal. 2002;2002(101):10.

[57] IBM solidDB; retrieved in March 2011. http://www-01.ibm.com/software/ data/soliddb/.

[58] Team T. In-memory data management for consumer transactions the timesten approach. ACM SIGMOD Record. 1999;28(2):528–529.

[59] Stonebraker M, Abadi D, Batkin A, Chen X, Cherniack M, Ferreira M, et al. C-Store: A Column Oriented DBMS. In: Proceedings of the International Conference on Very Large Databases (VLDB). Trondheim, Norway; 2005. p. 553–564.

[60] Abadi D, Madden S, Hachem N. Column-stores vs row-stores: how different are they really? In: Proceedings of the ACM International Conference on Management of Data (SIGMOD). Vancouver, Canada; 2008. pp. 967–980.

[61] Fox A, Gribble SD, Chawathe Y, Brewer EA, Gauthier P. Cluster-based scalable network services. In: Proceedings of Symposium on Operating System Principles (SOSP). Saint Malo, France; 1997. pp. 78–91.

[62] Shen K, Yang T, Chu L. Clustering support and replication management for scalable network services. IEEE Transactions on Parallel and Distributed Systems (TPDS). 2003;14(11):1168–1179.

[63] Dean J, Ghemawat S. MapReduce: simplified data processing on large clusters. In: Proceedings of the USENIX Symposium on Operating System Design and Implementation (OSDI). San Francisco, CA; 2004. pp. 137–150.

[64] Apache Hadoop; retrieved in March 2011. http://hadoop.apache.org/.

[65] Goldberg B. Functional programming languages. ACM Computing Surveys. 1996;28(1):249–251.

[66] Catozzi J, Rabinovici S. Operating system extensions for the teradata parallel VLDB. In: Proceedings of the International Conference on Very Large Databases (VLDB). Rome, Italy; 2001. pp. 679–682.

[67] Kurç T, Lee F, Agrawal G, Çatalyürek Ü, Ferreira R, Saltz J. Optimizing reduction computations in a distributed environment. In: Proceedings of the International Conference for High Performance Computing, Networking, Storage and Analysis (SC). Phoenix, AZ; 2003. p. 9.

[68] Page L, Brin S, Motwani R, Winograd T. The PageRank Citation Ranking: Bringing Order to the Web. Stanford InfoLab; 1999. SIDL-WP-1999-0120.

[69] Shneiderman B, Plaisant C, Cohen M, Jacobs S. Designing the User Interface: Strategies for Effective Human–Computer Interaction. Pearson Education; 2010.

[70] Shneiderman B. Software Psychology: Human Factors in Computer and Information Systems. Winthrop Publishers; 1980.

[71] Robles-De-La-Torre G. Principles of Haptic Perception in Virtual Environments. Birkhauser Verlag; 2008.

[72] Dix A, Finlay J, Abowd G, Beale R. Human–Computer Interaction. 3rd edn. Pearson and Prentice Hall; 2004.

[73] Sears A, Jacko JA. Human–Computer Interaction Handbook. CRC Press; 2007.

[74] Ahlberg C, Shneiderman B. Visual information seeking: tight coupling of dynamic query filters with starfield displays. In: Proceedings of the ACM Conference on Human Factors in Computing Systems (CHI). Boston, MA; 1994. pp. 313–317.

[75] Asahi T, Turo D, Shneiderman B. Using treemaps to visualize the analytic hierarchy process. Information Systems Research. 1995;6(4):357–375.

[76] Adobe Flex; retrieved in September 2010. http://www.adobe.com/products/flex/.

[77] JFreeChart; retrieved in September 2010. http://www.jfree.org/jfreechart/.

[78] Tibco SpotFire; retrieved in September 2012. http://spotfire.tibco.com.

[79] TPC. TPC Benchmark E – Standard Specification – Version 1.12.0. Transaction Processing Performance Council (TPC); 2011. TPCE-v1.12.0.

[80] Chang C, Moon B, Acharya A, Shock C, Sussman A, Saltz J. Titan: a high-performance remote sensing database. In: Proceedings of the IEEE International Conference on Data Engineering (ICDE). Birmingham, UK; 1997. pp. 375–384.

[81] DeWitt D, Gray J. Parallel database systems: the future of high performance database systems. Communications of the ACM (CACM). 1992;35(6):85–98.

[82] Zou Q, Wang H, Soulé R, Andrade H, Gedik B, Wu KL. From a stream of relational queries to distributed stream processing. In: Proceedings of the International Conference on Very Large Databases (VLDB). Singapore; 2010. pp. 1394–1405.

[83] Cocke J, Slotnick D. Use of Parallelism in Numerical Calculations. IBM Research; 1958. RC-55.

[84] Dongarra J, Duff I, Sorensen D, van der Vorst H, editors. Numerical Linear Algebra for High Performance Computers. Society for Industrial and Applied Mathematics; 1998.

[85] Levitin A. The Design and Analysis of Algorithms. Pearson Education; 2003.

[86] Rubinstein R, Kroese D. Simulation and the Monte Carlo Method. John Wiley & Sons, Inc.; 2008.

[87] Shoch JF, Dalal YK, Redell DD, Crane RC. Evolution of the Ethernet local computer network. IEEE Computer. 1982;15(8):10–27.

[88] Tanenbaum A, Wetherall D. Computer Networks. 5th edn. Prentice Hall; 2011.

[89] OpenMP ARB. OpenMP Application Program Interface – Version 3.0. OpenMP Architecture Review Board (OpenMP ARB); 2008. spec-30.

[90] Feo J, Cann DC, Oldehoeft RR. A report on the Sisal language project. Journal of Parallel and Distributed Computing (JPDC). 1990;10(4):349–366.

[91] The GHC Team. The Glorious Glasgow Haskell Compilation System User Guide. The Glasgow Haskell Compiler (GHC) Group; 2010. document-version-7.0.1.

[92] Trinder PW, Hammond K, Loidl HW, Peyton-Jones SL. Algorithm + strategy = parallelism. Journal of Functional Programming. 1998;8(1):23–60.

[93] Mitrionics. Mitrion Users' Guide. Mitrionics AB; 2009.

[94] Geist A, Beguelin A, Dongarra J, Jiang W, Mancheck R, Sunderam V. PVM: Parallel Virtual Machine. A Users' Guide and Tutorial for Networked Parallel Computing. MIT Press; 1994.

[95] Gropp W, Lusk E, Skjellum A. Using MPI: Portable Parallel Programming with Message-Passing Interface. MIT Press; 1999.

[96] Du W, Ferreira R, Agrawal G. Compiler support for exploiting coarse-grained pipelined parallelism. In: Proceedings of the International Conference for High Performance Computing, Networking, Storage and Analysis (SC). Phoenix, AZ; 2003. p. 8.

[97] Ferreira R, Agrawal G, Saltz J. Compiler supported high-level abstractions for sparse disk-resident datasets. In: Proceedings of the ACM International Conference on Super-computing (ICS). São Paulo, Brazil; 2002. pp. 241–251.

[98] Saraswat V, Sarkar V, von Praun C. X10: Concurrent programming for modern architectures. In: Proceedings of the ACM Symposium on Principles and Practice of Parallel Programming (PPoPP). San Jose, CA; 2007. p. 271.

[99] Toomey L, Plachy E, Scarborough R, Sahulka R, Shaw J, Shannon A. IBM Parallel FORTRAN. IBM Systems Journal. 1988;27(4):416–435.

[100] Yelick K, Hilfinger P, Graham S, Bonachea D, Su J, Kamil A, et al. Parallel languages and compilers: perspective from the Titanium experience. International Journal of High Performance Computing Applications. 2007;21(3):266–290.

[101] Sun Microsystems. RPC: Remote Procedure Call Protocol Specification Version 2. The Internet Engineering Task Force (IETF); 1988. RFC 1050.

[102] The Object Management Group (OMG), Corba; retrieved in September 2010. http://www.corba.org/.

[103] Grisby D, Lo SL, Riddoch D. The omniORB User's Guide. Apasphere and AT&T Laboratories Cambridge; 2009. document-version-4.1.x.

[104] OMG. IDL to C++ Language Mapping, Version 1.2. Object Management Group (OMG); 2008. formal/2008-01-09.

[105] OMG. IDL to Java Language Mapping, Version 1.3. Object Management Group (OMG); 2008. formal/2008-01-11.

[106] Microsoft. Distributed Component Object Model (DCOM) Remote Protocol Specification (Revision 13.0). Microsoft Corporation; 2012. cc226801.

[107] Newcomer E. Understanding Web Services: XML, WSDL, SOAP, and UDDI. Addison Wesley; 2002.

[108] Smith S. The Scientist and Engineer's Guide to Digital Signal Processing. California Technical Publishing; 1999.

[109] Brigham E. Fast Fourier Transform and Its Applications. Prentice Hall; 1988.

[110] Guyon I, Gunn S, Nikravesh M, Zadeh L, editors. Feature Extraction, Foundations and Applications. Physica-Verlag – Springer; 2006.

[111] Quatieri T. Discrete Time Speech Signal Processing – Principles and Practice. Prentice Hall; 2001.

[112] Blanchet G, Charbit M, editors. Digital Signal and Image Processing using MATLAB. John Wiley & Sons, Inc and ISTE; 2006.

[113] Zhu Y, Shasha D. Efficient elastic burst detection in data streams. In: Proceedings of the ACM International Conference on Knowledge Discovery and Data Mining (KDD). Washington, DC; 2003. pp. 336–345.

[114] Zhu Y. High Performance Data Mining in Time Series: Techniques and Case Studies [Ph.D. Thesis]. New York University; 2004.

[115] Kay S. Fundamentals of Statistical Signal Processing: Estimation Theory. Prentice Hall; 1993.

[116] Devore J. Probability and Statistics for Engineering and the Sciences. Brooks/Cole Publishing Company; 1995.

[117] Data Mining and Statistics: What's the Connection?; published November 1997; retrieved in April, 2011. http://www-stat.stanford.edu/~jhf/ftp/dm-stat.pdf.

[118] Data Mining and Statistics: What is the Connection?; published October 2004; retrieved in April, 2011. The Data Administration Newsletter – http://www.tdan.com/view-articles/5226.

[119] Zaki M, Meira W. Data Mining and Analaysis: Fundamental Concepts and Algorithms. Cambridge University Press; 2014.

[120] Fayyad U, Piatetsky-Shapiro G, Smyth P, Uthurusamy R, editors. Advances in Knowledge Discovery and Data Mining. AAAI Press and MIT Press; 1996.

[121] Yang Y. The Online Customer: New Data Mining and Marketing Approaches. Cambria Press; 2006.

[122] Choudhary A, Harding J, Tiwari M. Data mining in manufacturing: a review based on the kind of knowledge. Journal of Intelligent Manufacturing. 2009;20(5): 501–521.

[123] Harding J, Shahbaz M, Srinivas S, Kusiak A. Data mining in manufacturing: a review. Journal of Manufacturing Science and Engineering. 2006;128(4):969–976.

[124] Obenshain MK. Application of data mining techniques to healthcare data. Infection Control and Hospital Epidemiology. 2004;25(8):690–695.

[125] Aggarwal C, editor. Social Network Data Analytics. Springer; 2011.

[126] Aggarwal C, Wang H, editors. Managing and Mining Graph Data. Springer; 2010.

[127] Guazzelli A, Lin W, Jena T. PMML in Action: Unleashing the Power of Open Standards for Data Mining and Predictive Analytics. CreativeSpace; 2010.

[128] Guazzelli A, Zeller M, Chen W, Williams G. PMML: An open standard for sharing models. The R Journal. 2009;1(1):60–65.

[129] The Data Mining Group; retrieved in September 2010. http://www.dmg.org/.

[130] Park H, Turaga DS, Verscheure O, van der Schaar M. Tree configuration games for distributed stream mining systems. In: Proceedings of the International Conference on Acoustics, Speech, and Signal Processing (ICASSP). Taipei, Taiwan; 2009. pp. 1773–1776.

[131] Aggarwal C, editor. Data Streams: Models and Algorithms. Springer; 2007.

[132] Joshi M, Moudgalya K. Optimization: Theory and Practice. Alpha Science International; 2004.

[133] Petrova S, Solov'ev A. The Origin of the Method of Steepest Descent. Historia Mathematica. 1997;24(4):361–375.

[134] Bradley S, Hax A, Magnanti T. Applied Mathematical Programming. Addison Wesley; 1977.

[135] Lipák B, editor. Instrument Engineer's Handbook – Process Control and Optimization. 4th edn. CRC Press and Taylor & Francis; 2006.

[136] Avriel M. Nonlinear Programming – Analysis and Methods. Dover Publications; 2003.

[137] Křížek M, Neittaanmäki P, Glowinski R, Korotov S, editors. Conjugate Gradient Algorithms and Finite Element Methods. Berlin, Germany: Springer; 2004.

[138] Wagner H. Principles of Operations Research, with Applications to Managerial Decisions. Prentice Hall; 1975.

[139] Sarker R, Mohammadian M, Yao X, editors. Evolutionary Optimization. Kluwer Academic Publishers; 2003.

[140] Goldberg D. Genetic Algorithms in Search, Optimization, and Machine Learning. Addison Wesley; 1989.

[141] De Jong K. Evolutionary Computation – A Unified Approach. MIT Press; 2002.

[142] Kennedy J, Eberhart R, Shi Y. Swarm Intelligence. Morgan Kaufmann; 2001.

[143] Kirkpatrick S, Gelatt CD, Jr, Vecchi MP. Optimization by simulated annealing. Science. 1983;220(4598):671–680.

[144] Bellman R. Dynamic Programming. Princeton University Press; 1957.

[145] Held M, Karp R. The traveling-salesman problem and minimum spanning trees. Operations Research. 1970;18(6):1138–1162.

[146] Cormen T, Leiserson C, Rivest R, Stein C. Introduction to Algorithms. 3rd edn. MIT Press; 2009.

[147] Luenberger D. Investment Science. Oxford University Press; 1998.

[148] Fourer R. Software for optimization: a buyer's guide (part I). INFORMS Computer Science Technical Section Newsletter. 1996;17(1):14–17.

[149] Fourer R. Software for optimization: a buyer's guide (part II). INFORMS Computer Science Technical Section Newsletter. 1996;17(2):3–4, 9–10.

[150] Moré J, Wright S. Optimization Software Guide. Society for Industrial and Applied Mathematics; 1993.

[151] Abadi D, Ahmad Y, Balazinska M, Çetintemel U, Cherniack M, Hwang JH, et al. The design of the Borealis stream processing engine. In: Proceedings of the Innovative Data Systems Research Conference (CIDR). Asilomar, CA; 2005. pp. 277–289.

[152] Fu F, Turaga D, Verscheure O, van der Schaar M, Amini L. Configuring competing classifier chains in distributed stream mining systems. IEEE Journal on Selected Topics in Signal Processing (J-STSP). 2007;1(4):548–563.

[153] Turaga D, Foo B, Verscheure O, Yan R. Configuring topologies of distributed semantic concept classifiers for continuous multimedia stream processing. In: Proceedings of the ACM Multimedia Conference. Vancouver, Canada; 2008. pp. 289–298.

[154] Wolf J, Bansal N, Hildrum K, Parekh S, Rajan D, Wagle R, et al. SODA: An optimizing scheduler for large-scale stream-based distributed computer systems. In: Proceedings of the ACM/IFIP/USENIX International Middleware Conference (Middleware). Leuven, Belgium; 2008. pp. 306–325.

[155] Wolf J, Khandekar R, Hildrum K, Parekh S, Rajan D, Wu KL, et al. COLA: Optimizing stream processing applications via graph partitioning. In: Proceedings of the ACM/IFIP/USENIX International Middleware Conference (Middleware). Urbana, IL; 2009. pp. 308–327.

[156] Andrade H, Kurç T, Sussman A, Saltz J. Exploiting functional decomposition for efficient parallel processing of multiple data analysis queries. In: Proceedings of the IEEE International Conference on Parallel and Distributed Processing Systems (IPDPS). Nice, France; 2003. p. 81.

[157] Gedik B, Andrade H, Wu KL. A code generation approach to optimizing high-performance distributed data stream processing. In: Proceedings of the ACM International Conference on Information and Knowledge Management (CIKM). Hong Kong, China; 2009. pp. 847–856.

[158] Caltech. Sensing and responding – Mani Chandy's biologically inspired approach to crisis management. ENGenious – Caltech Division of Engineering and Applied Sciences. 2003;Winter(3).

[159] Boyer S. SCADA: Supervisory Control and Data Acquisition. 2nd edn. Instrument Society of America; 1999.

[160] Arasu A, Babcock B, Babu S, Datar M, Ito K, Motwani R, *et al.* STREAM: the Stanford Stream data manager. IEEE Data Engineering Bulletin. 2003;26(1):665.

[161] Balakrishnan H, Balazinska M, Carney D, Çetintemel U, Cherniack M, Convey C, *et al.* Retrospective on Aurora. Very Large Databases Journal (VLDBJ). 2004;13(4):370–383.

[162] Chandrasekaran S, Cooper O, Deshpande A, Franklin M, Hellerstein J, Hong W, *et al.* TelegraphCQ: continuous dataflow processing. In: Proceedings of the ACM International Conference on Management of Data (SIGMOD). San Diego, CA; 2003. pp. 329–338.

[163] Thies W, Karczmarek M, Amarasinghe S. StreamIt: a language for streaming applications. In: Proceedings of the International Conference on Compiler Construction (CC). Grenoble, France; 2002. pp. 179–196.

[164] Upadhyaya G, Pai VS, Midkiff SP. Expressing and exploiting concurrency in networked applications with Aspen. In: Proceedings of the ACM Symposium on Principles and Practice of Parallel Programming (PPoPP). San Jose, CA; 2007. pp. 13–23.

[165] IBM InfoSphere Streams; retrieved in March 2011. http://www-01.ibm.com/software/data/infosphere/streams/.

[166] StreamBase Systems; retrieved in April 2011. http://www.streambase.com/.

2 Introduction to stream processing

2.1 Overview

Stream processing has been a very active and diverse area of research, commercial, and open-source development. This diversity of technologies has brought along new terminology, concepts, and infrastructure necessary for designing and implementing sophisticated applications.

We start this chapter by describing some of the application domains where stream processing technologies have been successfully employed (Section 2.2), focusing on the distinctive characteristics of these applications that make them suitable for the use of stream processing technologies.

These application scenarios allow us to illustrate the motivating requirements that led to the development of multiple *information flow processing* systems, a class that groups multiple technical approaches to continuous data processing (Section 2.3). We discuss some of its broad subcategories, including active databases, Continuous Query (CQ) systems, publish–subscribe systems, and Complex Event Processing (CEP) systems. All of them are precursors that have helped shape the stream processing paradigm.

We then switch the focus to the conceptual foundations and the architectural support behind the stream processing technology, and the applications it supports (Section 2.4). We also include an overview of *analytics*, i.e., the algorithms and knowledge discovery techniques, that form the basis of most innovative Stream Processing Applications (SPAs) and a historical perspective on the research that led to the development of several Stream Processing Systems (SPSs).

Finally we include a survey of academic, open-source, and commercially available SPS implementations, and describe their different characteristics. Section 2.5 concludes this chapter and provides a starting point for the discussion of the practical design and implementation aspects of SPAs carried out in Part II of this book.

2.2 Stream Processing Applications

In this section, we discuss scenarios that motivate the need for SPAs. Examples can be found in domains ranging from financial markets, manufacturing, healthcare, traffic systems, large-scale infrastructure monitoring and security, to scientific and environmental

Stock market
- Impact of weather on securities prices
- Analyze market data at ultra-low latencies

Natural systems
- Wildfire management
- Water management

Transportation
- Intelligent traffic management

Manufacturing
- Process control for microchip fabrication

Health and life sciences
- Neonatal ICU monitoring
- Epidemic early warning system
- Remote healthcare monitoring

Law enforcement, defense and cyber security
- Real-time multimodal surveillance
- Situational awareness
- Cyber security detection

Fraud prevention
- Multi-party fraud detection
- Real-time fraud prevention

e-Science
- Space weather prediction
- Detection of transient events
- Synchrotron atomic research

Other
- Smart Grid
- Text Analysis
- Who's Talking to Whom?
- ERP for Commodities
- FPGA Acceleration

Telephony
- CDR processing
- Social analysis
- Churn prediction
- Geomapping

Figure 2.1 Stream processing applications.

domains like radio astronomy, water management, and wildlife and natural resource monitoring. Consider the set of sample applications shown in Figure 2.1.

As can be seen, these applications include stock transaction analysis for market making, process control for manufacturing, analysis of various sensors streams in natural and physical sciences, multi-modal surveillance for law enforcement, fraud detection and prevention for e-commerce, physiological sensor monitoring for healthcare, Call Detail Record (CDR) processing for telecommunications, and many others.

In the rest of this section we describe three application scenarios in more detail. These scenarios include applications in cybersecurity, transportation, and healthcare. Each one of them demonstrates certain unique data processing and analytical characteristics.

2.2.1   Network monitoring for cybersecurity

Application context

The security of data networks is of vital importance to the operations of many institutions. However, the vulnerability of these networks and the threat landscape is changing rapidly. The sophistication of attackers as well as the scope of their activities has been increasing, leading to data loss and theft, which in many cases results in financial and reputational risks.

Cybersecurity threats have traditionally arisen from individuals and informal groups that use *malware* to interfere with or steal information, including online account and credit card numbers. These exploits make use of known vulnerabilities in a host's Operating System (OS) or in one of the applications it runs (e.g., web servers and DBMSs). In some cases, these attacks can be extremely disruptive not only to individual organizations, but also to the Internet ecosystem.

In more recent times, however, a new class of threats have emerged from members of well-funded and organized groups that target the theft of sensitive, high-value information. These groups may exploit zero-day vulnerabilities or use *social engineering* mechanisms and credentials of trusted insiders to evade detection [1].

The goal of cybersecurity applications is to combat these different types of threats by *continuously* analyzing network data, identifying and neutralizing vulnerabilities, and ideally preventing such attacks. Consider a specific type of cyber threat posed by *botnets*. In this scenario a host infected with malware becomes a *bot*, i.e., a host controllable by a *botmaster* using a Command and Control (CC) infrastructure. Botnets may be used to launch Distributed Denial of Service (DDoS) attacks or to deliver spam. In fact, Conficker [2] and other recent botnets have grown to include several millions of infected bots, posing a serious threat to networks worldwide.

Identifying a botnet involves locating botmaster domains and malware infected hosts that are part of the CC infrastructure. There are several sources of information about the network, ranging from Domain Name System (DNS) queries and responses, to network traffic flow and raw packet traces. These sources of information must be analyzed in real-time to model and detect malicious behavior.

Application requirements

Consider the sample cybersecurity application shown in Figure 2.2. The sources of information pertinent to botnet identification contain streaming data with different underlying data types and formats, spanning a range of network protocols and time granularities.

These traditional data sources may also be supplemented with additional information from the Internet community, including expert and user-collated lists of suspicious domains as well as information describing different types of bot behavior.

This problem requires continuous data processing as a detection system must monitor several sources of streaming data from potentially distributed locations. Additionally, there are several requirements on performance and scalability, with desired latencies for detection in the order of milliseconds (to successfully prevent host infection) and

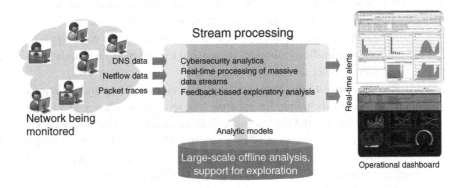

Figure 2.2 Stream processing in cybersecurity.

with aggregated data rates approaching several tens to hundreds of gigabits per second (Gbps) for a reasonably large corporate network.

Unlike with traditional malware that can be detected with signature matching, detecting and preventing the growth of botnets requires *exploratory analysis*. This type of analysis includes dynamic decision making with respect to several aspects of the processing. These decisions typically include which sources to ingest data from and what fraction of the data to consider. They also include algorithmic choices and on-the-fly adapation with both supervised (e.g., classification, regression, and signature detection) as well as unsupervised (e.g., clustering, anomaly detection) analytic techniques.

Finally, the exploration must account for adversarial moves, where botmasters attempt to obfuscate their behavior dynamically in response to designed detection and prevention strategies. This adds to the complexity of the analysis task and might require combining streaming analytics with offline analysis (with potentially multiple passes on stored and aggregated data) performed on historical data. Hence, the streaming portion of one such application must also be able to interact with other external analytic algorithms and platforms, requiring different types of data adaptation and analytic integration. This aspect is also depicted in Figure 2.2.

2.2.2   Transportation grid monitoring and optimization

Application context
Road networks and vehicles are becoming more instrumented with several different types of sensors generating continuous data that can be analyzed by offline applications as well as SPAs. For instance, road networks are instrumented with loop sensors, which are embedded under the pavement to compute *traffic volumetrics*, with sensors on toll booths, as well as cameras at different entry and exit points and traffic intersections. Similarly a majority of the public transport as well as commercial fleets are nowadays instrumented with Global Positioning System (GPS) sensors that report location and travel information periodically.

Each of these sensors acts as a streaming data source with the individual streams consisting of sequences of timestamped numeric and non-numeric measurements, capturing data in various formats, and sampled at varying time intervals.

Stream processing applied to this data can enable various novel applications ranging from real-time traffic monitoring, to personal trip advisors and location-based services. These scenarios are illustrated in Figure 2.3.

Real-time monitoring of the traffic grid involves building and updating a representation of its state including, for example, the level of occupancy and the average speed of different links in the road network. This aggregated information can be used to identify events of interest including congestion, accidents, and other types of infrastructure problems. Moreover, the availability of this type of *live* information can also allow the public infrastructure authorities to make decisions that can, in many cases, mitigate traffic problems.

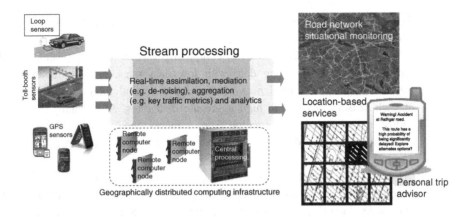

Figure 2.3 Stream processing in transportation.

These decisions can include measures such as disseminating information to provide real-time digital signage updates sent directly to the public, as well as directives to make operational changes to the public transport system.

The real-time availability of traffic information can also be the enabler for the development of personalized travel information. This type of application can provide real-time advice as part of trip planning as well as dynamic route modification and updates on the availability of public transport.

Finally, this same type of information can be used in numerous other types of commercial applications to provide location-based services, including the delivery of better targeted content, as well as the announcement of new products and deals to potential customers. More sophisticated applications of this sort can also combine traffic information with other sources of data (e.g., live weather, seasonal patterns, maintenance schedules) to provide information and forecasts that are even better tailored to consumers.

Application requirements

To support these transportation applications, the latency between when the data is generated and when action can be taken is in the order of minutes. Clearly, the latency is not as stringent as in network monitoring applications, but *near* real-time processing is necessary.

Additionally, several custom stream analytics are required for different tasks ranging from parsing and extracting information from the source data streams, to synchronizing the data, accounting for missing values, de-noising, correlation, tracking, and forecasting.

There is also a need for spatio-temporal analytics to combine GPS and other location readings onto maps, estimating direction and speed of travel, and computing shortest paths. Integrating these different types of information across vehicles, road segments, sensor types, and supporting these complex tasks also require the use of multi-modal (i.e., representing different types of data) and multi-variate (i.e., representing multiple random variables) statistical analytics.

While some of the individual low-complexity sensors (e.g., loop sensors, toll booth sensors) generate streaming data with low data rates, around hundreds of bytes per second, a typical road network may have several thousands or even millions of such sensors, leading to high aggregate data rates. Additionally, with the availability of multi-modal information from cameras, the streaming data also includes unstructured video and image signals. Hence these applications must be scalable and adaptive.

Finally, a unique characteristic of applications in the transportation domain is the geographical distribution of the data sources and the use of potentially unreliable network links. In such cases, it is often beneficial to make use of a distributed processing application layout by placing computation near the sensors, and partitioning the analysis hierarchically into local and central processing, as depicted by Figure 2.3.

2.2.3   Healthcare and patient monitoring

Application context

Care of critically ill patients is dependent upon the capture, analysis, aggregation, and interpretation of large volumes of data. The rapid and accurate integration of this data is essential for high-quality, evidence-based decision making in the context of a modern Intensive Care Unit (ICU). Yet, despite the increasing presence of new Information Technology (IT) artifacts in the ICU, most doctors, nurses, and other healthcare providers perform many data analysis tasks manually [3].

Originally, patient monitoring in an ICU was restricted to clinical examinations and observations of a limited number of physiological readings, such as heart and respiratory rate, blood pressure, oxygen saturation, and body temperature. With advances in sensor technologies, continuous monitoring of complex physiological data has became possible, enabling the dynamic tracking of changes in a patient's state.

The data from these sensors provides physicians with more information than they can (manually) analyze and it is not necessarily integrated with other important information, such as laboratory test results and admission records. As a result, the physicians and other caregivers are often required to integrate all these disparate sets of information as best as they can, before they can determine the state of their patients and prescribe an appropriate course of treatment.

Application requirements

There are several difficulties associated with manual analysis and integration of healthcare data. First, the inflow of data can be both voluminous and too detailed for humans to process without data analysis tools. Second, simply performing periodic reviews of the data can add significant latency between the occurrence of a medically significant event, and the (possibly necessary) reaction by a caregiver. Third, manual analysis can miss relevant subtleties in the data, particularly when an actionable event is the result of a combination of factors spread across several data streams.

As a result, in most cases, medical care delivered in ICUs tends to be reactive. Physicians are often responding to events that occurred, potentially, several hours or days earlier. Clearly, decreasing the latency to react can result in improved patient outcomes.

Finally, a modern ICU is a busy, complex, and stressful environment. The margin for error associated with certain medical procedures is small. The work flow and processes can sometimes be both archaic and complex. Hence, the amount of time healthcare providers can dedicate to data analysis is often limited.

In essence, the combination of all these factors results in the available information originating from sophisticated instrumentation not being used to its fullest potential in real-time, to aid in clinical management and, on a longer time scale, in clinical research.

Hence, the use of continuous applications and stream processing can improve the delivery of healthcare along several axes, including the automated capture of physiological data streams and other data from patients in an ICU as well as its real-time use by online analytics put in place to identify and predict medically significant events. These applications have several operational requirements.

The first requirement relates to the diverse nature of the signals and of the data that must be continuously captured and processed. This data includes periodically sampled signals with rates of tens to hundreds of kilobits per second (kbps) per patient, as well as asynchronous events resulting from information extracted from the admission interview, from test results, and from medical interventions. Therefore, the ingestion of the data by a continuous application requires the design of cleaning, filtering, and data alignment operations as well as multi-variate statistical techniques.

The second requirement relates to the design of medically relevant analytics, combining the expertise of medical doctors (with domain knowledge) and computer scientists (with continuous data analysis knowledge).

The third requirement relates to the implementation of mechanisms to capture and maintain audit trails of relevant historical physiological data, along with features extracted from the data, analytic results, and patient outcomes. This is essential because medical procedures are often re-examined to improve clinical care.

The fourth requirement relates to fault tolerance. Healthcare SPAs must provide continuous uptime with no data loss, due to their direct impact on patient well-being and treatment outcomes.

The fifth requirement corresponds to data privacy. In the clinical setting, data privacy must be protected to prevent data and result leakage to unauthorized parties. This requirement translates into a need for appropriate authorization and authentication mechanisms, data encryption, and anonymization.

The final requirement concerns the aggregation and management of information across multiple patients over extended time horizons. This translates into the need for the infrastructure and the analytics to scale gracefully up in order to make use of an increased corpus of accumulated data.

All of these requirements are expressed in the ICU scenario depicted by Figure 2.4. As is shown, multiple real-time physiological streams generated by sensors embedded

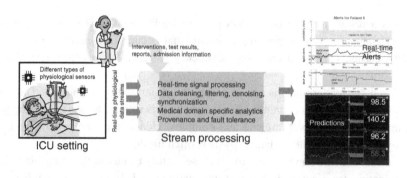

Figure 2.4 Stream processing in healthcare.

on the patient are combined with information from test results, interventions, and medications to facilitate the continuous monitoring and forecasting of her state. A similar application, developed using the InfoSphere Streams SPS, has been deployed in the SickKids Hospital in Toronto [4].

2.2.4   Discussion

Across these applications the analytic requirements include the need for streaming capabilities in data ingest and analysis, along with the ability to continuously tune the analytics in response to changes in data and in the knowledge that has been accumulated so far. From a system infrastructure point of view, the requirements include the ability to adapt the processing and change the physical structure of the application to scale up, and delivering high performance and fault tolerance, while providing controlled and auditable access to the data being processed and the results extracted by the analytical process.

Moving forward we will discuss the information flow technologies that bring us progressively closer to making this possible, and describe how we can design and build analytics as well as production-grade applications to address these requirements.

2.3   Information flow processing technologies

The need to process large volumes of data in real-time and to produce continuous information with *actionable* intelligence has attracted significant attention from the research community [5]. There have been several attempts at building technologies for *information flow processing*, which paved the way for current SPSs. The early set of such information flow processing systems can be grouped into four broad classes: active databases, Continuous Query (CQ) systems, publish–subscribe systems, and Complex Event Processing (CEP) systems. In the rest of this section, we describe these technologies in more detail to provide historical context for the emergence of stream processing technologies.

2.3.1   Active databases

Active database systems have been implemented as database extensions aimed at providing a framework for reacting to continuous changes in the data managed by a database. In general, active databases are used to support applications that implement continuous data transformation and event detection tasks.

Several active database systems have been developed, each with different architectures and with different target application domains. Nevertheless, a common characteristic of active databases is their reliance on Event-Condition-Action (ECA) rules, which capture events, the conditions surrounding these events, and the actions to be triggered when these conditions are met.

An *event* represents a change in the database associated with either a time or a time interval. For instance, the insertion of a new row into a database table is an event. Events can be primitive or composite. Primitive events are elementary occurrences and can be mapped directly to point in time changes. Composite events are formed by combining multiple primitive events or even multiple other composite events.

A *condition* describes the state in which the database should be for an action to execute once a relevant event is detected. An *action* represents the reaction to an event and is executed when the relevant condition holds, after an event is detected. An action can contain arbitrary database transactions. In fact, the execution of actions can cause the triggering of additional ECA rules. An action can also dynamically activate and deactivate other rules and perform external interactions like, for example, send an email message.

Active databases can be categorized as either *closed database* (e.g., Ode [6] and HiPac [7]) or *open database* systems (e.g., Samos [8] and Snoop [9]).

In closed database systems, ECA rules are applied to the data stored in the database and external event sources are not allowed. In open database systems events from inside the database as well as from external sources are allowed. Hence, open active database systems are generally more appropriate for continuous processing of streaming data. Several current relational Data Base Management Systems (DBMSs) have implemented some form of ECA rules using Structured Query Language (SQL) *triggers* [10].

While active databases can be used to implement certain types of stream processing, they also have several shortcomings that make them unsuitable for these applications. First, active databases were not designed to handle data rates associated with large-scale data intensive scenarios where data sources can generate hundreds of thousands or millions of tuples per second. Second, while most active databases provide a flexible programming model, which allows the specification of several types of data processing, they have limitations when it comes to supporting user-defined operations. Third, these systems tend to provide a centralized runtime environment, which makes it more difficult to tackle distributed monitoring and analysis applications.

Notwithstanding these shortcomings, the concepts embedded in the programming models of active databases have had a large influence in the design of current event and stream processing programming models.

2.3.2   Continuous queries

CQs are used to express information monitoring requests. A CQ is a *standing* query (as opposed to a snapshot query) that monitors information updates and returns results whenever these updates reach a user-specified threshold.

A CQ has three main components: a *query*, a *trigger*, and a *stop condition*. Whenever the trigger condition becomes true, the query is executed and the part of the query result that is different from the result of a previous execution (if any) is returned. The stop condition specifies when a particular CQ will cease to compute updates.

An example CQ can capture and implement the following English statement: "monitor the price of 10 megapixel (MP) digital cameras in the next two months and notify me when one with a price less than $100 becomes available." In this example, "the price of 10 MP digital cameras" is the query, "price less than $100" is the trigger, and "next two months" is the stop condition.

Clearly, when contrasted to a traditional relational query, a CQ has a number of distinguishing characteristics. First, a CQ is evaluated continuously over changing data sources and provides a *stream* of results. Second, a CQ provides users with a push-based model for accessing information as opposed to a pull-based (or, request-response) model implemented by a traditional relational query.

Naturally, an application that makes use of CQs (and of a system supporting them) for its data processing needs, must also implement adaptive mechanisms to cope with variations in the update rates at runtime.

The research and development behind CQ systems has been highly influenced by techniques employed by active databases as well as by query processing on append-only databases [11]. CQ systems including the University of Wisconsin's NiagaraCQ [12] and Georgia Tech's OpenCQ [13] have also served as precursors to today's SPSs.

In essence, CQ systems depend on the repeated evaluation of queries over continuously changing databases, relying on incremental query processing techniques to accelerate the re-evaluation of outstanding queries. SPSs take this mode of operation one step further by pushing the data through the queries in a continuous manner. In other words, in SPSs the data is continuously run through the queries, rather than the queries being continuously run through the data.

2.3.3   Publish–subscribe systems

A publish–subscribe (or pub–sub, for short) system is an event-driven system that decouples data producers from data consumers. In these systems, a *publisher* is a data producer. It generates data in the form of messages that are called *publications*. A *subscriber* is a data consumer and specifies its topics of interest using a *subscription*.

Publishers and subscribers do not know each other, i.e., each party is oblivious to where the others are located or even to whether they exist. The delivery of relevant publications to interested subscribers is achieved through a *broker network*, often formed by a set of broker nodes on a Wide Area Network (WAN). Pub–sub systems can be broadly divided into two categories: *topic-based* pub–sub systems and *content-based* pub–sub systems.

In a topic-based pub–sub system, each publication can have one or more topics associated with it and a subscription specifies one or more topics of interest. In this case, the matching between subscriptions and publications is straightforward, based on topic equality. As a consequence, the algorithms used for routing within the broker network can be simple.

In a content-based pub–sub system, a publication contains a set of properties and a subscription is expressed using conditions defined on these properties. Such conditions are usually made up of predicate expressions. For instance, a subscription may be looking for all publications where there is a *price* property whose value is less than "$1000" and a *name* property whose value is equal to "laptop."

Most content-based pub–sub systems support *atomic* subscriptions that are defined on individual publications. This is the case for IBM Research's Gryphon [14] and the University of Colorado's Siena [15] systems. Some content-based pub–sub systems support *composite* subscriptions that are defined on a series of publications using concepts such as sequences and repetition. An example of this type of system is the University of Toronto's Padres [16]. Supporting such subscriptions in an efficient manner is more challenging, since the brokers' routing algorithm must be capable of filtering publications as early and as fast as possible to ensure good performance.

In fact, CEP systems, which we discuss next, are built on the concept of *complex events*, which are very similar to composite subscriptions in pub–sub systems. However, CEP systems usually implement such functionality in a centralized system.

As is the case with other types of earlier information flow systems, some of the pub–sub heritage spills over into the design of SPSs. SPSs are similar to pub–sub systems in terms of their event-driven architecture, where the publication of new messages results in a stream of notifications being generated. Yet, at the system architecture-level, there are major differences.

First, in a SPS, events (akin to publications) are assumed to be flowing through an application, which can be seen as a set of standing queries (akin to subscriptions), to produce real-time results.

Second, while SPSs are often distributed, their processing core components commonly do not rely on a WAN. Instead, they typically work on a cluster or on a cloud environment, with high-speed networks interconnecting the computational infrastructure to deliver high throughput and low latency to applications.

2.3.4   Complex event processing systems

CEP systems were designed as computing platforms where events generated by multiple sources across an organization can be collected, filtered, aggregated, combined, correlated, and dispatched to analytic applications in real-time. In this way, applications can process and react to information embedded in the incoming data on a continuous basis.

These systems typically associate precise semantics to the processing of data items, based on a set of rules or patterns expressing the event detection tasks. For instance, temporal logic can be used to express the nature of the events to be detected and processed [17].

CEP systems typically support the detection of complex events, which are derived from simpler events as a result of additional processing and analysis. These include temporal sequences (e.g., event A following event B), negation (e.g., event A not appearing in a sequence), and Kleene closure (e.g., event A appearing repeatedly a number of times). These patterns can also be specified with respect to a temporal window (e.g., the occurrence of sequence of events happening within a given time frame).

Indeed, most CEP systems including Oracle CEP [18], TIBCO BusinessEvents [19], IBM Websphere Business Events [20], EsperTech's Esper [21], Cornell University's Cayuga [22], and the University of Massachusetts' SASE [23] focus heavily on rule-based detection tasks, expressed in terms of complex patterns defined as a function of event sequences, logic conditions, and uncertainty models.

In summary, CEP systems provide sophisticated support for rule-based complex event detection over data streams. Yet, they also have certain limitations. First, many of these systems make use of a centralized runtime, which limits scaling when applications must cope with large data volumes. Second, the programming models associated with CEP systems are not particularly adequate for applications implementing exploratory tasks, primarily because they lack the capability for dynamically modifying how the incoming data is processed in response to changes in the task, or in the data itself. They also provide only very limited support for handling unstructured data and complex analytics beyond rule-based logic.

2.3.5   ETL and SCADA systems

In addition to the different systems discussed so far, there are other types of technologies, for instance, Extract/Transform/Load (ETL) as well as Supervisory Control and Data Acquisition (SCADA) systems, that are designed as platforms to ingest and analyze high-volume data flows. Because they are geared towards specialized types of applications, we will only provide a very brief description of these systems.

ETL systems are used to implement moderately simple data transformation tasks (see Section 2.4.2). A distinctive feature of these systems is that they typically operate on offline stored data, producing output that will also be stored. Hence, in these tasks, there is no real need for streaming analysis or support for low end-to-end processing latency event when an ETL task is carried out in incremental steps.

On the other hand, SCADA systems support *monitoring* applications, implementing detection tasks (Section 2.4.2) to identify failures in the equipment used to run utilities, telecommunication, and manufacturing operations. While many SCADA platforms have limited support for event correlation, they tend to not provide in-depth analytics of the sort SPSs focus on, and usually handle much lower rate data inflows compared to the scenarios outlined in Section 2.2.

2.4   Stream Processing Systems

The earlier information flow systems can support some of the processing and analytical requirements associated with continuous data processing applications. In general, these technologies are sufficient to implement simple and small-scale streaming data analysis with low performance and fault tolerance needs.

As discussed in Section 2.2, SPAs usually implement refined, adaptive, and evolving algorithms as part of their analytical task. Moreover, these applications must be flexible, scalable, and fault-tolerant to accommodate large and, in many cases, growing workloads, with support for low-latency or real-time processing.

These requirements led to the emergence of the *stream processing* computational paradigm and SPSs. While SPSs were developed by incorporating many of the ideas of the technologies that pre-dated them, they also required several advancements to the state-of-the-art in algorithms, analytics, and distributed processing software infrastructure.

To address analytical limitations of earlier systems, SPSs include sophisticated and extensible programming models (Chapters 3–6), allowing continuous, incremental, and adaptive algorithms (Chapters 10 and 11) to be developed. To address scalability limitations, SPSs provide distributed, fault-tolerant, and enterprise-ready infrastructure (Chapters 7 and 8).

In the rest of this section, we will examine the stream processing paradigm, starting with the type of data usually processed by SPAs, as well as introduce its basic concepts and constructs.

2.4.1   Data

A streaming *data source* is a producer of streaming data, usually broken down into a sequence of individual data items (or *tuples*), that can be consumed and analyzed by an application.

Examples of such sources include different types of sensors, ranging in complexity from simple data motes,[1] to complex physical sensors such as medical devices [24] and radio telescopes [25], to instrumented infrastructure equipment and systems such as network and telecommunication switches [26], Automated Test Equipments (ATEs) used in manufacturing settings [27], and stock exchange market feeds [28].

In many cases, a source may also be a data repository, including a relational database, a data warehouse, or simply a plain text file, from which tuples are streamed and, often, *paced* according to explicit or implicit timestamps.

Analogously, a stream *data sink* is a consumer of results, also expressed in the form of a collection of individual tuples, generated by an application.

Examples of sinks include different types of control, visualization and monitoring applications, repositories as, for example, file systems or relational databases, message queues, and publish–subscribe interfaces.

---

[1] A *mote* is a small, usually battery-operated, sensor participating as a node in a wireless sensor network.

As mentioned, any continuous data source typically produces a sequence of tuples, which can be manipulated by an application. The next three concepts describe the characteristics of the stream processing *data model*.

The stream processing data model can be traced back to data models supporting database technologies, primarily the relational model. Yet, to accommodate the richer set of requirements associated with continuous processing, two key distinctions exist: first, the addition of the *stream* as a primary abstraction and, second, the inclusion of a richer and extensible set of data types, allowing greater flexibility in what an application can manipulate. With this in mind, the key components of this data model are briefly described next.

A *data tuple* is the fundamental, or atomic data item, embedded in a data stream and processed by an application. A tuple is similar to a database row in that it has a set of named and typed attributes. Each instance of an attribute is associated with a value.

Note that, a tuple differs from an *object* in Java or C++ in that it does not have *methods*. Thus, a tuple is *not* an Abstract Data Type (ADT).

A *data schema* is the type specification for a tuple and for its attributes. The structure of a tuple is, therefore, defined by its schema. An attribute of a tuple might itself be recursively defined by a yet another schema, thus making it possible to create structurally complex tuples.

Given the definition of a tuple and of its schema, a *data stream* is a potentially infinite sequence of tuples sharing a common schema, which is also the *stream schema*. Each tuple in a stream is typically associated with a certain time step (or timestamp) related to when the tuple was captured or generated, either in terms of an arrival time, a creation timestamp, a time-to-live threshold, or, simply, a sequence number.

Structured and unstructured data

In general, streaming data can be categorized into three broad classes: *structured*, *semi-structured*, and *unstructured*.

Structured streaming data includes data whose schema (or structure) is known a priori and its data items are organized into name/type/value triples.

Semi-structured streaming data includes data whose complete schema (or structure) is either nonexistant or not available. Yet, the data or some of its components are associated with self-describing tags and markers that distinguish its semantic elements, and enforce a hierarchy of records and attributes of its components. Operationally, it means that additional parsing and analysis must be undertaken to extract the parts of the data relevant to an application. For instance, semi-structured data expressed using markup languages (e.g., HyperText Markup Language (HTML) and eXtensible Markup Language (XML)) must be parsed so its components can be properly extracted and, subsequently, processed.

In contrast, unstructured streaming data consists of data in custom or proprietary formats, in many cases, binary-encoded and may include video, audio, text, or imagery data.

The data consumed by a SPA is often a mixture of structured, semi-structured, and unstructured data, where the unstructured and semi-structured components are progressively processed and their inherent structure is teased out, step-by-step, through the analytic operations performed by the application.

An application example

To make the concepts described so far concrete, we consider an application example that processes both structured and unstructured data. Consider a hypothetical application designed to analyze Voice over IP (VoIP) traffic [29] for improving marketing efforts as well as for surveillance. Its aim is to catalog *interesting* conversations and identify the people carrying out these conversations, Such an application may be used to identify social networks in support of either marketing or advertising campaigns, or to enforce corporate policies such as Chinese walls[2] that are present in many investment, accounting, and consulting firms.

In this setting, individual users of the phone system may be viewed as data sources that generate a sequence of tuples, in this case network packets with voice data that are transmitted across the network to other users. Specifically, the data consists of packets created using commercial VoIP codecs (G.729 [30] or Global System for Mobile Communications (GSM) [31]), which both encode as well as encrypt the analog voice data, for instance, using 256-bit Advanced Encryption Standard (AES) schemes [32]. This data is then further encapsulated and transported by Real-time Transport Protocol (RTP) [33] packets. Hence, the data to be processed by a streaming application consists of a sequence of RTP packets, where each packet includes a 12-byte non-encrypted header followed by 20 ms of encrypted and compressed voice.

The various RTP header fields provide the necessary information for network traffic analysis implemented by such an application. In particular, the *Payload Type* (PT) field identifies VoIP streams to be separated from other types of streams, the *Synchronization Source* (SSRC) field uniquely identifies individual streams of packets, and the *Timestamp* (TS) field reflects the sampling timestamp of the first byte in the RTP payload. In such a setting, our application might employ one or more network sniffers as data sources to extract the data packets for analysis.

Before the data can be analyzed, however, the individual packets must be converted into tuples through a process called *edge adaptation*. This process consists of extracting relevant information from the raw data items and populating a tuple's attributes (Section 9.2.1). In the case of these RTP network packets, the mapping is fairly obvious, each RTP header field can be converted into corresponding tuple attributes.

---

[2] In an organization, a Chinese wall (also referred to as a *firewall*) is an information barrier implemented to separate groups of people as a means of segregating access to different types of information. For instance, in investment firms, personnel allocated to making investment decisions regarding a particular stock or bond cannot interface with personnel acting as advisors and consultants who possess *material non-public* information regarding the firm issuing that stock or bond. The information barrier exists so investment personnel are not in an advantageous position with respect to regular investors who don't have access to that information.

Each SPS defines a tuple using its own syntax. For example, employing the notation used by Streams and its SPL programming language (Chapter 3), we can have data tuples with the typed attributes described by the following tuple schema definition:

```
tuple < timestamp TS, // the time stamp
        uint8 PT, // the payload type
        int64 SSRC, // the synchronization source
        list<uint8> voicePayload // a sequence of unsigned 8-bit integers
    >
```

This tuple definition has both structured as well as unstructured components. The structured portion includes the attributes TS, whose type is a timestamp, as well as the PT and the SSRC attributes, represented respectively as a uint8 (an unsigned 8-bit value) and int64 (a signed 64-bit signed value) numbers. The unstructured portion, represented by the voicePayload attribute, transports the encrypted voice payload and is represented by a list of uint8 values, which, simply put, is a sequence of bytes.

Naturally, this sequence of bytes is *opaque*. In other words, operations on tuples with this schema must have a priori knowledge of what type of data is expressed in the payload before it can be decrypted and parsed.

In general, the attribute types used in this example have one-to-one mappings and corresponding bindings to types in programming languages including C++ and Java, which, in the case of the SPL language, allows for a straightforward integration with processing logic written in these languages.

The process of edge adaptation is illustrated in Figure 2.5, where we can see multiple VoIP data sources communicating over the network as well as the network sniffers used to send the RTP packets to the application for analysis. In this application, an edge adapter is used to transform individual RTP packets into tuples, before further analysis can be performed on the data.

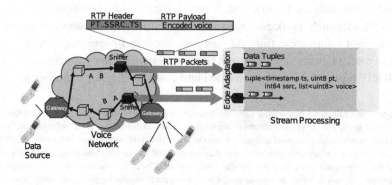

Figure 2.5 Data sources, streams, and tuples.

2.4.2 Processing

Once the data is available as tuples, it can be processed by the application's analytic components. In this section, we will outline basic stream processing concepts and terminology.

Operator

An *operator* is the basic functional unit in an application. In the most general form, an operator processes input tuples from incoming streams, applies an arbitrary function to them, and outputs tuples to outgoing streams. However, it is also possible that an operator might not process any incoming streams or generate any outgoing streams when acting as a data source or sink, respectively.

An operator may perform many different tasks:

(a) *Edge adaptation*, a task that consists of converting data from external sources into tuples that can be consumed by downstream analytics implemented by other operators.

(b) *Aggregation*, a task that consists of collecting and summarizing a subset of tuples from one or more streams, bounded by different logical, physical, or temporal conditions.

(c) *Splitting*, a task that consists of partitioning a stream into multiple streams for better exploiting data or task parallelism opportunities as well as for addressing other application-specific requirements.

(d) *Merging*, a task that consists of combining multiple input streams with potentially different schemas based on a matching or alignment condition including temporal or other explicitly defined affinity condition.

(e) *Logical* and *mathematical* operations, tasks that consist of applying different logical processing, relational processing, and mathematical functions to tuple attributes.

(f) *Sequence manipulation*, a task that consists of reordering, delaying, or altering the temporal properties of a stream.

(g) *Custom* data manipulations, tasks that consist of applying data mining, machine learning, or unstructured data processing techniques, potentially, combined with additional user-defined processing.

As will be seen in Chapters 10 and 11, many *custom* analytics include techniques borrowed from signal processing, statistics, and data mining as well as optimization theory. Irrespective of the task it performs, an operator generally ingests incoming data from other operators or external sources and produces new data. Incoming and outbound data is channeled in and out of an operator as separate streams. Each stream is associated with a *port*, which acts as an endpoint for receiving or producing tuples.

An *input port* is the logical communication interface that allows tuples to be received. Similarly, an *output port* allows tuples to be transmitted by an operator. An operator may have zero or more input ports as well as zero or more output ports.

A *source operator* has no regular input ports, even though it ingests external data, acting as an edge adapter to transform raw data into tuples. For instance, the source operator in our previously defined example extracts information from RTP packets to populate the corresponding tuple attributes.

A *sink operator* has no regular output ports, even though it emits data to external consumers, acting as an edge adapter to transform incoming tuples into data encoded for external consumption. For instance, a sink can convert tuple attributes into RTP packets.

As discussed, an operator typically produces new data as a result of its internal analytical or data manipulation task. In general, an internal analytical task has a long-lived *continuous* processing requirement, as opposed to a *process-to-completion* model found in traditional data processing tasks. Moreover, the analytical task is generally temporarily constrained, either because specific tasks must be completed in a given amount of time or simply because the execution of the task must be able to keep up with the incoming flow of data.

Stream

Streaming data in the form of a sequence of tuples is transported through a *stream connection* between the output port of an upstream operator and the input port of a downstream operator.

A stream connection may be physically implemented using a transport layer (Section 7.3.5), providing a common interface that hides the low-level transport mechanism.

Stream Processing Application

An operator typically works in concert with others by sharing tuples via streams. A set of operators and stream connections, organized into a data flow graph, form the backbone of a *Stream Processing Application* (SPA).

A SPA[3] is usually designed to solve a specific continuous analytical task. Hence, raw data flows into its source operators, which continuously convert them into tuples and send them for further processing by the rest of the operators in the application flow graph, until results eventually are sent by sink operators to external consumers.

The structure of the flow graph represents the *logical view* of the application, in contrast to its *physical view* (Section 5.3.1). The flow graph may be arbitrarily complex in its structure[4] with multiple branches, fan-ins, and fan-outs as well as possible feedback loops, which are typically used for providing control information back to the earlier stages of processing.

Stream processing application tasks

Stream Processing Applications (SPAs) are different from regular procedural applications. There are several fundamental distinctions that arise from the type and nature of the data processing carried out by them.

---

[3] From this point forward, we will simply use the term *application* when it is clear from the context that we are referring to a SPA.

[4] This book's cover depicts the flow graph for a very large, synthetic, application.

First, a SPA must cope with the *temporally* and *sequentially* paced nature of the incoming data. In other words, an application must be designed to cope with *when* and *how* quickly the data arrives, as well as to handle the data in the order it arrives.

Second, a SPA generally makes use of complex processing logic applied to continuously changing data. This is because of the *in-motion* data processing required to fulfill its analytical task. In most cases, random access to the data is not possible, unless a buffering mechanism is used, incurring additional computational and resource utilization costs.

Moreover, the analytical task must be designed to perform a *single pass* over the incoming data, as multiple passes are usually detrimental to the application's performance. Furthermore, the implemented data analysis must be able to *cope* and *adapt* continuously to changes in data, resource availability, and goals as the application runs.

The analytical tasks can have different latency and time-sensitivity requirements, may require managing different amounts of state, and may perform different types of analysis. Stream processing analytical tasks include different degrees of time-sensitivity, as a function of how quickly information must be processed and how quickly it must lead to a reaction (Section 2.2).

SPAs are also distinct along two other dimensions: whether their internal processing task is *stateful* or *stateless*, and also in the nature of these tasks. *Stateless* tasks do not maintain state and process each tuple independently of prior history, or even from the order of arrival of tuples. *Stateful* processing, on the other hand, involves maintaining information across different tuples to detect complex patterns. Note that stateful processing also creates more stringent fault tolerance requirements for the application, as state must be reliably maintained in spite of the occurrence of failures.

The nature and complexity of the analytical task is determined by the function it performs. Common functions include *transformation*, *detection*, and *exploration*.

A *transformation task* is often the first step in any data analysis undertaking. It includes several data pre-processing steps, such as cleaning, filtering, standardization, format conversion, and de-duplication. In addition to pre-processing the data, a transformation task might also include different types of business logic that must also be applied to the incoming data before it is ready for downstream consumption. For the most part, transformation tasks are stateless, although some of them, for instance, de-duplication, are stateful.

A *detection task* involves the identification of certain pre-defined patterns in the data. This type of task exploits the known structure of the data as well as of the sequence in which data items might appear in the stream. The patterns may be codified as *rules*, with a *logical predicate* representing the condition of interest on an individual data item or on a sequence of items. These rules may be implemented by different procedures and, often, by data mining models.

An *exploration task* focuses on the identification of previously unknown patterns or items of "interest" in the data using *unsupervised* analytics and *incremental* machine learning techniques. Exploration tasks require dynamic adaptation of the processing in response to different variations in data characteristics and resources, and based on *what* is discovered in the data. These tasks often use *feedback-driven adaptation* mechanisms. Since exploration tasks require sophisticated analysis and processing, they also may need to interact with external non-streaming offline sources of data and analytical processing placed outside the SPS.

Revisiting the VoIP application

We now revisit the example from Section 2.4.1, considering two possible analytical applications. The goal of the first application, named *Profile*, is to analyze the qualitative characteristics of the incoming VoIP data. This application publishes results at periodic intervals to a visual interface with information that includes the aggregate as well as the instantaneous rates of traffic, latency, and jitter[5] experienced by each user. This application implements a detection task.

The goal of the second application, named Who is Talking To Whom (WTTW) [34], is to identify pairs of conversations from the multiple simultaneous VoIP streams. The results produced by this application are used to update an evolving *social network* [35] capturing the interactions between people in a constantly evolving community. This application transmits its results to a social network visualization engine, which is able to display information about newly detected users of the VoIP system as well as interconnections between them.

Each of these applications has several subcomponents (Figure 2.6). For instance, the WTTW application must not only pair RTP data packets reconstructing the phone conversations, but also identify individual speakers to make sure that the social network can be updated accurately.

In the figure, there are two distinct logical application flow graphs representing the two applications. Each of them receives tuples from common source operators, which perform edge adaptation tasks, consisting of transforming the RTP packets into tuples.

The application flow graphs have different structures including operators with multiple input and output ports. Structurally, we can also observe *fan-in* (multiple streams

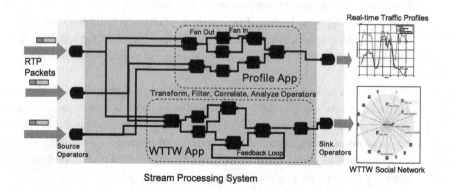

Figure 2.6 Operators, ports, and streams in the *Profile* and *WTTW* applications.

---

[5] Jitter is a deviation from the assumed periodicity of a signal.

connected to an operator's input port) and *fan-out* (a single stream being routed to multiple downstream operators) configurations as well as *feedback loops*.

Both applications employ different sink operators to send results to their respective external visualization interfaces. In *Profile*'s case, a real-time Graphical User Interface (GUI) for depicting the VoIP flow profiling results and, in WTTW's case, a social network visualizer.

Figure 2.6 showcases the structure for both applications in the form of their *logical* (and, in this case, contrived) flow graphs delineating each application's analytical components. Both of these flow graphs have *physical* counterparts that correspond to how these applications are instantiated on the computing environment where they run. The physical application flow graph is a collection of distributed and interconnected processes with individual operators mapped to the physical processes with potentially multiple operators hosted by the same process.

2.4.3   System architecture

A Stream Processing System (SPS) provides the *middleware*[6] on which streaming data is ingested, analyzed, and output by SPAs. SPSs typically have two main components: an application development environment and an application runtime environment.

Application development environment
The *application development environment* defines the framework and tooling used for implementing applications. This environment typically comprises a (1) *programming model*, organized around a programming or query language to express the application's internal processing logic and structure, and (2) a development environment where an application can be interactively implemented, tested, debugged, visualized, and fine-tuned.

In general, the continuous nature of streaming data and of the analytics implemented by a SPA are best expressed by specialized stream processing languages and programming models where abstractions for operators, ports, streams, and stream connections exist. Yet, it is also possible to write the code required by a SPA using conventional procedural programming and relational query languages, at the expense of some semantic impedance, that is, a mismatch between what needs to be expressed by an application and what can be expressed naturally and elegantly by a programming language.

Thus, the programming model embodied by a *stream processing language* provides constructs for manipulating data as well as for implementing, connecting subcomponents, and configuring an application. As discussed in Chapter 1, the data management literature [36] employs two terms to distinguish between the languages used for these tasks: Data Definition Language (DDL) and Data Manipulation Language (DML). In many cases, however, they are simply combined into a single one.

Fundamentally, a stream processing language includes constructs for (1) defining a *type system*, which captures the structure of the data an application manipulates;

---

[6] Middleware is the term used to designate the software that provides services to other software applications beyond those available from the operating system.

(2) defining the application as a *flow graph*, which translates the analytic requirements into a runnable program; (3) implementing mechanisms to manipulate data, which capture the necessary buffering and subsequent operations to be carried out; (4) creating processing operators, which embody an analytic algorithm and allow the modularization of an application as well as the reuse of common functionality across applications; and, finally, (5) defining configuration and extensibility parameters, which enable an application to be customized for a particular runtime environment.

There are four different classes of programming languages to support the implementation of SPAs: declarative, imperative, visual, and pattern-based.

Declarative languages express the analytic processing and the expected results of a computation without describing the execution or control flow necessary to implement it. This is the paradigm behind SQL-inspired languages commonly used in traditional relational databases. Their stream processing cohorts include the necessary extensions required for implementing continuous processing. Examples of such languages include StreamBase's StreamSQL [37] and STREAM's Continuous Query Language (CQL) [38].

Imperative languages explicitly specify the analytic processing execution plan in the form of an application, by wiring up operators that process the data stream and thus prescribe exactly how the data is to be processed and how the distributed state of an application is maintained. Imperative languages are also often combined with visual tools used to create an application by piecing together and configuring its compoents. An example of such languages is Massachusetts Institute of Technology (MIT)'s StreamIt [39].

Visual programming languages specify the analytic processing implemented by an application using a visual composition mechanism and a drag-and-drop interface to wire up widgets representing different computational steps. An example of such languages is Aurora's Stream Query Algebra (SQuAl) [40].

Pattern-based languages use *rules* to specify the analytic processing to be performed on the incoming data. Typically this processing includes identifying patterns in the data and detecting *interesting* occurrences in a monitoring task. These languages often include conditions that define when a rule *fires* as a function of logical, temporal, and sequencing-based pre-conditions. As seen in Sections 2.3.1 and 2.3.4, rules are also associated with *actions* to be undertaken once the pre-conditions for a rule are satisfied.

Finally, some languages combine elements from these different programming language classes. IBM's SPL is one example of such languages and so is the Event-Flow/StreamSQL employed by StreamBase.

Example
Consider the examples, adapted from an earlier survey [5] depicted by Figure 2.7, showing these different programming language classes.

The example demonstrates how a *transformation rule* that requires merging tuples from streams F1 and F2 with a specific windowing configuration can be implemented.

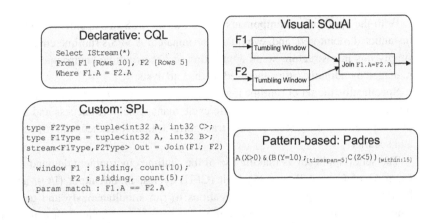

Figure 2.7 Classes of languages for stream processing.

The condition in this case states that the previous ten tuples from F1 and the previous five tuples from F2 should be merged if (and only if) the attribute A from an F1 tuple and the attribute A from an F2 tuple in the streams' respective windows match. Note that this rule makes use of temporal and logical patterns.

The declarative language example shows the CQL version, while the visual example depicts the SQuAl version. The SPL version corresponds to what a hybrid language might look like with elements from both declarative as well as imperative languages.

Finally, the figure also shows an example of a detection rule expressed in a pattern-based language, in this case, Padres. Note that its formalism is substantially different when contrasted with the other languages. Specifically, A, B, and C are different item types, with individual attributes X, Y, Z. This Padres rule is designed to trigger when an item of type A enters the system with an attribute X whose value is greater than zero, an item of type B whose attribute Y equals 10, and an item of type C whose attribute Z is less than 5 within 5 to 15 time steps after the arrival of an item of type B.

In many cases, the task of writing applications in any of these languages is made simpler by Integrated Development Environments (IDEs) [41, 42], where quick prototyping, testing, debugging, and, in some cases, deploying applications onto a test runtime environment is supported.

In Chapter 3, we will revisit this discussion when we switch the focus to the development of applications and highlight how the characteristics of SPAs shape stream processing programming models and languages.

Application runtime environment
The application runtime environment comprises the infrastructure that provides the support and services for running one or more SPAs and managing their execution life cycle. The environment consists of a software layer, potentially distributed across a multi-host environment on which the applications are instantiated, along with the environment's own management components.

Both the application components as well as the SPS management services must be instantiated monitored, and, eventually, terminated. A SPS's runtime environment must supply this infrastructure so an application and the runtime environment itself can be managed by administrators and used by data analysts.

Specifically, the set of runtime management interfaces must provide support for capabilities that include (a) application life cycle management, (b) access and management of computational resources, (c) data transport, (d) fault tolerance, and (e) authentication and security.

The set of applications making use of the runtime environment demand *resources* in the form of Central Processing Unit (CPU) cycles, Input/Output (I/O), and network bandwidth, allowing multiple applications to run simultaneously and to collaborate among themselves, sharing the common computational resources drawn from a pool of computational nodes.

The *resource management* service provided by the runtime environment ensures that the computational resources are continuously available to the applications by monitoring them and by ascertaining that they are well-apportioned across the applications. This is accomplished by employing scheduling and load balancing policies. Indeed, the more sophisticated SPSs include the ability to autonomically monitor their own distributed software components, ensuring their availability even in the presence of transient failures.

Moving tuples from one operator or from one application to another is a fundamental function of a SPS. A SPS *data transport* comprises the set of mechanisms for reliable movement of tuples from one operator to another as well as from a producer application to a consumer one. The data transport design in a SPS's runtime environment shares many commonalities with communication substrates developed to support modern network-based applications [43]. Yet, the transport layer provided by SPSs must also address stream processing-specific requirements in terms of Quality of Service (QoS) levels, fault tolerance, as well as in terms of the continuous nature and low-latency demands of many SPAs.

Finally, enterprise-grade SPSs must also be capable of supporting the concurrent execution of multiple applications as well as of a community of users. Hence, the ability to authenticate users and applications, provide audit trails, as well as security barriers between applications are all part of a SPS's *authentication* and *security* infrastructure.

Not every SPS includes all of the capabilities discussed above. Even when some of these capabilities are present, there is a reasonably large amount of variability in how they are engineered. In Chapter 7, we will revisit this discussion focusing on the architectural building blocks supporting a conceptual, full-fledged runtime environment.

2.4.4   Implementations

The combination of a more sophisticated programming model and runtime environment makes SPSs uniquely suited to address the requirements of workloads and analytics of the type discussed in Section 1.2. There are several implementations of SPSs by different research and open-source groups as well as commercial enterprises, that realize different engineering solutions and capabilities. This diversity makes it worth describing these systems, not only to provide a historic perspective, but also to showcase the different technical solutions and system infrastructure they provide to application developers. This is the focus of the rest of this section.

2.4.4.1   Academic systems

The development of full-fledged, commercial-grade SPSs can be traced back to a series of trailblazer prototypes produced in academia. We will highlight three of the earliest systems, Telegraph Continuous Queries (TelegraphCQ), STREAM, as well as Aurora/Borealis, and provide a very brief survey of other important academic systems.

Telegraph Continuous Queries

TelegraphCQ [44] is one of the earliest SPSs and was developed at the University of California, Berkeley. Its design roots come from Telegraph [45], an earlier system built for adaptive dataflow processing, and its implementation borrowed heavily from PostgreSQL [46]. TelegraphCQ's development was relatively short-lived, but led to a commercial implementation and the creation of a startup, Truviso, later acquired by Cisco.

TelegraphCQ was designed to support continuous queries over a combination of tables and data streams. Its programming model relied on a declarative programming language, StreaQuel, with operators designed for stream relational processing directly inspired by SQL counterparts and extended with a rich windowing mechanism used to buffer and operate on groups of tuples.

TelegraphCQ also included a simple version of edge adaptation in the form of a user-defined wrapper designed to ingest data from external sources. StreaQuel's query plans were executed by a TelegraphCQ's distributed runtime with adaptive logic used to route data efficiently across operators and the distributed nodes of the runtime.

The runtime environment was eventually optimized to handle queries combining real-time with historical data by employing a framework in which multiple resolutions of summarization and sampling could be simultaneously generated [47]. This approach can be looked at as a precursor to many of the techniques used to speed up the online analytics that SPAs rely on (more in Chapters 9 and 10).

STREAM

STREAM [48] was developed at Stanford University. Its early prototype had targeted environments with streams that may be fast-paced and have load characteristics that may vary over time. Due to this variability, the STREAM system was designed to work under severe resource constraints and to support continuous adaptivity as an application runs.

The STREAM programming model employs CQL [38], a declarative language with relational and stream processing operators whose abstract semantics is based on two basic constructs: *streams*, which are bags (collections) of tuple–timestamp pairs, and *relations*, which are time-varying bags of tuples.

CQL operators are particularly interesting because they were partitioned into three classes according to the input and output they operate on and generate as a result: *relation-to-relation*, *stream-to-relation*, and *relation-to-stream*. The first type, relation-to-relation operators, is derived directly from SQL counterparts and these operators are used to perform mostly transformational tasks. The second type, stream-to-relation operators, adds window constructs on top of the corresponding relational operators to enable temporal analysis. Finally, the third type, relation-to-stream operators, takes one or more relations as input and produces a stream as output.

In the STREAM runtime environment, applications are expressed in terms of query plans. Owing to its database architectural roots, when a continuous query is registered with the STREAM system, a *plan*, comprising a tree of operators, is generated. This plan indicates how the actual processing is carried out and it employs *queues* to buffer tuples (or references to them) as they move between operators, and *synopses* to store the operator state as the execution progresses.

When a query plan is executed, a *scheduler* selects operators to execute based on the (explicit) timestamps of the tuples they are supposed to process, ensuring that the order of execution has no effect on the query result. As will be seen later in this book, this determinism is not shared by all SPSs as, in some cases, it is not even necessary. Determinism, however, does impose additional computational costs during the execution of the query plan.

STREAM's runtime employs a set of interesting performance optimizations to speed up the execution of query plans. For instance, *synopsis sharing* is used to reduce memory utilization when multiple synopses in a query plan materialize nearly identical relations; *constraints exploitation* is used when data or properties of its arrival can be harnessed to reduce the size of a synopsis; and *global operator scheduling* is used to reduce memory utilization, particularly with queries that manipulate bursty input streams.

The STREAM runtime environment also includes a monitoring and adaptive query processing infrastructure, StreaMon. It manages the query plan, collects and maintains statistics about the streams and plan characteristics, and, more importantly, can re-optimize the query plan and the in-memory structures used in its execution. STREAM also makes use of load shedding techniques to probabilistically drop tuples under severe resource constraints.

STREAM includes a GUI-based query and system visualizer, the STREAM Visualizer, which allows system administrators and developers to inspect the system while it is running, as well as to experiment with adjustments.

The early STREAM prototype did not implement any support for distributed processing nor did it provide any fault tolerance mechanisms, but the STREAM researchers clearly identified these components as critical to a production-grade SPS.

Aurora

Aurora [40, 49] was developed jointly by a collaboration between Brandeis University, Brown University, and MIT. The development of Aurora is particularly notable because its architectural underpinnings [49] brought along a series of ground-breaking innovations in the design of both the programming model as well as in the runtime environment of SPSs.

Aurora's programming model is built on an imperative and visual programming language named SQuAl. SQuAl uses a *boxes and arrows* paradigm, common in workflow and process flow systems, to specify applications as Directed Acyclic Graphs (DAGs), where arrows represent stream connections between boxes representing operators.

SQuAl's operators are drawn from the relational algebra and include `filter`, `split`, `map`, `union` as well as order-sensitive operators such as `resample`, `bsort`, `join`, and `aggregate`.

Aurora's runtime system includes a scheduler [49], a QoS monitor, a load shedder [50], high-availability support [51], and a storage manager. After the development of Medusa [52], a system designed to provide networking infrastructure for Aurora operations, an application running on the system, called an Aurora *network*, could be spread across any number of hosts. Indeed, these hosts can either be under the control of one entity or can be organized as a loosely coupled federation with different autonomous participants.

Medusa provided a distributed naming scheme so that the output of an application from a host can be connected to another on a different host. It also provided a collection of distributed catalogs used to store the pieces of the workflow described by an application and its interconnections.

Finally, Medusa provided a distributed data transport layer, allowing the efficient delivery of messages between hosts. All of these functions now exist in commercial SPSs (Chapter 7).

When in execution, an Aurora network can accept input data, perform data filtering, computation, aggregation, and correlation, and, eventually, deliver output messages to client applications. In this network, every output can be optionally tagged with a QoS specification, indicating how much latency the connected application can tolerate as well as information on what to do if adequate responsiveness cannot be assured.

Using the QoS tag, Aurora's optimizer can rearrange the physical structure of the network to meet the required QoS requirements. Along the same lines, at runtime, Aurora's scheduler makes decisions based on the form of the network, the QoS specifications as well as the length of the various queues. These decisions include reordering flows, delaying or storing parts of the streaming data, as well as combining operators to minimize communication and processing overheads.

After substantial groundwork was in place, the Aurora platform led to the creation of StreamBase (Section 2.4.4.2), which developed one of the earliest commercial SPSs.

Borealis

Borealis [53] is a research offshoot created to address some of the limitations in Aurora's design. Borealis inherited the core stream processing functionality from Aurora as well as the distributed functionality from Medusa. Due to its distributed architecture, a collection of queries submitted to the system can be seen as a network of operators whose processing occurs at multiple sites. To support this model, the Borealis's runtime added infrastructure to allow the dynamic deployment and adaptation of the applications running on the system as well as capabilities to dynamically revise derived results, reflecting additional information gleaned from delayed data.

Many of these capabilities have been included in modern commercial SPSs. As described in Aurora's retrospective paper [54], support for historical data, synchronization capabilities, resilience to unpredictable stream behavior, XML and other feed format edge adaptation, programmatic interfaces, and globally accessible catalogs as well as an unrelenting focus on performance and scalability have all become commonplace in commercial SPSs (Chapters 7 and 8).

Finally, in addition to implementing two of the first SPSs, the Aurora/Borealis research team was also the first to propose a set of rigorous performance tests in the form of their Linear Road benchmark [55], allowing the performance of different SPSs to be compared [38, 56, 57].

While the Aurora/Borealis systems proposed multiple ground-breaking ideas, substantial advances in theory, programming models, distributed programming, and system support have also been made by other academic, open-source, and commercial institutions leading to the development of current SPSs. Some of these systems are outlined later in this chapter.

Other systems

The AT&T Labs Gigascope [58] is a SPS specifically designed for network traffic analysis and performance monitoring with an SQL-based programming language named Gigascope SQL (GSQL). Stream Mill [59], a general-purpose SPS developed at the University of California, Los Angeles (UCLA), also employs an SQL-based language to allow its operators to be combined with ad hoc functions that perform custom aggregation. The Public Infrastructure for Processing and Exploring Streams (PIPES) [60] project from Philipps Universität Marburg was built not as a full-blown system, but as a set of building blocks that can be used to assemble a SPS. This project has a programming model with an operator algebra along with well-defined deterministic temporal semantics implemented by the XXL library [61]. Other academic SPSs include NiagaraST [62] and the Global Sensor Networks (GSN) [63, 64].

2.4.4.2   Commercial systems

Commercial SPSs provide production-grade middleware with regular maintenance, training, and the plethora of services provided by their respective vendors. In this section, we will first provide a brief overview of two commercial systems that, while not implementing all of the features expected from a SPS can still be used for developing certain types of SPAs: TIBCO BusinessEvents and Oracle CEP.

Next, we will describe two commercial systems closely linked to earlier research efforts: StreamBase, based on the Aurora/Borealis systems, and IBM InfoSphere Streams, based on the System S research prototype.

Finally, another notable system is SAP's Sybase Event Stream Processor (ESP) [65, 66], which isn't further discussed here, but also shares certain commonalities with both

TIBCO's and Oracle's offerings, positioning itself as a CEP system with a few technical similarities to SPSs.

TIBCO BusinessEvents

TIBCO BusinessEvents [19] is a CEP system that provides reasonably sophisticated query and event pattern matching capabilities. TIBCO BusinessEvents has been used, for example, to process data from financial data feeds as well as website monitoring feeds.

As a CEP platform, TIBCO BusinessEvents lacks many of the distributed processing, extensibility, and fault tolerance capabilities provided by SPSs. Yet, simpler transformation and detection tasks can easily be implemented on it. As a commercial platform, its major strengths are in its programming model and in its integration with messaging infrastructure, including other TIBCO software products and other vendors providing the relatively common Java Message Service (JMS) interface.

The TIBCO BusinessEvents programming model employs Unified Modeling Language (UML)-based models to describe applications, services, and servers. These models are used to define the causal and temporal relationships between events as well as actions to be taken when a particular event occurs. The modeling task is performed using a rule editor, which is part of its TIBCO BusinessEvents Studio.

TIBCO BusinessEvents supports applications written using two different interfaces: a continuous, windowed query interface as well as a pattern matcher interface. The former is where this platform shares some commonalities with more sophisticated SPSs.

The SQL-like query language [67] allows developers to craft both *snapshot* as well as *continuous* queries. These queries are executed by agents deployed on the platform's runtime environment. An agent can execute the query against a previously primed cache or listen to an event stream as if the data was hosted by regular database tables.

On the other hand, its pattern description language [68] is simple and shares some similarities with both SQL and with regular expression-based languages. It includes constructs for defining patterns and configuring them for deployment using the platform's pattern matcher. The language allows the implementation of correlation operations across event streams, the definition of temporal event sequence recognition operations, the suppression of event duplicates, and the implementation of simple store-and-forward tasks, where data can be accumulated in time windows before additional processing.

Oracle CEP

Oracle Complex Event Processing (CEP) [18, 69] also includes a few features found in more sophisticated SPSs. Its primary analytical goals include filtering, processing, and correlating events in real-time. Its main application development framework is an Event Processing Network (EPN), where the analytics are implemented as a combination of data adapters and continuous queries described in Oracle's CQL [70]. CQL is fully compliant with the American National Standards Institute (ANSI) SQL 99 standard [71], but also adds constructs to support streaming data.

Typically an EPN makes use of three operational constructs: *event sources, event sinks*, and *processors*. An *event source* is a data producer on which CQL queries operate. Practically, an event source can be an *adapter* connecting to a JMS messaging interface, a HyperText Transfer Protocol (HTTP) server, a file, a stream (or *channel*), a *processor*, a *cache*,[7] or a *relational database table*. Conversely, an *event sink* is a consumer of query results. An event sink can be an *adapter*, a *channel*, a *processor*, or a *cache*.

More interestingly, a *processor* is where the analytical portion of an application resides, but a processor can also act as either an event source or sink. A processor is associated with one or more CQL queries that operate on events arriving from multiple *incoming* channels. A processor is usually connected to one or more *output* channels where the results are written. To make results available to external applications, each EPN output channel is normally connected to a user-written Plain Old Java Object (POJO) that takes user-defined actions based on the events it receives.

Oracle CEP's programming model is supported by a visual development environment that is also integrated with conventional Java development tooling. This environment is used to manage all of the development tasks, supporting syntax highlighting, application validation, graphical EPN editing, as well as management of component and application assembly.

Oracle CEP's query execution engine works in-memory and continuous queries listening on a stream are able to combine streaming events with data that has been previously stored. Oracle CEP's runtime environment hosts a server where containers for the EPN applications are used to manage their life cycle. This environment consists of a real-time JVM with deterministic garbage collection, providing services that include security, database access through a Java Data Base Connectivity (JDBC) interface [72], web connectivity via HTTP interfaces, connectivity to publish–subscribe systems, as well as logging, and debugging infrastructure.

Concerning scalability and distribution, applications can be designed such that the incoming data flow can be partitioned at the point of ingress or inside an EPN, providing a basic mechanism for segmenting the workload.

Furthermore, Oracle CEP supports application *hot standby* replicas, where each replica of a server processes an identical stream of events. The use of this fault-tolerant capability requires an EPN to make use of a High Availability (HA) adapter implementing additional infrastructure to ensure continuous operation under faulty conditions.

As part of the runtime management infrastructure, additional tooling in the form of the GUI-based Oracle CEP Visualizer provides an environment designed to support runtime configuration and the administration of applications, queries, caching, and clustering.

---

[7] In this case, a cache is a temporary storage area for events, created to improve the performance of an application. Oracle CEP includes its own caching implementation associated with a local Java Virtual Machine (JVM) and making use of in-memory storage. Other kinds of cache, for instance using disk storage, can also be implemented and plugged-in.

StreamBase

StreamBase [73] is a commercial SPS whose lineage can be traced to the Aurora and Borealis SPS projects. It has been used in numerous application domains, including real-time algorithmic trading as well as order execution and routing in financial engineering applications.

StreamBase's programming model has two language facets: EventFlow [74], a visual language, and StreamSQL [37], a combined declarative SQL-like DDL and DML. Both languages, which in some ways are interchangeable,[8] include relational operators extended to be used with time- or event-based windows when manipulating streaming data.

These operators perform a variety of functions including filtering, aggregation, merging, and correlation of data from multiple streams as well as carrying out complex analytics. Due to its SQL roots, StreamSQL can seamlessly reference database-stored data. A StreamSQL query can also make use of external data sources via *adapters*, including a variety of market data feeds and financial data messaging interfaces.

The set of StreamSQL operators can be augmented with custom Java-implemented operators [75]. Likewise, StreamSQL can also be extended with new custom adapters (used to convert data to and from the StreamBase tuple protocol), functions, and *client* applications.[9]

Finally, StreamSQL also includes a Drools operator [37], which relies on a Java-based rule management engine to support a simple pattern matching language that can be used to implement typical transformation and detection tasks. Interestingly, this capability demonstrates an approach by which CEP-like functionality can be made available as part of a SPS.

The StreamBase runtime environment [76] consists of one or more StreamBase *servers*, called sbd. A server, that in addition to hosting applications, provides tracing, debugging, logging, as well as profiling and monitoring services. Multiple sbd instances can be executed on a host.

StreamBase defines the notion of *containers* [77] for organizing and connecting multiple applications. Each container runs a single application and serves as a handle for an application's resources. Hence, it is possible to run multiple simultaneous applications in a server by employing multiple *containers*. The container abstraction also makes it possible to establish and maintain connections, including dynamic ones, between streams from different applications as well as to tap into JMS-based messaging infrastructure [78] for exchanging information with other components in a larger IT environment.

An application hosted by StreamBase can scale up by making use of multiple servers. Yet, this approach requires that the developer first breaks down their application into smaller components. Consequently, the workload partitioning, the data distribution, as well as the coordination of the multiple components are all handled manually.

---

[8] Limited conversion is possible between EventFlow and StreamSQL [37].

[9] In StreamBase terminology, a client application is used to feed and consume data to and from a StreamBase server.

StreamBase's runtime environment provides fault tolerance by relying on a redundant *hot standby* replica. Both replicas execute the same application, one running as the primary and the other as a secondary backup.[10] StreamBase's high-availability (HA) infrastructure can be integrated with third-party components to make use of virtualization as well as of Storage Area Network (SAN)-managed data repositories.

Applications are developed using StreamBase Studio, a Rapid Application Development (RAD), Eclipse-based [79] visual IDE, which provides tooling to manage all stages of an application's development process, including design, test, and deployment.

Finally, as part of StreamBase's infrastructure ecosystem, StreamBase LiveView [80] provides *live* data warehousing-like capabilities against which continuous queries can be executed. From an application's standpoint, LiveView provides data preprocessing, continuous query processing, data aggregation capabilities, alerting and notification, as well as desktop and client connectivity.

System S/InfoSphere Streams

System S, also one of the earliest SPSs, and its commercial counterpart InfoSphere Streams were developed by IBM (Section 8.2). We omit a more detailed description of these systems in this section as the upcoming chapters will use InfoSphere Streams to illustrate the process of developing and implementing SPAs (Chapters 3–6), as well as to discuss the internal architecture of a SPS (Chapters 7 and 8).

2.4.4.3   Open-source systems

In parallel with the development of academic and commercial SPSs, two open-source systems, Yahoo S4 and Twitter Storm, are, at the time of this writing, under active development. We discuss them briefly in this section.

Yahoo S4

S4 [81, 82] is a general-purpose, distributed, scalable, partially fault-tolerant, pluggable platform that allows programmers to develop applications for processing streaming data. S4 was initially proposed at Yahoo as a framework to process user feedback related to online advertisement placement. One of its first applications consisted of tuning parameters of a search advertising system using the live traffic generated by queries posted via a web search engine.

Its design is inspired by both the MapReduce framework [83] as well as by the Stream Processing Core (or SPC) [84], the original processing engine used by System S. The S4 architecture provides encapsulation and location transparency, which, in similar fashion to InfoSphere Streams, allows applications to be concurrent, while exposing a simple programming model to application developers. A limitation of its runtime, however, is the lack of reliable data delivery, which can be a problem for several types of applications.

---

[10] StreamBase also supports a mode where a *warm* replica is used. This type of replica differs from a *hot* one in the sense that it does not send tuples downstream until a fault occurs when it becomes the primary.

Applications are written in terms of small execution units, or Processing Elements (PEs), a term which they share with System S, written in Java. PEs are meant to be generic, reusable and configurable and can be employed across various applications. Hence, S4's programming model is also extensive, but lacks the high-level support of a composition language such as Streams' SPL.

After S4's initial release by Yahoo in 2010, it became an Apache Incubator project and is still under active development.

Twitter Storm

Storm [85, 86] was originally implemented by BackType, a small company later acquired by Twitter. Its original design was to provide the ability to process the stream of *tweets* produced by users of the Twitter [87] social networking service, in real-time.

Storm is a general-purpose SPS engineered to execute continuous queries over streaming data with a runtime environment that provides reliable data delivery, robustness, and application-level fault tolerance.

Storm's distributed runtime includes a master component, called *nimbus*, used to distribute the application work across a cluster, and worker components, called *supervisors*, which manage the local *worker* processes that implement (a subset of) a Storm application, called a *topology*.

Storm's primitive operators are either a *spout*, a generator of streams connecting directly to an external data source, or a *bolt*, a consumer and/or a producer of streams that encapsulates arbitrary processing logic. Multiple *spouts* and *bolts* can be packaged in a *topology* to construct a SPA.

Storm's programming model is language-agnostic and processing logic can be implemented in any language, although Java is the natural choice, it being the language in which Storm was implemented. When non-JVM byte code languages are employed, the internal processing logic is hosted by a subprocess that communicates with its managing *worker* by exchanging JavaScript Object Notation (JSON) messages.

A *topology* can specify the level of parallelism it desires by stating how many workers it wants. Each of its *spouts* and *bolts* can also set the number of threads they need. To further aid with scalability, tuples produced by *spouts* and *bolts* can be distributed using stream *groupings*, which allows for a round-robin (*shuffle grouping*) or partitioned (*fields grouping*) tuple distribution for downstream processing.

Storm's fundamental application fault tolerance mechanism consists of ensuring that each message coming off a *spout* is fully processed and that *workers* remain active. If any *worker* goes down abruptly, it gets reassigned elsewhere, ensuring the continuity of the application's execution.

The specification of the application *topology* as well as its deployment and management is done using specially built Java Application Programming Interfaces (APIs). It is also possible to employ a *local* mode of execution, specifically designed for simplifying the task of testing an application, where no distributed resources are used.

Clearly, Storm's runtime environment and programming model provide a degree of flexibility and extensibility similar to what is available in commercial SPSs.

Table 2.1 Feature overview – commercial systems.

System	Programming model	Extensibility	Scalability	Analytics
StreamBase	SQL-based language and visual drag-and-drop interface	Advanced with wizard support to incorporate new custom Java operators	Multi-host support, manually managed	Large collection of built-in adapters and operators
TIBCO Business Events	Traditional CEP with SQL-based language and visual drag-and-drop interface	Limited	Limited	Built-in operators and pattern matching support
Oracle CEP	Traditional CEP with SQL-based language and visual drag-and-drop interface	Moderate with Java integration	Limited multi-host support, manually managed	Built-in operators and integration with code written in Java
Streams	SPL language and visual drag-and-drop interface	Very advanced, including semantic description and integration with SPL compiler	Extensive multi-host support, partially automated	Several domain-specific built-in toolkits with operators and edge adapters, plus direct integration with Java and C++ libraries

Nevertheless, the lack of a query or application composition language is a notable shortcoming.

2.4.5   Discussion

As indicated by the overview in this section, there are a large number of academic, commercial, and open-source SPSs. The design space explored by these platforms is extremely diverse in terms of the programming models as well as the runtime environments they provide. As expected, the development effort to write the same application on these different platforms can vary widely.

More importantly, however, the availability of commercial SPSs provides solid environments that can be used for implementing production-grade SPAs. Yet, these platforms have programming models and runtime environments that are substantially different. At the time of this writing, their main features can be summarized across a few dimensions. We list these in Tables 2.1 and 2.2.

Table 2.2 Feature overview – open-source systems.

System	Programming model	Extensibility	Scalability	Analytics
Yahoo S4	Java	Advanced, but no high-level program-ming support	Advanced, implicit and manually managed	No toolkits, user-developed
Twitter Storm	Primarily Java with mechanism to integrate with other languages	Advanced, but no high-level program-ming support	Very advanced, implicit and manually managed	No toolkits, user-developed

From a programming standpoint, StreamBase relies on SQL-like semantics, while InfoSphere Streams provides SPL, a flexible application composition programming language. Both systems have extensible programming models that can be used to implement custom operations and functions.

Nevertheless the StreamBase programming model is constrained by its SQL roots, while SPL was designed from the ground up with numerous mechanisms for language extensions (Chapter 5), including: (1) an extensible type system with support for reflection; (2) a common infrastructure for both system-provided and user-developed operators, which are built using the same infrastructure; (3) the availability of *thunks*[11] to operators, allowing their parameterization with arbitrary expressions and aggregations; (4) a common window-management library with well-defined tuple management semantics; (5) the support for developing operators with arbitrary number of inputs and outputs; (6) dynamic composition capabilities allowing the integration of multiple applications sharing the same runtime environment; and (7) the support for higher-order[12] composite operators.

When evaluated from a distributed computing standpoint, InfoSphere Streams focuses on mid-to-large-scale clusters able to support a gamut of applications ranging in complexity from small-scale, single-node analytics to a fully distributed large-scale, fault-tolerant *solution*. A single logical multi-component SPL application implementation can be mapped to different physical layouts using a simple re-compilation step (Section 5.3.1). Conversely, in StreamBase the wiring of a distributed application is a manual process as is its re-tuning when the application is being prepared to be scaled up.

Finally, while academic and open-source SPSs are now available they lack some features of commercial systems, including integrated tooling for development, large libraries of operators and edge adapters, support for debugging, and optimization as

---

[11] In SPL we use the term *thunk* to denote a construct that allows certain operator aspects to be replaced at code generation time once these aspects are extracted from an application's source code.

[12] Operators that take other operators as parameters.

well as sophisticated application and system management environments. Nevertheless, open-source SPSs rely on implicit parallelism where the middleware manages the distribution of work and the execution of the distributed application, facilitating the process of prototyping a new application.

Therefore, these SPSs can be very useful for practitioners to test and evaluate stream processing concepts as well as to design real-world applications. On the other hand, the process of deploying and the long-term management of a production-grade application can be complex, particularly in a large-scale environment where this application might have to be integrated with other pieces of a large IT infrastructure.

2.5   Concluding remarks

There are several application domains where the use of continuous analytics and SPAs are required. The need for stream processing stems from the widespread availability of data that needs to be monitored and analyzed and that is continuously produced by sensors implemented as part of software and hardware systems. This data is available in both structured and unstructured formats, requiring pre-processing logic to make it consumable by the data analysis segments of an application.

The data analysis typically includes data mining and machine learning algorithms designed to infer and make forecasts based on the information that is gleaned from the data.

The applications hosting these analytics face stringent performance requirements in terms of latency as well as throughput to ensure that not only are they able to keep up with the data flow, but also that they can react quickly when an action is warranted. Moreover, these applications must be scalable, adaptive, and fault-tolerant, creating a combination of requirements that is unique and characteristic of stream processing.

While the specific domains and applications we described earlier in this chapter were not meant to be comprehensive, they clearly highlighted some of these different application requirements.

In the cybersecurity domain, we observed the need for real-time analysis in an adversarial and exploratory setting. In the transportation domain, we highlighted the need for specialized parsing and analysis in a geographically distributed setting. In the healthcare domain, we emphasized the need to integrate stream data with semi-static contextual information as well as the need for data provenance tracking and fault tolerance. Like these applications, many others can benefit from this new technology and we discuss additional usage examples in Chapters 9 and 12.

The development of stream processing as a computing paradigm can be seen as an evolutionary step in the set of technologies referred to as information flow processing. Indeed, the programming and system infrastructure available to support continuous applications has become increasingly more sophisticated, starting with active databases, continuous query systems, publish–subscribe systems, CEP systems, and culminating with the development of SPSs.

In this chapter, we also provided definitions and illustrative examples of the fundamental concepts of stream processing, ranging from its building blocks (e.g., streams, operators, and applications) to the infrastructure that enables building and running SPAs, a SPS's programming model and its application runtime environment.

We closed this chapter with an overview of academic, commercial, and open-source systems to provide a global perspective of what has been explored in the system design space and the capabilities that are now available to those interested in writing new applications or converting existing ones.

In the next few chapters we will switch to the topic of application development where many of the ideas discussed here will take form.

2.6   Exercises

1. **Finance engineering and algorithmic trading.** Consider several data sources streaming live trade transactions from a *derivatives market* as well as from a *stock market.*

   Consider also the existence of a data warehouse with the historic profile of each of the derivative underliers (an underlier is a security or commodity which is subject to delivery upon the exercising of an option contract or convertible security) as well as with broad market measures.

   Design a SPA as a block diagram that assesses whether the prices for the options are *correct* by employing a price modeler that considers both live and historic information.

   While the specific algorithms are not important to your solution, the design must include a sketch of the stream schemas for the live data sources as well as an Entity Relationship (ER) diagram for the static data sources, along with a refined flow graph that connects the data sources, the processing logic components, and the data sinks.

   As a result the application must annotate each market transaction with additional attributes indicating the fair price and a recommendation to buy or sell. As part of this exercise, you are also expected to research the finance engineering literature and provide a design that is as realistic as possible.

2. **Real-time operations monitoring.** Consider a manufacturing plant that consists of a number of production lines. Assume that each line encompasses a number of processing stations. Each station generates a live feed with log information in the form of a data stream.

   Each log stream contains a continuous feed of events, corresponding to a manufactured item's arrival and departure to and from a processing station. Consider the existence of a database that contains additional information about the manufacturing lines, including the expected processing times for different stations.

   Design a SPA in the form of a block diagram that monitors the manufacturing plant, computes global and local statistics including averages of the processing times and the production rates, detects performance problems (e.g., sudden increases in the processing times), and identifies quality problems (e.g., items that have not been through an assembly step).

While the specific algorithms are not important to your solution, you are expected to research the literature on manufacturing automation and provide a design that is as realistic as possible. Minimally, it must include a sketch of the stream schemas for the live data sources as well as an ER diagram for the static data sources, along with a detailed flow graph that interconnects sources, the processing logic, and the consumers of the application results.

References

[1] Andress J, Winterfeld S. Cyber Warfare: Techniques, Tactics and Tools for Security Practitioners. Syngress; 2011.

[2] Worm Infects Millions of Computers Worldwide; published January 2009; retrieved in July, 2011. http://www.nytimes.com/2009/01/23/technology/internet/23worm.html.

[3] Amarasingham R, Pronovost PJ, Diener-West M, Goeschel C, Dorman T, Thiemann DR, *et al.* Measuring clinical information technology in the ICU setting: application in a quality improvement collaborative. Journal of American Medical Informatics Association (JAMIA). 2007;14(3):288–294.

[4] Blount M, Ebling M, Eklund M, James A, McGregor C, Percival N, *et al.* Real-time analysis for intensive care: development and deployment of the Artemis analytic system. IEEE Engineering in Medicine and Biology Magazine. 2010;29(2):110–118.

[5] Cugola G, Margara A. Processing flows of information: from data stream to complex event processing. ACM Computing Surveys. 2012;44(3):15.

[6] Lieuwen D, Gehani N, Arlein R. The Ode active database: trigger semantics and implementation. In: Proceedings of the IEEE International Conference on Data Engineering (ICDE). Washington, DC; 1996. pp. 412–420.

[7] Dayal U, Blaustein B, Buchmann A, Chakravarthy U, Hsu M, Ledin R, *et al.* The HiPAC project: combining active databases and timing constraints. ACM SIGMOD Record. 1988;17(1):51–70.

[8] Gatziu S, Dittrich K. Events in an active object-oriented database system. In: Proceedings of the International Workshop on Rules in Database Systems (RIDS). Edinburgh, UK; 1993. p. 23–39.

[9] Chakravarthy S, Mishra D. Snoop: an expressive event specification language for active databases. Elsevier Data & Knowledge Engineering. 1994;14(1):1–26.

[10] Act-Net Consortium. The active database management system manifesto: a rulebase of ADBMS features. ACM SIGMOD Record. 1996;25(3):40–49.

[11] Terry D, Goldberg D, Nichols D, Oki B. Continuous queries over append-only databases. In: Proceedings of the ACM International Conference on Management of Data (SIGMOD). San Diego, CA; 1992. pp. 321–330.

[12] Chen J, DeWitt DJ, Tian F, Wang Y. NiagaraCQ: a scalable continuous query system for Internet databases. In: Proceedings of the ACM International Conference on Management of Data (SIGMOD). Dallas, TX; 2000. pp. 379–390.

[13] Liu L, Pu C, Tang W. Continual queries for Internet scale event-driven information delivery. IEEE Transactions on Data and Knowledge Engineering (TKDE). 1999;11(4): 610–628.

[14] Strom RE, Banavar G, Chandra TD, Kaplan M, Miller K, Mukherjee B, *et al.* Gryphon: An Information Flow Based Approach to Message Brokering. The Computing Research Repository (CoRR); 1998. cs.DC/9810019.

[15] Carzaniga A, Rosenblum DS, Wolf AL. Design and evaluation of a wide-area event notification service. ACM Transactions on Computer Systems. 2001;19(3):332–383.

[16] Jacobsen HA, Cheung AKY, Li G, Maniymaran B, Muthusamy V, Kazemzadeh RS. The PADRES publish/subscribe system. In: Hinze A, Buchmann AP, editors. Principles and Applications of Distributed Event-Based Systems. IGI Global; 2010. pp. 164–205.

[17] Muhl G, Fiege L, Pietzuch P. Distributed Event-Based Systems. Springer; 2006.

[18] Oracle CEP; retrieved in October 2012. http://www.oracle.com/technetwork/middleware/complex-event-processing/overview/.

[19] TIBCO BusinessEvents; retrieved in October 2012. http://www.tibco.com/products/event-processing/complex-event-processing/businessevents/.

[20] IBM Websphere Business Events; retrieved in April 2011. http://www-01.ibm.com/software/integration/wbe/.

[21] Team E, Inc E. Esper Reference (Version 4.7.0). EsperTech Inc.; 2012.

[22] Brenna L, Demers A, Gehrke J, Hong M, Ossher J, Panda B, *et al.* Cayuga: a high-performance event processing engine. In: Proceedings of the ACM International Conference on Management of Data (SIGMOD). Beijing, China; 2007. pp. 1100–1102.

[23] Wu E, Diao Y, Rizvi S. High-performance complex event processing over streams. In: Proceedings of the ACM International Conference on Management of Data (SIGMOD). Chicago, IL; 2006. pp. 407–418.

[24] Turaga D, Verscheure O, Sow D, Amini L. Adaptative signal sampling and sample quantization for resource-constrained stream processing. In: Proceedings of the International Conference on Biomedical Electronics and Devices (BIOSIGNALS). Funchal, Madeira, Portugal; 2008. pp. 96–103.

[25] Biem A, Elmegreen B, Verscheure O, Turaga D, Andrade H, Cornwell T. A streaming approach to radio astronomy imaging. In: Proceedings of the International Conference on Acoustics, Speech, and Signal Processing (ICASSP). Dallas, TX; 2010. pp. 1654–1657.

[26] Bouillet E, Kothari R, Kumar V, Mignet L, Nathan S, Ranganathan A, *et al.* Processing 6 billion CDRs/day: from research to production. In: Proceedings of the ACM International Conference on Distributed Event Based Systems (DEBS). Berlin, Germany; 2012. pp. 264–267.

[27] Turaga DS, Verscheure O, Wong J, Amini L, Yocum G, Begle E, *et al.* Online FDC control limit tuning with yield prediction using incremental decision tree learning. In: 2007 Sematech Advanced Equipment Control / Advanced Process Control Symposium (AEC/APC). Indian Wells, CA; 2007. pp. 53–54.

[28] Park Y, King R, Nathan S, Most W, Andrade H. Evaluation of a high-volume, low-latency market data processing sytem implemented with IBM middleware. Software: Practice & Experience. 2012;42(1):37–56.

[29] Hersent O. IP Telephony – Deploying VoIP Protocols and IMS Infrastructure. 2nd edn. John Wiley & Sons, Inc.; 2011.

[30] ITU. G.729: Coding of Speech at 8 kbit/s Using Conjugate-Structure Algebraic-Code-Excited Linear Prediction (CS-ACELP). International Telecommunication Union (ITU); 2007. G.729.

[31] Eberspächer J, Vögel HJ, Bettstetter C, Hartmann C. GSM – Architecture, Protocols and Services. 3rd edn. John Wiley & Sons, Inc.; 2009.

[32] NIST. Advanced Encryption Standard (AES). National Institute of Standards and Technology (NIST); 2001. FIPS-PUB-197.

[33] Schulzrinne H, Casner S, Frederick R, Jacobson V. RTP: A Transport Protocol for Real-Time Applications. The Internet Engineering Task Force (IETF); 1996. RFC 1889.

[34] Verscheure O, Vlachos M, Anagnostopoulos A, Frossard P, Bouillet E, Yu PS. Finding "who is talking to whom" in VoIP networks via progressive stream clustering. In: Proceedings of the IEEE International Conference on Data Mining (ICDM). Hong Kong, China; 2006. pp. 667–677.

[35] Knoke D, Yang S. Social Network Analysis. 2nd edn. Sage Publications; 2008.

[36] Ullman J. Principles of Database and Knowledge-Base Systems. Computer Science Press; 1988.

[37] StreamSQL; retrieved in October 2012. http://www.streambase.com/developers/docs/latest/streamsql/.

[38] Arasu A, Babu S, Widom J. The CQL continuous query language: semantic foundations and query execution. Very Large Databases Journal (VLDBJ). 2006;15(2):121–142.

[39] Thies W, Karczmarek M, Amarasinghe S. StreamIt: a language for streaming applications. In: Proceedings of the International Conference on Compiler Construction (CC). Grenoble, France; 2002. pp. 179–196.

[40] Abadi D, Carney D, Çetintemel U, Cherniack M, Convey C, Lee S, et al. Aurora: a new model and architecture for data stream management. Very Large Databases Journal (VLDBJ). 2003;12(2):120–139.

[41] Gedik B, Andrade H, Frenkiel A, De Pauw W, Pfeifer M, Allen P, et al. Debugging tools and strategies for distributed stream processing applications. Software: Practice & Experience. 2009;39(16):1347–1376.

[42] Reyes JC. A Graph Editing Framework for the StreamIt Language [Masters Thesis]. Massachusetts Institute of Technology; 2004.

[43] Tanenbaum A, Wetherall D. Computer Networks. 5th edn. Prentice Hall; 2011.

[44] Chandrasekaran S, Cooper O, Deshpande A, Franklin M, Hellerstein J, Hong W, et al. TelegraphCQ: continuous dataflow processing. In: Proceedings of the ACM International Conference on Management of Data (SIGMOD). San Diego, CA; 2003. pp. 329–338.

[45] Shah MA, Madden S, Franklin MJ, Hellerstein JM. Java support for data-intensive systems: experiences building the Telegraph dataflow system. ACM SIGMOD Record. 2001;30(4):103–114.

[46] PostgreSQL; retrieved in July 2011. http://www.postgresql.org/.

[47] Chandrasekaran S, Franklin M. Remembrance of streams past: overload-sensitive management of archived streams. In: Proceedings of the International Conference on Very Large Databases (VLDB). Toronto, Canada; 2004. pp. 348–359.

[48] Arasu A, Babcock B, Babu S, Datar M, Ito K, Motwani R, et al. STREAM: the Stanford Stream data manager. IEEE Data Engineering Bulletin. 2003;26(1):665.

[49] Carney D, Çetintemel U, Cherniack M, Convey C, Lee S, Seidman G, et al. Monitoring streams – a new class of data management applications. In: Proceedings of the International Conference on Very Large Databases (VLDB). Hong Kong, China; 2002. pp. 215–226.

[50] Tatbul N, Çetintemel U, Zdonik SB. Staying FIT: efficient load shedding techniques for distributed stream processing. In: Proceedings of the International Conference on Very Large Databases (VLDB). Vienna, Austria; 2007. pp. 159–170.

[51] Cherniack M, Balakrishnan H, Balazinska M, Carney D, Çetintemel U, Xing Y, *et al.* Scalable distributed stream processing. In: Proceedings of the Innovative Data Systems Research Conference (CIDR). Asilomar, CA, USA; 2003. pp. 257–268.

[52] Zdonik SB, Stonebraker M, Cherniack M, Çetintemel U, Balazinska M, Balakrishnan H. The Aurora and Medusa projects. IEEE Data Engineering Bulletin. 2003;26(1):3–10.

[53] Abadi D, Ahmad Y, Balazinska M, Çetintemel U, Cherniack M, Hwang JH, *et al.* The design of the Borealis stream processing engine. In: Proceedings of the Innovative Data Systems Research Conference (CIDR). Asilomar, CA; 2005. pp. 277–289.

[54] Balakrishnan H, Balazinska M, Carney D, Çetintemel U, Cherniack M, Convey C, *et al.* Retrospective on Aurora. Very Large Databases Journal (VLDBJ). 2004;13(4):370–383.

[55] Arasu A, Cherniak M, Galvez E, Maier D, Maskey A, Ryvkina E, *et al.* Linear Road: a stream data management benchmark. In: Proceedings of the International Conference on Very Large Databases (VLDB). Toronto, Canada; 2004. pp. 480–491.

[56] Jain N, Amini L, Andrade H, King R, Park Y, Selo P, *et al.* Design, implementation, and evaluation of the Linear Road benchmark on the stream processing core. In: Proceedings of the ACM International Conference on Management of Data (SIGMOD). Chicago, IL; 2006. pp. 431–442.

[57] Zeitler E, Risch T. Massive scale-out of expensive continuous queries. Proceedings of the VLDB Endowment. 2011;4(11):1181–1188.

[58] Cranor C, Johnson T, Spataschek O, Shkapenyuk V. Gigascope: a stream database for network applications. In: Proceedings of the ACM International Conference on Management of Data (SIGMOD). San Diego, CA; 2003. pp. 647–651.

[59] Bai Y, Thakkar H, Wang H, Luo C, Zaniolo C. A data stream language and system designed for power and extensibility. In: Proceedings of the ACM International Conference on Information and Knowledge Management (CIKM). Arlington, VA; 2006. pp. 337–346.

[60] Kräer J, Seeger B. PIPES: a public infrastructure for processing and exploring streams. In: Proceedings of the ACM International Conference on Management of Data (SIGMOD). Paris, France; 2004. pp. 925–926.

[61] Cammert M, Heinz C, Krämer J, Schneider M, Seeger B. A status report on XXL – a software infrastructure for efficient query processing. IEEE Data Engineering Bulletin. 2003;26(2):12–18.

[62] Naughton JF, DeWitt DJ, Maier D, Aboulnaga A, Chen J, Galanis L, *et al.* The Niagara Internet query system. IEEE Data Engineering Bulletin. 2001;24(2):27–33.

[63] Aberer K, Hauswirth M, Salehi A. A middleware for fast and flexible sensor network deployment. In: Proceedings of the International Conference on Very Large Databases (VLDB). Seoul, Korea; 2006. pp. 1199–1202.

[64] The GSN Project; retrieved in October 2012. `http://sourceforge.net/apps/trac/gsn/`.

[65] Sybase ESP; retrieved in October 2012. `http://www.sybase.com/products/financialservicessolutions/complex-event-processing/`.

[66] Sybase. Sybase Event Stream Processor – Programmers Reference (Version 5.1). Sybase, an SAP Company; 2012.

[67] TIBCO. TIBCO BusinessEvents – Event Stream Processing – Query Developer's Guide (Release 5.0). TIBCO; 2011.

[68] TIBCO. TIBCO BusinessEvents – Event Stream Processing – Pattern Matcher Developer's Guide (Release 5.0). TIBCO; 2011.

[69] Oracle. Oracle Complex Event Processing: Lightweight Modular Application Event Stream Processing in the Real World. Oracle white paper; 2009.

[70] Oracle CQL Language Reference 11g Release 1 (11.1.1); retrieved in October 2012. http://docs.oracle.com/cd/E15523_01/doc.1111/e12048.pdf.

[71] ISO. Information Technology – Database Languages – SQL. International Organization for Standardization (ISO); 1999. ISO/IEC 9075-[1-5]:1999.

[72] Andersen L. JDBC API Specification 4.1 (Maintenance Release). Oracle; 2011. JSR-000221.

[73] StreamBase Systems; retrieved in April 2011. http://www.streambase.com/.

[74] StreamBase EventFlow; retrieved in October 2012. http://www.streambase.com/developers/docs/latest/authoring/.

[75] StreamBase API Guide; retrieved in October 2012. http://www.streambase.com/developers/docs/latest/apiguide/.

[76] StreamBase Test/Debug Guide; retrieved in October 2012. http://www.streambase.com/developers/docs/latest/testdebug/.

[77] StreamBase Administration Guide; retrieved in October 2012. http://www.streambase.com/developers/docs/latest/admin/.

[78] Deakin N. Java Message Service – Version 2.0 (Early Draft). Oracle; 2012.

[79] Clayberg E, Rubel D. Eclipse Plug-ins. 3rd edn. Addison Wesley; 2008.

[80] StreamBase LiveView; retrieved in October 2012. http://www.streambase.com/products/liveview/.

[81] Neumeyer L, Robbins B, Nair A, Kesari A. S4: distributed stream computing platform. In: Proceedings of the International Workshop on Knowledge Discovery Using Cloud and Distributed Computing Platforms (KDDCloud). Sydney, Australia; 2010. pp. 170–177.

[82] S4 – Distributed Stream Computing Platform; retrieved in October 2012. http://incubator.apache.org/s4/.

[83] Dean J, Ghemawat S. MapReduce: simplified data processing on large clusters. In: Proceedings of the USENIX Symposium on Operating System Design and Implementation (OSDI). San Francisco, CA; 2004. p. 137–150.

[84] Amini L, Andrade H, Bhagwan R, Eskesen F, King R, Selo P, et al. SPC: a distributed, scalable platform for data mining. In: Proceedings of the Workshop on Data Mining Standards, Services and Platforms (DM-SSP). Philadelphia, PA; 2006. pp. 27–37.

[85] Storm Tutorial; retrieved in October 2012. https://github.com/nathanmarz/storm/wiki/Tutorial/.

[86] The Storm Project; retrieved in October 2012. http://storm-project.net/.

[87] Twitter; retrieved in March, 2011. http://www.twitter.com/.

Part II

Application development

3 Application development – the basics

3.1 Overview

In this chapter we introduce the fundamentals of application development using the stream processing paradigm. It is organized as follows. In Section 3.2, we provide an overview of the main characteristics of Stream Processing Applications (SPAs), focusing on how they affect the development process. In Section 3.3, we review the existing stream processing languages, deepening the discussion started in Chapter 2. In Section 3.4, we provide an introduction to the SPL language which will allow the readers to follow the examples provided in the rest of the book. In Section 3.5, we provide a list of commonly used stream processing operators and discuss their roles by looking into their implementation as provided by the SPL standard toolkit.

From this point onwards, we will pair the technical exposition with actual code examples. These examples are provided using the Streams platform and the SPL language. We encourage the readers to try them out and to modify them as they go along.

3.2 Characteristics of SPAs

We first look at the characteristics of SPAs and discuss their impact on application development. These characteristics not only influence their design and implementation, but also permeate the design of the programming model and of the application runtime features provided by SPSs.

Data-in-motion analytics
SPAs implement *data-in-motion* analytics. These analytics ingest data from live, streaming sources and perform their processing as the data flows through the application. As we have discussed in Chapters 1 and 2, this on-the-fly model of computation has many advantages, providing the ability to ingest and analyze large volumes of data at rates that are not possible to achieve with traditional *data-at-rest* analytics, to adapt to changes and to derive insights in real-time.

Structuring data-in-motion analytics as a SPA usually leads to an elegant *data flow graph* representation as well as to an efficient and scalable implementation. Data flow graphs represent computation as a processing flow where operations are applied on the data as it flows through a graph of analytic *operators*. These operators implement generic, reusable analytic functions. Instances of these operators are combined into data flow graphs via stream connections to form complete applications or application fragments.

Edge adaptation

Since SPAs ingest data from live sources and transmit results to different end-systems and users, integration with external systems is an integral part of the application development process and requires both source as well as sink *adapters*.

The process of adapting the data coming from sources or transforming the streaming data to be sent to sinks is referred to as *edge adaptation*. This process might involve data cleaning and transformation as well as protocol conversions and rate adjustments to match the needs of the external systems.

Long-running analytics

An important characteristic of SPAs is their long-running nature. This is due to data sources continuously producing the data that drives these applications. As a result, SPAs must keep up with the rate of the input data they process. Since data workloads can be dynamic, applications must be designed to handle peaks as well as variations in the workload. This means that an application as well as the runtime environment must implement adaptive resource management techniques including *load balancing* and *load shedding* [1, 2, 3, 4, 5, 6] (Section 9.3.3), which influences not only the internal design of a SPS, but also the language used to implement applications.

Another important implication of long-running analytics is the need for updating applications without enforcing scheduled downtime. While updating and evolving SPAs require a certain level of support from the application runtime environment, they also require applications to be designed in a way that eases the piecemeal update and expansion of their data flow graphs.

The long-running nature of streaming applications also creates a need for mechanisms to monitor the health of individual applications as well as of the application runtime environment itself. In both cases, the health of the applications and of the runtime system is defined in terms of quality metrics. These monitoring mechanisms are used by external parties to perform administrative tasks, as well as from within the applications themselves to implement self-healing and self-adapting behaviors.

In summary, SPAs need to be designed and developed to be autonomic and to cope with changes in the workload as well as in resource availability.

Reactive and proactive analytics

A common goal of SPAs is to detect patterns of interest in streaming data and generate actions accordingly. In other words, these applications need to be *reactive*. This requirement implies that SPAs must maintain state and use this state to detect patterns on a continuous basis. Due to the infinite nature of streams, such state is commonly maintained only for the most recent data in the form of a *window*, whose contents are updated as the data flows. As a result, the notion of *windows over streams* is a fundamental concept in SPAs.

In addition to being reactive, many SPAs are also *proactive*, particularly when they implement exploration tasks as described in Section 2.4.2. This means that these applications use the current and past history to build models that will predict future outcomes. Both reactive and proactive analytics must adapt the internal processing logic of the application based on the newly detected or predicted events. This requires the ability to propagate control information from later stages of a flow graph to earlier ones. As such, flow graphs representing SPAs may contain *feedback loops* for control information, and are not necessarily acyclic.

In general the feedback control information can also trigger topological modifications to an application's data flow graph. Examples of such changes include the use of a different set of sources and sinks, the addition of new analytical pieces, or the removal of existing segments from a running application.

Multi-modal analytics

SPAs are multi-modal, i.e., they use multiple modes of input/output. This is a direct reflection of the variety of the data sources that are used for data exploration applications and the increasing need to correlate information across heterogeneous data sources. Structured information such as relational database records, unstructured information such as audio, video, and text, and semi-structured information described in eXtensible Markup Language (XML) [7] and JavaScript Object Notation (JSON) [8] are a few examples of the types of data processed by a SPA.

One implication of this diversity of data sources is that a rich type system must be part of the programming model so that the analytics employed by SPAs can be properly expressed.

The variety of the analytics used in multi-modal SPAs demands a high level of *extensibility* from a SPS. At a minimum, a set of cross-domain operators must be provided by the system to facilitate rapid development. More importantly, domain-specific operators must also be supported through extension mechanisms, which enable application developers to pick the right set of building blocks for their applications.

Extensibility goes hand-in-hand with *reusability*. Since applications often share functionality, the flexibility of reusing analytic operators, whether they are provided by the platform or are extensions developed by third parties, is critical in addressing the software engineering challenges faced during the development of an application.

High performance and scalable analytics

High throughput and low latency are common requirements to several types of SPAs. Therefore, delivering high performance, while providing high-level abstractions that ease the development of applications, is a challenge faced by SPSs.

A flow graph representation is good both for expressing the internal structure of a SPA in a natural way, as well as for exposing its components that might function independently in parallel. Nevertheless, the flow graph representation used during application development is a *logical* abstraction, whereas the flow graph that gets executed on a set of computational hosts represents its *physical* organization. In many cases, to facilitate the process of tuning high-performance SPAs, configuration directives (when available) can be used to influence the mapping between the logical and the physical representations.

SPAs must be scalable to handle the vast volume of data they analyze. This requires both *vertical scalability* (also known as *scale-up*) and *horizontal scalability* (also known as *scale-out*) [9]. Vertical scalability consists of delivering more performance as the computational capacity of the hosts where an application runs increases, when additional CPU cores, network bandwidth, or memory are added to an existing system. Horizontal scalability consists of delivering more performance as the number of hosts used by an application increases.

Providing abstractions that help expressing *data parallelism*, *task parallelism*, and *pipeline parallelism* [10] in the data flow graph representation of SPAs is important to achieve both horizontal and vertical scalability. These abstractions generally provide the ability to decompose an application in a way that exposes the different forms of parallelism inherent in it. Moreover, the runtime environment must also provide an execution model that takes advantage of parallelization.

3.3   Stream processing languages

In Section 2.4.3 we discussed the make up of a stream processing development environment and provided an overview of programming models and languages developed for SPSs. In this section, we pick up from where we left off, providing an in-depth look into language characteristics and describing the foundations behind SPL, the programming language that will be used throughout the rest of this book.

As previously discussed, SPAs are developed using programming tools and languages that provide abstractions that match the characteristics and needs of these applications. The fundamental concepts associated with stream processing are unique enough that existing, general purpose programming languages and tools are either insufficient or inadequate to construct SPAs with ease. For instance, as seen in Section 2.4.2, operators, ports, streams, and stream connections are first class abstractions in stream processing.

3.3.1   Features of stream processing languages

We now look at the fundamental features that are provided by stream processing languages.

Compositional elements

A stream processing language is expected to offer compositional features targeted at constructing data flow graphs. At a high level, the fundamental task of an application developer in stream processing is to map its computation to a flow graph. The language should facilitate this by providing mechanisms to define, instantiate, and link operators to each other with stream connections. This form of composition is not limited to application development time, but includes also dynamic composition that may take place at runtime.

Another important aspect of composition is to be able to represent fragments of flow graphs as *logical*, coarse-grain operators and compose them into higher-level flow graphs, which can simplify the development of large-scale applications through modularity.

Declarative operators

Analytic operators form the basic building blocks of SPAs. Operators are expected to provide a declarative interface,[1] so that application developers can focus on the application's overall structure and minimize the effort devoted to understanding *how* operators accomplish their tasks. Instead, application developers can focus on understanding *what* operators do and use that knowledge to construct the flows for their application.

Declarative stream processing operators are reusable, as they can work with different stream schemas, support different parameterizations, and allow the customization of the results they produce.

Expression language

An *expression language* must be embedded in a stream processing language to express basic flow manipulations and to parameterize declarative operators with logical conditions.

For instance, if an operator is to be used to join or correlate items across two streams, the application developer must be able to specify a condition for the match that defines the correlation.

Type system

A stream processing language must provide a rich set of types encapsulated in a *type system* to facilitate the development of flexible operators and analytics.

As seen in Section 2.4.1, a stream is a time-ordered sequence of tuples where each tuple's internal structure is defined by a *schema*, represented by a *type*.

Having types associated with streams enables operators to define a clear interface that must be adhered to while assembling an application's flow graph. For instance, a stream produced by an operator can only be consumed by another operator if the type of the stream expected by the consuming operator matches the one emitted by the producing operator.

---

[1] An operator's *declarative interface* consists of its parameters and other configurations (input and output ports, windowing, and output tuple attribute assignments) along with its name and namespace. Section 5.2 provides a discussion of these in the context of the SPL language.

In other words, having types associated with streams makes it possible to define a compile time *contract* between operators. Furthermore, the same type system can be used to maintain and manipulate the operator's internal state.

Windowing

Owing to the continuous nature of streaming analytics, it is often infeasible to maintain the entire history of the input data. There are two main approaches to tackle this challenge.

First, a running *data synopsis* can be created and maintained, where this synopsis stores an aggregated summary metric (e.g., a moving average). In this case, individual tuples are no longer available after the aggregation operation has taken place.

Alternatively, a *window* over streaming data can be created, buffering individual tuples. Windows can be built according to a multitude of parameters that define what to buffer, resulting in many window variations. These variations differ in their policies with respect to evicting old data that should no longer be buffered, as well as in when to process the data that is already in the window.

Many stateful stream processing operators are designed to work on *windows* of tuples, making it a fundamental concept in stream processing. Therefore, a stream processing language must have rich windowing semantics to support the large diversity in how SPAs can consume data on a continuous basis.

Standard operators and adapters

Having a set of generic and reusable operators and adapters at the disposal of application developers is important to increase the productivity in designing and implementing SPAs.

While applications may have very specialized requirements in terms of the type of operators they require, there is a set of well-understood operators that are found in almost all applications. Similarly, while applications may connect to custom, proprietary data sources and sinks, there is a large number of standard protocols and devices for which general purpose edge adapters can be created. Ideally, a stream processing language must provide a *standard* set of operators and edge adapters to enable rapid application development.

Modularity

SPAs are full-fledged programs. Therefore, software engineering principles including code *reuse*, *maintainability*, and *serviceability* are important concerns.

Consequently, stream processing languages must support modular design principles and include mechanisms for partitioning an application into multiple compilation units, organizing code into reusable components, providing a notion of name spaces, and managing libraries, as well as for versioning and dependency management.

Configuration

A SPA can be large and consist of a set of distributed components. These components may execute on a cluster of compute hosts. Thus, a given application can be mapped to the set of available hosts and processors in different ways.

While applying automatic system-level optimizations is a desired way of hiding some of the complexity associated with this mapping, user-defined application-level constraints must be taken into account in doing so. These constraints include placing a certain component on a particular host (e.g., because the component might make use of specialized hardware), ensuring that two components are placed together (e.g., because they operate synergistically), or ensuring that two critical components are not co-located (e.g., because it creates a single point of failure in the application).

As a result, a stream processing language should enable developers to influence the mapping of logical components onto physical processors, or to express constraints that state undesirable placement characteristics.

In addition, configurations can be used to specify other non-functional aspects of applications including the inter-process communication substrate to be used, the logging and debugging aids that should be enabled, checkpointing, and settings related to fault tolerance.

Extensibility

A general-purpose stream processing language must be extensible so it becomes possible to support domain-specific SPAs with their own set of operators. Moreover, there is a large body of existing libraries and analytics written in C++, Java, and Fortran, representing intellectual and capital investment that organizations typically want to integrate in their SPAs.

Therefore, a well-designed stream processing language must also provide the necessary mechanism to facilitate the reuse of these assets.

3.3.2   Approaches to stream processing language design

We have discussed characteristics of SPAs and looked at the language features that facilitate their development. We now discuss the design choices made by different stream processing languages.

SQL-based languages

The Structured Query Language (SQL) is not a stream processing language as it was originally designed to manipulate data from relational tables.

While multiple relational operations may need to be performed on streaming data, the continuous nature of a stream must be taken into consideration. In general, a table-oriented implementation of a relational operator is not directly usable for a stream-oriented task. For instance, some of the SQL relational operators, e.g., group-by aggregations, are blocking, in the sense that they need to consume the entire input before producing an output. Blocking is generally undesirable in the stream processing context.

Clearly, SQL cannot be used as is for implementing continuous processing operations. Still, a common approach has been to adapt SQL for stream processing, enhancing the language with windowing concepts as well as with non-blocking versions of relational operators.

These languages (e.g., Continuous Query Language (CQL) [12] and Event Processing Language (EPL) [13]) have streaming semantics, and queries implemented in them are referred to as *continuous queries* [14].

There are several advantages to SQL-based languages. First, SQL is a well-established language and stream processing languages based on it have a familiar look-and-feel.

Second, since SQL was designed for manipulating *data-at-rest*, extending it to managing *data-in-motion* simplifies the development of applications that integrate these two forms of computation.

Third, SQL is a standard [15] adopted by virtually all relational database systems. As a result, parsers, visualizers, and composers for SQL are readily available. Relying on existing tooling and experience is a cost-effective way of building a stream processing language.

Fourth, SQL and the relational algebra lie at the heart of relational databases. Extending existing relational database engines to support streaming is a possible incremental approach that can be used to quickly build a stream processing engine.

Nevertheless, an SQL-based stream processing language does not necessarily require a relational engine as its runtime environment. In general, a SPA can achieve better performance when decoupled from relational database technologies, which are typically optimized for disk-based processing.

A fundamental disadvantage of SQL-based stream processing languages is their lack of support for advanced flow composition. In SQL, both the operators and the overall physical structure of the query are hidden behind clauses (e.g., select, from, group by). This creates a mismatch between the need to express the data flow structure of the computation to be performed by a SPA and the syntax offered by the programming language.

Additionally, the non-explicit nature of flow composition in SQL makes it difficult to have language support for influencing the mapping from the logical flow graph to the physical one when an application is going to run on multiple distributed hosts.

Finally, relational algebra has a fixed number of operators that provide relational completeness [16] for SQL queries. There is no such theoretical result for stream processing, particularly because the underlying data model is more fluid, expanding beyond simpler relational tuples, as discussed in Section 2.4.1.

As a conclusion, extending SQL to provide stream processing capabilities and extensibility mechanisms to support the development of domain-specific operators requires substantial syntactic improvements, which clearly weakens the argument for using a SQL-based language as is.

Visual languages
Visual stream processing languages provide a *boxes and arrows* notation to create data flow graphs. The programming environment associated with these languages usually provides a canvas editor paired with a palette of available operators, which can be used to instantiate, configure, and connect operators to form data flow graphs [17, 18, 19].

The key advantage of visual languages is their ease of use. They do not require strong programming skills and enable Rapid Application Development (RAD) for a broad group of developers.

Visual editing frameworks are common in other development domains as well, including Electronic Design Automation (EDA), model-based simulation, robot programming, Graphical User Interface (GUI) design and implementation, and database design. Nevertheless, there is no established standard for visual languages across these different domains since the nature of the development tasks vastly different across them.

Visual stream programming is typically composition-oriented, where the manipulation of the flow graph structure is a central piece of many application development workflows. Unlike textual languages, which are limited by the linear nature of text, visual languages can better represent the application analytic processing topology. This characteristic helps a programmer, potentially not the one who wrote the application in the first place, to better understand its structure.

On the flip side, a weakness of visual programming languages is their lack of scripting and automation support. For instance, textual languages can easily be paired with code generation frameworks to automate the creation of applications or application fragments. This is not generally possible with visual languages, except in cases where pre-defined templates exist. Furthermore, while visual languages are well-suited for representing the data flow structure of a SPA, they are not as adequate for representing control flow. Examples include loops that create replicated segments, conditional statements that turn segments on and off, or higher order composite operators, i.e., operators that take other operators as parameters and embed them into an application's flow graph.

Another limitation of visual languages is the representation of free-form code, as, for example, expressions and statements used for configuring or defining new operators, as well as for writing custom functions used by operators. In general, these operations involve programming tasks that cannot be easily represented through visual metaphors.

One possible approach to overcome some of these limitations is to augment a textual language with visual capabilities, leveraging the flexibility afforded by textual languages while making use of visual metaphors that reduce the cognitive load on developers.

For instance, an Integrated Development Environment (IDE) can provide a read-only view of the application being implemented. In this way, the representation is synchronized with the textual representation, but changes cannot be made directly to the visual representation. This approach enhances the development process when employing a textual language, but still requires programming skills to build SPAs. Engineering such a solution requires developing a fast mapping between textual and visual representations, by incrementally capturing changes in the application code and converting these changes into modifications to the visual representation.

A next step in this approach is to provide an environment where developers can use either representation interchangeably to make modifications to the application implementation. Engineering such an environment is more challenging, but some SPSs, including InfoSphere Streams, provide this capability.

Custom domain-specific languages
A Domain-Specific Language (DSL) is a programming language targeted at a particular problem domain. Many existing languages can be seen as DSLs. For instance, SQL can be considered a DSL for relational data management where the problem representation consists of relations and tuples. Likewise, in the context of SPAs, a natural problem representation is a data flow graph. Therefore, the visual and textual languages we have discussed are also DSLs.

An important advantage of these languages is their ability to provide features that suit specific needs of a particular application domain. For example, when using a stream processing DSL, applications can be expressed with more appropriate abstractions from the stream processing domain. This results in succinct, idiomatic code that is easy to develop and to maintain.

In contrast, however, DSLs require additional learning effort from the developers. While this aspect may slow down a beginner, the long-term benefits in terms of ease of development and maintenance often outweigh the initial learning curve. An example of a DSL used for stream processing is the SPL language.

3.4   Introduction to SPL

In this section we give a brief introduction to SPL. The goal of this discussion is to familiarize the reader with the SPL syntax so that the various stream processing concepts, techniques, and code examples presented in the rest of this book can be understood.

3.4.1   Language origins

SPL is a programming language for creating distributed SPAs that run on IBM InfoSphere Streams [20].

Its development can be traced back to the implementation of the System S stream processing platform [21]. System S was developed as part of a research project at the IBM T. J. Watson Research Center, beginning in 2003. This project aimed at investigating runtime, language, and analytics for distributed large-scale stream processing.

System S's SPADE language [22][2] is a precursor to the SPL language [23] and shares many of its fundamental design principles. Nevertheless, SPL was designed from scratch with substantial syntactic improvements as well as mechanisms for large-scale software development and a renewed focus on extensibility.

SPL offers a stream-centric and operator-based programming model. When writing SPL code, an application developer can quickly translate a back-of-the-envelope prototype into an application skeleton by simply listing its data flows, the operators that

---

[2] SPADE stands for Stream Processing Application Declarative Engine.

generate them, and the connections between these operators. With SPL's focus on operators, developers can design their applications by reasoning about the right set of building blocks that are necessary to deliver the computation their application is supposed to perform.

The SPL language embodies several features:

- A *composition language* targeted at creating large-scale data flow graphs.
- A *statement language*[3] used for operator parameterization and customization as well as for defining new operators and functions.
- A *standard toolkit* of operators that provide basic flow manipulation capabilities, including stream relational capabilities, in addition to common source and sink edge adapters.
- Configuration support to influence how applications are mapped and executed on a distributed application runtime.
- Runtime and code generation interfaces for extending SPL with generic, declarative operators using general purpose programming languages.

In the remainder of this chapter we explore these features in more detail and also discuss debugging, profiling, and visualization, which are intimately connected to SPL's programming model.

3.4.2   A "Hello World" application in SPL

A typical application contains three main high-level components: *data ingest, analytical processing*, and *data egress*.

We consider a minimal "Hello World" application that has these three components and whose source code is shown in Figure 3.1. On the ingest side, this application employs a *source operator*, which, in this contrived example, is also a data generator. This source operator produces a fixed number (10) of tuples, each a second apart in time, containing random identifiers (which are values between 0 and 9).

Typically, in a real-world application, a source operator continuously ingests data produced by an external source, for instance, a Transmission Control Protocol (TCP) socket connection, a HyperText Transfer Protocol (HTTP) feed, a device driver interface, or a dynamic set of input files.

The analytical processing component of the "Hello World" application consists of an operator that populates each tuple it receives from the source with an additional attribute that contains the string "Hello World!" preceded by the random identifier from the tuple produced by the source operator.

In a real application the analytical processing component is usually expressed by a combination of several operators that work in concert to perform a complex analytical task, as discussed in Section 2.4.2.

---

[3] A statement language is used to put together expressions to create larger programs that manipulate variables using assignments and control flow structures.

```
1   namespace sample;
2
3   composite HelloWorld {
4     graph
5       stream<uint32 id> Beat = Beacon() {
6         param
7           period: 1.0; // seconds
8           iterations: 10u;
9         output
10          Beat: id = (uint32) (10.0*random());
11      }
12      stream<rstring greeting, uint32 id> Message = Functor(Beat) {
13        output
14          Message: greeting = "Hello World! (" + (rstring) id + ")";
15      }
16      () as Sink = FileSink(Message) {
17        param
18          file: "/dev/stdout";
19          format: txt;
20          flush: 1u;
21      }
22  }
```

Figure 3.1 A "Hello World" application in SPL.

```
1   namespace sample;
2
3   composite HelloWorld {
4     graph
5       ...
6   }
```

Figure 3.2 An application – the basics.

Finally, the application's egress component simply writes the augmented tuples in textual form, resulting in messages consisting of a greeting and an associated identifier, which are displayed in the application's standard output. This task is accomplished using a *sink operator*, in this case, a file sink that employs the Unix /dev/stdout device.

In a more realistic application, one or more sink operators can produce and route the application's results to external consumers that might be a TCP socket endpoint, a message queue, or a database.

The "Hello World" application contains three operators in total, connected in a simple pipeline topology. Real-world SPAs contain many more operators organized in a data flow graph topology that is significantly more complex than in this example.

De-constructing the "Hello World" application

We now go over the "Hello World" application piece-by-piece. We first look at the outer-most layer of the source code depicted by Figure 3.2.

The first line of the code is a *namespace* declaration. It states that this program is part of the sample namespace. A namespace creates a scope for the different types of entities that are used in an SPL program.

The next construct seen in the code is a *composite operator* definition named HelloWorld. Composite operators are discussed in detail in Section 4.2. For now it suffices to know that this composite is the main entry point of the program.

```
1   stream<uint32 id> Beat = Beacon() {
2     param
3       period: 1.0; // seconds
4       iterations: 10u;
5     output
6       Beat: id = (uint32) (10.0*random());
7   }
```

Figure 3.3 The first operator invocation.

The subsequent `graph` clause precedes the definition of the application's flow graph, which is specified as a series of operator invocations and stream connections that link them.

In SPL, an operator invocation is used to instantiate an operator and to define the output streams produced by it. As part of the operator invocation, a list of output streams and their types, and a list of input streams to be consumed are specified. Either list can be empty when source and sink operators are used. In addition, several options can also be specified for customizing an operator. They include parameters, output assignments, and window configurations.

Next, we look at the first operator invocation inside the `graph` clause depicted by Figure 3.3. This excerpt includes the invocation of the `Beacon` operator. This operator is a tuple generator that belongs to SPL's standard toolkit. It is also a source operator and does not have input ports.

The `Beacon` operator has a single output port, which generates a stream of tuples. In this particular invocation, the output stream created by this operator is named `Beat`.

The `Beat` stream generates tuples whose *schema* is defined as `tuple<uint32 id>`. In other words, each tuple in this stream contains an `id` *attribute* of type `uint32` (a 32-bit unsigned integer in SPL).

This invocation of the operator `Beacon` also specifies several parameters using the `param` clause. The first is the `period` parameter, which represents the amount of time between successive tuples generated by the operator; 1 second in this case. The second one is the `iterations` parameter, which represents the number of tuples to be generated; 10 in this case (u is a type suffix in SPL, identifying an unsigned integer literal).

Finally, the operator invocation also specifies an `output` clause. This clause is used to describe how the attributes in a tuple produced by one of this operator's output ports is assigned values. In this case the `id` attribute is populated with a uniformly distributed random integer in the range `[0,10)`.

By default, operator instances in SPL are named using the name of the first stream they produce. As a result, in this example the name of the `Beacon` operator instance is `Beat`. Alternatively, the operator can be named explicitly as shown in the following code excerpt:

```
(stream<uint32 id> Beat) as MySource = Beacon() { ... }
```

We now look at the second operator in the application, depicted by Figure 3.4, which is responsible for processing the data produced by the `Beat` operator. This invocation is similar to the first one we have seen with a few notable differences.

First, the Message operator is a Functor. This operator is used to perform basic tuple and attribute manipulations and is part of SPL's standard toolkit.

Second, this operator has one input port. This input port is an endpoint for the Beat stream, enabling the operator to consume tuples. This operator also has one output port, which generates the Message stream.

Third, this operator's output port has a different tuple schema defined as follows: tuple<rstring greeting, uint32 id>. This schema has an additional attribute greeting of type rstring (a raw string[4] in SPL).

Finally, this operator has an output clause, which describes how the greeting attribute is populated. In this case, it is given the value "Hello World!" followed by the id value enclosed in parentheses. For example, greeting will have the value "Hello World! (5)" when the id attribute has the value 5. In this case it is implicit in the assignment expression "Hello World! (" + (rstring) id + ")" that the id attribute from the incoming Beat tuple will be used to populate the attribute greeting in the outgoing Message tuple. The assignment expression can also be written as "Hello World! (" + (rstring) Beat.id + ")" to explicitly state that the id attribute value to be used comes from the Beat stream.

Note that an assignment for the id attribute for the Message tuples is not specified in the output clause. Consequently, the Functor operator *forwards* the value of the id attribute from the input tuple to the output one by automatically making a copy of the incoming value.

We now look at the last operator invocation, depicted by Figure 3.5, which employs a FileSink operator that consumes the application results.

The FileSink operator is also part of SPL's standard toolkit and is used to serialize the contents of a stream to a file. This operator has one input port, but no output ports.

```
1  stream<rstring greeting, uint32 id> Message = Functor(Beat) {
2    output
3      Message: greeting = "Hello World! (" + (rstring) Beat.id + ")";
4  }
```

Figure 3.4 The second operator invocation.

```
1  () as Sink = FileSink(Message) {
2    param
3      file: "/dev/stdout";
4      format: txt;
5      flush: 1u;
6  }
```

Figure 3.5 The third operation invocation.

---

[4] A raw string (rstring) in SPL is a sequence of bytes. SPL also supports Unicode strings (ustring), which are sequences of logical characters.

This particular `FileSink` operator instance processes tuples from the `Message` stream. It is parameterized to write the tuples it receives to the *standard output* as specified by the `file` parameter.

This operator also makes use of the `format` parameter to specify how the results should be formatted. In this case, the `txt` setting indicates that SPL's built-in text format should be used. Note that this operator is also configured with the `flush` parameter, which indicates that results should be printed immediately after each incoming tuple is received, *flushing* the underlying buffer associated with the `/dev/stdout` Unix device. The absence of this parameter causes the operator to print results either when the output buffer is full or when the application terminates.

Figure 3.6 depicts the data flow graph of the "Hello World" application. The gray boxes in the figure represent operator instances and the connections between them represent stream connections. Operator ports are indicated using rectangular boxes on either side of the operator, input ports on the left and output ports on the right.

Building the "Hello World" application

We now briefly describe the process of writing the code, compiling, and running the "Hello World" application.

The initial step is to write the code for the application. First, an application directory, `HelloWorldApp`, for example, is created. Under this directory, a subdirectory named after the namespace used by the application is created. In our example, the namespace is `sample` and the name of the subdirectory is `HelloWorldApp/sample`.

Next, in that subdirectory, we store the `HelloWorld.spl` file, which holds the code from Figure 3.1. Any text-based editor can be used for this purpose.

The specific steps executed on a Unix command shell using only I/O redirection are as follows:

```
~> mkdir HelloWorldApp
~> cd HelloWorldApp
~/HelloWorldApp> mkdir sample
~/HelloWorldApp> cd sample
~/HelloWorldApp/sample> cat > HelloWorld.spl
# populate the file with the SPL code and press Ctrl+D to save
```

The second step is to compile the application, transforming it into an executable. For this purpose, we use the Streams `sc` compiler, which should be invoked from the application directory (`HelloWorldApp`).

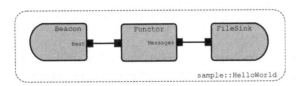

Figure 3.6 A visual representation of the "Hello World" application from Figure 3.1.

We will make use of two compiler options. The first one is `--main-composite`, which specifies the top-level operator defining the entry point for the application. In this case, it takes the value `sample::HelloWorld`.

The second compiler option is `--standalone-application`, which instructs the compiler to build a *standalone application*. Standalone applications are regular executables that can be launched from the shell command line, making use of only the local host. Typically, real-world, large-scale SPAs written in SPL are executed on a cluster of hosts, using the Streams distributed runtime. However, simple examples and test versions of real applications can be compiled as standalone applications for testing and debugging.

The specific compilation step executed on a Unix shell is as follows:

```
~/HelloWorldApp>sc --main-composite sample::HelloWorld --standalone-application
```

After this step, the compiler will have produced an executable (in the `output/bin` folder) that can be used to run the application. The resulting standalone application can be executed as follows:

```
~/HelloWorldApp>./output/bin/standalone
```

This program runs for around 10 seconds and produces an output similar to the following:

```
{greeting="Hello World! (8)",id=8}
{greeting="Hello World! (8)",id=8}
{greeting="Hello World! (2)",id=2}
{greeting="Hello World! (6)",id=6}
{greeting="Hello World! (8)",id=8}
{greeting="Hello World! (6)",id=6}
{greeting="Hello World! (9)",id=9}
{greeting="Hello World! (9)",id=9}
{greeting="Hello World! (4)",id=4}
{greeting="Hello World! (3)",id=3}
```

In the next sections we use SPL language examples to supplement our discussion and, when necessary, introduce additional language features. For a complete treatment of the SPL language we refer the interested reader to IBM's product documentation [24].

3.5   Common stream processing operators

Most SPSs provide a set of built-in operators that can be used to create applications. These operators can be grouped into three main categories: *stream relational*, *utility*, and *adapter*.

3.5.1   Stream relational operators

The set of stream relational operators implements operations similar to their counterparts in relational algebra. In general, each of these operators has an internal processing logic that provides a continuous and non-blocking version of the corresponding operation as implemented in relational databases.

Projection

The *projection operator* is a stateless operator (Section 4.3.1) used to add, remove, and update attributes of an inbound tuple producing a new modified outbound tuple.

In SPL, the `Functor` operator can be used to perform projection operations. The following code excerpt demonstrates this capability:

```
1  stream<rstring name, float64 weight, float64 volume> Src = MySource() {}
2  stream<rstring name, float64 density> Res = Functor(Src) {
3    output
4      Res: density = weight / volume;
5  }
```

In this example, the `Functor` operator is used to remove the `weight` and `volume` attributes and to add a new attribute `density` whose value is the result of evaluating the expression `weight/volume`. The values for `weight` and `volume` are extracted from the inbound tuple. Hence, the stream produced by this operator (`Res`) has tuples with only the `name` attribute (whose value is the same as in the corresponding inbound tuple) as well as the `density` attribute which holds the result of the calculation.

Selection

The *selection operator* is a stateless operator used to filter tuples. On the arrival of each tuple, a boolean predicate is evaluated on its attributes and a determination is made as to whether it should be filtered out or not.

In SPL, the `Filter` operator can be used to perform selection operations. The following code excerpt demonstrates this capability:

```
1  stream<rstring name, float64 density> Src = MySource() {}
2  stream<rstring name, float64 density> Res = Filter(Src) {
3    param
4      filter: density > 1.0;
5  }
```

In this case the `Filter` operator is used to eliminate tuples from the inbound stream (`Src`) whose value for the `density` attribute is not greater than 1, producing an outbound stream (`Res`).

Aggregation

The *aggregation operator* is a stateful windowed operator (Section 4.3.1) used to group a set of inbound tuples and compute aggregations on their attributes, similarly to the `group-by` clause in a SQL statement.

The actual aggregations are computed continuously and results are produced every time the window *trigger policy* fires.

In SPL, the `Aggregate` operator can be used to perform several types of aggregations. The following code excerpt demonstrates the use of this operator when computing the average value of an attribute, while grouping tuples based on the value of another:

```
1   stream<rstring name, rstring department, int32 age> Src = MySource() {}
2   stream<rstring department, int32 avgAge> Res = Aggregate(Src) {
3     window
4       Src: tumbling, count(100);
5     param
6       groubBy: department;
7     output
8       Res: avgAge = Average(age);
9   }
```

In this example, the `Aggregate` operator is used to group every inbound set of 100 tuples, separating them into groups based on the unique values for the `department` attribute. For every group, this operator instance outputs a tuple that has its `department` attribute value set to that of the group's and the `avgAge` attribute value set to the average `age` value for the tuples within the group.

In the general case, the `Aggregate` operator can be configured in many different ways including: the specific attributes used to create the aggregation groups, the specific aggregation functions to apply to the attributes from the inbound tuples, the type of window mechanism as well as the specific window management policy to use, and, finally, whether partitioning of the inflow traffic should take place before operating on the inbound tuples.

Sort

The *sort operator* is a stateful and windowed operator used to first group inbound tuples and then sort them based on the value of a *key* assembled from a set of attributes. After tuples are sorted, results are produced every time the window *trigger policy* fires.

In SPL, the `Sort` operator performs this task. The following code excerpt demonstrates this capability:

```
1   stream<rstring name, rstring department> Src = MySource() {}
2   stream<rstring name, rstring department> Res = Sort(Src) {
3     window
4       Src: tumbling, count(100);
5     param
6       sortBy: department, name;
7   }
```

In this case a `Sort` operator instance is used to group every inbound set of 100 tuples and sort them based on a key formed by the concatenation of the values for the `department` and `name` attributes.

In the general case, the `Sort` operator can be configured in many different ways, including: the specific attributes used to create the sorting key, whether tuples should be sorted in ascending or descending order, the type of window mechanism as well as the specific window management policy to use, and, finally, whether partitioning of the inflow traffic should take place before sorting.

Many Stream Processing Systems (SPSs) also provide a *progressive* version of the sort operator. In a progressive sort, the smallest tuple is output whenever a new tuple is inserted   into   a   sliding window.

Join

The *join operator* is a stateful and windowed operator used to correlate tuples from two streams[5] based on a particular matching condition.

The operator's internal processing logic is triggered by the arrival of an inbound tuple to either one of the operator's input ports. First, the tuple is placed in that input port's window buffer. Second, this tuple is *matched* against all tuples stored in the window of the opposing input port. If the match predicate, a boolean expression, evaluates to true, an output tuple is produced with the correlation.

In SPL, the `Join` operator can be used to perform correlations between two streams as seen in the following code excerpt:

```
1  stream<rstring id, uint32 value> Values = MyValueSource() {}
2  stream<rstring id, rstring name> Names = MyNameSource() {}
3  stream<rstring id, rstring name, uint32 value> Res = Join(Values; Names) {
4    window
5      Values: sliding, time(10.0), count(1);
6      Names: sliding, time(10.0), count(1);
7    param
8      match: Values.id == Names.id;
9  }
```

In the example, this `Join` operator keeps track of tuples received in the last 10 seconds from the `Values` and `Names` streams, employing a sliding window on both input ports. Thus, each newly received tuple is matched against the tuples stored in the window buffer belonging to the other port using the expression `Values.id == Names.id`. This particular matching condition brings together tuples from the `Names` and `Values` streams which share the same value for the `id` attribute.

In the general case, a `Join` operator can be configured in different ways, including different join algorithms (e.g., an equi-join) as well as different forms of outer joins, different eviction policies, window partitioning, and different output port configurations for separating matching and non-matching results from an outer-join operation.

An equi-join can have one or more conjunctive terms in its match condition, where each term is an equality. Such joins can be more efficiently implemented via hash-based indices that make it possible to locate matching tuples from a window in constant time.

Outer-joins are used to output tuples that do not match any other tuples during their entire stay in the join window. When such a tuple is evicted from its window, it is emitted on a separate output port reserved for non-matching tuples. The following example illustrates an outer equi-join operation:

```
1  stream<rstring id, uint64 region, uint32 value> Values = MyValueSource() {}
2  stream<rstring id, uint64 region, rstring name> Names = MyNameSource() {}
3  (stream<Values, Names> Res;
```

---

[5] A stream relational `Join` operator's internal algorithm is substantially different from a standard join algorithm employed by a relational database. Cost estimation and optimization strategies employed by a relational database's join operator assume finite input relations, differently from, potentially, infinite streams. Moreover, a long-running stream relational join operator must be adaptive to react to changes and fluctuations in the data and in resource availability. It must also manage its internal state as, at no point in time, the algorithm has complete access to all of the data it will eventually correlate. For the interested reader, we recommend an earlier survey [25] covering the different join processing techniques available for SPAs.

```
4    stream<Values> UnmatchedValues;
5    stream<Names> UnmatchedNames) = Join(Values; Names)
6    {
7      window
8        Values: sliding, time(10.0), count(1);
9        Names: sliding, time(10.0), count(1);
10     param
11       algorithm: outer;
12       match: value < 10;
13       matchLHS: Values.id, Values.region;
14       matchRHS: Names.id, Names.region;
15   }
```

In this example, the Join operator matches the id and region attributes
from tuples that arrive in its left input port (Values) with the same attributes
from tuples arriving at its right input port (Names). It also has an inequality condition, defined by the expression value<10. This combined join condition is functionally equivalent to having match: Values.id==Names.id &&
Values.region==Names.region && value < 10, but can be implemented
more efficiently when expressed using the matchLHS and matchRHS parameters.

Another interesting aspect of this example is that it specifies outer as the join algorithm to use, which allows the use of two additional output ports. The first one has the
same schema as the first input port, and the second one has the same schema as the
second input port. These additional ports are used to emit tuples that have never been
involved in a match during their stay in the join windows.

3.5.2   Utility operators

The set of utility operators provides numerous data flow management functions typically demanded by the continuous nature of SPAs. These operators deal with tasks such
as separating multiplexed flows into multiple physically independent flows, multiplexing related flows into a single stream, delaying and throttling the tuple flow, compression
and decompression, parsing and formatting, as well as synchronizing streams under certain conditions. In the rest of this section, we will describe some of these operators as
well as their implementation in the SPL language standard toolkit.

Split
The *split operator* is a stateless operator used to divide an inbound stream into multiple streams, demultiplexing the incoming tuples according to an expression whose
evaluation indicates which outbound streams will transport each of these tuples.

In SPL, the Split operator can be used to demultiplex a stream into a (pre-
determined) set of outbound streams as seen in the following code excerpt:

```
1    stream<rstring id, rstring name> Src = MySource() {}
2    (stream<Names> SmallIds; stream<Names> LargeIds) = Split(Src) {
3      param
4        index: (id<100)? 0: 1; // output port to forward the tuple
5    }
```

In this example, an instance of the Split operator is used to divide the inbound
stream into two outbound streams. Incoming tuples whose id attribute value is less

than 100 are routed to the `SmallIds` stream and the remaining tuples are routed to the `LargeIds` stream. Note that the port indices start from 0 and follow the stream declaration order.

Barrier

The *barrier operator* is a stateful operator used to synchronize streams. Unlike the join, the barrier does not have a match condition, and ensures that it receives one tuple for each incoming stream before it emits an outbound tuple. The inbound tuples can be combined through simple data manipulations to produce an outbound tuple.

In SPL, the `Barrier` operator can be used for this task. The following code excerpt illustrates this operation:

```
1  stream<rstring id, blob image> Images = MySource() {}
2  stream<rstring id, rstring background> Backgrounds = BackgroundFinder(Images) {}
3  stream<rstring id, list<rstring> objects> Objects = ObjectFinder(Images) {}
4  stream<Backgrounds, Objects> Results = Barrier(Backgrounds; Objects) {}
```

In this example, consider a scenario where images from a source operator are streamed as tuples to image analysis operators. One of the operators extracts the background from a still image, producing an outbound tuple with that information transported by the `Backgrounds` stream. The other downstream operator detects objects in the image, producing the `Objects` stream. In this scenario, an instance of the `Barrier` operator is used to combine one tuple with the background information with the corresponding one containing information on the objects that have been detected. As a result, for each source image, the `Barrier` operator outputs a tuple that contains both the background information and the list of objects that have been located in each image.

3.5.3 Edge adapter operators

The set of edge adapter operators provides the tools used to convert data from external sources into a format that can be used by an application as well as to convert streams from this application into a format suitable for consumption by other data processing systems.

Most SPSs provide large sets of such adapters. In the rest of this section, we provide an overview of commonly found adapters, as well as their implementation in the SPL language.

File adapters

A *file adapter* is used to read and write data to and from files. File adapters are very effective, in particular, as an integration mechanism for legacy systems with the need to interoperate with SPAs.

When used as a source, file adapters provide three types of service. First, they can extract data from a file and stream it out. While a file has a fixed size, in contrast to a potentially infinite stream, the use of a file adapter to channel its contents into an application can, for example, enable consistent and repeatable scenarios that can be used to test and validate the data analysis carried out by a SPA.

Second, file adapters can be used to stream out the contents of a file to bootstrap an application with configuration data in the form of lookup tables or pre-computed data mining models.

Third, file adapters can be used to ingest and analyze data produced by an external (usually, legacy) source that produces files or to add new content to these files in a continuous mode.[6] When used in this way, a source file adapter typically consumes files deposited in a directory, as they are generated, and removes them as their content is streamed out.

When used as a sink, file adapters provide some very similar services. First, the output of a SPA running in test mode can be streamed out to a file, as a means of validating the implemented analytics.

Second, a file sink adapter can write to one or more files, which can in turn be consumed by another application or by a periodic batch process, providing a simple yet effective inter-process communication mechanism for external applications that inter-operate with a SPA.

The following code excerpt illustrates the use of a file source operator to read files from a directory:

```
1   stream<rstring file> Files = DirectoryScan() {
2     param
3       directory: "/tmp/work";
4       pattern: "^work.*";
5   }
6   stream<uint64 id, rstring name, list<uint32> values> Data = FileSource(Files){
7     param
8       format: csv;
9       compression: gzip;
10      deleteFile: true; // delete the file when processing is finished
11  }
```

In this example, an instance of the `DirectoryScan` operator is used to look for files in a given directory whose names match a particular pattern, specified as a regular expression. Each time a new file appears in this directory as a result of an external system depositing it in that location,[7] this new file is detected and its name is emitted as a tuple. The `DirectoryScan` operator also belongs to the SPL standard toolkit and is often used in conjunction with the `FileSource` operator when more than one file is to be read.

An instance of the `FileSource` operator is used to receive the stream that contains the names of files to be processed. This operator reads the contents of each file and emits tuples corresponding to this data. In the example, this operator is configured with a `format` parameter indicating that the tuples in the source file are laid out in the

---

[6] We refer to files that are updated continuously as *hot* files.

[7] It is important that the file is made available atomically, i.e., that all of its contents are present when its name appears. This is often achieved by moving the file to the destination directory with a temporary file name and performing an *atomic* renaming operation once the move is complete.

Comma-Separated Value (CSV) format. A `compression` parameter is also used to indicate that the file is to be decompressed using the `gzip` compression format. Finally, the `deleteFile` parameter is used to instruct the source operator to delete the file once it is fully processed.

TCP and UDP adapters

A TCP adapter is used to connect to external data sources and sinks via a TCP connection.

When acting as a data source, a TCP adapter can link to a remote data provider implemented as a TCP server, or provide a TCP server which a remote data provider can connect to, as a client. In both cases the external data provider pushes data into the TCP source adapter.

When acting as a data sink, a TCP adapter can link to a remote consumer providing a TCP server or alternatively, it can provide a TCP server which a remote consumer can connect to, as a client. In both cases the sink adapter pushes data into the remote consumer.

A User Datagram Protocol (UDP) adapter is similar to a TCP one, but employs the UDP protocol instead.

The following code excerpt illustrates the use of a matching pair of TCP source and sink operators in SPL:

```
1   // in one application
2   stream<uint64 id, rstring name> Data = MyDataSource() {}
3   () as Sink = TCPSink(Data) {
4     param
5       role: client;
6       port: 23145u;
7       address: "mynode.mydomain";
8       format: bin;
9   }
10  // in another application running on host "mynode.mydomain"
11  stream<uint64 id, rstring name> Src = TCPSource() {
12    param
13      role: server;
14      port: 23145u;
15      format: bin;
16  }
```

In this example, we consider a scenario where two different applications are communicating through a TCP connection. The code only shows the relevant parts of each application. In the first one we use a `TCPSink` operator to externalize the `Data` stream. This TCP sink is configured in the client mode via the `role` parameter, which means that it will establish a connection to a TCP server. The port and the address of the server it will contact are specified via the `port` and the `address` parameters, respectively. In the second application, we use a `TCPSource` operator to process the incoming TCP data and to output it as the `Src` stream. This TCP source is configured as a server, which means that it will accept a connection from a client. It also specifies the port on which the connection will be accepted. In this example, both the source and the sink operators use a `format` parameter to specify that the incoming data is binary encoded according to SPL's internal protocol.

The file, TCP, and UDP adapters provided by SPL's standard toolkit support different data formats. These adapters also support additional options including compression, character encoding, and encryption.

Other adapters

There are also other types of adapters usually provided by SPSs:

(a) Adapters to interface with *message queues*, which provide asynchronous communication between loosely coupled systems, with messages placed into a queue and stored until the recipients retrieve them at a later time. For instance, a SPA may interact with a queue[8] to make its results available to external systems acting as consumers. Conversely, a SPA may use a queue to consume results produced by external systems that act as producers. In Streams, operators to receive (XMSSource) and send (XMSSink) messages are available in the messaging toolkit (in the com.ibm.streams.messaging namespace).

(b) Adapters to interoperate with *relational databases*, which allow data to be read and written to relational tables. A database adapter usually defines a mapping between the types supported by the DBMS and the SPS. This mapping is used to convert database tuples into streaming tuples, and vice-versa. In addition to sources and sinks, database adapters designed to perform *enrichment* operations are also usually available. These adapters query a database system and use the results to add additional information to an existing stream. The queries can be customized based on the contents of the stream to be enriched, on a per-tuple basis. A common use case is to perform a lookup operation on a database table and use the result to populate an attribute of an outgoing tuple. In Streams, Open Data Base Connectivity (ODBC)-based operators are provided for reading (ODBCSource) and for appending (ODBCAppend) data from/to a database, for enriching tuples using data read from a database (ODBCEnrich), and for running an arbitrary SQL query (ODBCRun). These operators are available in SPL via the *db toolkit*[9] (in the com.ibm.streams.db namespace).

(c) Adapters to interact with Internet-based services, which support the use of HTTP-, File Transfer Protocol (FTP)-, and Simple Mail Transfer Protocol (SMTP)-based exchanges. For instance, these adapters can be used to read web pages, post to websites, read Really Simple Syndication (RSS) feeds, send email messages, and upload as well as download files to and from an FTP server. In SPL, the InetSource operator can be used to read data from Internet sources and is available in the *inet toolkit* (in the com.ibm.streams.inet namespace).

(d) Adapters to interact with *web services*, which are used to integrate SPAs with the existing software ecosystem in an enterprise environment. A common use case is to perform automatic actions as a result of the streaming analysis by invoking a web service interface that exposes a set of services. The core functionality provided by

---

[8] Popular implementations include Active MQ from the Apache Foundation [26] and WebSphere MQ from IBM [27].

[9] The db toolkit also contains several other vendor specific operators.

a web service adapter is to make Simple Object Access Protocol (SOAP) calls on a web service interface defined using the Web Services Description Language (WSDL).

Many other types of adapters have also been implemented, including adapters for processing large datasets in *big data*-type applications (e.g., Streams' *big data toolkit*), for data mining (e.g., Streams' *mining toolkit*), for integrating with financial data feeds (e.g., Streams' *financial services toolkit*), and for processing text (e.g., Streams' *text toolkit*).

3.6 Concluding remarks

In this chapter, we started exploring the process of building SPAs. We began with a summary of the distinguishing characteristics of SPAs (Section 3.2) and followed with a discussion of language features specifically designed to facilitate the development of these applications (Section 3.3). We introduced the IBM SPL language (Section 3.4) and used SPL code to create a simple application containing a data source, a small processing core, and a data sink. Finally, we introduced a set of operators that are commonly used in SPAs (Section 3.5). In the next chapter, we will make use of some of these operators and start exploring the mechanisms to manipulate and compose the flows of data that drive SPAs.

3.7 Programming exercises

The following exercises require the reader to become familiar with the SPL language and the Streams platform. We recommend accessing the product documentation available online [24] and using it as an additional source of information.

1. **File and socket manipulation, command-line processing, and window management.** Consider the Unix wc utility. It displays the number of lines, words, and bytes contained in a file. A line is defined as a string of characters delimited by a *newline* character. A word is defined as a string of characters delimited by *white space* characters.
   (a) Write a SPA that reads a file, counts the total number of lines, words, and characters, and prints out the result.
   (b) Extend your application so that it performs the counting over the last 100 lines of text and generates a stream of tuples where each one includes the word and character count for the last 100 lines.
   (c) Make the number of lines to be used (i.e., the last *n* lines) a submission time argument. Change your application such that it accepts data from an external TCP client rather than reading data from a file. Test your application using the telnet utility.

2. **Edge adaptation and data aggregation.** Many computational systems provide a logging service so system administrators can keep tabs on a system's activity and diagnose problems based on knowledge gleaned from the logs. Continuous and automated processing of log entries can speed up the identification of problems. Web servers are examples of such systems. A web server log contains information regarding HTTP-based access to web pages hosted by the server, including the Internet Protocol (IP) address of the host accessing the page, the time of a particular access, the number of bytes sent by the server, and the specific page addresses that have been served. For real examples of web server logs, consult the Internet Traffic Archive [28].

   (a) Write an application that processes the logs produced by a web server. For each unique IP address, compute the total number of bytes sent to it.

   (b) Extend your application such that the log entries are fed to the system as a stream, rather than being read from a file. Add an option to your application that allows the log entries to be replayed with the same inter-arrival time associated with the actual log entries, considering the timestamps of each individual log message.

   (c) Every time a log entry associated with a particular IP address is received, update the total number of bytes served to that IP address and output the new value. You should discard IP addresses for which no activity was detected during the last hour.

   (d) Extend your application with an additional stream whose individual tuples contain queries. Each query tuple contains a k value. Every time a query is received, the top k IP addresses with the highest number of bytes served should be output into a result stream.

3. **Edge adaptation and data aggregation.** The National Oceanic and Atmospheric Administration (NOAA) weather service website (http://weather.noaa.gov) provides live weather information for all Zone Improvement Plan (ZIP) codes in the United States. The information is readily available through a web browser. We now want to transform this information into input to a SPA.

   (a) Design an application to process the weather information such that a specific website (e.g., NOAA's) and any other necessary arguments (such as the specific ZIP code, the type of information to be retrieved, and the periodicity) to extract it are provided as submission time arguments. Test this application by outputting the results to a file.

   (b) Use the collected data to compute average metrics for different time scales as well as for different geographic regions (represented as an arbitrary group of ZIP codes). For instance, for each region, find the average humidity level over the last 1, 2, and 4 hours based on the timestamps associated with the data.

   (c) Compute the same information as an aggregate average for all available ZIP codes.

4. **Workload generation, command-line processing, composite operators, and stream relational operators.** Consider a scenario with two data sources: a stream with *sensor readings* and a stream with *queries* on the sensor data. The tuples storing

the sensor readings contain a sequence number, a timestamp, a location, and a sensor reading value attribute. The tuples in the query stream specify a threshold value and a region of interest. These queries aim at locating *recent* sensor readings that have a value that is *close* to the query threshold value and originate in a sensor whose location is within the query region.

(a) Write an application that reads these two streams (sensor and query) from two input files and replays them based on delays specified in the files. For each query received, the application must locate the matching sensor readings that occurred in the last *t* seconds (based on the tuple's timestamp attribute) and output them. The application should write the results to the console as well as to a file.

(b) Replace the file sources with an on-the-fly data generator, which should also be instantiated as part of the application.

(c) Replace the application's sink with a verifier sink that checks the correctness of the results, making sure the result tuples have sensor readings and queries that match.

(d) Change the data generator so that it outputs out-of-order sensor readings. The data generator should be configurable and periodically output tuples whose timestamps are out-of-order. Assume that the distance between two out-of-order tuples seen in a sequence is bounded by a threshold δ whose value is provided as a submission time argument (i.e., if two tuples with timestamps t_1 and t_2 appear out-of-order in the stream, then we have $|t_1 - t_2| \leq \delta$).

(e) Write an application segment (as a composite operator) that sorts a partially *ordered* stream. The order of a stream is established by the timestamp attribute of the tuples in this stream.

(f) Now assume you have two such unordered sensor readings streams. Update your application so that it makes use of the composite operator you just created, sorting each stream individually and merging them into a single fully ordered stream. Test your application using two generator sensor sources, one generator query source, and the verifier sink.

References

[1] Babcock B, Datar M, Motwani R. Load shedding in data stream systems. In: Aggarwal C, editor. Data Streams: Models and Algorithms. Springer; 2007. pp. 127–146.

[2] Chi Y, Yu PS, Wang H, Muntz RR. LoadStar: a load shedding scheme for classifying data streams. In: Proceedings of the SIAM Conference on Data Mining (SDM). Newport Beach, CA; 2005. pp. 346–357.

[3] Gedik B, Wu KL, Yu PS. Efficient construction of compact source filters for adaptive load shedding in data stream processing. In: Proceedings of the IEEE International Conference on Data Engineering (ICDE). Cancun, Mexico; 2008. pp. 396–405.

[4] Gedik B, Wu KL, Yu PS, Liu L. GrubJoin: an adaptive, multi-way, windowed stream join with time correlation-aware CPU load shedding. IEEE Transactions on Data and Knowledge Engineering (TKDE). 2007;19(10):1363–1380.

[5] Tatbul N, Çetintemel U, Zdonik SB, Cherniack M, Stonebraker M. Load shedding in a data stream manager. In: Proceedings of the International Conference on Very Large Databases (VLDB). Berlin, Germany; 2003. pp. 309–320.

[6] Tatbul N, Zdonik SB. Dealing with overload in distributed stream processing systems. In: Proceedings of the IEEE International Workshop on Networking Meets Databases (NetDB). Atlanta, GA; 2006. p. 24.

[7] Bray T, Paoli J, Sperberg-McQueen CM, Maler E, Yergeau F. Extensible Markup Language (XML) 1.0 – Fifth Edition. World Wide Web Consortium (W3C); 2008. http://www.w3.org/TR/REC-xml/.

[8] Crockford D. The application/json Media Type for JavaScript Object Notation (JSON). The Internet Engineering Task Force (IETF); 2006. RFC 4627.

[9] Michael M, Moreira JE, Wisniewski DSRW. Scale-up x scale-out: a case study using Nutch/Lucene. In: Proceedings of the IEEE International Conference on Parallel and Distributed Processing Systems (IPDPS). Long Beach, CA; 2007. pp. 1–8.

[10] Hennessy JL, Patterson DA. Computer Architecture: A Quantitative Approach. 2nd edn. Morgan Kaufmann; 1996.

[11] Aho AV, Ullman JD. Universality of data retrieval languages. In: Proceedings of the ACM Symposium on Principles of Programming Languages (POPL). Chicago, IL; 1979. pp. 110–119.

[12] Arasu A, Babu S, Widom J. The CQL continuous query language: semantic foundations and query execution. Very Large Databases Journal (VLDBJ). 2006;15(2):121–142.

[13] Oracle BEA. EPL Reference Guide. Oracle; 2011. http://docs.oracle.com/cd/E13157_01/wlevs/docs30/epl_guide/index.html.

[14] Terry D, Goldberg D, Nichols D, Oki B. Continuous queries over append-only databases. In: Proceedings of the ACM International Conference on Management of Data (SIGMOD). San Diego, CA; 1992. pp. 321–330.

[15] ISO. Information Technology – Database Languages – SQL. International Organization for Standardization (ISO); 2011. ISO/IEC 9075.

[16] Codd EF. Relational completeness of data base sublanguages. In: Rustin R, editor. Database Systems. Prentice Hall; 1972. pp. 65–98.

[17] StreamBase Systems; retrieved in April 2011. http://www.streambase.com/.

[18] Sybase Aleri Streaming Platform; retrieved March 2011. http://www.sybase.com/products/financialservicessolutions/aleristreamingplatform/.

[19] Abadi D, Ahmad Y, Balazinska M, Çetintemel U, Cherniack M, Hwang JH, et al. The design of the Borealis stream processing engine. In: Proceedings of the Innovative Data Systems Research Conference (CIDR). Asilomar, CA; 2005. pp. 277–289.

[20] IBM InfoSphere Streams; retrieved in March 2011. http://www-01.ibm.com/software/data/infosphere/streams/.

[21] Jain N, Amini L, Andrade H, King R, Park Y, Selo P, et al. Design, implementation, and evaluation of the Linear Road benchmark on the Stream Processing Core. In: Proceedings of the ACM International Conference on Management of Data (SIGMOD). Chicago, IL; 2006. pp. 431–442.

[22] Gedik B, Andrade H, Wu KL, Yu PS, Doo M. SPADE: the System S declarative stream processing engine. In: Proceedings of the ACM International Conference on Management of Data (SIGMOD). Vancouver, Canada; 2008. pp. 1123–1134.

[23] Hirzel M, Andrade H, Gedik B, Kumar V, Losa G, Mendell M, et al. Streams Processing Language Specification. IBM Research; 2009. RC-24897.

[24] IBM InfoSphere Streams Version 3.0 Information Center; retrieved June 2011. `http://publib.boulder.ibm.com/infocenter/streams/v3r0/index.jsp`.

[25] Aggarwal C, editor. Data Streams: Models and Algorithms. Springer; 2007.

[26] Apache Active MQ; retrieved in April 2011. `http://activemq.apache.org/`.

[27] IBM WebSphere MQ; retrieved April 2011. `http://www.ibm.com/software/integration/wmq/`.

[28] Internet Traffic Archive; retrieved in July 2012. `http://ita.ee.lbl.gov`.

4 Application development – data flow programming

4.1 Overview

In this chapter we study data flow programming, including *flow composition* and *flow manipulation*. Flow composition focuses on techniques used for creating the topology associated with the flow graph for an application, while flow manipulation covers the use of operators to perform transformations on these flows.

We start by introducing different forms of flow composition in Section 4.2. In Section 4.3, we discuss flow manipulation and the properties of stream processing operators, including their internal state, selectivity, and arity, as well as parameterization, output assignments and functions, punctuations, and windowing configuration used to perform such manipulation.

4.2 Flow composition

Flow composition patterns fall into three main categories: *static*, *dynamic*, and *nested* composition.

Static composition is used to create the parts of the application topology that are known at development time. For instance, consider an application that has a source operator consuming data from a specific Internet source, for example, a Twitter [1] feed. Let's assume that the stream generated by the source is to be processed by a specific operator that analyzes this data, for example, a sentiment analysis operator [2] that probes the messages for positive or negative tone. In this case, the connection between the source operator and the analysis operator is known at development time and thus can be explicitly created by connecting the output port of the source operator to the input port of the analysis operator. This is referred to as *static composition*, because the connection between these two operators is made at development time.

Dynamic composition is used to create the segments of an application topology that are not *fully* known at development time. For instance, consider an application containing a video analysis operator that can consume multiple video streams. Let's assume that the operator is interested in all video streams from a specific location (e.g., satellite imagery of certain sections of the planet). Further assume that the video sources are dynamic and appear and disappear based on availability (e.g., satellite cameras might be re-oriented and pointed to other regions of the planet).

In this case, the connection between the analysis operator and the video sources can be specified implicitly at development time. In other words, the analysis operator can indicate the properties of the streams it is interested in consuming. However, it cannot explicitly name the connections, as the connections depend on the runtime availability of source operators. This is an example of dynamic composition, because the connections between these operators are transient and not known until runtime.

Another interesting aspect of dynamic composition is the possibility of topology self-modification in an application's flow graph – one of the features that makes the implementation of exploration tasks possible (see Section 2.4.2). In other words, the data flow graph topology of an application employing dynamic composition might change at runtime as a result of changes exerted by the application itself.

These modifications occur either as a result of an application changing the *subscription* describing the data it is interested in, or as a result of data sources altering the *properties* of the data they generate. For instance, the location of interest for the video analysis operator in our earlier example can change over time, which means that the set of streams it is interested in receiving gets updated at runtime, potentially severing existing stream connections and creating new ones. Likewise, data sources associated with satellite cameras might tag the images with a different geographical label as they are re-oriented, which might also sever existing stream connections as well as create new ones.

Finally, *nested composition* addresses the *modularity* problem in designing large-scale flow graphs. The fundamental notion is to group a subset of the flow graph as if it were a regular operator. These operators are called *composite operators*. They simply hide internal implementation details of flow subgraphs. Yet, the internals of a composite operator can be inspected by drilling down on particular composite operator instances using visualization techniques. This layered representation makes it possible to produce smaller, more manageable views of the overall data flow graph as well as to functionally break down large applications.

For example, SPAs typically have three high-level components as seen in the "Hello World" application (Figure 3.1): data ingest, data processing, and data egress. While in that case each component was a single operator, in real-world applications these functional components are more complex, involving a larger number of operators and more sophisticated flow graph topologies. Yet, these functional blocks can often be grouped within composite operators, producing a layered representation of the application topology.

Another facet of nested composition is *reuse*. The flow graph topology embodied by a composite operator can be instantiated multiple times, possibly with different configuration parameters. This capability can minimize the amount of code repetition.

Consider an application where multiple replicas of a very similar processing flow must be instantiated to perform the same analytics on different data partitions. A composite operator can be used to implement the processing logic once and be reused multiple times; once per data partition.

In the following sections we discuss how flow composition is implemented in SPL.

Table 4.1 Operator port configurations in SPL.

Type	Diagram	SPL code
Source		`stream`<StrType1> SrcStr1 = MySource() {}
Multi-port source		(`stream`<StrType1> SrcStr1; `stream`<StrType2> SrcStr2) = MySource() {}
Sink		() `as` Sink = MySink(SrcStr1) {}
Multi-port sink		() `as` Sink = MySink(SrcStr1; SrcStr2) {}
Simple		`stream`<StrType1> Str1 = MySink(SrcStr1) {}
Splitter		(`stream`<StrType1> Str1; `stream`<StrType2> Str2) = MyOper(SrcStr1) {}
Merger		`stream`<StrType1> Str1 = MyOper(SrcStr1; SrcStr2) {}
Complex		(`stream`<StrType1> Str1; `stream`<StrType2> Str2) = MyOper(SrcStr1; SrcStr2) {}

4.2.1   Static composition

Static composition consists of connecting the output port of an operator, which produces a stream, to the input port of another operator. This linkage creates a *static stream connection*.

Table 4.1 depicts the SPL operator types based on their port configurations. Code segments are provided to illustrate the syntactic constructs used to represent these different configurations.

Source operators

A source operator generally has one or more output ports. Since an output port produces a single stream, a multi-output source operator produces one stream per output port.

Having multiple output ports associated with a source operator is useful in two situations. First, when streaming data is being demultiplexed, i.e., the raw stream tapped by the source operator is being separated into distinct streams for the independent consumption of downstream operators. Second, when an operator produces multiple streams with different schemas, again to demultiplex more complex data embedded in the raw input.

Note that, in SPL, multiple output ports are represented in an operator invocation using the semicolon character (;) as a separator in the list of stream definitions. Furthermore, parentheses are used to enclose the list of stream definitions. They are mandatory only when the list includes more than a single stream definition.

Sink operators

A sink operator generally has one or more input ports. Multi-input sink operators are useful when streaming data consisting of tuples with different schemas must be consumed by the same sink.

As with output ports, multiple input ports are represented in an operator invocation using the semicolon character (;) as a separator in the list specifying the streams that should be statically routed to that operator's input ports. Each group of streams, separated by a semicolon, corresponds to one operator input port. If multiple streams are routed to the same input port, each name is then separated by a comma character (,).

Analytic operators

Simple analytic operators have a single input port and a single output port. The SPL standard toolkit includes many operators of this type performing operations that include filtering, projection, aggregation, and sorting, among others.

As is the case with source and sink operators, many analytic operators make use of additional input and output ports. For example, several streaming operations require *merging* or *correlating* data from multiple incoming streams. Specific operations include joining data from different streams, barrier synchronization across streams, and merging sorted streams. All of these operations have corresponding operators in the SPL standard toolkit.

Also, there are cases where an operator employs an additional input port as a mechanism to receive *control* directives. For instance, a data mining operator that *scores* incoming data against a particular model might have a control input port used to periodically receive an updated model.

Similarly, certain streaming operations require splitting the resulting data across multiple streams, which, in turn, requires multiple output ports. For instance, an analytic operator might produce *diagnostic measurements* associated with its computed results. This might occur when an operator performing *parsing* of incoming data, must also generate a diagnostic stream whose tuples encode statistics on the number of successful and failed matches produced by its parser.

Static connections

Static connections between operator instances come in different variations. Table 4.2 provides a summary alongside code segments to demonstrate the specific syntactic constructs used to represent them. In our discussion, we use two terms to describe specific topological configurations: downstream and upstream operators.

The term *downstream operators* refers to the set of operators that directly or indirectly receive data from a particular output port. Hence, when referring to the downstream operators of a particular operator, we are indicating the union of all the downstream operators connected directly to one of that operator's output ports as well as the ones indirectly connected through intermediate operators, recursively.

Similarly, the term *upstream operators* refers to the operators that are directly or indirectly sending data to a particular input port. Hence, when referring to the upstream

Table 4.2 Static connection variations in SPL.

Type	Diagram	SPL code
Direct	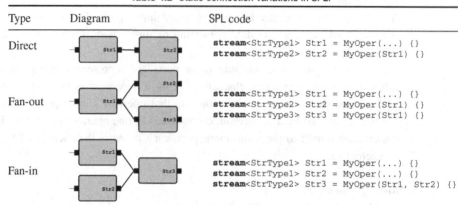	`stream<StrType1> Str1 = MyOper(...) {}` `stream<StrType2> Str2 = MyOper(Str1) {}`

Fan-out		`stream<StrType1> Str1 = MyOper(...) {}`
		`stream<StrType2> Str2 = MyOper(Str1) {}`
		`stream<StrType3> Str3 = MyOper(Str1) {}`

Fan-in		`stream<StrType1> Str1 = MyOper(...) {}`
		`stream<StrType1> Str2 = MyOper(...) {}`
		`stream<StrType2> Str3 = MyOper(Str1, Str2) {}`

Note: For brevity, ellipses (. . .) are used to represent the upstream part of the flow that is not shown in the figure. This is not a valid SPL syntax.

operators of an operator, we are indicating the union of all the upstream operators connected directly to one of that operator's input ports as well as the ones indirectly connected through intermediate operators, recursively.

Direct static connection

A *direct connection* is one involving a single output port connected to a single input port. When an operator submits a tuple to its output port, the tuple is delivered to the input port of the *downstream operator* directly connected to the receiving end of the connection.

Fan-out static connection

A fan-out connection is one involving a single output port connected to more than one input port, potentially located on different downstream operators.

When an operator submits a tuple to its output port, the tuple is delivered (or fanned out) to each one of the input ports of the downstream operators on the receiving end of the fan-out connections.

All of these input ports must be defined with the same schema, matching the output port on the sending side of the connection. This constraint is enforced by the SPL compiler, which identifies violations when an application is built.

Fan-in static connection

A fan-in connection is one where multiple output ports, associated with one or more *upstream* operators, are connected to a single input port.

Tuples received by an input port on the receiving end of a fan-in connection may have been produced by any of the upstream output ports whose streams are connected to it. The incoming tuples from different streams are multiplexed[1] on arrival to the input port.

[1] In SPL, this multiplexing happens according to the tuple arrival order.

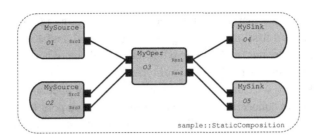

Figure 4.1 A visual representation of the application from Figure 4.2, illustrating various forms of static connections.

```
1  namespace sample;
2  composite StaticComposition {
3    graph
4      (stream<MyType1> Src1) as O1 = MySource() {}
5      (stream<MyType1> Src2; stream<MyType2> Src3) as O2 = MySource(){}
6      (stream<MyType3> Res1; stream<MyType4> Res2) as O3 = MyOper(Src1,Src2; Src3) {}
7      () as O4 = MySink(Res1) {}
8      () as O5 = MySink(Res1; Res2) {}
9  }
```

Figure 4.2 Static composition showcasing multiple types of static connections.

As is the case with fan-out connections, the schemas associated with the output ports producing the streams that are part of the fan-in connection must be the same.

An application flow graph example

An application flow graph is the representation of an application consisting of operators (vertices) interlinked by stream connections (edges). This representation can be arbitrarily complex.

Figure 4.1 illustrates one such application flow graph with a combination of fan-in and fan-out connections, employing operators of different types. The equivalent SPL source code can be seen in Figure 4.2.

Using this flow graph, we can make a few observations. First, operator instances are named either with the name of the stream produced by their first output port (the default) or with an explicit name indicated by the as clause. In this case, we can see that all of the operator instances have been explicitly named.

The first instance of the MySource operator, with a single output port, is named O1. The second instance of the MySource operator, with two output ports, is named O2. Note that these *operator instances* employ the same *operator type* with different configurations.

Similarly, the single instance of the MyOper operator is named O3. The first instance of the MySink operator, with a single input port, is named O4, and its second instance, with two input ports, is named O5.

Second, we can see that the first input port of operator O3 is one of the endpoints of a fan-in connection, where the streams Src1 and Src2 are connected. Stream Src1 is produced by the output port of the operator O1, whereas stream Src2 is produced by the first output port of the operator O2, and these two streams have the same schema: MyType1.

Note also that the syntax used for specifying the fan-in in line 6 indicates that this operator's first input port ingests streams Src1 and Src2, whereas the second port ingests stream Src3. Observe that the second output port in this operator has a different schema since stream Src3 is of type MyType2.

Finally, we can see that the first output port of operator O3 is one of the endpoints of a fan-out connection. The stream produced by this port, Res1, is connected to the input port of operator O4 as well as to the first input port of operator O5.

Flow graphs with feedback loops

Data flow graphs representing SPAs can have cycles. These cycles often represent feedback loops where a downstream operator sends control information back to an upstream operator, supplying information that can be used to dynamically modify the application behavior. This structure is one of the cornerstones of exploratory tasks, as discussed in Section 2.4.2.

Consider, for example, an application whose data flow graph is organized as a pipeline used to perform semantic searches [3] over an online data source. Assume that the first operator in the pipeline interacts with an external source to setup filters based on current topics of interest whereas downstream operators in the pipeline perform text analytics to determine if any interesting information can be extracted from the current feed. This process can be made adaptive by having a downstream operator notify the upstream filtering operator to adjust its filters based on the current results, enlarging or constraining the search dynamically, as needed.

Figure 4.3 shows the data flow graph for this application and Figure 4.4 provides the corresponding source code. The notable part in this example is in line 5, where the stream Fdb1 from operator Op4 is connected to an upstream operator (Op2).

4.2.2   Dynamic composition

Dynamic composition is established by specifying *stream subscriptions* and associating them with an operator's input port; and by specifying *stream properties* and associating them with an operator's output port.

Stream subscriptions are used to express *interest* on specific types and sources of data. These subscriptions describe the streams that an operator desires to receive on a particular input port, although the exact identities of the streams that might carry the data matching a subscription are not known at compile time. Instead, the subscriptions are used to describe predicates on the *properties* of streams to identify them dynamically, at runtime.

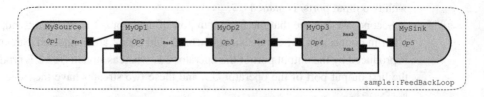

Figure 4.3 A visual representation of the application from Figure 4.4 showcasing a feedback loop.

```
1   namespace sample;
2   composite FeedbackLoop {
3     graph
4       (stream<MyType1> Src1) as Op1 = MySource() {}
5       (stream<MyType1> Res1) as Op2 = MyOp1(Src1; Fdb1) {}
6       (stream<MyType1> Res2) as Op3 = MyOp2(Res1){}
7       (stream<MyType1> Res3, stream<MyType2> Fdb1) as Op4 = MyOp3(Res2){}
8       () as Op5 = MySink(Res3) {}
9   }
```

Figure 4.4 A sample application in SPL illustrating feedback loops.

On the sending side, a stream produced by an output port can have a list of properties associated with it. These properties can be extended, removed, and modified at runtime. Input port subscriptions are evaluated against these properties to determine the set of dynamic connections to be created at runtime.

Similar to static connections, the output port and the input port involved in a dynamic connection must have matching schemas. Unlike static connections, dynamic connections are established at runtime, but can change as a result of any of these events:

1. A new application instance that includes operators generating streams with associated stream properties or consuming streams based on stream subscriptions is submitted for execution and starts running.
2. An existing application instance that includes operators generating streams with associated stream properties or consuming streams based on stream subscriptions is canceled and stops running.
3. The stream properties of an output port in an existing application instance are modified at runtime, possibly as a result of changes in the characteristics of the data transported by this stream.
4. The stream subscription of an input port in an existing application instance is modified at runtime by an operator, possibly as a result of changes in the data of interest to this operator.

In Streams, SPAs are instantiated by submitting them to a *runtime instance*. An *application instance* is called a *job*. Items 1 and 2 above result in updates in the dynamic connections of the flow graph when new jobs are submitted and existing ones are canceled.

As will be described in Section 5.2, the processing logic embodied by SPL operators can be written in general purpose programming languages that make use of Streams' runtime APIs or in SPL using Custom operators. Relevant to the present discussion is the fact that these programming interfaces include specific support for changing stream properties and subscriptions at runtime. Items 3 and 4 from the list above occur when an operator from one of the existing jobs makes use of those APIs.

Dynamic connections can be used to implement certain functionalities common to exploration tasks, where application behavior needs to change at runtime as a result of changes in the data, as well as to support certain administrative tasks where, an application or parts of it must be taken offline for maintenance or replacement. We discuss these use cases next.

Incremental deployment of an application

Dynamic connections enable application instances to connect to each other at any time during execution. Using this mechanism, larger applications can be deployed incrementally. As new pieces are constructed, they can be launched to augment the currently running application, providing additional or specialized analytical capabilities not available in the core application.

Incremental deployment facilitates phased development and deployment of large-scale applications, where the core functionality can be developed and deployed first and the extensions can be deployed later, when necessary. Moreover, it is also possible to create and sever connections between operator instances as a result of changes in the characteristics of the workload or as a shift in focus of the analytics, possibly as a result of refining the application goal from data that has been analyzed earlier.

Clearly, when dynamic connections are used by an application to interface with another, it also becomes necessary to engineer mechanisms to ensure that their life cycles are properly aligned in such a way that producer/consumer relationships are managed appropriately. For instance, if a particular application must be running and available before another one starts operating, the overall multi-application solution must include orchestrating mechanisms to ensure that such conditions hold.

Application maintenance

SPAs tend to run continuously for long periods of time and frequently include critical monitoring tasks that must operate non-stop. Hence, in many cases, application maintenance tasks must be designed to accommodate continuous operation.

Here, the ability to create dynamic connections can be used to quiesce one segment of a larger application, modify it, and bring it back up online later, without disturbing other parts of the application.

Similarly, the ability to create and sever dynamic connections can be used to replace an existing application segment by swapping it with an alternative, without introducing a lengthy downtime. Swapping segments in this form is a way to address bugs that might be present in that segment or to add functionality.

Source and sink discovery

The ability to establish dynamic connections can be used by an application to discover new sources of data as the application instances that generate these sources are brought up online. Alternately, these other applications might, at some point, *tag* results that they produce with information that might be relevant to data consumers, who might become interested or, alternatively, lose interest in that data.

In both cases, *exploratory* applications can be written without prior knowledge of the exact identities of the data sources, but can specify type and property related requirements, which will determine the set of data sources they will be connected to at runtime.

Self-evolving applications

Exploration tasks performed by a SPA require, in certain cases, that the data analysis *evolve*. Evolution, in this case, can take the form of a refinement or a recalibration of a data mining model employed by an application. It can also be more drastic and include the replacement of a portion of the analytics altogether. In either case, evolving the application is something that the application itself can command, as a function of discoveries it makes, changes in the data, or changes in the focus of the exploratory task, and it happens dynamically, as the application is running.

The ability to establish dynamic connections clearly facilitates the creation of a self-evolving application ecosystem. For instance, based on the results of the analysis performed by an application, a new set of applications can be deployed or the existing ones can be removed. In this way, applications can transform the global data flow graph based on their own findings.

Employing dynamic composition
In Streams, dynamic connections are supported through the `Export` and `Import` SPL operators.

An `Export` operator is used to *export* a stream produced by an operator's output port by specifying a set of properties describing this stream. As previously discussed, these properties are used for the *dynamic connection matching* process that takes place at runtime.[2]

An `Import` operator is used to *import* streams to be dynamically routed to an operator's input port. The operator specifies a subscription expression that is used during the *dynamic connection matching*. Exported streams whose properties satisfy the subscription expression will induce the creation of a dynamic connection between the output ports producing those streams and that input port.

`Import` and `Export` are pseudo-operators in SPL. They are simply syntactic constructs used to specify input port subscriptions and output port stream properties.

Table 4.3 depicts their usage, alongside with SPL code segments to illustrate the specific syntactic constructs used to establish a dynamic connection.

Table 4.3 Stream properties and input port subscription specifications in SPL.

Type	Diagram	SPL code
Import		```stream<MyType> Src = Import() { param subscription: color=="red" || size>3; } stream<MyType> Res = MyOper(Src) {}```
Export		```stream<MyType> Res = MyOper(...) {} () as Sink = Export(Res) { param properties: { color="red", size=1 }; }```

[2] Only exported streams can be part of a dynamic connection and the non-exported streams are not considered during the matching process.

An `Import` operator specifies a stream subscription expression associated with its input port. In this example, the tuples resulting from the subscription funnel into the stream `Src`, which, at any given point, is the result of multiplexing all the exported streams that matched. This stream can then be consumed normally by one or more downstream operator instances. In this excerpt, only one consumer (`MyOper`) is present.

The subscription expressions are specified using the `subscription` parameter of the `Import` operator and take the form of predicate expressions in SPL.

Finally, the `Export` pseudo-operator is used to specify the stream properties associated with the stream (in this excerpt, `Res`) originating in its upstream operator (in this excerpt, the operator instance `MyOper`) that it is going to export. The stream properties are specified using the `properties` clause of the `Export` operator and take the form of tuple literals, i.e., a list of attribute names and values.

A working example

We now consider a scenario where an application needs to process video from several camera feeds. The camera feeds each have a *location* and a *priority*. We assume that camera feeds from different locations with different priorities can be added and removed, dynamically. These changes might occur because cameras are moved (pan, zoom, tilt) or dynamically turned on and off.

In this scenario, the application analytic segment, which processes the video, does not know a priori the complete set of cameras that are currently active at any given time, but keeps track of the locations and priority levels it is interested in.

Unlike the diagrams we have seen so far, the flow graph in Figure 4.5 is not a compile-time view of a single application, but instead it is a *dynamic* view of a set of applications. In other words, multiple jobs are deployed corresponding to the two applications in the diagram, namely `sample::CameraFeed` and `sample::Analytic`. Each one has been instantiated three times. The *job identifiers* (shown in the top corner) are used to differentiate among instances of the same application.

As previously discussed, stream subscriptions and properties can be updated at runtime. In the present example, we assume, for simplicity,[3] that the `Camera` operator instances update the properties of the streams they produce based on a configuration specified at application instantiation time. Similarly, the instances of the `Analyzer` operator update their subscription expressions based on a configuration specified at application instantiation time. As a result, it can be seen in Figure 4.5 that the camera feed from job 1 is only connected to analyzers in jobs 3 and 4 because the properties `loc="Front"`, `priority=2` satisfy the expression `loc=="Front" &&` `priority<5` and `priority>0`, but not `loc=="Back"`.

Figure 4.6 provides the code that produces the flow graph in Figure 4.5 and it has several notable characteristics.

First, it contains two application *entry points* in the form of two composite operators: `sample::CameraFeed` and `sample::Analytic` that correspond to two applications.

---

[3] In real applications changes in properties and subscriptions occur over time.

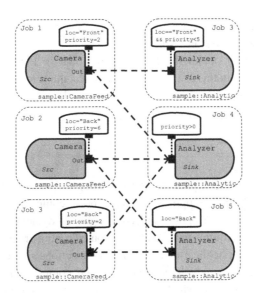

Figure 4.5 A live view of the dynamic connections that might be established when running the code seen in Figure 4.6.

```
1   namespace sample;
2   composite CameraFeed {
3     graph
4       (stream<MyType> Out) as Src = Camera() {
5         param
6           configuration: getSubmissionTimeValue("camera.conf");
7       }
8       () as Sink = Export(Out) {
9         param
10          properties: { loc="", priority=0 };
11      }
12  }
13  composite Analytic {
14      stream<MyType> Src = Import() {
15        param
16          subscription: loc=="none";
17      }
18      () as Sink = Analyzer(Src) {
19        param
20          configuration: getSubmissionTimeValue("analyzer.conf");
21      }
22  }
```

Figure 4.6 Establishing a properties-based dynamic connection.

Second, lines 6 and 20 in Figure 4.6 specify `configuration` parameters to both the `Camera` and `Analyzer` operators, where the value of the parameter is set at job submission time. Retrieving the actual value at runtime can be done by employing the `getSubmissionTimeValue` SPL built-in function. As will be described shortly, the instances of these operators can be configured with different values for these parameters during application instantiation. It is also possible to define and configure a stream subscription and stream properties dynamically at runtime as part of an operator's internal logic (Section 9.2.3).

Third, the `Camera` and `Analyzer` operators are user-defined and, therefore, not part of any of the SPL toolkits.

Finally, to obtain the flow graph shown in Figure 4.6 we must instantiate the applications in this example multiple times, thus creating six jobs.

Before this is done, however, these applications must be compiled using the `sc` compiler, as follows:

```
~/CameraApp>sc --main-composite sample::Analytic --output-directory=output/Analytic
~/CameraApp>sc --main-composite sample::CameraFeed --output-directory=output/CameraFeed
```

Unlike the compilation step used for the "Hello World" application from Section 3.4.2, these applications have been compiled without the `--standalone -application` option. Indeed, the use of dynamic connections between different application requires the Streams distributed runtime to manage the different jobs as well as to orchestrate the establishment of connections as needed.

As a result of building this code, the compiler creates an Application Description Language (ADL) file, which can be used as a handle to submit instances of the application to the distributed runtime, creating new jobs in the execution environment.

In Streams, this task is carried out by the `streamtool` command-line utility (Section 8.4.4), which interacts with the runtime environment to, among other tasks, submit and cancel jobs (Section 8.5.1).

During the job submission process, certain configurable parameters can be employed to specialize the application instances as described earlier. In the current scenario, different configuration files are used to populate operator-specific parameters, which operators will then use to define their own stream properties and subscriptions. The job submission along with each job's specific configuration is carried out as follows:

```
~/CameraApp>streamtool submitjob --instance-id StreamsSampleInstance
    -P analyzer.conf="analyzer1-conf.txt" output/Analytic/sample::Analytic.adl
~/CameraApp>streamtool submitjob --instance-id StreamsSampleInstance
    -P analyzer.conf="analyzer2-conf.txt" output/Analytic/sample::Analytic.adl
~/CameraApp>streamtool submitjob --instance-id StreamsSampleInstance
    -P analyzer.conf="analyzer3-conf.txt" output/Analytic/sample::Analytic.adl

~/CameraApp>streamtool submitjob --instance-id StreamsSampleInstance
    -P camera.conf="camera1-conf.txt" output/Analytic/sample::CameraFeed.adl
~/CameraApp>streamtool submitjob --instance-id StreamsSampleInstance
    -P camera.conf="camera2-conf.txt" output/Analytic/sample::CameraFeed.adl
~/CameraApp>streamtool submitjob --instance-id StreamsSampleInstance
    -P camera.conf="camera3-conf.txt" output/Analytic/sample::CameraFeed.adl
```

This example assumes that a Streams runtime instance called `StreamsSampleInstance` has already been created and is operational. All the jobs are then submitted to this runtime instance. In general, the Streams runtime can be instantiated multiple times, providing different and isolated execution environments. This feature is further described in Chapter 8.

The first three commands create three jobs, each with a unique job identifier. Each one of them is an instance of the `sample::Analytic` application, but parameterized differently. The last three commands are used to create three instances of the `sample::CameraFeed` application.

In each case, the -P option employed by streamtool is used to pass the configuration file names, whereas the --instance-id option is used to specify the Streams instance on which the jobs are being created.

The last argument given to the streamtool command is the ADL file that describes the application to the runtime environment.

Revisiting the incremental deployment use case
We have looked at dynamic connections in their most generic form, where the identities of the connections were not at all known at development time. In certain scenarios, the set of desired dynamic connections is known at development time, but multiple applications with dynamic connections between them may still be preferable to a single application with static connections.

For instance, when incrementally deploying an application, it is necessary to establish dynamic connections across multiple applications, even though the actual endpoints of the stream connections are known.

In these cases, *name-based* dynamic connections can be established by specifying an identifier on the exporting end and an identifier plus an application name on the importing end. Applications that include name-based exports cannot be instantiated more than once,[4] thus the application name and the identifier specified on the importing end unambiguously identify the exporting end of a dynamic connection.

Figure 4.7 provides the code for a scenario that makes use of a name-based dynamic connection. This excerpt also includes two composite operators. The first is sample::MainApp, which stands in as the core application and exports a stream by name. The second is sample::ExtensionApp, which stands in as the extension application and consumes, via a dynamic connection, the exported stream.

The Res stream from the main application is exported using the streamId parameter. A matching value for the streamId parameter is specified on the importing

```
1   namespace sample;
2   composite MainApp {
3     graph
4       stream<MyType> Res = MyMainApp() {}
5       () as Sink = Export(Res) {
6         param
7           streamId: "main-results";
8       }
9   }
10  composite ExtensionApp {
11    graph
12      stream<MyType> Src = Import() {
13        param
14          streamId: "main-results";
15          application: "sample::MainApp";
16      }
17      () as Sink = MyExtensionApp(Src) {}
18  }
```

Figure 4.7 Establishing a name-based dynamic connection.

---

[4] Streams supports *application scopes*, which are used to create runtime namespaces that apply to dynamic connections. Different instances of an application that export by name can be present in different application scopes.

side. Also, the name of the main composite is specified on the importing side via the `application` parameter.

Dynamic composition and content filtering

So far we have looked at dynamic composition at the stream level, where stream subscriptions and properties are used to match exported streams to its potential consumers. When an exported stream is matched to an import subscription, the content of the exported stream is delivered to the input port with a matching subscription.

In some application scenarios additional upstream filtering might be desired or needed. Even though unwanted tuples can always be filtered on the importing side, this often results in wasted bandwidth resources, because tuples will be sent from the exporting to the importing side only to be discarded later on. When the filtering to be performed on the importing side is highly selective in nature, the amount of bandwidth wasted can be significant.

As an example, consider a scenario where a news parsing application is exporting several streams, one for each main topic of news, for example, *politics*, *business*, *technology*, and *sports*. Assume that in this application each one of these exported streams carry tuples containing the full text of a news item as well as the source of the news (e.g., the name of the news agency). Consider a client application that is interested in receiving news items related to *business* and *technology*, but only from a specific news agency, e.g., *World News Press*. Naturally, this scenario can be implemented using stream subscriptions and properties with additional filtering on the importer side.

Figure 4.8 shows the code for implementing this idea. In this example, we see an application called `NewsExporter` that exports four different streams, each corresponding to a different news topic. The name of the topic is indicated as a property associated with the stream being exported.

We also see the `MyNewsImporter` application which uses an `Import` operator with a subscription expression indicating its interests on the `business` and `technology` topics. Finally, a `Filter` operator is used on the importing application side to screen for only those news items whose source is `World News Press`.

Alternatively, this type of dynamic composition can be enhanced with *content filters*. These filters are defined on the content of the matched streams, rather than on the stream properties using a subscription expression. In that sense, they operate at a different level than stream subscriptions. A stream subscription is evaluated against the stream properties of exported streams to determine the list of matching streams to be imported. In contrast, a content filter is defined on the attributes of the tuples flowing through the streams that matched a subscription.

In our example, we can modify the news importer application to explicitly indicate that filtering should be applied on the contents of the matched streams such that only news items from *World News Press* should be transmitted. While the end result is the same, this latter approach is more efficient in terms of communication costs.

```
1   composite NewsExporter {
2     type
3       NewsT = tuple<rstring source, list<rstring> text>;
4     graph
5       stream<NewsT> Politics = NewsAggregator() {
6         param topic: "politics";
7       }
8       () as PoliticsExporter = Export (Politics) {
9         param properties: { topic = "politics" };
10      }
11      stream<NewsT> Business = NewsAggregator() {
12        param topic: "business";
13      }
14      () as BusinessExporter = Export (Business) {
15        param properties: { topic = "business" };
16      }
17      stream<NewsT> Technology = NewsAggregator() {
18        param topic: "technology";
19      }
20      () as TechnologyExporter = Export (Technology) {
21        param properties: { topic = "technology" };
22      }
23      stream<NewsT> Sports = NewsAggregator() {
24        param topic: "sports";
25      }
26      () as SportsExporter = Export (Sports) {
27        param properties: { topic = "sports" };
28      }
29  }
30
31  composite MyNewsImporter {
32    type
33      NewsT = tuple<rstring source, list<rstring> text>;
34    graph
35      stream<NewsT> InterestingNews = Import() {
36        param
37          subscription : topic == "business" || topic == technology";
38      }
39      stream<NewsT> FilteredInterestingNews = Filter(InterestingNews) {
40        param
41          filter: source == "World News Press";
42      }
43  }
```

Figure 4.8 Filtering imported streams on the consumer side.

```
1   composite MyNewsImporter {
2     type
3       NewsT = tuple<rstring source, list<rstring> text>;
4     graph
5       stream<NewsT> InterestingNews = Import() {
6         param
7           subscription : topic == "business" || topic == technology";
8           filter : source == "World News Press";
9       }
10  }
```

Figure 4.9 Filtering imported streams using content filters.

Figure 4.9 shows the source code for this alternative. Note that we eliminated the Filter operator used in the original implementation. Instead, we specified a filter parameter in the Import operator. Internally, the runtime environment installs these filters on the exporting side, before any data is transmitted.

4.2.3   Nested composition

Nested composition consists of defining a flow subgraph as a composite operator, which is so named because, in contrast to primitive operators, it can host an internal flow graph.

In fact, we employed composite operators in all of the preceding examples, as any SPL application must have a top-level composite, enclosing the entire application flow graph. This specific composite is referred to as an application's *main* composite and is the entry point for an application. Note that a main composite operator does not have input ports or output ports.

When a composite operator is defined, the set of input ports and output ports it exposes are attached to some of the input and output ports that belong to the operators in its internal flow graph. As any primitive operator, composite operators can have various input and output ports and can serve as sources and sinks.

Table 4.4 summarizes the different composite operator variations, along with SPL code illustrating the specific syntactic constructs. Several notable aspects can be seen in this table.

The definition of a composite operator starts with the `composite` keyword, followed by the declaration of output and input ports, which are marked, respectively, by the `output` and `input` keywords. In each group, the individual ports are separated via the comma character (,), whereas a semicolon (;) is used to separate the list of input ports from the list of output ports. Each port definition is (optionally) associated with a stream schema as shown in the source code examples.

The body of a composite operator, which is delimited by curly braces, includes the `graph` clause marking the beginning of the topological description of the internal flow subgraph. This graph represents the operator's processing logic as a function of the collection of operators it embeds.

The invocation of a composite operator is syntactically similar to a primitive operator invocation, but its implementation is fully defined by SPL source code.

Composite operators address three main use cases: *encapsulation*, *reuse*, and *templatization*, in the form of higher-order composites.

Encapsulation

A composite operator can encapsulate large flow graphs and this capability can be seen as a modularization mechanism.

Composite operator definitions can use *other* composite operators as well, enabling the creation of hierarchical flow graphs. This capability usually reduces the engineering effort of assembling large-scale applications, as composite operators can be developed and can function as independent modules.

Figure 4.10 showcases one such example, where a subgraph employing four operators is encapsulated by a composite operator. Figure 4.11 provides the equivalent SPL source code. Note that the figure does *not* depict an application (in other words, the `ComplexComposite` is not a *main*). Also, in this code excerpt, this composite operator is merely *defined* and never *instantiated*, demonstrating how one can separate the definition of composite operators from the applications that use them, as a means to organize large source code bases.

Table 4.4 Composite operator definitions and uses.

Type	Diagram	SPL code
Source (*def*)	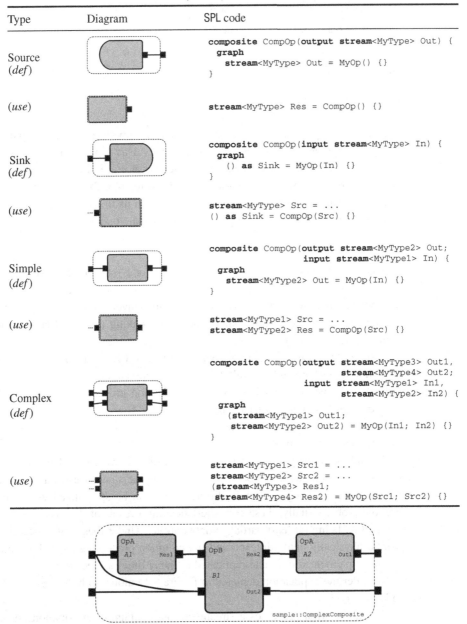	```composite CompOp(output stream<MyType> Out) {` ` graph` ` stream<MyType> Out = MyOp() {}` `}```
(*use*)		`stream<MyType> Res = CompOp() {}`
Sink (*def*)		```composite CompOp(input stream<MyType> In) {` ` graph` ` () as Sink = MyOp(In) {}` `}```
(*use*)		`stream<MyType> Src = ...` `() as Sink = CompOp(Src) {}`
Simple (*def*)		```composite CompOp(output stream<MyType2> Out;` ` input stream<MyType1> In) {` ` graph` ` stream<MyType2> Out = MyOp(In) {}` `}```
(*use*)		`stream<MyType1> Src = ...` `stream<MyType2> Res = CompOp(Src) {}`
Complex (*def*)		```composite CompOp(output stream<MyType3> Out1,` ` stream<MyType4> Out2;` ` input stream<MyType1> In1,` ` stream<MyType2> In2) {` ` graph` ` (stream<MyType1> Out1;` ` stream<MyType2> Out2) = MyOp(In1; In2) {}` `}```

(*use*)		`stream<MyType1> Src1 = ...`
		`stream<MyType2> Src2 = ...`
		`(stream<MyType3> Res1;`
		` stream<MyType4> Res2) = MyOp(Src1; Src2) {}`

Figure 4.10 A visual representation of the composite operator from Figure 4.11 demonstrating encapsulation.

This example also demonstrates how composite operators can ingest streaming data. Specifically, a composite's input ports are connected directly to input ports of the operator instances in its internal subgraph. For instance, operator instance A1 uses stream `In1` as input, where `In1` is also the name of the first input port of the composite

```
1   namespace sample;
2   composite ComplexComposite(output Out1, Out2; input In1, In2) {
3     graph
4       (stream<MyType1> Res1) as A1 = OpA(In1) {}
5       (stream<MyType2> Res2;
6        stream<MyType3> Out2) as B1 = OpB(Res1; In1, In2) {}
7       (stream<MyType3> Out1) as A2 = OpA(Res2) {}
8   }
```

Figure 4.11 Encapsulation via composite operators.

operator. As expected, this establishes a connection between the first input port of the composite operator and that of operator instance A1 as depicted in Figure 4.10.

Similarly, the output ports of a composite operator are directly connected to output ports belonging to operators that are part of the internal flow subgraph. For example, operator instance A2 produces the stream Out1 on its first output port, where Out1 is also the name of the first output port of the composite operator.

Reuse

The composite operator abstraction is also a code *reuse* mechanism. As discussed before, a composite operator can be instantiated multiple times, in different sites in an application flow graph. Moreover, individual invocations of composite operators can be customized, on a per instance basis, by making use of different values for its configuration parameters.

Code reuse is useful in many scenarios both as a way of factoring out common code as well as a mechanism for parallelization as discussed next.

First, in some cases applications apply common processing to different data, processed in different locations in an application's flow graph. In these situations the common processing can be encapsulated in a composite operator, which is then instantiated multiple times.

Second, streaming data can sometimes be demultiplexed and processed independently. In certain cases, the actual processing is the same irrespective of the data. In others, the processing is slightly different, but can be specialized via simple configuration knobs. Usually, these characteristics are a result of the task performed by an application being (at least partially) data parallel. Under these circumstances, the data can be physically demultiplexed and the common processing carried out on this data can be implemented once and replicated as many times as necessary.

Consider the application depicted in Figure 4.12 along with the equivalent SPL code in Figure 4.13.

This application processes a stream of data containing information on individuals such as their name, id, and age. The application's objective is to filter the data stream, preserving only information related to individuals who are known to be either *teachers* or *students* and separating the data into two streams corresponding to each of these two categories. The application also ensures that only tuples for individuals within a specific age range (different for each category) are considered.

The classification is carried out based on information maintained in two databases, one for *teachers* and another for *students*.

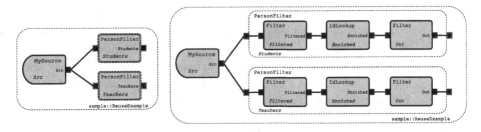

Figure 4.12 A visual representation of the application from Figure 4.13, illustrating code reuse via composite operators. The unexpanded graph is shown on the left and the expanded one on the right.

```
1   namespace sample;
2   composite PersonFilter(output Out; input In) {
3     param
4       expression<list<int32>> $ageRange;
5       expression<rstring> $table;
6     graph
7       stream<In> Filtered = Filter(In) {
8         param
9           filter: In.age >= $ageRange[0] && In.age <= $ageRange[1];
10      }
11      stream<In, tuple<boolean found>> Enriched = IdLookup(Filtered) {
12        param
13          table: $table;
14        output
15          Enriched: found = Lookup(Filtered.id);
16      }
17      stream<In> Out = Filter(Enriched) { // projects out the 'found' attribute
18        param
19          filter: In.found == true;
20      }
21  }
22  composite ReuseExample {
23    graph
24      stream<int32 id, rstring name, int32 age> Src = MySource() {}
25      stream<Src> Students = PersonFilter(Src) {
26        param
27          ageRange: [6, 12];
28          table: "student-db";
29      }
30      stream<Src> Teachers = PersonFilter(Src) {
31        param
32          ageRange: [24, 65];
33          table: "teacher-db";
34      }
35  }
```

Figure 4.13 Composite operator reuse.

To perform this filtering, the application employs a composite operator that encapsulates a pre-screening and a lookup step, each of them implemented by a separate primitive operator. Specifically, the application's *main* ReuseExample invokes the composite operator PersonFilter twice in lines 25 and 30 to extract information related to *teachers* and *students*.

In each invocation, the age range to use in the filter predicate as well as the database to use for the lookups are specified as configuration parameters, via the ageRange and the table parameters, respectively.

The assembly of all these different pieces requires additional SPL syntax.

The set of configuration parameters for the `PersonFinder` composite operator can be seen in lines 4 and 5. The $ sign is used as a *sigil*[5] for the formal parameter names.

The `PersonFilter` composite operator makes use of a `Filter` operator, instantiated in line 7. This `Filter` uses the `$ageRange` parameter from the composite operator to define the actual filter predicate to be applied to the incoming `In` stream.

Note that the stream produced by the `Filter` operator has the same schema as the input stream it consumes. This relationship is established using the `stream<In>` syntax, which indicates that the `Filtered` stream has the same schema as that of the `In` stream.

Next, in line 11, the `PersonFilter` operator employs the `IdLookup` operator, a user-defined operator and not part of the standard toolkit. This operator is used to look for an individual's `id` in a particular database table defined by the `$table` parameter from the `PersonFilter` operator. The schema for its output port is an extension of the one associated with its input port, where an additional `found` attribute of type `boolean` is added (using the syntax `stream<In, tuple<boolean found>>`).

The `IdLookup` operator includes an `output` clause to specify the assignment of values to the attributes of its outgoing tuples. In this particular case, the `found` attribute will have the result of the lookup operation as computed by the `Lookup` function using the value of the `id` attribute from each tuple in the `Filtered` stream.[6]

Finally, the `PersonFilter` operator invokes a second `Filter` operator in line 17 to retain only the individuals for which the lookup succeeded. This operator also drops the `found` attribute from the final output.

Figure 4.12 shows both the unexpanded and the expanded data flow graphs for this application. Both instances of the `PersonFilter` operator, `Students` and `Teachers`, include a similar internal flow graph. Thus, both of their graphs include an operator instance named `Filtered`, but each one is individually configured and performs a different operation. To differentiate among these instances, one can use their *fully qualified* operator instance names: `Students.Filtered` and `Teachers.Filtered`.

Higher-order composites

As we have seen, a composite operator can be parameterized through configuration parameters. A composite operator parameter can be used for several purposes, some already seen in earlier examples. One of the most interesting, however, is to use a parameter to provide the name of an operator (primitive or composite) to be instantiated as part of a composite operator's internal data flow subgraph.

---

[5] In the programming language literature, a sigil is a symbol attached to a variable name, showing the variable's data type or scope. In SPL, it is used to indicate that the variable is a parameter whose value, which can be an expression, will be expanded in-place wherever the variable is used.

[6] In most operators, the value given to attributes that are not explicitly assigned in a newly produced tuple is taken from the incoming tuple processed to produce this new tuple.

```
1   namespace sample;
2   composite Process(output Out; input In) {
3     param
4       operator $sampler;
5       expression<float64> $rate;
6     graph
7       stream<In> Filtered = MyFilter(In) {}
8       stream<In> Sampled = $sampler(Filtered) {
9         param
10          rate: $rate;
11      }
12      stream<In> Out = MyReducer(Sampled) {}
13  }
14  composite RandomSampler(output Out; input In) {
15    graph
16      param
17        expression<float64> $rate;
18      stream<In> Out = Filter(In) {
19        param
20          filter: random() < $rate;
21      }
22  }
23  composite Main {
24    graph
25      stream<int32 id> Src = Beacon() {}
26      stream<int32 id> Sampled1 = Process(Src) {
27        param
28          rate: 0.2;
29          sampler: sample::RandomSampler;
30      }
31      stream<int32 id> Sampled2 = Process(Src) {
32        param
33          rate: 0.2;
34          sampler: some.other::Sampler;
35      }
36  }
```

Figure 4.14 Higher-order composite operators.

This type of *higher-order* composite provides *code generation* capabilities as its instantiation can produce different flow graphs resulting from the values given to its parameters, which can include other composite operators, as well as functions, types, expressions, and attributes.

The code in Figure 4.14 demonstrates the use of a higher-order composite. In this example the Process operator is used to perform three operations on a stream: filtering, sampling, and data reduction.

While the filtering and the data reduction segments in Process' flow graph are fixed, the sampling piece is configurable via the sampler parameter and different sampling operators (primitive or composite) can be used.

Lines 29 and 34 show two different uses of the Process operator where different sampling operators are passed as arguments, the RandomSampler (whose definition can be seen in line 14) and some.other::Sampler (whose definition is not shown).

Line 4 shows how the formal parameter intended to be configured with an operator name is defined. Specifically, the operator keyword is used to indicate the meta-type[7] for the parameter $sampler. Line 8 shows how it is used in the body of the Process operator.

---

[7] SPL supports other parameter meta-types, including expressions, functions, and attributes.

4.3   Flow manipulation

The topology of a SPA flow graph is determined by the combination of all subflows created by the different types of composition employed by the application. Yet, to understand the semantics of an application, it is necessary to look at how the streaming data is manipulated by the different operator instances in the flow graph.

As discussed before, an operator consumes streams received by its input ports and generates a new stream on each of its output ports. The operator's internal processing core is responsible for consuming the incoming data and for producing the results that are eventually emitted through its output ports.

The processing logic implemented by an operator not only depends on its internal algorithm, but also on its specific invocation in a particular SPA, which configures its ports, parameters, and other runtime properties.

In the SPL language, the syntax for invoking an operator provides the mechanisms to specify and customize the desired functionality for each instantiation of the operator. Thus, the possible ways in which an operator can be invoked defines the *declarative interface* of the operator. Many operator properties are reflected in the operator's declarative interface and affect its internal implementation, whereas certain other operator properties are traits of the operator's internal implementation and are not visible at the declarative interface level.

Primitive operators are implemented in general purpose programming languages and their implementation details are not visible in the SPL source code. The various syntactic and semantic properties of a primitive operator are defined in an *operator model* that describes it. This model is further discussed in Section 5.2.

The following discussion looks at operator properties, including those that are visible at the declarative interface level as well as those that are characteristics of its internal implementation, and their impact on the syntax and semantics of an operator.

4.3.1   Operator state

A stream processing operator can either maintain internal state across tuples while processing them, or process tuples independently of each other. Formally, operators can be categorized with respect to *state management* into three groups: *stateless*, *stateful*, and *partitioned stateful*.

A *stateless operator*, as the name indicates, does not maintain any internal state. An operator of this kind performs computation on a tuple-by-tuple basis without storing or accessing data structures created as a result of processing earlier data.

For instance, the *projection* relational operator implements a stateless operation. This operator is used to remove attributes from a tuple, from a table (in a relational database query), or from a stream (in a SPA). Clearly, removing an attribute from a tuple requires no additional data beyond the tuple being operated on.

In SPL the `Functor` operator can be used to perform a projection. In the following code segment, it is used to remove the `value` attribute from the tuples in the `Src` stream:

```
1  stream<rstring name, int32 value> Src = MySource() {}
2  stream<rstring name> Res = Functor(Src) {}
```

The simplicity of stateless operators can be leveraged by the application runtime environment to optimize their execution. First, a stateless operator can be parallelized easily as it processes individual tuples independent of other tuples. Moreover, a stateless operator can function without synchronization in a multi-threaded context, where multiple upstream operators might be sending tuples to it concurrently. Finally, a stateless operator can be restarted upon failures, without the need of any recovery procedure. These issues are further discussed in Chapter 8.

In contrast to stateless operators, a *stateful operator* creates and maintains state as it process the incoming tuples. Such state, along with its internal algorithm, affects the results the operator produces.

SPAs are continuous in nature and the input data they process is potentially unbounded. Hence, the state maintained by an operator is usually a *synopsis* of the tuples received so far, a subset of recent tuples kept in a window buffer, a collection of handles to external data (from non-stream sources), or a combination of all these. State accumulation is necessary because streaming analytics such as an aggregation operation, a pattern detection algorithm, or a method for time series analysis must inspect data over either a period of time or until a certain number of samples have been seen.

As an example of a stateful operator, consider the `DeDuplicate` operator, which is part of SPL's standard toolkit. This operator filters out tuples that are deemed duplicates of previously received tuples. With this operator, a tuple is considered a duplicate if it shares the same *key* with a previously seen tuple within a pre-defined period of time. Clearly, the implementation of such operation requires maintaining an internal data structure as well as the logic to expire data that is no longer relevant.

In the following code segment, the `DeDuplicate` operator is used to keep the last minute's worth of data, which is then used to screen the incoming tuples for duplicates:

```
1  stream<rstring name, int32 value> Src = MySource() {}
2  stream<rstring name, int32 value> Res = DeDuplicate(Src) {
3    param
4      key: name;
5      timeOut: 60.0; // only remember the last 60 seconds
6  }
```

The runtime support for a stateful operator is considerably more complex than for stateless operators. It includes management of the inbound tuple flow as well as specialized parallelization and fault tolerance support.

More specifically a stateful operator requires synchronization to protect its internal state from concurrent access in a multi-threaded execution context. In other words, upstream traffic from multiple streams might trigger simultaneous access to its internal data structures, which become *critical section*s where operations must be properly coordinated to ensure runtime correctness.

Moreover the processing logic carried out by a stateful operator is not as easily parallelizable. Its internal state creates dependencies between the processing of individual tuples as each tuple might alter the operator's state.

When it comes to fault tolerance, a stateful operator must employ some form of state persistence (e.g., a checkpoint infrastructure, a transactional log, or replicas) or strategies for rebuilding state (e.g., *priming* the operator with additional data for a certain amount of time) to ensure that its internal state can be recovered after a failure requiring an operator restart.

Finally, a *partitioned stateful operator* is an important special case of stateful operators. While this type of operator manages an internal state, the data structures supporting the state can be separated into independent *partitions*. Each partition corresponds to an independent segment of the input data. In other words, the streaming data processed by one such operator can be seen as a collection of substreams multiplexed together, but with an element (e.g., a key or a unique combination of the tuple attributes) that allows the operator to perform the necessary data segmentation.

As an example, consider a stream that carries trading transactions from a stock exchange, where each transaction relates to a particular stock symbol. Specifically, each transaction tuple contains an attribute `ticker`, which is a symbol that identifies the company whose share is being sold or bought (e.g., "IBM," "AAPL" for Apple, "GOOG" for Google). Let us now consider an application that is designed to compute the Volume Weighted Average Price (VWAP) for each type of stock over the last 10 transactions on that stock.

In such an example, the VWAP is the state that must be maintained while the application runs. To compute a particular stock's VWAP, separate tallies for the average of each company's stock price must be maintained. The incoming data flow can be viewed as a stream that multiplexes several substreams (one per company), each transporting tuples related to a single stock ticker symbol. Similarly, the state maintained by the operator can be divided into partitions, where each partition of the state (i.e., the individual tallies for each ticker symbol) only depends on the data from that specific substream.

Figure 4.15 demonstrates this scenario as an SPL application. The running tallies are computed by an `Aggregate` operator instance, which computes different types of aggregations, for example, `Max`, `Min`, and `Sum`.

```
1   namespace sample;
2   composite VWAP {
3     graph
4       stream<rstring ticker, decimal32 price, decimal32 volume> Src = TickerSource() {}
5       stream<rstring ticker, decimal32 weight, decimal32 volume> Res = Aggregate(Src) {
6         window
7           Src: tumbling, count(10);
8         param
9           partitionBy: ticker;
10        output
11          Res: weight = Sum(volume*price), volume = Sum(volume);
12      }
13      stream<rstring ticker,decimal32 vwap> Out = Functor(Res) {
14        output
15          Out: vwap = weight / volume;
16      }
17  }
```

Figure 4.15 Computing the volume weighted average price for stock trading transactions.

The `partitionBy` parameter, seen in line 9, is used to indicate that the internal state, defined as a function of a window of 10 tuples, should be segmented based on the value of the `ticker` attribute. In other words, independent aggregation operations take place on the resulting substreams created *after* demultiplexing the `Src` stream based on the values of the `ticker` attribute. As a result, this `Aggregate` instance computes summations over the last ten tuples for each partition, independently. The downstream `Functor` is used to compute the final volume weighted average price.

The runtime support for a partitioned stateful operator can also be specialized. Specifically, a partitioned stateful operator is more efficiently parallelizable provided that coordinated access to the different partitions is guaranteed. In fact, the logical partitioned substreams can be physically split into actual streams and processed independently by multiple operator instances, as long as the tuples belonging to the same partition are kept as part of the same stream. Naturally, the fault tolerance support provided by the runtime environment can also be refined and segmented based on the existence of different partitions, a topic further discussed in Chapter 8.

4.3.2   Selectivity and arity

A non-source operator generally executes its internal processing logic in response to the arrival of a tuple to one of its input ports. The result of the execution of this processing logic *might* cause the operator to emit one or more tuples. Two operator properties govern the relationship between data received and produced by an operator: *selectivity* and *arity*.

The *selectivity* of an operator captures the relationship between the number of tuples produced by its internal processing logic as a function of the number of tuples it ingested. Broadly, operators can be divided into two selectivity categories: *fixed* selectivity and *variable* selectivity.

A fixed selectivity operator consumes a fixed number of inbound tuples and generates a fixed number of outbound tuples.

The SPL standard toolkit's `Barrier` is one such operator and it is used to synchronize the data flow from multiple inbound streams. It operates by consuming one tuple at a time from each of its (configurable) n input ports such that one tuple from each port is processed together, in lock-step. As a result, it produces a single outbound tuple.

A variable selectivity operator exhibits a non-fixed relationship between the number of tuples consumed and produced.

The SPL standard toolkit's `Join` operator is one such operator and it is used to correlate tuples buffered in windows associated to its two inbound streams. For each tuple consumed from either input port, there could be zero, one, or more tuples generated based on the number of tuples satisfying the operator's matching condition. In other words, there is no fixed relationship between the number of tuples consumed and produced as this relationship is purely dependent on the characteristics of the incoming data.

The *arity* of an operator refers to the number of ports it has. The term arity can be applied to both the input or output side of an operator. A particularly important category of operators is the one comprising operators with an arity of 1 for both input and output ports or, simply, *single input/single output* operators. Such configuration is commonly encountered in practice and their selectivity can be one of the following:

- A *one-to-one* operator (1 : 1) produces an outbound tuple for each inbound tuple it receives, which is usually the result of applying a simple *map* operation to the data inflow.
- A *one-to-at-most-one* operator (1 : [0, 1]) produces either no outbound tuples or a single one for each inbound tuple it receives, usually as a result of applying a *filter* to the data inflow.
- A *one-to-many* operator (1 : N) produces more than one outbound tuple for each inbound tuple it receives, usually as a result of performing *data expansion*.
- A *many-to-one* operator (M : 1) produces an outbound tuple for every M inbound tuples it receives, usually as a result of performing *data reduction*.
- A *many-to-many* operator (M : N) produces zero or more outbound tuples as a result of consuming zero or more inbound tuples. This case covers all other possibilities beyond the ones listed above.

Understanding the arity and selectivity of an operator helps reasoning about the overall flow rates within a SPA. For instance, operators that perform filtering and data reduction typically reduce the outflow data rates, in contrast to operators performing data expansion, which increase the outflow data rates.

4.3.3 Using parameters

As seen in earlier examples, operators are parameterizable. An operator parameter is used to customize an operator instance and can either be optional or mandatory. An optional parameter may or may not be specified by an operator invocation, whereas a mandatory parameter must always be specified.

As an example, consider the `Functor` operator which can perform two (stream) relational operations, *projection* and *selection*. Depending on where it is used, it can be instantiated to perform either or both of these operations. This flexibility is possible because its `filter` parameter is optional and the setting that controls what attributes to project out can also be tweaked.[8]

An operator parameter can optionally be typed. If a type is defined it can be fixed or selected from a set of types. As an example of untyped parameters, the `groupBy` parameter of an `Aggregate` operator can be of any type, since it is used to divide tuples into groups, which are uniquely defined by the value of a specific `groupBy` attribute.

An example of a fixed typed parameter is the `filter` condition of a `Functor` operator, which must result in a `boolean` value. In other words, applying a filtering

---

[8] The `Functor` operator decides which attributes to project out based on the schema associated with its output port.

predicate to a tuple must result in a *true* or *false* result, indicating whether the tuple meets the predicate.

An example of the second category of typed parameters is the `sortBy` parameter in the `Sort` operator which must be associated with a value from a type that can produce a *strict weak ordering* of values (e.g., one of SPL's numeric types) as it will be used for comparing tuples based on the values associated with the `sortBy` attribute.

Parameters can have bounded or unbounded cardinality. A *bounded cardinality* parameter must be configured with a fixed number of settings, whereas an *unbounded cardinality* can be configured with a potentially unbounded number of settings. Here, we use the term *setting* to denote the expressions specified as part of the operator invocation.

For instance, the `filter` parameter of a `Functor` operator has cardinality 1, meaning that the operator expects a single setting for that parameter. In contrast, the `sortBy` (mandatory) parameter of a `Sort` operator may accept one or more settings for that parameter, in a sequence where each of them is separated by a comma (`,`). The following code excerpt illustrates these cases:

```
1   stream<rstring id, int32 width, int32 height> Src = MySource() {}
2   stream<rstring id, int32 width, int32 height> Filtered = Filter(Src) {
3     param
4       filter: id != ""; // mandatory, fixed cardinality (=1) parameter
5   }
6   stream<rstring id, int32 width, int32 height> Sorted = Sort(Filtered) {
7     window
8       Filtered: tumbling, count(4);
9     param
10      sortBy: width, height; // mandatory, unbounded cardinality parameter
11  }
```

A parameter can also be associated with an *expression mode*. The expression mode prescribes the valid forms of an expression that can be used as a setting for a parameter. Three broad categories are commonly used: *attribute-free*, *attribute-only*, and *unconstrained*.

An *attribute-free* expression must not reference any attributes from the schemas associated with one of the operator's input ports. Such a constraint is put in place either because the operator does not have input ports or because the setting for this parameter must be independent of the current tuple or tuples being processed.

As as an example, the `file` parameter in the standard toolkit's `FileSource`[9] is attribute-free as it specifies the name of a file to be read by the operator.

An *attribute-only* expression must contain a single attribute from one of the schemas associated with one of the operator's input ports. As an example, the `DeDuplicate` operator has a parameter named `key`, consisting of one or more attribute-only expressions. Thus one or more of such expressions, each with a single attribute, can be used as the keys to detect duplicates.

As an example, the `DeDuplicate` operator has a parameter whose setting specifies the attribute or the set of attributes to be used as keys to detect duplicates.

An *unconstrained* expression may reference attributes associated with the schemas defining the tuple formats from multiple input ports. As an example, the `Join` operator

---

[9] The `FileSource` operator is used to read a file and stream it out.

has a `match` parameter that takes a *boolean* expression as a value, where this expression states the conditions under which two tuples (coming from its two different input ports) are considered a match. The following code exemplifies such use of an unconstrained expression parameter:

```
1   stream<int32 id, int32 value> Values = MyValueSource() {}
2   stream<int32 id, rstring name> Names = MyNameSource() {
3   stream<Values, Names> Joined = Join(Values; Names) {
4     window
5       Values: sliding, time(5.0);
6       Names: sliding, time(5.0);
7     param
8       match: Values.id == Names.id && value > 0;
9   }
```

In this invocation of the `Join` operator, the setting for the `match` parameter is an unconstrained expression used in the operator's internal processing logic to compare the `id` attribute value for a tuple from the `Values` stream with the one from the `Names` stream as well as to check if the `value` attribute (also from the `Values` stream[10]) is greater than zero.

4.3.4   Output assignments and output functions

An operator invocation's *output assignment* specifies how the attributes of outbound tuples are to be populated. An attribute can either be populated by assigning it a value directly (i.e., a *plain assignment*) or by computing a value (i.e., an *assignment with an output function*).

A *plain* assignment employs an expression whose computed value is assigned to a tuple attribute. Similar to an expression used to set the value for a parameter, the output assignment expressions can also have different modes, for instance, attribute-free, attribute-only, and unconstrained.

As an example of a plain output assignment, consider the following use of the `Functor` operator:

```
1   stream<rstring name, float64 distance, float64 duration> Src = MySource() {}
2   stream<rstring name, float64 speed> Res = Functor(Src) {
3     output
4       Res: speed = distance / duration;
5   }
```

In this code excerpt, the expression `distance / duration` is used to compute the value to be assigned to the `speed` attribute in the tuples transported by the `Res` stream.

In general, when assignments for output attributes are omitted, an operator instance assigns them using the value of a matching attribute from the inbound tuple that triggered the processing. Hence, in the preceding example, the original code is equivalent to declaring `Res: name = Src.name, speed = distance / duration;` since there isn't an explicit assignment for the `name` attribute.

An assignment with an *output* function is syntactically similar to a plain assignment, but rather than specifying an expression, it specifies an *output* function that can take

---

[10] An unambiguous attribute name can be used without stating which stream it belongs to.

zero or more expressions as arguments. The value produced by this function is then used to populate the attribute of an outbound tuple. The need for an output function occurs in two situations.

First, when each tuple produced by certain type of operators is the result of processing multiple inbound tuples. Such operators specify a *reduction function* to be applied to a collection of values. As an example, consider the following code segment with an invocation of the `Aggregate` operator:

```
1   stream<rstring name, int32 age> Src = MySource() {}
2   stream<rstring name, int32 age> Res = Aggregate(Src) {
3     window
4       Src: tumbling, count(10);
5     output
6       Res: name = ArgMax(age, name), age = Max(age);
7   }
```

In this case, the attribute `age` in an outbound tuple is assigned the maximum value observed for that attribute in the set of inbound tuples from the `Src` stream that have been accumulated in the operator's window, before the computation is carried out. The maximum value is computed by the `Max` function. Note that the attribute `name` in the outbound tuples is assigned the value of the `name` attribute belonging to the inbound tuple with maximum value for the `age` attribute. Such result is obtained by applying the `ArgMax` function to the inbound tuples.

Second, a source operator not having any input ports may need to specify how the attributes of its outbound tuples are to be populated. The following code snippet illustrates this case:

```
1   stream<blob image, int32 w, int32 h> Src = MyImageSource() {
2     output
3       Src: image = Image(), w = Width(), h = Height();
4   }
```

In such a situation, a hypothetical `MyImageSource` operator producing tuples containing images makes use of three functions: the `Image` function to produce the actual image as a sequence of bytes (whose type is a `blob`, one of the supported types in the SPL type system), a `Width` function to extract the image's width, and a `Height` function to extract the image's height. These functions[11] are invoked every time the operator produces an outbound tuple, and their values are used to populate the tuple attributes `image`, `w`, and `h`.

For operators requiring attribute assignment with an output function, the assignment can still be omitted. In this case, the actual value assignment is made using the result of applying a *default* function invoked with the matching attribute value (if any) from the inbound tuple. Each operator's implementation can define what its default output assignment functions is.[12] A common approach is to define a function called `Last` that returns the value of the attribute from the last tuple among the set of tuples over which the output function is evaluated. As an example, consider the following code segment:

---

[11] The prototypes of output functions supported by an operator are defined in its operator model.
[12] The default output function is defined in the operator model.

```
1   stream<rstring id, int32 value> Src = MySource() {}
2   stream<rstring id, int32 valueSum> Res = Aggregate(Src) {
3     window
4       Src: sliding, count(10), count(1);
5     output
6       Res: valueSum = Sum(value);
7   }
```

In this example, an `Aggregate` operator with a sliding window (Section 4.3.6) is used to output the `Sum` of the last ten `value` attributes for each new tuple received. Note that the `id` attribute does not have an output assignment specified, which is equivalent to defining `Res: id=Last(id)`, `valueSum=Sum(value);`. In this case, the `id` value of the tuple that triggers the processing (i.e., the last tuple in the window) is assigned to the outbound tuple's `id` attribute.

4.3.5   Punctuations

A *punctuation* is a marker that appears along with the tuple flow in a stream. It is used to segment or delimit the data flow, and as a control signal for operators. Unlike a tuple, punctuations do not have a schema as they do not transport data.

In the SPL language two types of punctuations are defined: *window punctuations* and *final punctuations*.

A window punctuation is used to segment a stream into logical groups of tuples. Thus, a punctuation signals the end of a group, comprising the tuples that have been part of the flow so far, as well as the beginning of a new group, comprising all tuples up to the next punctuation.

The intent of segmenting a stream with punctuations is to group together sequences of related tuples. The boundaries created by consecutive window punctuations can be used by an operator to define the set of tuples to be collectively processed as a unit.

As an example, consider again the standard toolkit `Join` operator. A tuple arriving at one of the operator's input ports can result in zero or more outbound tuples being emitted depending on whether the inbound tuple matches existing tuples in the other input port's window. When there are matches, the set of outbound tuples generated by the operator are logically related, as they are the result of the same processing step. In this case, the `Join` operator outputs a window punctuation after the outbound tuples are emitted to indicate that these tuples form a logical group.

The signal provided by a punctuation can be ignored, but it is often useful to a downstream operator's internal processing. For instance, an `Aggregate` operator that follows a `Join` can treat a punctuated set of inbound tuples as a processing window, and perform an aggregation operation over them.

The operator's semantics are also determined by how it behaves in the presence of inbound punctuations, and whether and under which circumstances its own internal processing produces them. For instance, an operator that has a punctuation-based window defined on a particular input port can only work with an input stream that carries window punctuations. In other words, its input port is a *window punctuation expecting* port.

```
1    namespace sample;
2    composite SensorQuery {
3      graph
4        stream<rstring id, uint64 value, float32 locX, float32 locY>
5          Sensor = SensorSource() {}
6        stream<float32 locX, float32 locY, float32 distance> Query=QuerySource(){}
7        stream<rstring id, int32 value> Res = Join(Sensor; Query) {
8          window
9            Sensor: sliding, time(60);
10           Query: sliding, count(0);
11         param
12           match:sqrt(pow(Sensor.locX-Query.locX, 2)+pow(Sensor.locY-Query.locY,2))
13               <= Query.distance;
14       }
15       stream<rstring id, int32 value> FltRes = Filter(Res) {
16         param
17           filter: value > 0;
18       }
19       stream<rstring id, int32 maxValue> Summary = Aggregate(FltRes) {
20         window
21           FltRes: tumbling, punct();
22         output:
23           Summary: id = ArgMax(value, id), maxValue = Max(value);
24       }
25   }
```

Figure 4.16 The use of punctuation-based windows.

Along the same lines, an operator's output port can create either a *window punctuation free* stream or a *window punctuated* stream. A window punctuated stream is generated either because the operator has an output port producing punctuations or it has an output port that can preserve window punctuations received by the operator. The former means that the operator issues its own punctuations, as is the case with SPL's `Join` operator. The latter means that the operator *preserves* window punctuations received from an upstream operator and forwards them as part of its outbound flow. The `Filter` operator in SPL exhibits such behavior.

Figure 4.16 depicts an application that employs window punctuations. In this hypothetical example, the `Sensor` stream, with tuples representing readings from a physical sensor, is joined with the `Query` stream whose tuples represent queries on the sensor data.

The actual correlation is performed as follows. For every tuple in the `Query` stream, the set of sensors readings from the last minute, within the spatial range specified by the `Query` tuple, is output as the result of correlations computed by the `Join` operator. This sequence of outbound tuples is followed by a window punctuation.

Downstream from the `Join`, a `Filter` operator instance is used to remove sensor readings whose values are less than or equal to zero. The `Filter` operator's output port *preserves* punctuations. Thus, the input to the `Aggregate` operator instance, which defines a punctuation-based window (note the `punct()` function in line 21), is segmented and each group of tuples is kept together in a window, which is then used to search and compute the highest value for the sensor reading.

While SPAs are continuous in nature and streams are typically infinite, there are cases where a stream comes to an end. Two of these situations are common. First, a stream can be created by an operator as a result of converting static data from an external finite source (e.g., a file or a database table). Second, a stream belonging to an application instance can come to an end as a result of that application instance being terminated.

To indicate that no additional tuples will be generated by an output port as well as to enable other operators to react to such events, a second type of punctuation is necessary. In SPL, this is called a *final* punctuation.

Thus, when an operator exhausts its external finite source of data, or during an application instance shutdown, final punctuations are issued and operators receiving them can perform their own procedure towards finalizing their internal logic and clean-up.

4.3.6   Windowing

A *window* is a buffer associated with an input port to retain previously received tuples. The window *policies* define the operational semantics of a window, for example, when tuples are removed from it as well as when it becomes ready for processing.

The first parameter of the windowing policy is called the *eviction policy* and determines the properties of the tuples that can be held in the buffer. In certain cases, its behavior is governed by a property of the window itself, such as its buffer's maximum capacity. In other cases, the buffer size fluctuates as the number of earlier tuples, whose characteristics match the properties defined by the eviction policy, also fluctuates.

The second component of the windowing policy is called the *trigger policy* and it determines how often the data buffered in the window gets processed by the operator internal logic. For instance, for an `Aggregate` operator, it indicates when the computation of the specific aggregation operation is carried out, and for a `Sort` operator it indicates when the tuples buffered in the window are sorted and the outbound results are produced.

Most SPSs support two types of windows, referred to as *tumbling window* and *sliding window*. These two types of windows both store tuples in the order they arrive, but differ in how they implement the eviction and trigger policies.

A *tumbling window* supports batch operations. When a tumbling window fills up (according to the rules set by its eviction policy), all the tuples in the window are evicted from the buffer by a window *flush* operation.

In contrast to that, a *sliding window* is designed to support incremental operations. When a sliding window fills up (again, according to the rules set by its eviction policy), older tuples in the window must be evicted to make up space for the new ones.

These two types of windows also differ in how they initiate (or trigger) an operator's processing logic.

A tumbling window's trigger policy is simple. Once the conditions are met for a window flush operation, but before the actual eviction of the tuples, the operator's internal logic is executed and the tuples are processed.[13] In other words, the trigger and batch eviction happen one after another when the window gets full. As a result, tumbling windows need not specify a separate trigger policy.

Many types of eviction policies exist and these are discussed in detail later in this section. Figure 4.17 depicts one of them, a four-tuple *count-based* eviction policy.

---

[13] This can also be performed incrementally, as each tuple arrives into the tumbling window

Figure 4.17 The evolution of a tumbling window with a four-tuple count-based eviction policy. Tuples arrive and are buffered until the limit is reached, after which the buffer is emptied.

Figure 4.18 The evolution of a sliding window with a four-tuple count-based eviction policy. Once four tuples are in the buffer, the arrival of a new one implies the eviction of the oldest one in the buffer.

Count-based eviction policies are very simple. In the case of a tumbling window, they define the maximum number of tuples that can be buffered. Once the number of buffered tuples reaches this limit, it triggers the processing of all of the data in the window as well as a window flush, which empties the buffer.

The following code excerpt demonstrates this scenario in the context of an `Aggregate` operator:

```
1  stream<rstring name, uint32 size> Src = MySource() {}
2  stream<rstring name, uint32 maxSize> Res = Aggregate(Src) {
3    window
4      Src: tumbling, count(4);
5    output
6      Res: name = ArgMax(size, name), maxSize = Max(size);
7  }
```

In this example, the `Aggregate` operator outputs the tuple with the maximum value for the attribute `size` amongst the last four inbound tuples from the `Src` stream.

In contrast to the behavior of a tumbling window, a sliding window's trigger policy works independently from its eviction policy. Figure 4.18 depicts a sliding window, again with a four-tuple count-based eviction policy, but also with a two-tuple count-based trigger policy. In this figure, the symbol $\star$ is used to indicate when the operator logic is triggered. Again, as is the case for eviction policies, different types of trigger policies exist and they are discussed later in this section.

The *count-based* policy used in the example defines the number of tuples that must be received since the last time the operator processing was triggered. Once this limit is reached, the operator's internal logic goes over the data buffered in the window and processes it, potentially generating outbound tuples with the results of the computation. The following code excerpt shows how this type of policy can be used:

```
1  stream<rstring name, uint32 size> Src = MySource() {}
2  stream<rstring name, uint32 maxSize> Res = Aggregate(Src) {
3    window
4      Src: sliding, count(4), count(2);
5    output
6      Res: name = ArgMax(size, name), maxSize = Max(size);
7  }
```

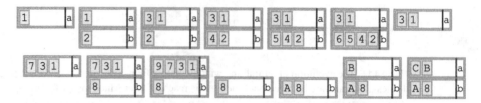

Figure 4.19 A partitioned tumbling window example, with a four-tuple count-based eviction policy. The tuples are inserted in the following order: $(1, a), (2, b), (3, a), (4, b), (5, b), (6, b),$ $(7, a), (8, b), (9, a), (A, b), (B, a), (C, a)$. The partition attribute is the second one in the tuple, whose value is either a or b.

In this example, an instance of the `Aggregate` operator is used to locate the tuple with the maximum value for the `size` attribute amongst the last four tuples from the `Src` stream, continuously, after every two new tuples are received.

Both sliding and tumbling windows can be *partitioned*. Inbound tuples are assigned to a partition based on an operator-specific condition. For instance, a hypothetical `id` attribute in a tuple can be used to create a separate partition, one for each of the unique values seen for the `id` attribute. Each partition acts as a separate logical window. In other words, the eviction and trigger policies are independently applied to each individual partition.

Figure 4.19 depicts a partitioned tumbling window with a four-tuple count-based eviction policy. The value of the second attribute in the tuples is used to determine the partition. As seen in the figure, the partitions get populated and tuples are evicted independently when a partition gets full. For instance, the arrival of the tuple $(6, b)$ results in the logical window associated with the b partition getting full, causing it to be flushed.

The following code excerpt demonstrates the use of a partitioned tumbling window:

```
1   stream<rstring name, rstring kind, uint32 size> Src = MySource() {}
2   stream<rstring name, rstring kind, uint32 maxSize> Res = Aggregate(Src) {
3     window
4       Src: tumbling, count(4), partitioned;
5     param
6       partitionBy: kind;
7     output
8       Res: name = ArgMax(size, name), maxSize = Max(size);
9   }
```

In this example, an instance of the `Aggregate` operator is used to output the tuple with the maximum value for the `size` attribute amongst the last four tuples, for each individual value of the `kind` attribute, i.e., after four tuples are accumulated in a partition.

Figure 4.20 depicts a partitioned sliding window with a four-tuple count-based eviction policy and a two-tuple count-based trigger policy. The second attribute in the tuples is used to determine the partition assignment. Again, the logical windows corresponding to different partitions operate as if they were separate windows.

The following SPL code excerpt demonstrates the use of a partitioned sliding window:

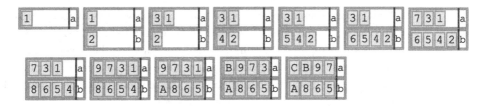

Figure 4.20 A partitioned sliding window example, with a four-tuple count-based eviction policy. The tuples are inserted in the following order: $(1, a)$, $(2, b)$, $(3, a)$, $(4, b)$, $(5, b)$, $(6, b)$, $(7, a)$, $(8, b)$, $(9, a)$, (A, b), (B, a), (C, a). The partition attribute is the second one in the tuple, whose value is either a or b.

```
1  stream<rstring name, rstring kind, uint32 size> Src = MySource() {}
2  stream<rstring name, rstring kind, uint32 maxSize> Res = Aggregate(Src) {
3    window
4      Src: sliding, count(4), count(2), partitioned;
5    param
6      partitionBy: kind;
7    output
8      Res: name = ArgMax(size, name), maxSize = Max(size);
9  }
```

In this example, an instance of the `Aggregate` operator is used to output, for each unique value of the `kind` attribute, the tuple from the partition with the maximum value for the `size` attribute amongst the last four tuples accumulated in that partition. This operation is triggered after two new tuples are accumulated in the partition, according to the configuration for the count-based trigger policy employed by this operator.

In a partitioned sliding window, a logical window associated with a partition will never become empty once a tuple is buffered in that partition. This has important implications for memory utilization. For example, consider that the attribute used to partition the window is drawn from a very large domain as, for example, an Internet Protocol (IP) address extracted from a network packet. In this case, it is conceivable that one partition for every possible IP address will eventually be created as time progresses, resulting in extensive memory consumption.

To counter this effect, a partitioned window can optionally specify a *partition expiration policy*, which states when stale partitions should be eliminated.

Windowing management policies

This section provides an in-depth look at specific types of eviction and trigger policies. These policy types can be grouped into four main categories based on what property forms the basis for deciding which/how many tuples will be part of the window at any given point: *count-based*, *delta-based*, *time-based*, and *punctuation-based*.

A *count-based policy* is characterized by the maximum number of tuples a window buffer can hold. When a count-based policy is used with a tumbling window as its eviction policy, it defines the maximum number of tuples its buffer can hold. On the arrival of a new tuple, this tuple is inserted into the window. If the number of tuples is equal to the size specified by the count-based eviction policy, then the window is flushed.

When a count-based policy is used with a sliding window as its eviction policy, it defines the maximum number of tuples to be kept in the window, as it *slides*. On the arrival of a new tuple, if the number of existing tuples in the window is already equal to the size specified by the eviction policy, the oldest tuple in the window is evicted. Once the eviction is carried out (if needed), the newly arrived tuple is inserted into the window.

When a count-based policy is used with a sliding window as its trigger policy, it is termed a count-based trigger policy. In this case, it defines the number of tuples that must be buffered until an operator's internal processing is triggered. Before the operator logic runs, a check is made to decide whether any tuples need to be evicted and, if so, these evictions are performed *before* the operator performs its task. At this point, the counter is reset, restarting the cycle. For sliding windows with count-based trigger policies, the order of operation is given as: eviction, insertion, and trigger.[14]

A *delta-based policy* is specified using a *delta threshold value* and a tuple attribute referred to as the *delta attribute*. The *type* of the threshold value must match that of the delta attribute. As an exception to the rule, when the delta attribute is of type timestamp, the delta threshold can be of type float64 (to represent time duration in seconds).

Behaviorally, in contrast to capping the maximum window buffer size, the size of the buffer fluctuates dynamically and is determined based on the difference in the value of the earliest timestamp and that of the latest. Thus, with a *delta-based* policy all tuples whose values for their timestamp attribute are within the range, defined by the latest value minus the delta threshold value, are kept in the buffer. It is important to understand that delta-based windows are expected to be used when the value of the attribute (on which the delta is computed) increases over time, i.e., older tuples have smaller value for this attribute than newer tuples. A similar mechanism for the *delta-based* trigger policy is also used to determine when the operator internal logic is invoked.

In this way, if a *delta-based* eviction policy is used with a tumbling window, when a new tuple arrives, if the difference between the new tuple's value for the delta attribute and that of the oldest tuple in the window is larger than the delta threshold value, the window is processed and flushed and the newly arrived tuple is inserted into a new empty window buffer. Otherwise, the tuple is buffered normally.

Alternatively, if a delta-based eviction policy is used with a sliding window, when a new tuple arrives, the existing tuples in the buffer for which the difference between the newly received tuple's value for the delta attribute and that of their own attributes is greater than the delta threshold value are evicted. The newly received tuple is inserted into the window after the evictions take place (if they are performed at all).

When a sliding window makes use of a delta-based trigger policy, on the arrival of a new tuple, if the difference between the new tuple's value for the delta attribute and that of the last tuple that triggered the processing is greater than the delta threshold,

---

[14] This is true, except when a time-based eviction policy is involved, in which case eviction is independent of insertion and trigger.

the operator internal logic is executed again. If any tuples are also subject to eviction, these evictions are performed *after* the operator logic completes its execution cycle. For sliding windows with delta-based trigger policies, the order of operation is given as: trigger, eviction, insertion.[15]

A *time-based* window policy is specified using a wall-clock time period. In this way, eviction and operator logic trigger events can take place at regular time intervals.

If a tumbling window makes use of a *time-based* eviction policy, when a new tuple arrives, it is inserted into the window buffer. When the time elapsed since the last window flush exceeds the period specified by the eviction policy, the operator logic is invoked and the buffer is again flushed. Differently from other policies, in this case, the operator's internal logic might be invoked and, potentially, be followed by a window flush independently from a tuple arrival.

Alternatively, if a time-based eviction policy is used with a sliding window, as tuples arrive they are inserted into the window buffer and tuples that have been in the window longer than the policy's defined time period are evicted.

When a sliding window makes use of a time-based trigger policy, it defines the period of time between successive invocations of the operator's internal logic. As is the case for tumbling windows, the arrival of a new tuple is *not* necessary for the operator to process the tuples that have been buffered.

Finally, a *punctuation-based* window policy establishes that a window buffer becomes ready for processing every time a punctuation is received. This is essentially a mechanism for creating a user-defined specification of window boundaries. This windowing policy can only be used for tumbling windows, because punctuations, by their nature, define a sequence of non-overlapping windows. Hence, inbound tuples are buffered until a punctuation is received, which triggers the processing of the window buffer by the operator's internal logic, subsequently followed by a window flush, re-starting the cycle.

Partition management policies

In a regular partitioned tumbling window, a partition is eliminated only when it becomes empty as a result of a window flush operation specific to that partition. In a regular partitioned sliding window, partitions are *never* removed. To understand why that is, recall that this type of window *slides* as new tuples arrive, but the window never gets empty, as a new tuple simply takes the space previously used by an expired one.

Therefore, when partitions are used, the buffer space employed by a window can grow continuously and unboundedly. As discussed earlier, a partition expiration policy is put in place to establish the conditions under which an entire partition can be retired, expunging any tuples that might have been in that window partition buffer at that time.

A partition expiration policy can be used with both tumbling and sliding windows, but its use is optional. There are three common partition expiration policies: *partition-count-based*, *tuple-count-based*, and *partition-time-based*.

---

[15] This is true, except when a time-based eviction policy is involved, in which case eviction is independent of insertion and trigger.

A *partition-count-based* partition expiration policy specifies the maximum number of partitions that can be buffered in a window. If the number of partitions exceeds this threshold, then an existing partition is evicted to make space.

A *tuple-count-based* partition expiration policy specifies the maximum number of tuples that can be buffered in a window across all partitions. If the number of tuples exceeds such a threshold, then an existing partition is evicted to make space.

In both cases, an additional *partition expiration selection policy* must be employed to determine the specific partition to remove. Possible selection policies include: the Least Recently Updated Partition (LRUP), the Least Frequently Updated Partition (LFUP), and the Oldest Partition (OP).

Finally, a *time-based* partition expiration policy specifies the maximum amount of wall-clock time a partition can remain active without any updates. In other words, once this amount of time has been elapsed with the partition remaining unchanged, it, along with the tuples that might have been buffered so far, are expired and eliminated.

4.4 Concluding remarks

In this chapter, we studied flow composition to understand the structural aspects of stream processing (Section 4.2). We introduced the concepts of *static flow composition*, used to construct basic data flow graphs; *dynamic flow composition*, used to update an application's flow graph topology at runtime, and *nested flow composition*, used to manage large-scale application development in a hierarchical manner.

This chapter also introduced flow manipulation mechanisms (Section 4.3), focusing on the properties and configurations of operators used to customize their behavior. These properties include state, selectivity, and arity, and configurations covered parameterization, output configuration, punctuation modes, and windowing.

In the next chapter, we examine modularity, extensibility, and distribution, which complement the topics discussed in this chapter, providing the basis for the development of domain-specific and large-scale distributed applications.

4.5 Programming exercises

1. **A rule-based marketing platform**. Consider a phone company, Snooping Phone Services, whose management software gets a Call Detail Record (CDR) with information about each call placed using its network. Each CDR includes the caller and callee numbers as well as the time and duration of a phone call.

    Snooping's management software also has access to a database where phone numbers are categorized by the type of business such as banking and insurance institutions, electric and gas utilities, cable and Internet services, travel services and others.

    Snooping's chief marketing officer is now planning to develop a marketing platform to allow customers to opt-in to receive offers for services that match their calling profile, possibly in return for monthly discounts in their phone bill. For instance, if a customer has recently called an insurance company, the marketing platform can send additional insurance offers to him or her. Similarly, if a customer is placing multiple international calls, Snooping's marketing platform can send an offer with a special service package including long-distance calling deals.

In this exercise, you will implement a prototype for this marketing platform, the *SnoopMarkPlat* application. When analyzing the development steps outlined below, abide by two general software engineering principles:

(1) *Modularization*: design the main application components, for data ingest, data processing, and data egress, as separate composite operators.

(2) *Flow composition*: consider the portions of your application where static, dynamic, and nested composition are each particularly effective and incorporate their use in your design.

The design and implementation of the application should proceed through the following steps:

(a) Develop a CDR workload generator as a source operator. The workload generator should output the call records to be used for testing the marketing application. In this application, at least the following settings must be configurable: the number of customers, the rate at which calls are placed, and the distribution of the different types of calls a customer might place (e.g., international call to another person, call to a bank, call to an emergency number). Consider making use of a graph model of relationships between groups of people as well as people and businesses to make the workload generator as realistic as possible.

(b) Develop at least three templates to describe the types of rules to be supported by the marketing platform. A typical rule template will look similar to the following example: "if a *land line* makes more than X calls *per month* to *international numbers*, send a promotion for an *international call package*." In this template, the value for X can be configured at the time when the rule is activated and the rule segments typeset in italics can be customized with different categories (for instance, *per month* can be replaced by *per week* and *international* by *Indian phone number*).

(c) Implement *SnoopMarkPlatController*, a GUI-based application used by the marketing analysts to manage rules by customizing the template rules developed in the previous steps as well as by activating and deactivating them.

(d) Write a source operator that can interact with the *SnoopMarkPlatController* application to receive rule updates. For example, when a template rule is customized and activated, a notification indicating the existence of a new rule is generated by the application and received by the source operator, thus making the marketing platform aware of it. Similarly, when a rule is deactivated a notification indicating this change is also sent to the marketing platform.

(e) Develop *SnoopMarkPlat* as a SPA that takes as input two streams: the *CDR stream* from the CDR workload generator and the *rule stream* from the *SnoopMarkPlatController* application's source operator. *SnoopMarkPlat* should maintain a set of active rules and evaluate each incoming CDR tuple against them. This logic should be encapsulated by a composite operator named

SPSRuleEvaluator. This operator must output a tuple every time an incoming CDR matches a rule. The outgoing tuple schema must include attributes to describe the matching rule as well as the original CDR.

(f) Extend the *SnoopMarkPlatController* application to receive live updates from the SPA, depicting the phone number and the specific promotion to be sent to a customer.

2. **A query-based network monitoring application.** A network service provider, Zippy Network Services (ZNS), employs Deep Packet Inspection (DPI) techniques to extract a utilization profile of its networking resources. At its finest level, individual flows are identified and associated with an application protocol (as described in /etc/services from a Unix host), a source and destination IP addresses, as well as a set of network packets and their sizes.

ZNS is interested in providing live and continuous monitoring of its networking resources such that queries of the following type can be be answered:

- List protocols that are consuming more than H percent of the total external bandwidth over the last T time units.
- List the top-k most resource intensive protocols over the last T time units.
- List all protocols that are consuming more than X times the standard deviation of the average traffic consumption of all protocols over the last T time units.
- List IP addresses that are consuming more than H percent of the total external bandwidth over the last T time units.
- List the top-k most resource intensive IP addresses over the last T time units.
- List all IP addresses that are consuming more than X times the standard deviation of the average traffic consumption of all IP addresses over the last T time units.

In this exercise, you will prototype the software to support a continuous query-based network monitoring system as part of the *ZNSNetMon* application. When analyzing the development steps outlined below, abide by two general software engineering principles:

(1) *Modularization*: design the main application components, for data ingest, data processing, and data egress, as separate composite operators.

(2) *Flow composition*: consider the portions of your application where static, dynamic, and nested composition are each particularly effective and incorporate their use in your design.

The design and implementation of the application should proceed through the following steps:

(a) Develop a network flow workload generator as a source operator. This operator emulates the data that would normally come from a DPI module. Make at least the following settings configurable: the rate of incoming tuples summarizing an individual flow, the number of IP addresses and protocols involved in the flows, and the distribution of the network usage over the different kinds of protocols and IP addresses.

(b) Develop a GUI-based application, *ZNSQueryWorkbench*, that enables users to create queries that resemble the examples we listed above. Ensure that all of the query thresholds (shown as variables in the query examples) can be configured by a network analyst.

(c) Develop a source operator that can interact with the *ZNSQueryWorkbench* application so queries can be customized, activated, and deactivated. When a new query is added or removed using the GUI, the source operator should output a notification summarizing the operation performed by the data analyst.

(d) Develop a SPA that takes as input two streams: the *DPI stream* from the workload generator and the *query stream* from the *ZNSQueryWorkbench* application. Furthermore, the application must be configurable with information regarding the networking resources it manages, including the external connectivity links, its range of internal IP addresses, and a profile of resource consumption per type of protocol.

When running, the application must inspect the DPI stream based on the currently active set of queries. This logic, necessary to evaluate each of the active queries, must be encapsulated by an operator named ZNSQueryEvaluator, which is expected to output a stream whose tuples correspond to updates modifying the earlier state (if any) representing the result for each active query. For instance, when a protocol enters into the top-k list replacing one that was already in the list, two tuples should be sent regarding this query: one describing the addition of the new protocol into the top-k list and a second one describing the removal of the one to be replaced.

(e) Add a sink operator to the application so it can send the resulting stream to the *ZNSQueryWorkbench* application. Improve its GUI so that it can also provide visual representations for each active query displaying the updates received from the *ZNSNetMon* application.

References

[1] Twitter; retrieved in March 2011. `http://www.twitter.com/`.
[2] Liu B. Sentiment analysis and subjectivity. In: Dale R, Moisl H, Somers H, editors. Handbook of Natural Language Processing. 2nd edn. CRC Press; 2010. pp. 627–666.
[3] Guha R, McCool R, Miller E. Semantic search. In: Proceedings of the International Conference on World Wide Web (WWW). Budapest, Hungary; 2003. pp. 700–709.

5 Large-scale development – modularity, extensibility, and distribution

5.1 Overview

In this chapter, we study modularity, extensibility, and distribution. As part of modularity, we look at how types, functions, primitive, and composite operators can be organized into *toolkits* to facilitate the development of large-scale applications.

As part of extensibility, we discuss how to extend a SPS with new cross-domain and domain-specific operators. As part of distribution, we discuss mechanisms to manage and control how the logical view of a SPA maps to a corresponding physical view to be deployed on a set of distributed hosts.

This chapter is organized as follows. Section 5.2 discusses modularity and extensibility, providing the foundations for structuring large, complex, and multi-component applications. Section 5.3 tackles the issue of distributed programming and deployment.

5.2 Modularity and extensibility

Large-scale software engineering and system development in programming languages such as C++ and Java relies on creating a complete application from a collection of smaller building blocks. The design of complex SPAs is no different. Such a design and development approach is referred to as *modularity* [1].

The reliance on a *modular design* aims at mitigating engineering complexity by organizing an application into modules or components, where the relationship between these components is expressed in terms of interactions through well-defined component interfaces.

In earlier chapters we discussed how composite operators facilitate the creation of modular SPAs. Not only does a composite operator create an independent component by encapsulating a data flow processing subgraph, but its invocation in the context of an application specifies an external interface, determining how its corresponding subgraph interacts with other parts of an application through stream connections.

When implementing an application in SPL, modularity goes beyond the use of composite operators since its implementation makes use of other types of constructs, including types, functions, and primitive operators.

In SPL a group of functionally related constructs can be organized into a reusable package referred to as a *toolkit*. Just like a *library* is used to organize functions, classes, and static values in C++ applications, or a *package* is used to organize classes and

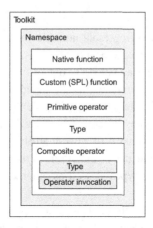

Figure 5.1 Structure of a toolkit.

interfaces in Java, a toolkit makes sharing pre-packaged components across applications possible. Figure 5.1 shows how an SPL toolkit is structured.

In a large project, modularization must be considered from the onset as should mechanisms for *namespace management*, to mitigate the possibility of naming collisions; *versioning*, to manage the addition of new software pieces as the applications evolve; and *dependency management*, to allow the integration of inter-related components.

Some of these requirements can also be addressed by organizing components into toolkits. For instance, in SPL, each toolkit has a name and a version. At the same time, a toolkit can specify its dependencies on other toolkits, using names and version ranges for the dependent toolkits. Toolkits can also contain documentation, sample code, and additional tools.

In the rest of this section, we will focus on the different types of components that can be grouped in a toolkit and used to construct applications. While discussing these specific components, we also look at mechanisms for extending SPL by adding new functions and operators to the language.

5.2.1   Types

Most SPSs come with a pre-defined set of types and associated basic operations and functions to manipulate values of these types. A *type* defines a stream schema, which in turn defines the domain for the set of values its tuples can have, the types of operations that can be performed on the stream, the size and encoding for the tuples the stream transports, and the set of candidate endpoints a stream can connect to.

A type can be primitive or composite. A primitive type is an indivisible singleton, whereas a composite type is defined in terms of other primitive or composite types.

Examples of primitive types include strings, as well as 64-bit versions of integers, decimal floating point numbers, and binary floating point numbers, which in SPL correspond to `ustring`, `int64`, `decimal64`, `float64`, and many other variations of these numeric types, where a different number of bits (i.e., the numeric suffix in the type name) is used to represent numeric values belonging to different ranges and with different precisions (for non-integer values).

A composite type can belong to one of two categories: *collections* or *structures*.

A collection type contains, as the name implies, a collection of values of a given type. Examples of collection types include lists, hash sets, and hash maps, which in SPL correspond to the types list, set, and map, respectively.

A structure type consists of a list of attributes, each defined as a *name* and a *type*. The type for each attribute can be either one of the primitive types or another composite type, including collection types as well as other structure types. An example of a structure type is the tuple.

When arbitrary levels of nesting are supported when defining a composite type, the list of specific types an application or a toolkit can make use of is virtually infinite.

Defining a *type* as part of a toolkit serves two main purposes. First, it defines a common and domain-specific vocabulary that establishes an interface for the use of the toolkit by other applications. For instance, an image processing toolkit can define an Image type to be employed by its clients when using the functions and operators provided by the toolkit. Second, the types used across related applications that cooperate in a larger analytic task can be factored out and organized in a common location. In this way, applications that talk to each other via dynamic connections can define the schemas for the streams that cross an application's boundary and place them in a shared toolkit used by both applications.

As an example of such capability, consider the code excerpt in Figure 5.2. It describes an image processing toolkit that defines a set of types in the my.image.lib namespace. The types are defined in the context of a composite operator called Types, effectively grouping all of them under a common namespace.

```
1    // file that defines the toolkit types
2    namespace my.image.lib;
3
4    public composite Types {
5      type
6        static ImgSize = tuple<uint32 width, uint32 height>;
7        static ImgDepth = enum {d8u, d8s, d16u, d16s, d32s, d32f, d64f};
8        static Image = tuple<blob data, ImgSize size, ImgDepth depth>;
9    }
```

```
1    // file that defines the application
2    namespace my.app;
3    use my.image.lib::*;
4
5    composite Main {
6      graph
7        stream<Types.Image> Images = ImageReader() {}
8        stream<Types.Image> SmallImages = Functor(Images) {
9          param
10           filter: size.width <= 64 && size.height <=64;
11       }
12   }
```

Figure 5.2 A hypothetical image processing operator toolkit and its use.

Three types are defined in this case. The `Types.ImgSize` type defines a *tuple* to represent the size of an image, including its `width` and `height`. The type `Types.ImgDepth` defines an *enumeration* to represent the depth of an image. Finally, the `Types.Image` type defines a tuple used to represent an image. This tuple comprises three attributes: `size` and `depth`, whose types are `ImgSize` and `ImgDepth`, respectively, as well as `data`, whose type is a `blob`, indicating that the `data` is a sequence of bytes.

The preceding example also includes a sample application that makes use of the types defined by the toolkit. This application, `my.app::Main`, uses the types from the toolkit both to declare its stream schemas, as in `stream<Types.Image>`, as well as to perform operations on tuples.

5.2.2 Functions

A *function* is a construct used to encapsulate a basic processing task. In stream processing, functions are generally used to configure operators, either as part of expressions used to specify parameter values, or to compute the value of attributes in outbound tuples. An example of the former is the value of the `match` parameter in a `Join` operator (Section 3.5.1), where a function can be used to determine the outcome of a match. An example of the latter is the outbound tuple attribute assignments in a `Functor` operator (Section 3.4.2), where the value of an attribute can be computed by a function applied to the values of the attributes of the incoming tuple.

Most SPSs support numerous built-in functions that operate on values of one of their built-in types. Nevertheless, an application might need additional functions that are not natively available.

New functions can be created by combining existing built-in functions or, alternatively, by leveraging third-party libraries where functions may have been originally written in languages such as C++ or Java. Existing SPSs differ substantially in the extension mechanisms and plumbing they provide to help incorporate external software assets.

SPL custom functions
In SPL, these two extension mechanisms are available. The first type of function that can be added to an application is a *custom function*.

Consider the code excerpt in Figure 5.3. In this example, the goal is to compute the total price for a sequence of customer orders. The internal structure of an `order` is defined by a type named `Order`. Hence, an instance of a customer `order` consists of an `id` attribute along with an `items` attribute used to store the list of items purchased by the customer. The `items` attribute is of type `Item`, a composite type, where each item has a `name` and a `price` attribute.

The calculation of the total price of an order is carried out by the `PriceCalculator` composite operator, which internally makes use of a `Functor` operator. This `Functor` produces an outbound tuple that, in addition to relaying the data from the incoming tuple, adds a new attribute called `total` to outbound tuples.

```
1   namespace my.trade.app;
2
3   type Item = tuple<rstring name, decimal32 price>; type Order =
4   tuple<uint64 id, list<Item> items>;
5
6   decimal32 computeTotalPrice(Order order) {
7     mutable decimal32 total = 0.0;
8     foreach (Item item in order.items) {
9       total += item.price;
10    }
11    return total;
12  }
13
14  composite PriceCalculator(input stream<Order> In; output Out) {
15    graph
16      stream<In, tuple<decimal32> total> Out = Functor(In) {
17        output
18          Out: total = computeTotalPrice(In);
19      }
20  }
```

Figure 5.3 Using the computeTotalPrice function to calculate the total price for a set of customer orders.

This attribute is computed by the computeTotalPrice custom function which is fully implemented in SPL.

SPL *native functions*

The second type of function is referred to as a *native function*. To support the implementation of this type of functions, SPL provides a programming interface that includes mechanisms for converting types between their SPL representation and the native language so data can be handed off to a native function and results can be shipped back to the SPL program.

Native functions must be *registered* with the SPL compiler. The registration involves specifying the *function signature*, containing its name, its set of formal parameter types, and the return type, as well as its list of library dependencies (as a list of Dynamically Shared Objects (DSOs)). The library dependencies include the library containing the function implementation as well as a list of any other libraries required for the function's operation.

The function signature allows the compiler to recognize a call to a native function and perform the proper syntactic validation, whereas the library dependency information enables the compiler and the application runtime environment to bind the function at compile-time and runtime, respectively.

Figure 5.4 illustrates how a native function implemented in C++ can be exposed to SPL. The figure shows the function model, an XML document used to describe the function so it can be understood by the compiler. This model is placed under a namespace directory [2], which defines the SPL namespace for registered functions (independent of the C++ namespace).

The function model specifies the name of the header file providing the C++ prototype (CorePinner.h, in this case) for the function (or, possibly, functions) to be registered (pinToCpuCore, in this case), its corresponding SPL signature as well as its free-form

```
1   <functionModel
2       xmlns="http://www.ibm.com/xmlns/prod/streams/spl/function"
3       xmlns:cmn="http://www.ibm.com/xmlns/prod/streams/spl/common"
4       xmlns:xsi="http://www.w3.org/2001/XMLSchema-instance"
5       xsi:schemaLocation="http://www.ibm.com/xmlns/prod/streams/spl/function␣
                functionModel.xsd">
6    <functionSet>
7     <headerFileName>CorePinner.h</headerFileName>
8     <functions>
9      <function>
10       <description>Pin to a given Cpu core</description>
11       <prototype><![CDATA[ stateful int32 pinToCpuCore(int32 cpu)]]></prototype>
12      </function>
13     </functions>
14     <dependencies>
15      <library>
16       <cmn:description>Thread pinning library</cmn:description>
17       <cmn:managedLibrary>
18        <cmn:includePath>..impl/include</cmn:includePath>
19       </cmn:managedLibrary>
20      </library>
21     </dependencies>
22    </functionSet>
23  </functionModel>
```

Figure 5.4 Registering the pinToCpuCore native function.

```
1   namespace sample;
2
3   composite ThreadPinner {
4    graph
5     stream<int8 dummy> Beat = Beacon() {
6      param iterations : 1u;
7     }
8     () as Sink = Custom(Beat) {
9      logic
10      onTuple Beat: {
11       int32 res = pinToCpuCore(1);
12       assert(res==0);
13       println("Pinned thread to CPU 1");
14      }
15     }
16   }
```

Figure 5.5 Using the pinToCpuCore native function.

human-readable description and, finally, its list of library dependencies, if any. In the case of pinToCpuCore, its implementation is inlined and fully described in a C++ header file.

The pinToCpuCore function uses OS APIs to pin the current thread to a specific core.[1] Once this native function is registered, it can be used just like any other function, as depicted by the code excerpt in Figure 5.5.

5.2.3 Primitive operators

In Section 3.5 we presented several primitive operators that are commonly provided as built-in operators by many SPSs. While there is substantial difference in the global set of operators provided by each of these systems, whatever operators this set might

---

[1] Its C++ implementation is not provided here for brevity, but it makes use of the sched_setaffinity system call available in the Linux OS.

contain is likely incomplete considering the variety of domains and applications that can be created.

Therefore, a desirable feature of a stream processing language is the ability to incorporate new primitive operators. When this is possible, the implementation of a new primitive operator must be carried out such that its semantics is both well-defined and malleable. In this way, the operator can be reused under different circumstances. This goal can be achieved by adhering to two design principles.

Design principles

A primitive operator should provide a *declarative interface*. Its syntax should enable a developer to specify *what* operation needs to be performed, while hiding *how* it is going to be performed.

Second, a primitive operator should provide a *generic interface*. It should not impose any constraints on the schema of streams it can consume and be able to handle a variable number of input and output ports, when relevant to the processing task the operator implements.

Looking more carefully at two operators in SPL's standard toolkit, one can see these two principles at work. First, consider the `Aggregate` operator (Section 3.5.1). As a generic operator, it can ingest streams with any schema. Furthermore, by being configurable with different types of windows and flexible output assignments, where different aggregation functions can be used, it allows an application developer to declare what aggregations she wants, while the details of how these aggregates are computed are not revealed.

Now consider the `Split` operator (Section 3.5.2). Not only can it process incoming streams with any schema, but it also supports n-way stream splitting, producing outbound tuples through n output ports, where n can be set to match the specific needs of an application.

Operator development API

In SPL, both built-in and user-defined primitive operators are developed in general purpose programming languages, allowing them to leverage the vast number of existing libraries available in both C++ and Java.

The development of new operators can be facilitated by an API that provides common services required by them. For instance, tasks such as the (1) ingestion of incoming tuples received by the ports of an operator; (2) submission of outbound tuples to the output ports of an operator; (3) processing of asynchronous events; (4) management of parameters, including parsing and validation; (5) management of output assignments; and (6) management of processing windows must be performed by most operators. Therefore, in SPL's operator development infrastructure these mechanisms are available in the form of C++ and Java APIs. Some of them are shown in Figure 5.6 as part of the abstract `Operator` class.

Tuple handling

On the input side of an operator, different variations of the `process` method (lines 3, 4, and 7 in Figure 5.6) must be overridden by the derived class implementing a new primitive operator.

```
1   class Operator {
2   public:
3     virtual void process(Tuple & tuple, int port);
4     virtual void process(Punctuation const & punct, int port);
5     void submit(Tuple & tuple, int port);
6     void submit(Punctuation const & punct, int port);
7     virtual void process(int index);
8     ...
9   };
```

Figure 5.6 C++ code illustrating a typical operator API.

The first two process methods are called by the *operator runtime* as a result of the arrival of a new inbound tuple (line 3) or of a punctuation (line 4) and they supply the operator-specific logic for processing them. The third process method (line 7) is used to perform *asynchronous* processing as it executes in the context of a separate thread (identified by index).

On the output side of an operator, the submit methods are used to emit tuples and punctuations using one of the operator's output ports (lines 5 and 6 in Figure 5.6). In general, but not always, the submit method is called in the context of the process methods, as new results are produced by the operator.

Note that these APIs employ a polymorphic Tuple type. As a result, a single process method can handle different types of tuples coming from different input ports (Section 5.2.3.1), making it possible to implement operators that are generic and reusable.

Parameter, output, and window handling
The parameter handling interfaces are used to allow the customization of an operator instance's behavior based on the settings supplied for it when the operator is instantiated in an application.

Similarly the output attribute handling interfaces are used to customize an operator instance's behavior concerning the values to be assigned to the attributes in the outbound tuples. Such assignments might be as simple as passing through a value from an attribute of an incoming tuple, but can also include the result of evaluating expressions and functions, the result of processing data held in window buffers, or a more complex combination of all these.

The windowing management interfaces supply the common processing logic that allows an operator to handle window-related operations such as evictions, triggers, and partition management. These interfaces are event-driven and handlers for particular windowing events are implemented according to the internal needs of the new operator. Each event is then delivered to the operator by the runtime where a specific handler is executed to react to it.

Life cycle management
There are several additional interfaces including the ones for managing the operator initialization and termination, start-up and shutdown, error and exception handling, thread management, checkpointing, metrics management and access, and dynamic connection

management. Many of them and their relationship to the Streams application runtime environment are described in Section 8.5.9.

5.2.3.1   Developing new operators

In this section, we will demonstrate how to develop and use new primitive operators by making use of the `sample::TextStats` application shown in Figure 5.7.

In this application, there are two new C++ primitive operators: `TextReader` and `WordCounter`. `TextReader` is a primitive operator that reads lines from a text file and streams out tuples corresponding to a single line of text as extracted by C++'s `getline` function. `WordCounter` is an operator that counts the number of characters, words, and lines in the text, outputting the results for each incoming tuple. The application also makes use of a `Custom` operator to print the results to the console.

The `TextReader` operator

The `TextReader` operator makes use of SPL's primitive operator development framework and its implementation comprises two C++ files: a header (Figure 5.8) and an implementation file (Figure 5.9).

The C++ header (lines 1 and 13) and implementation (lines 1 and 26) use pragmas to emit boiler-plate prologue and epilogue code as well as a fixed class name `MY_OPERATOR`, in accordance with SPL's operator development framework [2]. A new operator is automatically named based on its location in a toolkit.

The `TextReader` constructor is very simple and only initializes its member variable (`file_`, line 7 in Figure 5.9) used to keep the name of the file to be read, whose value is specified by the operator's `file` parameter and retrieved from its instantiation in SPL code (line 7 in Figure 5.7).

The operator also defines the `allPortsReady` method, a callback invoked by the SPL runtime when the operator is ready to begin processing. It indicates that all its ports are initialized and connected to downstream consumers and upstream producers. This method calls the SPL runtime's `createThreads` function, instructing it to create a new thread, which will provide the execution context to the operator's `process` method.

```
1   namespace sample;
2   composite TextStats
3   {
4     graph
5       stream<rstring line> Lines = TextReader() {
6         param
7           file : "input.txt";
8       }
9       stream<int32 nchars,int32 nwords,int32 nlines> Counts=WordCounter(Lines){}
10      () as Sink = Custom(Counts) {
11        logic
12          onTuple Counts:
13            println(Counts);
14      }
15  }
```

Figure 5.7 Sample application using C++ primitive operators

```
1   #pragma SPL_NON_GENERIC_OPERATOR_HEADER_PROLOGUE
2
3   class MY_OPERATOR : public MY_BASE_OPERATOR
4   {
5     public:
6       MY_OPERATOR();
7       void allPortsReady();
8       void process(uint32_t index);
9     private:
10      rstring file_;
11  };
12
13  #pragma SPL_NON_GENERIC_OPERATOR_HEADER_EPILOGUE
```

Figure 5.8 The C++ header file for the `TextReader` operator.

```
1   #pragma SPL_NON_GENERIC_OPERATOR_IMPLEMENTATION_PROLOGUE
2
3   #include <fstream>
4
5   MY_OPERATOR::MY_OPERATOR()
6   {
7     file_ = getParameter_file();
8   }
9
10  void MY_OPERATOR::allPortsReady()
11  {
12    createThreads(1);
13  }
14
15  void MY_OPERATOR::process(uint32_t index)
16  {
17    std::ifstream srcFile;
18    srcFile.open(file_.c_str());
19    OPort0Type otuple;
20    while(!getPE().getShutdownRequested() && srcFile) {
21      getline(srcFile, otuple.get_line());
22      submit(otuple, 0);
23    }
24  }
25
26  #pragma SPL_NON_GENERIC_OPERATOR_IMPLEMENTATION_EPILOGUE
```

Figure 5.9 The C++ implementation file for the `TextReader` operator.

`TextReader` is a source operator and, hence, it processes incoming data, but not incoming *tuples*. To perform this task, it employs the `process(uint32_t)` method, which runs asynchronously, in contrast to the synchronous `process(Tuple & tuple, int port)` method (Figure 5.6) used by non-source operators, which is driven by the arrival of tuples to an input port. In this function, the index argument identifies the asynchronous thread under which the method is running, allowing the method to customize its behavior on a per-thread basis, if necessary. In our example, the operator is single-threaded.

The asynchronous `process` method defines the main processing loop for this source operator. The loop executes indefinitely while the operator is not told to terminate and there is data to be ingested. To keep track of a termination request, the operator uses the SPL runtime's `getPE` method (Section 8.5.9) to obtain a reference to the application partition encompassing it, and this partition's `getShutdownRequested` method, to check whether this particular partition has been told to shutdown. Unless a shutdown is requested, the operator continues to processes the input file, line by line (line 21).

The operator's main loop reads each line into the `line` attribute of the output tuple whose type is `OPort0Type`. This type provides a C++ representation of the schema for the tuples output by the operator's first (its 0th) output port. Finally, the operator submits the populated tuple to the output port with index 0, using the `submit` method (line 22).

The `WordCounter` operator

The other operator we built, `WordCounter`, showcases the implementation of a simple `Functor`-like operator. Similarly to the implementation of the `TextReader` operator, `WordCounter`'s implementation comprises two files: a header (Figure 5.10) and an implementation file (Figure 5.11).

`WordCounter` is not a source operator, thus it must implement the two `process` methods that handle incoming tuples and punctuations (lines 3 and 14 in Figure 5.11, respectively).

The tuple-processing method makes use of a polymorphic cast to convert the incoming tuple reference held by the `tuple` parameter into the concrete tuple type `IPort0Type` held by the variable `ituple`. Here the type `IPort0Type` represents the schema of the tuples entering `WordCounter`'s first (and only) input port.

```
1   #pragma SPL_NON_GENERIC_OPERATOR_HEADER_PROLOGUE
2   class MY_OPERATOR : public MY_BASE_OPERATOR
3   {
4     public:
5       void process(Tuple const & tuple, uint32_t port);
6       void process(Punctuation const & punct, uint32_t port);
7   };
8
9   #pragma SPL_NON_GENERIC_OPERATOR_HEADER_EPILOGUE
```

Figure 5.10 The C++ header file definition for the `WordCounter` operator.

```
1   #pragma SPL_NON_GENERIC_OPERATOR_IMPLEMENTATION_PROLOGUE
2
3   void MY_OPERATOR::process(Tuple const & tuple, uint32_t port)
4   {
5     IPort0Type const & ituple = static_cast<IPort0Type const &>(tuple);
6     rstring const & line = ituple.get_line();
7     int32 nchars = line.size();
8     int32 nwords = SPL::Functions::String::tokenize(line, "_\t", false).size();
9     int32 nlines = 1;
10    OPort0Type otuple(nchars, nwords, nlines);
11    submit(otuple, 0);
12  }
13
14  void MY_OPERATOR::process(Punctuation const & punct, uint32_t port)
15  {
16    if(punct==Punctuation::WindowMarker)
17      submit(punct, 0);
18  }
19
20  #pragma SPL_NON_GENERIC_OPERATOR_IMPLEMENTATION_EPILOGUE
```

Figure 5.11 The C++ implementation for the `WordCounter` operator.

A reference to the line attribute is obtained from the ituple object using the get_line() method (line 6) and its contents are tokenized to extract the number of characters, words, and lines in the text it holds (line 8). Finally, an outbound tuple of type OPort0Type is created (line 10) and emitted through the operator's 0-th output port (line 11). WordCount's punctuation-processing method simply forwards all window punctuations from its single input port to its single output port (line 16).

Introspection and generic operators

SPSs face an important design decision concerning the *introspection* interface necessary to support the development of generic operators. Specifically, a generic operator must be able to process tuples as well as to produce tuples with different schemas as these schemas remain unspecified until the operator is instantiated in an application. Thus, the operator implementation must have mechanisms to extract and manipulate the values of the set of attributes belonging to inbound and outbound tuples, including ways to reference the attribute names and types.

An introspection interface can be used at runtime, at compile-time, or both. Performing all of the attribute handling operations at runtime often introduces execution overhead. On the other hand, performing these operations at compile-time requires the ability to infer all the schemas and specialize the data handling part of the operator, as an application is being compiled. The second approach is usually associated with lower execution overhead. In contrast to other SPSs, Streams provides compile-time introspection and code generation support [3] in addition to the more conventional runtime introspection interface.

In Streams, the compile-time introspection interface provides the information necessary for an operator to be fully customized as an application is being built. In this case, operators are developed as code generation templates. Operator templates are specialized as an application is compiled where they become operator code generators. Each generator is then used to emit highly specialized low-level code (in C++) for handling the specifics of an operator instance with its own parameters, ports, windowing, and output assignment configurations.

The code generation interfaces used in the operator templates are often provided in a scripting language suitable for text processing. In SPL's case, the compile-time introspection and code generation interfaces are provided in Perl [4] and are complemented by an additional runtime introspection interface provided in C++. The resulting operator code generator emits pure C++ code, which is then automatically compiled into a DSO or standalone executable.

Registering primitive operators

Adding user-defined primitive operators to a stream processing language requires providing its compiler (or interpreter) with syntactic and semantic information about these new operators. In this way the compiler can validate an operator invocation and produce error messages, when necessary. For instance, using this information a compiler can, for example, verify that an instantiation of the `Join` operator defines two input ports or that an instantiation of a `Filter` operator is associated with a `boolean` filter expression.

Furthermore, the compiler and the runtime environment can use the semantic properties of an operator to ensure that safety checks and optimizations are put in place according to the specific needs of an operator instance. For example, the compiler can perform checks such as verifying that punctuation-expecting input ports are only serving as endpoints to punctuated streams. Likewise, the compiler can make use of optimizations such as avoiding unnecessary locks and synchronization primitives when they are not necessary in a particular context.

Along the same lines, the runtime environment can reason about the execution conditions and enable dynamic optimizations such as the parallelization of an operator's internal logic when it has no state or when its internal state can be partitioned.

For these reasons, adding a new primitive operator to a toolkit generally requires a registration step where two artifacts must be provided: the operator's implementation in the form of a code generation template, along with a descriptive model capturing its syntactic and semantic characteristics. In SPL, this model is referred to as an *operator model* [3]. `TextReader`'s own model is depicted by Figure 5.12.

```
1  <operatorModel
2    xmlns="http://www.ibm.com/xmlns/prod/streams/spl/operator"
3    xmlns:cmn="http://www.ibm.com/xmlns/prod/streams/spl/common"
4    xmlns:xsi="http://www.w3.org/2001/XMLSchema-instance"
5    xsi:schemaLocation="http://www.ibm.com/xmlns/prod/streams/spl/operator␣
        operatorModel.xsd">
6    <cppOperatorModel>
7      <context>
8        <description>Reads lines from a text file</description>
9        <providesSingleThreadedContext>Always</providesSingleThreadedContext>
10     </context>
11     <parameters>
12       <allowAny>false</allowAny>
13       <parameter>
14         <name>file</name>
15         <description>Name of the file to read from</description>
16         <optional>false</optional>
17         <type>rstring</type>
18         <cardinality>1</cardinality>
19         ...
20       </parameter>
21     </parameters>
22     <inputPorts/>
23     <outputPorts>
24       <outputPortSet>
25         <expressionMode>Nonexistent</expressionMode>
26         <windowPunctuationOutputMode>Generating</windowPunctuationOutputMode>
27         <cardinality>1</cardinality>
28         <optional>false</optional>
29         ...
30       </outputPortSet>
31     </outputPorts>
32   </cppOperatorModel>
33 </operatorModel>
```

Figure 5.12 The operator model for the `TextReader` operator.

TextReader's operator model

The operator model describes several aspects of its behavior and user-facing characteristics. For instance, under the context element (line 7), the description element (line 8) provides a human-readable explanation of the operator's function and the providesSingleThreadedContext element (line 9) is used to indicate its threading semantics.[2]

Under the parameters element (line 11), the allowAny element (line 12) is used to indicate that this operator does not allow arbitrary command-line-style parameters. Furthermore, the parameter element is used to specify the non-optional file parameter (line 14) required by the operator, along with its type (rstring), human-readable description, and cardinality, stating that the value for this parameter consists of a single value.

Finally, the outputPorts element (line 23) spells out the characteristics of this operator's single output port. The expressionMode element (line 25) indicates that the operator cannot make assignments to its outbound tuples' attributes. The windowPunctuationOutputMode element (line 26) specifies that the operator generates window punctuations. The cardinality element (line 27) specifies that there is only a single output port and the optional element (line 28) states that this port is mandatory, in other words, when this operator is used in an application, it will produce a stream.

WordCounter's operator model

WordCounter's operator model (Figure 5.13) has many similarities to TextReader's, but it also differs in some important ways. First, it has a single (line 18), mandatory (line 19), windowless (line 16) input port, specified by the inputPorts element (line 14). Second, this operator's single (line 26), mandatory (line 27) output port preserves incoming punctuations (line 25), forwarding them through its output port, as was seen in Figure 5.11 (line 16).

5.2.4 Composite and custom operators

A composite operator, similarly to a primitive operator, can be included in a toolkit. In this way, applications that depend on the toolkit can use it, like any regular primitive operator.

Adding a composite operator is straightforward, but composites have inherent limitations. Notably, a developer can only rely on functionality already available in other primitive and composite operators. If additional customization is necessary, a different route must be pursued.

As we have discussed, one approach is to add a new primitive operator, implementing it in a general-purpose programming language. Yet, it is often more convenient to add a user-defined operator employing the statements and expressions provided by the stream

---

[2] An Always value for this setting stipulates that any instance of this operator is guaranteed to provide a single-threaded execution context. For a source operator, this information asserts that the operator will emit tuples from one sole execution thread.

```
1   <operatorModel
2     xmlns="http://www.ibm.com/xmlns/prod/streams/spl/operator"
3     xmlns:cmn="http://www.ibm.com/xmlns/prod/streams/spl/common"
4     xmlns:xsi="http://www.w3.org/2001/XMLSchema-instance"
5     xsi:schemaLocation="http://www.ibm.com/xmlns/prod/streams/spl/operator␣
           operatorModel.xsd">
6     <cppOperatorModel>
7       <context>
8         <description>Count words and characters in a line</description>
9         <providesSingleThreadedContext>Always</providesSingleThreadedContext>
10      </context>
11      <parameters>
12        <allowAny>false</allowAny>
13      </parameters>
14      <inputPorts>
15        <inputPortSet>
16          <windowingMode>NonWindowed</windowingMode>
17          <windowPunctuationInputMode>Oblivious</windowPunctuationInputMode>
18          <cardinality>1</cardinality>
19          <optional>false</optional>
20          ...
21        </inputPortSet>
22      </inputPorts>
23      <outputPorts>
24        <outputPortSet>
25          <expressionMode>Nonexistent</expressionMode>   <
                 windowPunctuationOutputMode>Preserving</windowPunctuationOutputMode>
26          <cardinality>1</cardinality>
27          <optional>false</optional>
28          ...
29        </outputPortSet>
30      </outputPorts>
31    </cppOperatorModel>
32  </operatorModel>
```

Figure 5.13 The operator model for the WordCounter operator.

processing language itself. While this approach may not be as flexible as developing the operator from scratch in a general-purpose programming language, it is usually speedier. We refer to an operator of this kind as a *custom operator*.

A custom operator is in fact just a different type of primitive operator. The fundamental differences are twofold. First, a custom operator is a singleton and cannot be directly reused. Second, while a conventional primitive operator *hides* its internal implementation in the form of a code generation template, a custom operator *exposes* it in the form of statements and expressions written in SPL itself.

Because a custom operator is a singleton, it does not provide a generic declarative interface and, hence, it cannot be reused as-is. This limitation can be easily addressed by using a composite operator to *wrap* the custom operator invocation, making it possible to include this combo in a toolkit. The code excerpt in Figure 5.14 demonstrates how this is can be done.

This example implements the CacheWithPeriodicUpdate composite operator whose function is to cache the latest inbound tuple and periodically send it out. This operator has a single input port and produces a single stream as output. It is configured with a period parameter. Its processing logic is implemented using SPL's standard toolkit Custom operator, which maintains the operator's internal state, performs processing upon receiving tuples from its input ports, and emits outbound tuples through its output ports.

```
1    namespace my.sample.ops;
2
3    composite CacheWithPeriodicUpdate(Input In; Output Out) {
4      param
5        expression<float64> $period: 10.0;
6      graph
7        stream<int8 dummy> Periodic = Beat() {
8          param
9            period: $period;
10       }
11       stream<In> Out = Custom(In; Periodic) {
12         logic
13          state: {
14             mutable bool seen = false;
15             mutable tuple<In> last = {};
16          }
17          onTuple In: {
18            seen = true;
19            last = In;
20          }
21          onTuple Periodic: {
22            if(seen)
23              submit(Out, last);
24          }
25       }
26    }
```

Figure 5.14 Wrapping a Custom operator in a composite operator.

```
1    namespace my.sample.app;
2    using my.sample.ops::*;
3    composite Main {
4      graph
5        stream<uint32 value> Values = ValueSource() {}
6        stream<uint32 value> PeriodicValue = CacheWithPeriodicUpdate(Values) {
7          param
8            period: 5.0;
9        }
10   }
```

Figure 5.15 Using a Custom operator wrapped by a composite operator.

In this particular example, the Custom operator keeps the last received tuple (line 13) and a boolean flag (seen) indicating whether an incoming tuple has reached this operator.

Every time a tuple is received by the CacheWithPeriodicUpdate's input port, it is routed to the first input port of Custom operator causing it to update its internal state. This action is specified via an onTuple clause (line 17).

A Beacon operator, connected to the Custom operator's second input port, sets the pace for sending out the last cached inbound tuple. The period parameter from the CacheWithPeriodicUpdate composite is passed to the Beacon operator for this purpose. That is, every time a tuple is received by the Custom operator's second input port, the cached tuple, if available, is sent out as specified by its other onTuple clause (line 21).

Figure 5.15 depicts how the newly defined CacheWithPeriodicUpdate operator can be used (line 6). Note that these two code fragments, the toolkit definition and the application, are placed in two different files, which makes it possible to separate and hide the implementation of the `CacheWithPeriodicUpdate` operator from the location where it is used.

5.3   Distributed programming

In this section, we discuss distributed programming in the context of SPAs. While one of the goals of the data flow graph abstraction is to shield the developers from the details of how an application is placed and executed on a set of distributed hosts, understanding how the operators are mapped to execution processes and hosts as well as how a transport layer shuttles tuples between operators are important design considerations.

Indeed, in many cases rather than relying on automated decisions made by the application runtime environment, a developer might choose to manually tweak some of these decisions to meet requirements such as resource utilization limits, performance goals as well as legal and auditing constraints. In the rest of this section, we discuss specific distributed programming configuration knobs available to developers.

5.3.1   Logical versus physical flow graphs

An application developer typically devises and implements an application as a *logical* data flow graph, where the vertices of the graph correspond to primitive and composite operator instances and the edges to stream connections.

In contrast to this logical representation, when a SPS deploys an application, it must map it to a corresponding *physical* data flow graph, where the vertices of the graph correspond to OS processes and the edges to transport connections that can move tuples between operators, potentially across hosts.

Figure 5.16(a) depicts a logical representation of an application, including the complete operator instance hierarchy. Due to the existence of the composite operators, different perspectives with different granularities of the logical view exist. For instance, Figure 5.16(b) shows the same application, but the composite operator instances are collapsed and only the top-level operator instance graph is shown. A logical view always corresponds to the topology constructed by the application developer, even though different perspectives of this topology with differing levels of composite operator expansion can be presented.

The physical topology of an application is more than just a different perspective on the logical flow graph. It is a *mapping* that can take different forms. This mapping is implemented in two steps.

First, the set of composite operators that make up an application are distilled to a corresponding set of primitive operators, which are broken into a number of execution partitions. An application execution partition *fuses* one or more operators in the context of an OS process.

Second, these partitions are assigned to hosts as part of a *scheduling* process carried out by the application runtime environment. Note that the physical data flow graph is a hierarchical graph. At the coarsest level, nodes are the hosts, at an intermediary

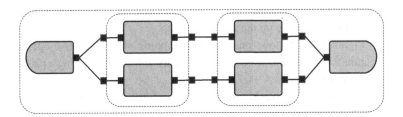

(a) Logical application view, showing the complete operator instance hierarchy.

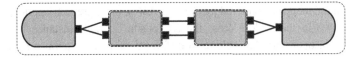

(b) Collapsed logical application view, showing the top-level operator instances.

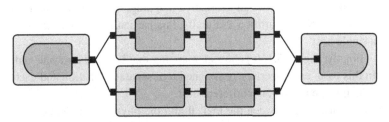

(c) One possible physical application topology, showing four partitions.

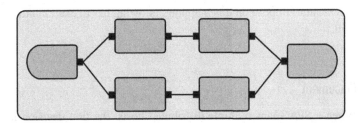

(d) Another possible physical application topology, showing a single partition with the operators fused together.

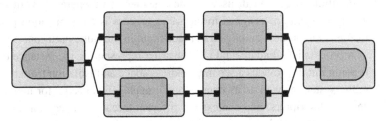

(e) Another possible physical application topology, showing one partition per primitive operator without fusion.

Figure 5.16 Logical and physical application topologies.

level, nodes are the application partitions, and at the finest level, nodes are the primitive operators that make up the application.

Figure 5.16(c) shows one possible mapping of the primitive operator instances from Figure 5.16(a) into a set of application partitions, thus yielding a partition-level physical data flow graph.

An interesting observation is that an application partition can cross the boundaries created by a composite operator. The latter is simply a logical grouping driven by concerns like encapsulation and code reuse, whereas the former is a physical grouping driven by concerns such as runtime performance and load balancing.

In the mapping depicted by Figure 5.16(c), there are four application partitions where the two in the middle cross the composite operator instance boundaries, since they contain operators that belong to different composite operator instances. This particular mapping can be distributed and executed on up to four hosts, potentially one per application partition.

Figure 5.16(d) shows an alternative mapping, where all operators are fused into one single application partition. Naturally, this particular mapping can only be executed on a single host.

Finally, Figure 5.16(e) shows a mapping that is at the opposite end of the spectrum. In this case, each operator is assigned to a separate partition and each partition can, in principle, be placed on a different host.

The mapping between the logical and the physical graph is often controlled by the Stream Processing System (SPS) to improve performance and to balance load as will be seen in Section 5.3.2. Yet, there are various constraints that an application designer may want to impose on this mapping, allowing her to exert control over this process as well.

5.3.2   Placement

When a SPA spans multiple distributed hosts, the first decision facing a developer is how to allocate different segments of an application to different hosts. This *placement* process takes two forms: *partition placement* and *host placement*.

Partition placement deals with the association of operators with OS *processes*. A group of operators assigned to the same process is an application partition. Operator fusion is the process employed to place them in the same partition.

A *process* is a fundamental OS construct representing the instance of a program that is being run. It provides a separate address space and isolation from the other processes in the system as well as its own resource management (e.g., for memory, CPU cycles, and file descriptors) and context for different threads of execution. As a result, when a partition is mapped to a process, it becomes the basic unit of execution in a distributed SPS as each process can be placed on a different host.

Host placement deals with the association of partitions with the hosts that make up the application runtime environment. The process of assigning application partitions to hosts is referred to as *scheduling*.

The combined process of creating application partitions along with scheduling brings about a great deal of deployment flexibility to a SPA. Indeed, the same application can be run on a variety of different environments, from a single host to large clusters of hosts without re-writing it.

While the decisions regarding both of these processes can be delegated to the compiler (partition placement) and to the application runtime environment (host placement), an application designer can define *constraints* and exert fine control over them, if so desired.

Partition placement constraints

The first type of partition placement constraints is referred to as partition *co-location* and is used to enforce the placement of two or more operators into the same application partition.

There are several motivations for co-locating two operators. First, co-location provides shared access to a process's common resources to all the operators fused into the same partition. Second, co-location provides a considerably more efficient means for cross-operator communication. In fact, compared to inter-process communication, intra-process communication enjoyed by operators in the same partition incurs considerably less overhead. Naturally, the existence of a common address space removes the need for data serialization as well as for network operations.

The second type of partition placement constraints is referred to as partition *ex-location* and is used to enforce the placement of one or more operators into different partitions. In other words, no two operators associated with a partition ex-location constraint can be part of the same partition.

The need for specifying a partition ex-location constraint arises when two or more operators employ a resource designed as a singleton. For example, certain embedded language interpreters as well as embedded databases are designed to have only a single instance within a process.

The third type of partition placement constraints is referred to as partition *isolation* and is used to ensure that a given operator does not share its partition with any other operators.

The need for specifying a partition isolation constraint usually stems from the need to provide memory protection to certain critical operators or to lower the amount of resource sharing. In the first case, an operator sharing its partition with others is subjected to critical failures unrelated to its own processing. Another operator's faulty behavior directly affects all other operators in the same partition. In certain cases, such failures can cause the whole partition to fail, irrespective of which operator triggered the failure, potentially leading to a hard crash. In the second case, all operators in a partition are constrained by the per-process resource limits, for instance, the total number of file descriptors and stack space that can be used. Hence, isolating an operator ensures that it is not subjected to failures other than its own as well as that it does not share resources and has access to the whole allocation quota available to its OS process.

```
1   composite Main {
2     graph
3       stream<MyType> Src = MySource() {
4         config
5           placement: partitionColocation("source"),
6                   partitionExlocation("source_sink");
7       }
8       stream<MyType> Flt = MyFilter() {
9         config
10          placement: partitionColocation("source");
11      }
12      stream<MyType> Agg = MyAgg(Flt) {
13        config
14          placement: partitionIsolation;
15      }
16      stream<MyType> Sink = MySink() {
17        config
18          placement: partitionExlocation("source_sink");
19      }
20  }
```

Figure 5.17 Using partition constraints.

Figure 5.17 demonstrates the use of the different partition placement constraints in SPL. The placement constraints are specified in an operator's optional `config` clause, using the `placement` directive.

The partition co-location and partition ex-location constraints are specified via the `partitionColocation` and `partitionExlocation` directives, respectively. These directives assign an *identifier* to specific constraints. Operators that share the same identifier for a partition co-location constraint are *fused* in the same application partition (e.g., `source` in lines 5 and 10). Operators that share the same identifier for a partition ex-location constraint are placed on different partitions (e.g., `source_sink` in lines 6 and 18).

As a result, in the example in Figure 5.17, the `Src` and `Flt` operators are fused into the same partition, which must be different from the partition where the `Sink` operator is placed.

Finally, a partition isolation constraint indicated by the `partitionIsolation` directive ensures that the `Agg` operator is placed in a partition of its own.

Host placement constraints

The first type of host placement constraints is referred to as *host co-location* and is used to enforce the placement of two or more operator instances on the same host.

This kind of co-location is used in scenarios where common access to per-host shared resources is required. Typical examples include access to a local disk as well as access to OS-managed shared memory segments. Also, co-locating operators on a host tends to reduce the communication costs between them as no network communication costs are incurred.

The second type of host placement constraints is referred to as *host ex-location* and is used to enforce the placement of one or more operator instances onto different hosts.

The need for host ex-location arises in two situations. First, as a means to improve the reliability of certain critical segments of an application. Specifically, different, but redundant copies, of the same operator instance can be deployed onto different hosts insulating that operator from single-host failures. Second, as a means to guard against per-host resource limits, such as physical memory or local disk space.

The third type of constraint is referred to as *host isolation* and is used to ensure that the partition hosting a given operator instance does not share its host with any other application partitions.

The main reason for establishing a host isolation constraint is to provide exclusive access (minus OS and the middleware's own overhead) to the resources from a host to the possibly multiple operators fused into the *single* application partition placed on that host.

It is important to note that partition and host placement constraints interact with each other. For instance, defining a partition co-location constraint implies also host co-location, since an OS process cannot cross host boundaries. Similarly, defining a host ex-location constraint implies also partition ex-location.

In SPL, host placement constraints are specified using a syntax similar to the partition placement constraints, except that different keywords are employed: `hostColocation`, `hostExlocation`, and `hostIsolation`.

Host tags

While host co-location, ex-location, and isolation constraints provide a lot of flexibility, these directives are not designed for host-level resource matching. As an example, consider an operator instance that requires access to a resource that is only accessible on a subset of the hosts belonging to a cluster such as a database, a specific piece of hardware, or a software license tied to specific host identifiers.

To address these cases, an operator can make use of placement directives that refer to a *host tag*, which is simply a label assigned to a host by a system administrator, creating a mapping between a specific host (or hosts) and the resources it has.

In InfoSphere Streams, the association between a host and its tags can be made as part of the creation of a runtime instance (Section 8.4.1).

With the use of host tags it is possible to avoid hard-coding host names in an application and instead request hosts with a required set of resources. This approach provides flexibility in deploying the application on different clusters of hosts as it can be executed on different groups of hosts as long as the resource requirements indicated by the tags are met.

Host pools

Host tags are used by host pools. A *host pool* defines a collection of hosts with similar resource configurations.

A host pool is characterized by three properties: an optional set of host tags each of its hosts must be associated with; an optional setting stating how many hosts a pool must

```
1    composite Main {
2      graph
3        stream<MyType> Src1 = DBReader() {
4          config
5            placement: host(poolA[0]);
6        }
7        stream<MyType> Src2 = DBReader() {
8          config
9            placement: host(poolA[1]);
10       }
11     config
12       hostPool: poolA = createPool({size=2, tags=["db"]}, Sys.Shared);
13   }
```

Figure 5.18 Using host pools with host tags.

have, which, if present, makes the pool a *fixed-size* pool; and a mandatory setting stating whether the pool is *exclusive* or *shared*. In contrast to a shared host pool, an exclusive one is reserved to a single application.

Once defined, a host pool can be referenced by using it in a host placement constraint. In this way, an operator can be placed on a specific pool or on a specific host in that pool by stipulating the index that identifies it in the host pool.

Figure 5.18 depicts a code excerpt demonstrating the use of host pools in SPL. Two instances of the DBReader operator are placed on the first and the second hosts of the poolA host pool. This pool is associated with the db host tag, which, in this case, indicates the availability of a database server in those hosts.

The placement is defined by first declaring the 2-host poolA shared host pool and stating that it employs hosts with the db tag (line 12). Once declared and configured, the host pool is used by referencing it in host placement constraints associated with the two operator instances (lines 5 and 9).

5.3.3 Transport

A stream transport layer is used to implement the physical connections supporting the logical stream connections used in an application's source code. The flexibility and efficiency of the transport layer is fundamental to enable the development of scalable distributed SPAs.

Two transport layer design aspects are particularly important: first, the ability to be malleable to address specific performance requirements such as a focus on latency or throughput and, second, the ability to employ different types of physical connections to provide the most performance while imposing the least amount of overhead, regardless of the physical layout employed by an application.

It is clear that the performance requirements of SPAs lie in a continuum (Section 3.2), yet it is important to take a look at the two extreme points in this continuum: *latency-sensitive* and *throughput-sensitive* applications.

A latency-sensitive application usually has little tolerance to buffering and batching as tuples must move between operators as quickly as possible, ensuring that fresh data is processed and fresh results, including intermediate ones, are output rapidly.

Conversely, a throughput-sensitive application is interested in producing as many final results as possible in as little time as possible. Thus, this type of application can receive a significant performance boost from tuple buffering and batching techniques, particularly in the early stages of its processing flow graph.

Ideally a *physical transport layer* supporting a stream connection must expose configuration options that can be used to adjust the trade-offs between latency and throughput. Tuning these settings is specially important for distributed SPAs, whose components run on a large number of hosts. For this reason, some SPSs expose these transport layer configuration options to application developers as well as provide automated methods to tune them, so that the settings that best match the needs of a particular application can be identified and used in production.

Partitioning an application

A small-scale SPA can be deployed either as a single application partition or a handful of them, sharing the same host. When possible, this layout is effective and simple to manage.

In many cases, a single-host deployment is not feasible and a SPA may be partitioned in a variety of ways. Medium-size SPAs might include multiple application partitions laid out on a small set of hosts. Large-scale SPAs take this type of organization to the extreme as they might employ a multitude of application partitions spanning tens or hundreds of hosts. In these cases, it is possible that specialty high-speed network fabrics such as InfiniBand [6] might be available to some of them and can be used to interconnect certain critical components of a distributed SPA.

Selecting a physical transport

Scalable SPSs such as Streams provide a variety of different physical transports (Section 7.3.5), optimized for different scenarios and with different inherent costs and performance trade-offs. For instance, stream connections between fused operators sharing the same application partition can be implemented very efficiently by function calls as these operators are part of the same OS process's address space.

Naturally, the same is not true for connections between operators placed on the same host, but on different application partitions. In this case, a physical connection corresponding to a stream can be implemented using one of several OS-supported Inter-Process Communication (IPC) abstractions: shared memory segments, named pipes, or network protocols, including TCP/IP as well as Remote Direct Memory Access (RDMA) [5] over either InfiniBand or Ethernet.

Despite the variety of ways by which tuples can be exchanged between operators, there are two invariants when it comes to managing the streaming data traffic: *in-order delivery* and *reliable delivery*. The first invariant implies the delivery of tuples and punctuations from a producing output port to a consuming input port in the same order as they were emitted. The second invariant means that the tuples and punctuations submitted to an output port are always delivered without loss.

In this way, in most cases an application developer does not need to worry about which physical transport is used when an application is deployed, except when they

```
1    composite Main {
2      graph
3        stream<MyType> LatencySensitiveSource = Source() {
4          config
5            placement: host(poolA[0]);
6            outputTransport: LLM_RUM_IB;
7        }
8        stream<MyType> LatencySensitiveResult = Process(LatencySensitiveSource) {
9          config
10           placement: host(poolA[1]);
11           inputTransport: LLM_RUM_IB;
12       }
13     config
14       hostPool: poolA = createPool({size=2, tags=["ib"]}, Sys.Exclusive);
15   }
```

Figure 5.19 Using different data transport options.

want to make this choice explicitly. For these cases, the code excerpt in Figure 5.19 illustrates how an alternative transport layer can be selected.

In this particular example, the developer of a hypothetical latency-sensitive application consisting of two operators wants each of them placed on a different host, exclusively. Since some of the hosts in the cluster where the application will be deployed are connected to an InfiniBand switched fabric, Streams' Low Latency Messaging (LLM) InfiniBand transport layer [7] can be used.

Two hosts with InfiniBand adapters can be requested at runtime by specifying a fixed-size host pool using the "ib" tag as well as the Sys.Exclusive marker (line 14). In this way, the application's Source operator can be configured with an outputTransport directive set to LLM_RUM_IB, indicating that the traffic generated by its outbound stream should be transported by Streams' reliable unicast messaging on InfiniBand (line 6). Similarly, the Sink operator is configured with a matching inputTransport directive also set to LLM_RUM_IB (line 11). In this way, the SPL compiler emits all of the necessary configuration settings allowing the tuple flow to be routed via the InfiniBand switched fabric at runtime.

Any operator that has an input port can be configured with an inputTransport directive and any operator that has an output port can be configured with an outputTransport directive. Because these are partition-level settings, connections with conflicting directives will be reported at compile-time.

5.4   Concluding remarks

In this chapter, we discussed the basic constructs that provide modularity and extensibility capabilities to SPAs, enabling the implementation of large-scale applications.

We discussed how user-defined functions and operators provide an extensible foundation so developers can elegantly address problems from various application domains (Section 5.2). As part of this discussion, we touched upon the basic application programming interfaces used to implement new operators using general-purpose programming languages and SPL.

As more complex and larger applications are developed, the need for distributed programming arises. Thus, we also looked at mechanisms for partitioning an application focusing on the specific needs of stream processing. We studied how the mapping between a logical data flow graph, representing an application, and a physical data flow graph, constituting the executable representation of this application, can be created and controlled by an application developer (Section 5.3).

Finally, we looked at various placement directives affecting the logical-to-physical application mapping process. We concluded with a discussion of the transport layer that implements the physical infrastructure used to manage the stream connections and its role in supporting distributed stream processing.

5.5   Programming exercises

1. **Application scale up: data partitioning and rule replication**. Consider the *SPS-MarkPlat* application described in the first exercise in Chapter 4. While Snooping's chief marketing officer was impressed by the small-scale demonstration using the application prototype you developed, concerns regarding how well the existing prototype would scale up were raised by the chief technology officer. She asked for additional improvements to address these concerns.

   During the subsequent discussion with other members of the technical team, the approach agreed upon to scale up the application consists of *partitioning* the flow of incoming CDR tuples and *replicating* the rule matching process across the different replicas.

   Thus, in this exercise, you will improve on the existing prototype by proceeding through the following steps:

   (a) Implement modifications to the application flow graph so it can partition and distribute the incoming *CDR stream* and rely on multiple replicas of its rule-matching analytics to simultaneously process a larger set of marketing rules:

      (i) Partition the incoming CDR tuple traffic. Instead of using a single instance of the `SnoopRuleEvaluator` operator, modify the application so it can use a configurable number p of instances of this operator to evaluate the marketing rules over the incoming *CDR stream*.

      (ii) Distribute the incoming CDR data over the p `SnoopRuleEvaluator` operator instances by making use of a `Split` operator. The data splitting strategy should be designed such that a hash function is applied to the caller identifier attribute from the incoming CDR tuples, yielding a balanced p-way data distribution schema.

      (iii) Employ a fan-out configuration to ensure that all instances of the `SnoopRuleEvaluator` operator receive the *rule stream* from the *SnoopMarkPlatController* application.

      (iv) Similarly, employ a fan-in configuration to merge the results from the p instances of the `SnoopRuleEvaluator` operator, delivering them back to the *SnoopMarkPlatController* application.

(v) Employ placement directives in the *SnoopMarkPlat* application so the different instances of the `SnoopRuleEvaluator` operator as well as of the splitting and merging logic can be allocated to different hosts of a multi-host cluster. Make sure that I/O operators (source and sink) can be allocated to different *I/O* hosts, while the operators implementing the application analytical part can be allocated to *compute* hosts.

(b) Study the performance of the new application by assembling the following experiments:

(i) **Baseline:** Run the original application you implemented as part of the first exercise in Chapter 4, connecting it to the workload generator.

While maintaining all of the other workload generator parameters fixed, particularly the number of rules $n = 100$, increase the number of customers c to the point of saturation (c_s), where the application can no longer cope with the inflow. Record the workload generation rate r_s, the single-host saturation rate measured in number of CDR/s, as the baseline performance for your subsequent experiments.

(ii) **Multiple partitions:** Run the new version of the application by varying p from 1 to 4. In each case, fuse the splitting logic with the workload generator source, deploying all of them on a single host $(h = 1)$. In other words, for $p = 4$, four instances of the `SnoopRuleEvaluator` operator should be used. For each setting, increase the data source rate and observe the system performance, recording the saturation rate $r_{h,p}$ for each setting (specifically, $r_{1,1}$, $r_{1,2}$, $r_{1,3}$, and $r_{1,4}$).

(iii) **Multiple partitions/multiple hosts:** On a 6-host cluster, place the workload generator and the *SnoopMarkPlatController* application's sink by themselves on their own hosts (the *I/O* hosts). Run the new version of the application by varying p and h together from 1 to 4, using the other four available hosts (the *compute* hosts). In each case, fuse the source operator and the operators implementing the splitting logic together. For each partition, ensure that the operators encapsulated by the `SnoopRuleEvaluator` composite operator are also fused and placed on a host of their own. In other words, for $p = h = 4$, one instance of the `SnoopRuleEvaluator` operator should be placed on each of four *compute* hosts. For each setting, increase the data source rate and observe the system performance, recording the saturation rate $r_{h,p}$ for each setting (specifically, $r_{1,1}$, $r_{2,2}$, $r_{3,3}$, and $r_{4,4}$).

(iv) **Computational costs of the analytics:** Investigate the impact of the number of rules on each of the configurations from items (i)–(iii) by repeating these experiments with n varying from 100 to 1000 (in increments of 100).

(v) **Analysis:** Report your results by analyzing the impact of data partitioning, distributed processing, and of the cost of matching CDRs to rules as the number of rules increases. Summarize these results in an *executive summary*, proposing an *ideal* configuration for different operating points of the *SnoopMarkPlat* application. Discuss why data partitioning along with the replication of the `SnoopRuleEvaluator` operator is a reasonable design from a scalability standpoint.

(vi) **Extended analysis:** The redesign of the application we considered consisted of partitioning the *CDR stream* and replicating the rules across multiple instances of the `SnoopRuleEvaluator` operator. Consider the following questions: How could the application be modified so it becomes possible to also partition the rules? What are the trade-offs, expressed as an analytical model, between the design you experimented with and the rule-partitioning configuration?

2. **Application scale up: data replication and rule partitioning**. Consider the *ZNSNetMon* application described in the second exercise in Chapter 4. Its initial prototype was well received by the technical manager overseeing its development, but before the prototype could be deployed in production, he pointed out that the application must first be scaled up so it can accommodate the network traffic already handled by ZNS as well as the future growth being anticipated.

He also observed that the work performed by the `ZNSQueryEvaluator` operator becomes increasingly more expensive as additional queries are being simultaneously evaluated. In light of this consideration, he articulated that it would be possible to scale up the application by replicating the data from the Deep Packet Inspection (DPI) module and partitioning the analytical queries, distributing them through multiple instances of the `ZNSQueryEvaluator` operator.

Following the technical discussion, it was agreed that two activities would take place to improve on the original prototype: a modification to incorporate his ideas as well as an empirical evaluation considering different workload characteristics and physical layouts of the application.

Thus, in this exercise, you will proceed through the following steps:

(a) Implement modifications to the application flow graph so the alternative proposed by the technical manager can be evaluated:

(i) Modify the application so that, instead of using a single instance of the `ZNSQueryEvaluator` operator, the number of instances can be configured.

(ii) Replicate the *DPI stream* by fanning it out so each instance of the `ZNSQueryEvaluator` operator receives all of the DPI data.

(iii) Implement a distribution strategy using a `Split` operator and a hash function to distribute the queries, which are transported as tuples in the *query stream*, to the different instances of the `ZNSQueryEvaluator` operator such that each instance gets, on the average, the same number of queries.

(iv) Modify the application, making use of placement directives, so that each instance of the `ZNSQueryEvaluator` operator can be allocated on a different host of a multi-host cluster, placing I/O-related operators on specific input and output *I/O* hosts and the analytics operators on their own *compute* host.

(b) Study the performance of the new application by carrying out the following experiments:

(i) **Baseline:** Run the original application you implemented as part of the second Chapter 4 exercise, connecting it to the workload generator. Define a fixed configuration for the workload generator, increase the rate of the *DPI stream* until it saturates the application, and record this rate.

(ii) **Multiple hosts:** On a six-host cluster, place the workload generator and the *ZNSQueryWorkbench* applications sink by themselves on their own hosts (the *I/O* hosts). Using the same workload generator configuration as in the prior experiment, increase the number of hosts (and, hence, the number of query partitions) from 1 to 4 and record the point where the rate of the *DPI stream* saturates the application.

(iii) **Computational costs of the analytics:** Investigate the impact of the overall number of queries on the baseline as well as on the multi-host configurations. By preserving the original workload settings you employed, carry out four new additional experiments where the number of queries increases by a factor of 8. In each case, compute the rate of the *DPI stream* that saturates the application.

(iv) **Analysis:** Report your results by analyzing the impact of data replication and query partitioning as the number of hosts as well as the number of rules increase. Summarize these results in an *executive summary*, considering both experimental knobs independently.

(v) **Extended analysis:** The redesigned application relies on a replication of the *DPI stream* so that each instance of the *ZNSQueryWorkbench* operator can compute the results for the queries it manages. In general, data replication incurs additional costs to move the tuples around the flow graph. Given this consideration, what modifications can be made such that both data and query partitioning can be used in the *ZNSNetMon* application? Analytically model the original and the new design alternative to identify the conditions (as a function of the workload characteristics) where each alternative might be superior.

References

[1] Gutknecht J, Weck W. Modular programming languages. In: Proceedings of the Joint Modular Languages Conference. Vol. 1204 of Lecture Notes in Computer Science. Linz, Austria: Springer; 1997. pp. 1–311.

[2] IBM InfoSphere Streams Version 3.0 Information Center; retrieved in June 2011. http://publib.boulder.ibm.com/infocenter/streams/v3r0/index.jsp.

[3] Gedik B, Andrade H. A model-based framework for building extensible, high performance stream processing middleware and programming language for IBM InfoSphere Streams. Software: Practice & Experience. 2012;42(11):1363–1391.

[4] Wall L. Programming Perl. 3rd edn. O'Reilly Media; 2000.

[5] RDMA Consortium; retrieved in November 2011. `http://www.rdmaconsortium.org/`.

[6] Infiniband; retrieved in November 2011. `http://www.infinibandta.org/`.

[7] IBM WebSphere MQ Low Latency Messaging; retrieved in September 2010. `http://www-01.ibm.com/software/integration/wmq/llm/`.

6 Visualization and debugging

6.1 Overview

In this chapter, we examine visualization and debugging as well as the relationship of these services and the infrastructure provided by a SPS. Visualization and debugging tools help developers and analysts to inspect and to understand the current state of an application and the data flow between its components, thus mitigating the cognitive and software engineering challenges associated with developing, optimizing, deploying, and managing SPAs, particularly the large-scale distributed ones.

On the one hand, visualization techniques are important at development time, where the ability to picture the application layout and its live data flows can aid in refining its design.

On the other hand, debugging techniques and tools, which are sometimes integrated with visualization tools, are important because the continuous and critical nature of some SPAs requires the ability to effectively diagnose and address problems before and after they reach a production stage, where disruptions can have serious consequences.

This chapter starts with a discussion of software visualization techniques for SPAs (Section 6.2), including the mechanisms to produce effective visual representations of an application's data flow graph topology, its performance metrics, and its live status.

Debugging is intimately related to visualization. Hence, the second half of this chapter focuses on the different types of debugging tasks used in stream processing (Section 6.3).

6.2 Visualization

Comprehensive visualization infrastructure is a fundamental tool to support the development, understanding, debugging, and optimization of SPAs. It is also effective in taming the complexity of a multi-application environment, particularly when the continuous data processing components are interconnected to other pieces of a large Information Technology (IT) infrastructure. In this section, we explore four different facets of the visualization problem and discuss how each of them can be addressed with the help of tools available in Streams.

6.2.1 Topology visualization

SPAs, due to their information flow processing structure, lend themselves to a natural visual representation in the form of a graph. Topology visualization focuses on providing the means and tools to examine and explore an application's data flow graph at compile-time as well as its properties after it is deployed on the SPS.

Despite the straightforward mapping between the data flow graph programming abstractions and visual metaphors, delivering an effective topology visualization framework requires providing certain fundamental capabilities and adhering to a few important design principles.

Hierarchy support

SPAs are naturally *hierarchical* owing to the fact that their corresponding logical and physical data flow graphs are both hierarchical.

An application's logical data flow graph may contain composite and primitive operator instances nested inside higher-level composite operator instances. This structure can be arbitrarily deep. Similarly, the physical data flow graphs may contain multiple primitive operator instances nested inside application partitions, multiple application partitions placed on a host, and multiple hosts organized in different computational pools.

Hence, providing a visualization mechanism to seamlessly display and integrate the hierarchical dimensions of an application is important. Not only can this approach portray an application's natural structure, but it can also be used to manage the complexity of larger applications, providing different perspectives ranging from a global overview to a local view, where an operator or a host becomes the focus.

As a result, visualization tools and the system services supporting them must expose mechanisms to manage and explore hierarchical views. For instance, the ability to adjust the visible levels of the hierarchy, including support for collapsing and expanding composite operator instances, grouping operators and application partitions in different ways as well as overlaying the visualization onto the physical, possibly multi-host, environment where applications run.

The management of the natural hierarchy present in a SPA coupled with well-designed navigational aids are particularly important in helping the visualization to be usable, understandable, and effective irrespective of the size and complexity of an application.

Many of these mechanisms are available in Streams through its IDE, the Streams Studio (Section 8.4.4), and its *Application Graph* development-time and *Instance Graph* run-time views. For instance, Figure 6.1 shows a development-time hierarchical visualization of an application where all of its elements are fully expanded. Figure 6.2 depicts the same application, but only the primitive operator instances and first-level composite operator instances are shown, providing a more compact representation.

Complexity management

An efficient visualization framework includes mechanisms to cope with the scale and complexity of large cooperating applications primarily by providing *scoping* controls.

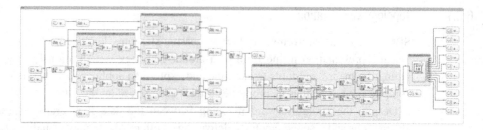

Figure 6.1 Hierarchical visualization of an SPL application using nested composition (fully expanded view).

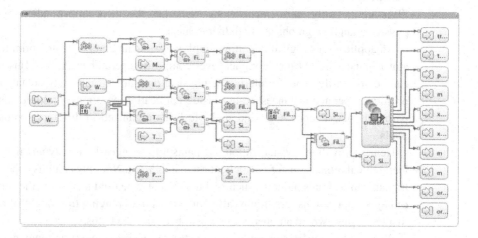

Figure 6.2 Hierarchical visualization of an SPL application using nested composition (partially collapsed view).

The most primitive form of scoping control consists of filtering based on different criteria where only elements (e.g., operators, application partitions, or streams) related to a particular application or job are displayed. The ability to highlight and locate elements based on their ids, names, and types as well as based on system administration criteria (Section 7.3.1 and 8.5.1) including job ownership and host where the element is located can be extremely useful. For instance, Figure 6.3 shows how an operator belonging to a large application can be located. In this case, text-based search is used to find all of the uses of the WaferLots operator. In the top diagram we see the instance of the operator highlighted by a bold border in the fully expanded application graph. In the bottom diagram we see the operator inside the composite operator that envelopes it.

Another form of scoping control consists of highlighting a rectangular area of an application, allowing an analyst to zoom in and out of a particular region of the overall topology, switching between a top-level global view and a detailed, localized view of an application segment. Figure 6.4 demonstrates how Streams Studio can be used in this capacity to explore a large-scale graph by employing the *Outline* view to focus on the

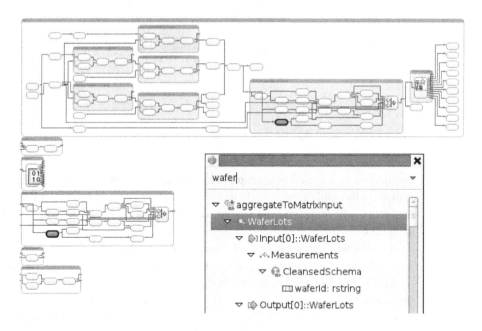

Figure 6.3 A Streams Studio view showing the operator search box.

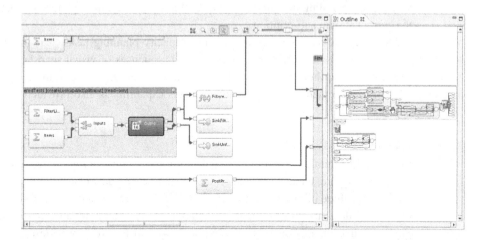

Figure 6.4 An outline view where zooming is used to explore local sites of a large application.

area of interest and, subsequently, the zooming capability available in the *Application Graph* view to portray a segment of the application with the desired level of detail.

Live and real-time
The continuous and long-running nature of SPAs demands that the visualization tools and system infrastructure include mechanism to extract and depict *live* snapshots of an

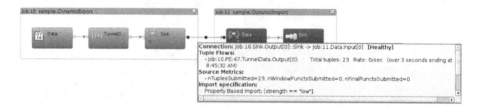

Figure 6.5 Inspecting the properties of a dynamic connection.

application as it runs. These snapshots must be continuously updated to reflect modifications in the application state. In this way, minute changes in an application are quickly and efficiently conveyed to an analyst, allowing her to observe the application behavior as it runs.

An effective visualization tool must be able to depict in real-time the arrival of new jobs and their interaction with existing ones as well as provide mechanism to inspect the instantaneous changes in data flows and resource consumption, including the application's use of network bandwidth and computing cycles.

Figure 6.5 depicts the Streams Studio's *Instance Graph* view rendering two jobs connected by a dynamic stream connection. This visualization also shows the properties of this connection including several live statistics culled from performance counters (Section 8.5.2), including the number of tuples that has flowed through the connection and the current stream subscription.

Moreover, in performance analysis and capacity planning tasks, it is often helpful to have access to tools that extract and record sequences of visual snapshots. In some cases, these snapshots may be synchronized with quantitative data from the application runtime environment for offline analysis. This capability is also provided by Streams Studio.

Interactivity

SPAs are live entities. Therefore, visualization tools must be *interactive*, enabling an analyst to explore, understand, and efficiently debug, profile, or monitor an application.

The prior three design principles are in many ways enablers to interactivity. The use of efficient data structures to manage the application's hierarchy and complexity, coupled with live interactions with the applications and the SPS, provide the sense that the visualization process itself is alive and is able to swiftly reflect changes in the application.

Interactivity is also important in a visual composition editor (Section 3.3.2) where an application is implemented by dragging-and-dropping components into a canvas. Generally, this type of environment uses visual aids when connections are created between application's components as well as when component's properties are configured.

Furthermore, a composition editor typically employs visual signals to indicate problems as the application is being assembled, prompting a developer to react and correct defects. This level of interactivity can make a development task more efficient as it propels the developer to continuously ensure that the application is syntactically correct.

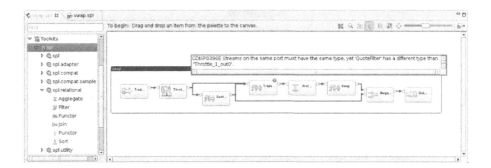

Figure 6.6 Interactive application construction in Streams Studio's visual composition editor.

Streams Studio's drag-and-drop visual composition editor has these capabilities. The editor links the visual representation to the corresponding SPL code and can seamlessly switch between the two. More importantly, it can provide on-the-fly validation of the code, regardless of the representation being used by the developer.

In Figure 6.6, the palette with the set of available operators is shown on the left and the current application flow graph is shown on the right. In this particular example, one of the operators has a configuration error, indicated by an X on its right-hand side. As the developer's mouse hovers over that operator, a tooltip is shown providing a diagnostic indicating, in this case, that one of the incoming stream connections has an incompatible schema.

Multi-perspective support

SPAs can be visualized using different *perspectives*, enabling a developer to tailor the visual representation to debugging, performance optimization, or capacity planning tasks. Two of the main visualization perspectives relate to whether an application is still being developed or whether it is already in execution.

A *development-time* visualization perspective is mostly static. Since the application is not yet instantiated, many of its dynamic aspects such as its dynamic stream connections, performance counters, and interconnection with other applications cannot be represented. Nevertheless, the visualization itself can be altered dynamically to use different layouts as well as to focus on different aspects of the application's flow graph. For instance, the visual tool can portray a flow graph consisting only of the application's smallest components, its primitive operators, grouped by the partitions that will eventually exist once it is deployed.

In contrast, an application's *runtime* perspective captures snapshots of its execution state as well as snapshots of the environment where it runs, providing substantially more information than the development-time perspective. The visualization tool can, for example, depict all of the applications that are currently sharing the runtime environment.

As is the case with the development-time perspective, the runtime perspective can incorporate hierarchical elements and present an application using different views,

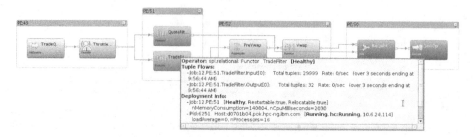

Figure 6.7 Runtime view depicting an application's set of partitions (PEs), each containing a data flow subgraph forming the larger application flow graph.

organized in terms of partitions, hosts, or jobs. For instance, Figure 6.7 shows an application organized as a set of partitions, or PEs (Section 7.2.2), where each partition hosts several operators. A tooltip is used to show the details about the state of one of the operators, where it is possible to see that it is healthy and running normally.

6.2.2 Metrics visualization

The continuous data processing performed by a SPA as well as the dynamic changes that affect it as a result of shifts in its workload or in its interactions with other applications creates challenges when it comes to understanding how the application is behaving.

Performance metrics, held by internal system performance counters, are the indicators used to capture raw system and application performance information and can be used directly or be further processed to synthesize other measures of interest to an analyst. Metrics are live by nature and reflect the effects of the workload imposed on an application as well as on the SPS supporting it, reporting memory, CPU, network, and storage utilization. In general, sophisticated SPSs are designed with instrumentation points and service interfaces designed to retrieve performance metrics (Sections 7.3.2 and 8.5.2) as are the operators that make up a large-scale SPA.

As a result visual tools that can help understand an application's resource utilization as well as the quantitative impact an application exerts on the runtime environment, are very important.

An effective way to visualize performance metrics is to overlay them on an application's flow graph. For instance, the changing tuple data rates for a stream connection can be visualized by representing this connection with a line whose varying thickness conveys the change in data volume, providing an instantaneous glimpse on the volume of data transported by a stream.

As an example, Figure 6.8 shows a fragment of a Streams Studio-extracted visualization where line thickness and color are used to indicate the rate at which tuples are flowing in a particular stream connection. The qualitative information implied by the visual cues can be augmented textually with a tooltip, as is the case in the figure, and relays the data rate instantaneously associated with that stream connection. Similarly,

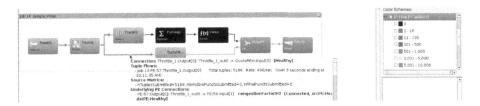

Figure 6.8 Line thickness and color (shown in shades of gray) as performance indicators for stream connections and operators.

the CPU load on a host or the CPU consumption associated with an operator instance can be visualized using different colors to convey how overloaded a host is and how demanding an operator instance is. Note that this data is continuously updated while the application runs.

6.2.3   Status visualization

Status visualization focuses on retrieving and summarizing the *health* and status indicators associated with an application as well as with the SPS that supports it. In general, SPSs provide both job management (Section 7.3.1 and 8.5.1) and resource management (Sections 7.3.2 and 8.5.2) interfaces used to expose this information so it can be visualized or queried textually.

As discussed earlier, an application might comprise a large number of partitions, spanning multiple hosts in a cluster. The SPS supporting one or more of these large applications must monitor and manage its own components (Sections 7.2 and 8.4) as well as the hardware resources used to support the platform's components and the applications.

Status visualization conveys the health and status of the system and of the applications it manages in real-time. In the former case, the status of a host indicates whether it is functional and in a *schedulable* state such that application partitions can be instantiated on it. In the latter case, the status of an application partition indicates whether the operator instances it manages are up and running. Similarly, the status associated with stream connections indicates whether the physical links associated with the connections are active and healthy.

Figure 6.9 depicts a Streams Studio-generated visualization that conveniently captures both of these facets. On the left-hand side of the figure, the list of hosts managed by a Streams instance (Section 8.4.1), the instance's components (Section 8.4.2), and their status can be seen. This snapshot also includes the list of applications currently in the system, their associated partitions (PEs), and indicators describing the PEs' states. Some of the same information is depicted pictorially in the flow graph shown in the middle of the figure. This perspective, the *host view*, is able to portray an application's data flow graph segmented by the hosts where it runs, providing a sense of its physical layout. Finally, on the right-hand side of the figure, the color schema used for the health, status, and performance indicators is shown.

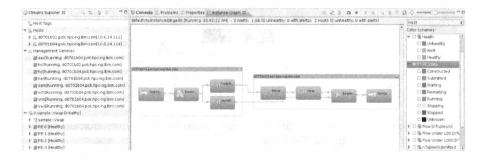

Figure 6.9 Combined system and application status visualization.

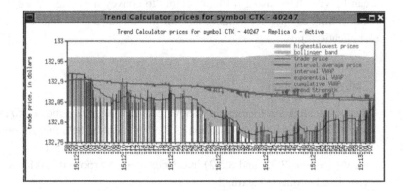

Figure 6.10 A snapshot of an application's live data visualization Graphical User Interface (GUI).

6.2.4   Data visualization

Presenting the results produced by a SPA in an effective form contributes directly to how user-friendly and empowering a data analysis application is. The large volume of data they process, the sophisticated machine learning and data mining techniques they employ (Chapter 10), as well as the continuous adaptation and evolving behavior embedded in their algorithms can only become an important data analysis asset if this analytical edge can be translated into consumable information directly useful to a decision-maker. A rich visual interface is, therefore, essential.

In general, the visualization interface provided by an application must usually rely on a comprehensive dashboard where charts and graphs representing both the raw data manipulated by an application as well as summarized metadata and results produced by the application can be inspected, manipulated, and interacted with.

As a consequence, the design of a SPA must consider how it will be integrated with one or more visual interfaces to accomplish these goals. In some cases, a straightforward chart, as the one in Figure 6.10, generated from the live contents of a stream externalized via a sink operator is sufficient. In this case, the application's GUI shows a snapshot of results produced while processing transactions from a stock exchange. While simple, this interface updates continuously as the application runs and processes more data, providing immediate feedback to its users. In other cases, considerably more effort must be devoted to creating a compelling and effective visual user interface.

While we won't provide a longer discussion on data visualization design principles and techniques in this book,[1] a key component of a SPA's visual interface is its ability to update continuously, directly reflecting the actual data flow and the current analytical results that have been computed. The ability to summarize and rank these results automatically, while retaining the capacity to drill down and inspect all of the results accumulated so far, is equally important (Section 12.4.5).

In general, more sophisticated visual user interfaces are usually handcrafted using User Experience (UX) toolkits with programmable data presentation frameworks. Examples include Adobe Flex [8], SpotFire [9], and gnuplot [10].

System-supplied data visualization
Designing a SPA's visual interface is a complex engineering problem in its own right. Yet, a certain degree of automation for simpler tasks can be provided by the tools supplied by the SPS. For instance, in an SPL application, a stream can be configured as a data source for a system-supported visualization interface by setting the boolean directive streamViewability in its operator's optional config clause as is the case for the PreVwap stream in Figure 6.11.

In this way, the Streams Console, the Streams' web-based administration interface (Section 8.4.2), can be used to automatically produce a visualization chart by defining a *data view* over a stream. Each data view is a window-based snapshot that is continuously updated as more data becomes available. Figure 6.12 depicts an example of a data view employing a time-based window where stock market transactions are used to compute the Volume Weighted Average Price (VWAP) as well as other pricing metrics for an IBM share as a function of time.

```
1   stream<rstring ticker, decimal64 vwap, decimal64, minprice,
2         decimal64 maxprice, decimal64 avgprice, decimal64 sumvolume>
3     PreVwap = Aggregate(TradeFilter)
4   {
5     window
6       TradeFilter: sliding, count(4), count(1), partitioned;
7     param
8       partitionBy: ticker;
9     output
10      PreVwap:
11        ticker = Any(ticker),
12        vwap = Sum(price * volume),
13        minprice = Min(price),
14        maxprice = Max(price),
15        avgprice = Average(price),
16        sumvolume = Sum(volume);
17    config
18      streamViewability: true;
19  }
```

Figure 6.11 Making an operator *viewable*.

[1] Extensive literature on the design of data visualization interfaces and dashboards is available [2, 3, 4, 5, 6, 7].

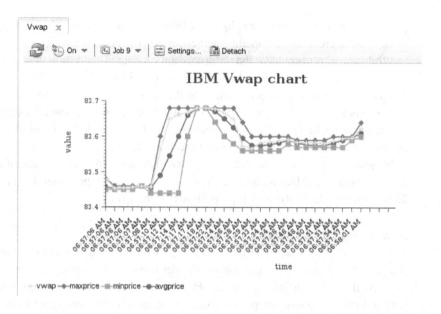

Figure 6.12 System-generated visualization of data.

6.3   Debugging

One of the stages of the software development life cycle is testing, when developers and quality assurance engineers ensure that an application is functionally correct as well as that it addresses operational and performance requirements set forth in its design specification.

A high-quality application is extensively tested along these dimensions during its development phase when the initial implementation defects are fixed. Later on, when the application evolves and incorporates new capabilities, it is also subjected to additional testing in the form of regression tests, which verify whether both older and newly implemented features work according to the specification.

Note that application correctness is a broad concept, including functional correctness, which asserts that an application performs the data processing tasks it was designed to carry out accurately, as well as non-functional correctness, which asserts that it delivers its results in accordance with pre-determined performance, fault tolerance, scalability, and usability goals.

When either of these two requirements is violated, the application developer's task is to find out the cause or causes of the failure to subsequently correct or improve the application's implementation. Identifying failures or sources of degradation typically requires the use of debugging tools, which, in our definition, includes both traditional debuggers as well as performance profilers.

Debugging SPAs can be particularly challenging due to the existence of asynchronous data flows and the distributed nature of the computation. As we have described in earlier chapters, the existence of asynchronous flows is simply a result of employing multiple execution contexts in the form of application partitions, which may happen to be distributed across several hosts.

Yet, despite these challenges, the well-defined structure of a SPA expressed as a data flow graph simplifies the understanding of the interactions that take place within the application and thus facilitates the analysis of data dependencies as well as of operator-to-operator interactions that might occur.

Interestingly, this structure contrasts favorably with the challenges in debugging applications built using other distributed processing platforms such as PVM [11] and MPI [12]. These types of applications usually employ free-form communication patterns, where one application component can, in principle, interact with any other, making it difficult to understand and visualize the data exchange flows without detailed analysis of the application source code.

Nevertheless, debugging a distributed SPA can still present formidable difficulties. For instance, when attempting to address defects arising from intra-operator interactions, which can be arbitrarily complex, as well as when tackling defects originating from performance slowdowns in a multi-component application spanning tens or hundreds of hosts.

Debugging tasks

In essence, debugging is a method for finding software defects in an application. When dealing with distributed SPAs, debugging comprises four different, but related tasks [13], which will be examined in more detail in the rest of this section:

- *Semantic debugging*, which includes the techniques and tools used to find errors in the application semantics by examining its data flow graph, including the application structure, the configuration of its operators, as well as the content of the stream traffic flow.
- *User-defined operator debugging*, which includes the techniques and tools used to find errors in the user-defined operator source code written in a general-purpose programming language.
- *Deployment debugging*, which includes the techniques and tools used to find errors in the runtime placement and configuration of an application.
- *Performance debugging*, which includes the techniques and tools used to locate inefficiencies and bottlenecks that prevent an application from meeting its performance goals.

6.3.1   Semantic debugging

Semantic debugging consists of identifying defects in three aspects of a SPA: its configuration, its compositional structure, and its data flow.

The first two aspects relate to reviewing the computational structure of an application (its configuration and composition), whereas the third one is specifically focused on the data and, in particular, on the intermediate results produced by an application.

6.3.1.1   Configuration correctness

In this step, the operator instances used in an application's data flow graph must be reviewed to ensure that they are properly configured. Clearly, employing the wrong operator, providing incorrect parameter values to it, making incorrect output assignments, and improperly configuring the windowing properties can result in semantic errors in an application. Unfortunately, however, these are just examples of a much larger set of configuration mishaps.

Debugging problems in an application's flow graph or in the configuration of its operators is best achieved using an IDE that is able to catch and report simpler problems at development time, but also able to represent the application graphically, providing intuitive visual metaphors corresponding to each of these debugging aspects as well as capabilities to quickly navigate the application.

As discussed in Section 6.2, the Streams Studio IDE can be used to inspect how operators have been configured and these configurations can also be edited (Figure 6.13). By linking the topological view (top left) to the source code containing the operator invocation (top right) as well as to a panel summarizing its configuration (bottom), the tasks of inspecting and, potentially, fixing an operator can be conveniently carried out.

6.3.1.2   Compositional correctness

A SPA expressed as a set of operators interlinked in a data flow graph must be reviewed to ensure that its overall structure is correct. Incorrectly specified stream connections, static or dynamic, are a common cause of semantic errors.

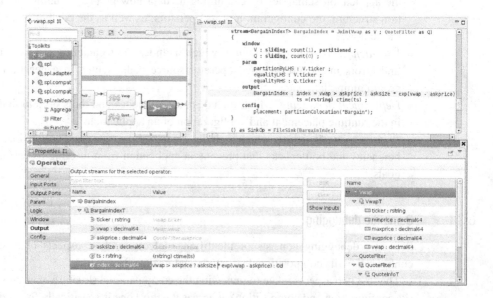

Figure 6.13 Viewing and editing an operator's properties.

The graphical representation generated by an IDE can be used to inspect the application structure. As is the case in Streams Studio, in many other development environments the visual representation employed by these tools provides either a direct link to the source code, a graphical drag-and-drop composition canvas, or both, allowing a developer to easily make the necessary corrections upon detecting a compositional error.

In this debugging task, a typical process consists of methodically exploring a SPA's flow graph starting from its outermost module (Section 5.2) and diving into each subcomponent, asserting its structural correctness, recursively.

Streams Studio's visualization capabilities provide numerous ways to navigate and inspect an application's flow graph, including its component hierarchy as previously discussed (Section 6.2.1).

6.3.1.3 Data flow correctness

The data transported by the multiple stream connections in an application must be reviewed to ensure that the transformations and processing applied to each operator's inbound streams are indeed producing the expected results. Unfortunately, errors in the data flow sometimes are only detected when the application is in user-acceptance testing (UAT), emulating production-like workloads, or already in production. Hence, a debugger must be able to connect to a running instance of that application such that its *live* global state can be probed with minimal impact to the application.

Specifically, a debugger must provide the means for examining the actual data flowing through its stream connections as well as the internal state of each operator. These tasks require traditional debugging techniques as well as new ones tailored to stream processing. In the rest of this section, we look at both dimensions of this problem, starting with an overview of traditional techniques used in non-streaming applications.

Break points and trace points

General-purpose debuggers include two categories of program execution control functions: break points and trace points. These two categories provide the basic tools used in semantic debugging [13]. The former is aimed at controlling the flow of execution of an application and the latter at inspecting its internal state.

Considering the data flow created by a SPA, a *break point* can be activated on an input or on an output port, causing the execution to temporarily stop. Such a break point is triggered every time a tuple or a punctuation hits the port in question. When this happens and the execution stops, the value of a newly arrived tuple or punctuation can be examined and modified. Furthermore, because the execution is stopped, the state of the operator can be inspected and, possibly, modified. Subsequently, the execution can be resumed by the user.

A break point can be made *conditional*, in which case an expression specifying a predicate that must be satisfied indicates when the break point should be triggered. Such a condition can include references to the tuple attributes and the values they must have, qualitatively expressing the tuples of interest to the debugger. This condition along with the location of the break point form its definition.

A *trace point* can also be defined on an input or on an output port, but it does not cause an application to stop. It gets triggered every time a tuple or punctuation hits that port. When this happens, the value of the tuple or of the punctuation is stored in a table associated with this particular trace point, which can later be inspected.

A trace point can also specify how much history to accumulate. When this limit is exceeded, the oldest tuples and punctuations that were stored earlier are discarded. By examining the traces, a developer can better understand the history of the tuple and punctuation traffic associated with a port to assess whether an operator's computation is being performed correctly. Similar to break points, trace points can also be conditional, so that only tuples matching their predicate are retained.

In Streams, the sdb debugger (Section 8.4.4) can be used to define break points and trace points, including their conditional variations, as well as to view and modify tuple contents, inject new tuples, and drop existing tuples from a flow.

In Figure 6.14 we see the console-based sdb interface. In this example, sdb is attached to one of the application partitions where two operators are running (SinkOp and BargainIndex). The actual list of operators can be retrieved using the o (list-(o)perators) command. Subsequently, a breakpoint can be defined using the b ((b)reakpoint) command. For instance, to define a breakpoint on the BargainIndex operator's first output port (denoted by o 0), the command b BargainIndex is used as seen in Figure 6.14.

The execution can be resumed by using the g ((g)o) command. As the application runs, we can see that whenever a tuple hits the breakpoint, the execution is again suspended and the tuple's contents are displayed. At this point, the user can decide whether to drop, modify, or continue with the tuple unchanged. The latter option requires using

```
 d0428b03 Default - 9 (on d0428b03.pok.hpc-ng.ibn  _ □ ✗
IBM Stream Debugger (SDB), pid: 17682
PE execution is suspended.
Set initial probe points, then run "g" command to continue execution.
(sdb) o
   #in  #out Operator                                  Class
    1    0   SinkOp                                     SinkOp
    2    1   BargainIndex                               BargainIndex
(sdb) b BargainIndex o 0
   Set +  0   Breakpoint  BargainIndex                      o   0
   stopped:false
(sdb) g
(sdb)
    +  0   Breakpoint  BargainIndex                          o   0
   dropped:false stopped:true
   ticker, "IBM", rstring
   vwap, 83.453333333333333, decimal64
   askprice, 83.46, decimal64
   asksize, 10, decimal64
   ts, "Tue Dec 27 14:30:19 2005", rstring
   index, 0, decimal64
(sdb) ▊
```

Figure 6.14 Inspecting a tuple's contents using breakpoints in sdb.

the (c)ontinue command. Alternatively, the user can also inject one or more tuples or punctuations into the stream before the execution is resumed.

Lineage debugging

While both trace and break points help pinpoint problems by allowing a controlled execution of an application as well as the inspection of the data flowing through the stream connections, these functions are not as effective when the location of a bug is unknown. A technique referred to as *lineage debugging* [14], specific to stream processing can provide additional assistance in such situations, as it allows the examination of data dependencies across the data flow graph.

When a debugger implements lineage debugging, it can provide information about the operations used to derive the value of a tuple's attributes as well as information about how a particular tuple affects others created downstream. In essence, a tuple's *backward lineage* includes information on all the tuples that have contributed to its production. Similarly, a tuple's *forward lineage* includes information on all the tuples whose production it has contributed to.

It is important to note that computing the lineage of a tuple requires tracking both the inter- and intra-operator data dependencies. Inter-operator data dependencies are naturally represented by the stream connections. In contrast, intra-operator data dependencies must be extracted from an operator's internal processing logic and, hence, are not immediately evident unless explicitly exposed, for instance, by the operator's model (Section 5.2.3.1).

The system infrastructure required to support lineage debugging must automatically track the history of tuples flowing through the input and output ports of all of the operators in an application and extract the data dependencies by identifying the set of inbound tuples used to produce an outbound tuple. These dependencies are then transitively extended to compute the forward and backward lineages.

The lineage information can be used in different ways. For instance, the forward lineage information is useful to perform *impact analysis* where the goal is to understand what type of influence a given tuple has on other downstream tuples as well as in the final results produced by an application. Conversely, the information extracted from the backward lineage relationships can be used to perform *root cause analysis*, where the goal is to understand what other tuples and operators have impacted the production of a given tuple.

Streams does not provide tooling to support lineage debugging on an application. Yet, it is possible to emulate certain aspects of this technique by adding instrumentation code to an application as well as additional sink operators to store information for future analysis. In this case, a debug operator is placed on the receiving end of a stream alongside the regular stream consumers in the application. Thus, the tuple traffic is tapped and accumulated for further on-screen or offline analysis.

Despite the lack of technical sophistication, the use of this method to output stream flows to a file, employing a FileSink operator, or to the console, employing a Custom operator, can be very effective, akin to the use of printing statements in a regular Java or C++ program.

6.3.2   User-defined operator debugging

User-defined operator debugging techniques and tooling are targeted at locating problems in the source code of primitive operators. As discussed before, these operators are written using general-purpose programming languages and, therefore, debugging their internal processing logic involves using native debuggers (e.g., Free Software Foundation (FSF)'s gdb [15]) that support the specific language that was used to implement an operator.

The internal processing implemented by an operator is usually driven by the arrival of external data, in the case of source operators, or tuples and punctuations, in the case of other types of operators. Hence, to diagnose and fix defects in the implementation of an operator, the development environment must provide a seamless integration between a conventional debugger and a stream processing debugger so the operator's behavior can be examined as it performs its analytical task.

A typical debugging workflow will consist of employing break points of the sort we discussed in Section 6.3.1 to initially home in on the tuples and ports related to this debugging task. Subsequently, once a break point gets triggered, the native debugger is automatically launched and the operator internal processing logic is stopped, allowing the stepwise tracing of its execution.

In Streams, the sdb debugger supports launching a gdb session window, using the (g)db command, when a tuple hits a break point on an input port. This command not only opens a gdb window but also sets a breakpoint on the operator's tuple and punctuation handling process methods (Section 5.2.3). When the operator execution is resumed using the c command, the gdb breakpoint will eventually be hit, making the gdb window active. At this point, the user can interact with gdb normally until the operator execution is again resumed and control is returned to the Streams runtime.

6.3.3   Deployment debugging

As discussed in Section 5.3, the logical data flow graph associated with a SPA can be mapped to one of many possible physical data flow graphs that will be deployed and, at some point, executed on a set of distributed hosts. Therefore, an application developer or a system administrator must have the means to verify whether the runtime layout and configuration of an application are correct.

Even though the semantic correctness of the applications may not necessarily depend on the runtime physical layout of an application, the specifics of the runtime layout are relevant to monitoring, load balancing, as well as data ingress and egress tasks.

Deployment debugging involves making sure that the physical layout of the application conforms to the placement constraints intended by the application developer.

This type of debugging task is best supported by tools that provide visual as well as text-based deployment views of the application runtime. These representations of an application must include a hierarchical decomposition of the application in terms of operator instances, partitions, hosts, and jobs as well as their association with the runtime resources they are currently using.

This task can also be carried out more productively if the tools used for debugging are able to link the logical and physical view of an application. In this way, developers can quickly examine the application and verify that the constraints they want to enforce on the mapping of the logical application components and their physical counterparts are consistent. This step also includes verifying that the distributed components of a SPA are correctly placed on the set of distributed hosts making up the application runtime.

Streams supports deployment debugging through Streams Studio. Specifically, through the real-time, live, and multi-perspective visualizations of the *Instance Graph* view shown in Section 6.2.

6.3.4   Performance debugging

Many SPAs are very performance-sensitive, a direct by-product of their continuous processing model and, frequently, the reason why these applications are developed in the first place (Section 3.2). Not surprisingly, identifying performance bottlenecks and aggressively optimizing these applications are an integral part of an application's life cycle.

The goal of performance debugging tools and techniques is to find problems that prevent the application from meeting its performance targets with respect to specific metrics laid out as part of the application specification.

Performance-driven debugging typically requires inspecting and monitoring various system-level and application-specific performance metrics associated with stream connections, operator instances, application partitions, and the hosts employed by the application runtime. These metrics might include stream data rates, CPU utilization, memory usage, along with a variety of user-defined and application-specific metrics as well as other derived metrics computed as a function of the low-level ones.

Starting with simpler, system-provided metrics, a fundamental measure of performance can be obtained by looking at a stream's data rate representing the number of tuples emitted per time unit. The rate of a stream can also be collected and reported as the number of bytes sent per time unit. In either case, these rates can be used to check whether the application meets its throughput-related performance requirements. Again, when debugging an SPL application, Streams Studio can be used to inspect and collect this information as seen in Figure 6.9.

Diagnosing back pressure
Stream data rates can be measured anywhere in the flow graph, but a particularly relevant instrumentation point is at the output port (or ports) of an application's source operators. These rates determine an application's throughput because an application's source operators naturally find an equilibrium between how quickly external data can be converted into tuples versus the rate at which an application can consume this data due to the *back pressure* the downstream processing exerts on the external sources (Section 9.2.1.1).

Back pressure is a condition that might occur in any multi-component application supported by information flow systems and, by consequence, in SPSs. This condition can be easily understood through an example. Consider a simple three-operator chain flow graph, where a source operator A sends data to B, which processes it and sends its results to a sink operator C. When operator C is slow in processing its incoming data, it eventually forces the operator B to block when it tries to emit additional tuples. If this blocking did not happen, the tuples output by B would have to be arbitrarily discarded until operator C frees up. Since this is not acceptable, B blocks. However, the blocking is not limited to B, as B starts waiting for C to free up, it is prevented from processing its incoming tuples. As a result, the pressure is felt by operator A, which also blocks. This condition may potentially block the original sensor producing the data, if it also employs a protocol where data cannot be dropped. In essence, back pressure slows down the entire application and, as a result, its throughput is limited by the slowest of the operators, C in our example.

This type of blockage happens on the tuple producing endpoint of a stream connection and is caused by buffering and flow management in a SPS's transport layer (Section 5.3.3). For instance, a stream connection that crosses two application partitions is usually transported by a TCP/IP connection. Since TCP/IP buffers are finite, if the consumer side cannot keep up with the producer side, these buffers eventually fill up and blockage occurs to prevent data loss.

Simply inspecting the stream rates of an application can provide a rough characterization of an application's throughput, but these rates alone are not directly useful to pinpoint the location of back pressure-induced bottlenecks. In fact, singling out the primary cause of a blockage and identifying the location of a bottleneck require measuring the stream data rates with segments of the application turned off. This methodical procedure starts with only the source operators, followed by gradually adding the additional downstream operators while, at each step, re-measuring the stream data rates.

Eventually, when the layer introducing the bottleneck is turned back on, the drop off in the stream rate will be detected. While this approach provides a clear understanding of which operator instance is causing throughput degradation, it is both labor- and time-intensive, particularly for large and distributed applications.

An equivalent, but speedier, approach consists of locating the stream connections that are experiencing back pressure with the help of instrumentation and tools for performance analysis. This method consists of inspecting the buffer occupancy associated with the producing endpoint of a stream connection. The higher this metric is, the more pressure the operator faces. From this location, one can follow the back pressure by tracing the data flow graph until a downstream operator that *does not* experience back pressure is located. This downstream operator *is* the source of the slowdown and it must be optimized.

Streams provides a *congestion factor* metric, which is a relative measure of the back pressure experienced by a stream. This metric can take values in the range [0, 100], where 0 indicates no back pressure on the link and 100 indicates a completely stopped flow. In the sequence of snapshots depicted by Figure 6.15, we show how an operator that creates a processing bottleneck can be located with the help of these measurements. In this example, an application used to process text and compute a few metrics for further analysis employs five operators organized in a pipeline. In this very simple scenario,

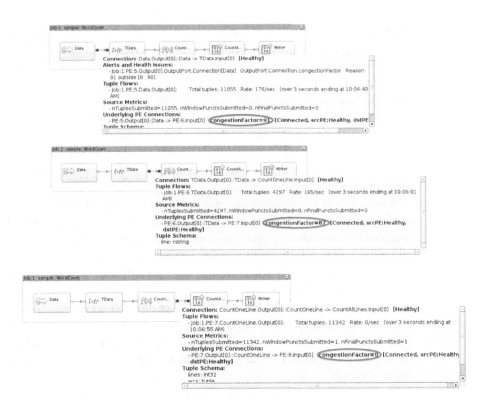

Figure 6.15 Locating a bottleneck using Streams Studio.

the application designer has inadvertently left a tracing statement in the third operator in the pipeline. As a result, the application, because of the extra tracing output it produces, does not achieve the expected throughput.

By inspecting the congestion factor values for each stream connection and observing that the first two operators are experiencing contention in their output ports with congestions factors of 91 and 87, respectively, and that the third one does not, it is easy to infer that the third operator is the site of the bottleneck in this application.

In this example, the slowdown is the result of a coding mistake, but frequently real performance problems are found. In such situations, multiple optimization approaches can be considered to mitigate the back pressure effect (Sections 9.2.1.1 and 9.3.2.2). In broad terms, the options consist of optimizing the operator's internal logic, replicating it, parallelizing it, or, in some cases, applying deeper changes to the structure of the application's data flow graph such that the workload can be better divided up and distributed.

Performance-monitoring tools

The steps sketched above require substantial tooling support, including runtime instrumentation and well-designed visualization tools. The visualization framework in

Streams relies on specific performance metrics obtained from instrumentation points automatically placed by the application runtime (Sections 7.3.2 and 8.5.2).

Beyond buffer occupancy and congestion factors, numerous other metrics can help locate the cause of blockages and other performance problems. Different SPSs collect and process different ones, which we survey next.

CPU utilization measures how much of the CPU computational resources available in a host are being used. This continuous measurement might aggregate the load by operator instance, by application partition (including all the operator instances it hosts), or by host, which naturally includes aggregating the load from all partitions as well as the OS background load.

When taken at a partition or operator instance-level, the magnitude of the CPU utilization can also be used for detecting back pressure. Clearly, a partition that depletes the available CPU resources is a bottleneck. Unfortunately, the reverse is not true. An operator instance may still be a bottleneck despite not consuming all of the CPU resources. For instance, in situations where an operator's performance is limited by other types of contention, including I/O (e.g., excessive disk or network activity) or other, perhaps external, resources (e.g., database, file system, or directory servers).

Note that, when aggregated at a host-level, the CPU utilization can be used to evaluate whether the load imposed by an application is well-balanced across the set of available hosts used by a distributed application, or whether there is enough slack to handle spikes in the workload.

Latency is another important metric of interest. Nevertheless, the presence of one-to-many (a data expansion) and many-to-one (a data reduction) relationships between inbound and outbound tuples processed by operators imply that measuring it is not a simple matter of tracking the amount of time it takes for a tuple to travel from one point to another within the flow graph. As a result, tracking latency is often performed at the application-level via user-defined metrics [16].

Another important point about latency relates to the issue of clock synchronization across the hosts in a distributed runtime environment. When it is measured across hosts, the result can be substantially inaccurate if the clock drift between the hosts is non-negligible relative to the actual measured latency. Indeed, addressing this issue typically requires hardware support or a software solution providing cluster-wide clock synchronization [17].

Therefore, defining the precise meaning of latency as well as of other metrics of interest for an operator requires specific knowledge of its internal processing logic. Thus, an important mechanism that can be provided by a SPS consists of a metric management subsystem, where an individual metric is associated with a *performance counter*.

In Streams, a performance counter, whether system-supported or user-defined, can be accessed at runtime by a developer as well as by performance analysts and system administrators (Section 6.2.1). Command-line and visualization tools can be used to look at instantaneous snapshots as well as to inspect past values of such metrics, allowing the detection of anomalies and the understanding of long-running performance trends.

Instrumentation and impact monitoring

Performance debugging techniques are central to locating the sites in a flow graph where slowdowns occur and, hence, provide the basis for fixing or improving the runtime behavior of an application.

Nevertheless, instrumentation and monitoring costs are often non-negligible and can become expensive in certain applications, impacting and distorting their runtime behavior. As an example, consider the collection of stream data rates in bytes per second. For a stream connection between fused operator instances, the tuple traffic might be delivered as simple function calls from the producing operator to the consumer (Section 8.5.4). In this case, given the absence of a serialization step to prepare a tuple for transmission over the network, simply computing the size of a tuple can become the dominating cost in an operator. Similarly, obtaining the CPU utilization on a per-operator instance basis, rather than on a per-partition basis, can also be computationally expensive. Specifically, it requires finding and teasing apart the individual computational demands associated with an operator instance, which might be sharing a single thread of execution with other operator instances.

In conclusion, enabling the collection of *all* performance metrics, including application runtime-based as well as the application-specific ones, can be too costly, if turned on by default. As a result, SPSs, such as Streams, often collect some basic metrics continuously and augment that with the support for an on-demand *profiling mode*, where compiler- or performance analyst-driven instrumentation is manually activated and deactivated, enabling the collection of additional performance metrics.

6.4   Concluding remarks

In this chapter, we looked at two critical software engineering aspects relevant to the development and management of large-scale and complex SPAs: visualization and debugging.

In the first half of this chapter, we discussed the visualization techniques and capabilities that can be used to better understand the dynamics of a SPA. These techniques rely on tools that make use of visual metaphors to depict an application's flow graph, its performance metrics, the health and status of its own components as well as of the SPS runtime components, and the in-flight streaming data associated with the applications. The relationship between the user-facing visualization tools and a SPS's runtime environment will be further discussed in Sections 7.3.10 and 8.5.11.

In the second half of this chapter, we examined how debugging is carried out in a streaming context and we focused on the data flow organization of a SPA, which creates additional challenges and requires other debugging tools in contrast to debugging regular computer programs. We discussed how semantic, user-defined operator, deployment, and performance debugging tasks can be carried out and how specific methods and tools can help a developer in performing them. In Section 7.3.9, and Section 8.5.10, where Streams' debugging service is examined in more detail, we will extend what was discussed in this chapter to show the relationship between the user-facing debugging tools introduced here and the services that support them.

Indeed, in the next two chapters, the focus will shift as we start examining the internal architecture of a SPS. The upcoming discussion will shed light on the structure and on the services provided by the distributed middleware used to support and manage the execution of SPAs.

References

[1] De Pauw W, Andrade H. Visualizing large-scale streaming applications. Information Visualization. 2009;8(2):87–106.

[2] Yau N. Visualize This: The FlowingData Guide to Design, Visualization, and Statistics. John Wiley & Sons, Inc.; 2011.

[3] Few S. Now You See It: Simple Visualization Techniques for Quantitative Analysis. Analytic Press; 2009.

[4] Steele J, Iliinsky N, editors. Beautiful Visualization: Looking at Data through the Eyes of Experts (Theory in Practice). O'Reilly Media; 2010.

[5] Telea AC. Data Visualization. A K Peters, Ltd; 2007.

[6] Ward MO, Grinstein G, Keim D. Interactive Data Visualization: Foundations, Techniques, and Applications. A K Peters, Ltd; 2010.

[7] Wong DM. The Wall Street Journal Guide to Information Graphics: The Dos and Don'ts of Presenting Data, Facts, and Figures. W. W. Norton & Company; 2010.

[8] Adobe. Using Adobe Flex 4.6. Adobe Systems; 2011.

[9] TIBCO Spotfire; retrieved in September 2012. http://spotfire.tibco.com.

[10] Janert PK. Gnuplot in Action: Understanding Data with Graphs. Manning Publications; 2009.

[11] Geist A, Beguelin A, Dongarra J, Jiang W, Mancheck R, Sunderam V. PVM: Parallel Virtual Machine. A Users' Guide and Tutorial for Networked Parallel Computing. MIT Press; 1994.

[12] Gropp W, Lusk E, Skjellum A. Using MPI: Portable Parallel Programming with Message-Passing Interface. MIT Press; 1999.

[13] Gedik B, Andrade H, Frenkiel A, De Pauw W, Pfeifer M, Allen P, et al. Debugging tools and strategies for distributed stream processing applications. Software: Practice & Experience. 2009;39(16):1347–1376.

[14] De Pauw W, Letia M, Gedik B, Andrade H, Frenkiel A, Pfeifer M, et al. Visual debugging for stream processing applications. In: Proceedings of the International Conference on Runtime Verification (RV). St Julians, Malta; 2010. pp. 18–35.

[15] Stallman RM, CygnusSupport. Debugging with GDB – The GNU Source-Level Debugger. Free Software Foundation; 1996.

[16] Park Y, King R, Nathan S, Most W, Andrade H. Evaluation of a high-volume, low-latency market data processing sytem implemented with IBM middleware. Software: Practice & Experience. 2012;42(1):37–56.

[17] Sobeih A, Hack M, Liu Z, Zhang L. Almost peer-to-peer clock synchronization. In: Proceedings of the IEEE International Conference on Parallel and Distributed Processing Systems (IPDPS). Long Beach, CA; 2007. pp. 1–10.

Part III

System architecture

7 Architecture of a stream processing system

7.1 Overview

After discussing how SPAs are developed, visualized, and debugged in Chapters 3 to 5, this chapter will focus primarily on describing the architectural underpinnings of a conceptual SPS application runtime environment.

Shifting the discussion to the middleware that supports stream processing provides an opportunity to discuss numerous aspects that affect how an application runs. These aspects include the support for resource management, distributed computing, security, fault tolerance as well as system management services and system-provided application services for logging and monitoring, built-in visualization, debugging, and state introspection.

This chapter is organized as follows. Section 7.2 presents the conceptual building blocks associated with a stream processing runtime: the computational environment and the entities that use it. The second half of this chapter, Sections 7.3 and 7.4, focuses on the multiple services that make up a SPS middleware and describes how they are integrated to provide a seamless execution environment to SPAs.

7.2 Architectural building blocks

Middleware is the term used to define a software layer that provides services to applications beyond what is commonly made available by an Operating System (OS), including user and resource management, scheduling, I/O services, among others. In general, middleware software provides an *improved* environment for applications to execute. This environment referred to as the *application runtime environment*, further isolates the application from the underlying computational resources.

Therefore, the fundamental role of any middleware is to supply additional infrastructure and services. This additional support is layered on top of a potentially disparate collection of distributed hardware and software. In this way, an application developer can focus more on developing the application rather than on the mechanisms that support it.

In the rest of this section we will describe the application runtime environment. We focus on the make up of the middleware software that supports it (Section 7.2.1), describing both the entities that cooperate to implement and maintain the applications sharing this environment (Section 7.2.2), and the basic services that are common across different SPSs (Section 7.2.3).

7.2.1   Computational environment

As described in Chapter 2, a SPS can run on a single computational host or across multiple, distributed hosts. In either case, it provides a layer, that includes one or more software components running on top of the host's OS, providing *services* that simplify the development and management of SPAs.

For the purposes of this chapter, we focus on stream processing middleware that manages multiple hosts. In other words, the primary middleware role is to provide the illusion of a single, cohesive, and scalable runtime.

Instance

A SPS, when instantiated, provides an application runtime environment. An *instance* of such a runtime environment is a self-contained isolated administration unit. In principle, multiple instances can coexist, sharing the same underlying computational resources.

Entities such as users and applications associated with an instance are generally isolated[1] from entities associated with other instances.

7.2.2   Entities

A SPS manages the interactions and requests of different entities that share the computational environment (Section 7.2.1) and interact with the middleware by making service requests (Section 7.2.3).

While different entities might be defined by different SPSs, we describe four entities of particular interest: user, application, job, and Processing Element (PE).

User

A *user* is an entity that employs a SPS as a client to perform specific analytical tasks. A user may interact with the system by playing different roles including *application developer, application analyst*, or *system administrator*.

A user can only interact with a system by presenting *user credentials* that can be authenticated and which are associated with certain capabilities (e.g., to submit new jobs, to retrieve the state of the system, among others) encoded in an Access Control List (ACL).

A user may correspond to a human user, manually interacting with the system, or to an *automated proxy* using a regular user credential, but acting on the behalf of an automated entity, e.g. another software platform making use of stream processing capabilities.

Interactions with a SPS as well as the ownership of assets managed by the application runtime environment are logged and associated with specific users. In other words, when

---

[1] It is possible to use adapters to create conduits for inter-operation of applications executed on different application runtime instances.

a service request is submitted to a SPS, in most cases the corresponding user credential must also be presented alongside with the request. This service request is then vetted against the capabilities associated with that credential before processing.

As will be seen in Section 7.2.3, certain service requests result in the creation of system-managed assets including jobs and other distributed processes. Once created, these assets are associated with a user entity for runtime management as well as for audit purposes.

Application

An *application* is the logical representation of an analytical task. An application is usually described using a query language or a domain-specific programming language, where the processing corresponding to an analytic task is fully spelled out and implemented.

Depending on the SPS in use, an application can be expressed as a *continuous query*, where continuous processing implied by a query is applied to streaming data. In this case, the query results produced from the processing executed by the individual operators employed by a query plan can themselves be output in the form of streaming data.

Alternatively, an application can be expressed as an explicit *data flow graph*, where operators (vertices) interconnected by streams (edges) are laid out as a graph through which a continuous data flow is transformed and results are produced.

In both of these cases, an application must be transformed from a logical query-based or data flow-based representation outlined above into a physical representation suitable for execution.

Different SPSs employ distinct approaches for carrying out this transformation including *interpretation, compilation,* as well as other techniques that might blend a little bit of each. In cases where standalone binaries are produced and eventually instantiated, the entity that is executed is referred to as a *job*.

Job

A *job* is a runtime entity corresponding to the instantiation of an application or of an application component.[2] Certain SPSs may further subdivide a job into smaller components. This mechanism of creating multiple, collaborating components is typically employed for a number of different reasons.

First, it permits associating a portion of a job (and, therefore, of the application it represents) with an OS process. Second, the existence of multiple of these job components (or PEs as we will refer to them from this point forward) provides a degree of isolation between different parts of an application. Such isolation is beneficial, for example, to insulate a PE from a fatal fault occurring in a different PE as well as to provide a security barrier between PEs, when the code or the data being manipulated is sensitive.

---

[2] In this chapter, we use the term *application* to define a self-contained entity that runs on a single host as a single OS process or as a multi-component entity where all of the *application components* work in concert to implement a data processing task. In the latter case, each application component runs in the context of different OS processes, potentially on different hosts.

Third, it provides a mechanism for distributing the computation across different processors or cores in a host as well as different hosts in a cluster, a fundamental mechanism for *scaling up* an application.

Processing element

A PE is a runtime entity corresponding to a portion of a job, and usually maps to a *process* that can be individually managed by the OS. A PE provides an execution context for a *portion* of the application, consisting, for example, of a portion of a query plan or of a portion of the data flow graph. In either case, this application portion usually comprises a collection of operators that cooperate to implement a fragment of the overall application.

A PE provides an *insulated* execution context for a portion of the application. Thus, a SPS can manage it individually when providing fault tolerance, resource allocation, and scheduling services.

Specifically, when it comes to fault tolerance, the checkpointing[3] mechanism to use as well as the steps required to restart a PE after a failure, are typically defined as policies specific to individual PEs.

Likewise, the use of memory, CPU cycles as well as disk and network bandwidth can be constrained, sometimes with the help of the OS, where specific allotments and resource utilization thresholds are enforced on particular PEs.

Finally, scheduling policies can be used by a SPS to, for example, allocate a specific PE to a particular host depending on the current workload distribution. More generally, PEs provide a degree of flexibility to the scheduling component of a SPS, as the jobs they represent can be adequately placed on the set of distributed resources in such a way that an application can run as efficiently as possible.

Thus the additional level of granularity provided by PEs is an important tool to ensure that hardware as well as other resources from the application runtime environment are efficiently utilized. Furthermore, certain compile-time and profile-driven optimizations can also make use of this mechanism as a means to coalesce or partition the components of an application, as required. Optimizations of this kind typically use as input the type and configuration of the underlying resources (e.g., hosts and networks) to generate a physical layout placement plan for an application.

7.2.3   Services

A SPS provides *services* to the different entities outlined in the previous section. These services are tailored to the needs of the specific entities. They range from supporting specific tasks such as application management, monitoring, enforcement of security and auditing policies, to resource scheduling and fault tolerance.

---

[3] Checkpointing refers to the process of extracting a snapshot of the internal state of an application (or application component) and storing it. In this way, the application (or application component) can be restarted later, following a failure, and leverage the work previously carried out and captured by the checkpointed state. This technique is typically used by long-running application where restarting the computation from start and reconstructing the state is computationally expensive.

The set of services made available by a SPS is usually accessible via APIs, which enable the construction of additional (and, possibly, third party) tools.

We discuss the scope of these fundamental services next, noting that different SPSs choose to implement distinct subsets of these capabilities.

7.3 Architecture overview

From a software architecture standpoint, a SPS can be seen as a collection of components implementing the different services it provides to applications. In this section, each of these services are explained with an overview of the specific support they provide to applications. While the implementation and packaging of these services and their scope varies greatly across commercial systems, the basic functions are generally present in all of them.

7.3.1 Job management

The fundamental role of a SPS is to manage one or more applications during their execution life cycle, from instantiation to termination. At runtime, an *application* is represented by one or more *jobs*, which *may* be deployed as a collection of PEs. Hence, the granularity of job management services and internal management data structures varies depending on the architecture of different platforms, i.e., whether a job is a singleton runtime unit or a collection of PEs. In cases where the platform makes use of PEs, each one can also be stopped, restarted, and potentially relocated from one host to another.

In general, the workflow associated with an application's *life cycle* starts with the launching of a set of jobs and their associated PEs. This may also involve the verification of the user's credentials against user rights to assess whether they can indeed run the application.

These PEs function as an execution context for operators and their input and output ports as well as the streams they consume and generate. Thus, the runtime's job management component must identify and track individual PEs, the jobs they belong to, and associate them with the user that instantiated them.

A SPS *job management component* includes both a *service interface* as well as a set of internal management data structures. Its service interface exposes methods for submitting requests to the component, while its management data structures keep track of the internal state of the component. This internal state is affected by requests made by clients as well as by changes in the application runtime environment.

Services associated with job management include the ability to instantiate, cancel, re-prioritize, monitor, visualize, and debug individual jobs.

As an application runs, the job management component typically makes available monitoring and management services that can be used by application developers and system administrators to keep tabs on the state of the SPS and the collection of applications that might, at any given time, be sharing the runtime environment.

These services are normally available via command-line utilities and web- or GUI-based interfaces.

A particular application's life cycle normally ends with a cancellation request. Upon receiving it, the job management component must ensure that the individual jobs and their PEs are all terminated and the resources associated with them are deallocated and returned to the system. These operations are followed by a corresponding cleanup of the internal management data structures.

Finally, requests for job management services as well as certain runtime changes in the state triggered by its operations may also be logged and tracked, creating an audit trail of runtime activity that can later be used for debugging as well as for tracing a user's activities in the system.

7.3.2   Resource management

A SPS also provides the capability to manage the computational resources used by the applications. The system's resource manager must keep track of hardware resources belonging to individual hosts, such as memory, network bandwidth, local disk space, and CPU cycles as well as distributed resources such as the distributed networking infrastructure and distributed file systems.

The resource manager also keeps track of software resources that are part of the SPS, including its own components as well as external infrastructure these components make use of, such as DBMSs and additional middleware.

In both cases, keeping track of resources involves two main tasks. First, the resource manager must ensure that resources are *available* and *alive*. This task can be implemented using a push model whereby the software and hardware resources report their availability through a heartbeat mechanism. Alternatively, a pull model can be employed where this task is implemented by actively probing the hardware and software resources periodically to report failures when they occur. A combination of both strategies is also possible.

Second, the resource manager collects metrics that can be used to account for how much of each resource is being utilized. This information can subsequently be used for detecting service degradation. This task is not too dissimilar from keeping track of the availability of each resource. Metric collection is also periodic and can use a push method, where metric data is automatically sent out to a metric collector by the entity producing it, a pull method where a metric collector polls the entity producing it, periodically or on demand, or a combination of both [1].

Resource management can be *passive* or *active*. In passive resource management, metrics are used for making scheduling decisions as a result of changes in the set of jobs running in the system (e.g., computing the placement for a newly submitted job) as well as changes in the availability of system resources (e.g., updating the placement of existing jobs when one or more hosts are removed from the system).

In active resource management, metrics are collected continuously and used to enforce specific thresholds and quotas (e.g., limiting the amount of memory or Central

Processing Unit (CPU) cycles a PE can use), sometimes through the use of additional middleware.[4]

SPSs approach resource management in different ways. For instance, some SPSs rely on lower-level resource management capabilities that are provided by generic distributed management platforms as well as virtualization solutions, such as Globus' Globus Resource Allocation Manager (GRAM) [3] and VMWare ESX Server [4]. When it comes to resources local to a host, many resource management tasks are typically carried out by the OS itself [5]. Thus, a SPS might elect to outsource the lower-level resource management tasks, making use of information and resource management services (e.g., job distribution and virtualization) provided by external software.

The resource manager provides the foundation to the two other services we describe next: scheduling and monitoring.

7.3.3   Scheduling

The execution environment provided by a SPS groups a potentially large set of disparate hardware and software resources providing a unified runtime environment. The scheduling service ensures that these resources are harnessed and used efficiently.

The basic scheduling goals are, first, to provide *load balancing* and, second, to ensure that applications are executed efficiently. Load balancing consists of ensuring that the overall workload is well distributed and balanced over the computational resources. Improving load balancing is important when the system is heavily utilized and imbalances can result in a performance toll on the applications sharing the computational cycles. Similarly, ensuring that the applications managed by the SPS are executed in their most performant configuration is important so they can meet or exceed their throughput and latency requirements.

Providing an optimal scheduling mechanism is generally difficult for several reasons. First, those two goals may sometimes be conflicting. For instance, given a single distributed application its performance may be optimized by minimizing communication costs between its components and placing them all on a single host, which results in an unbalanced load on the system resources. Second, as discussed in Section 1.2, SPAs are fundamentally dynamic. Hence, optimal scheduling, even if computable, needs to be periodically recomputed to account for changes in the workload, which is disruptive and often impractical.

Dynamically changing an application as a result of a scheduling decision may require changing its resource constraints. This includes enforcing lower quotas on computing cycles or memory and, possibly, a placement change, forcing the relocation of certain components of the application to less utilized hosts. While feasible, changes of the first type typically require a modification on how the data is processed, potentially reverting

---

[4] As an example, Class-based Kernel Resource Management (CKRM) [2] was used experimentally in an early version of System S to enforce CPU and network traffic consumption thresholds on PEs.

to a cheaper/faster/less precise algorithm, and changes of the second type require stopping a portion of the application and restarting it elsewhere. Naturally, if an application is to endure these changes, it must be built with these scenarios in mind. For instance, each of the operators used to implement an application must be specifically designed to adapt, choosing a suitable operating point, as well as be able to checkpoint its state and retrieve it, if relocated.

Given the nature of the problem, most SPSs provide best effort scheduling.[5] For SPSs employing a query formulation of an application, the scheduling problem is similar to query planning and scheduling in conventional database management systems [7, 8, 9, 10]. However, because a stream processing query employs continuous operators that might be distributed over a multi-host runtime environment [11, 12] these database management techniques are often not applicable, requiring instead adaptive query processing techniques [13].

For SPSs employing a data flow graph formulation of an application, the scheduling task includes two complementary sub-problems. First, given a multi-job application, it is necessary to determine the *assignment* of operators to PEs (subjected to application-specific co-location and ex-location constraints of the sort described in Section 5.3.2) that yields optimal application performance [14]. Second, considering such an assignment and the current conditions of the application runtime environment, it is necessary to determine the *physical placement* of PEs on the set of distributed computational resources that best utilizes them [15].

The scheduling problem can then be tackled in stages with offline (compile-time) assignment and online (runtime) placement. Addressing these two problems separately, while computationally tractable, is likely to result in sub-optimal solutions. In practice these solutions are combined with continuous adjustments while an application runs.

In Streams, computing the set of PEs from a high-level description of an application is a compile-time task, which can be continuously refined by making use of profiling information gathered from earlier runs, employing an inspector-executor approach [16]. Once the resulting application partitions and corresponding PEs are determined and the application is ready to be instantiated, the runtime scheduler is presented with the task of computing the best placement for the set of PEs. Subsequently, by making use of the monitoring service (described next), the placement of individual PEs can be tweaked to accommodate imbalances and workload spikes as the applications run.

7.3.4 Monitoring

The monitoring service captures metrics related to *system health* and *performance*. It actively oversees the SPS's components as well as exposes an interface that can provide a view on the state of the system.

---

[5] An interesting exception is found in StreamIt [6], where for an application subjected to several constraints, including stream rates and operator processing costs, certain scheduling guarantees, such as minimizing buffering requirements, can be provided.

The information collected and made available by the monitoring interface can be directly consumed by end-users, employing visualization tools to inspect the state of the system, as well as by other components of the system that might employ such information as part of their management workflow.

For instance, the monitoring service provides current performance and workload metrics to the scheduler. These measures can be used both for computing job placement at submission time as well as for dynamically modifying this placement at execution time.

Moreover, the monitoring service provides basic health indicators that can be used to ensure that both the SPS as a whole as well as individual applications are operating correctly. These indicators form the basis for *autonomic computing* [17], i.e., the ability to automatically react to failures as well as to events causing performance degradation.

Finally, the monitoring service may also provide an application-level interface to access information about the system state, including inspecting metrics about its own individual PEs. As indicated earlier, this capability allows applications to be adaptive and to react to changes in the application runtime environment (e.g., a sudden spike in a host's load) as well as to changes in the workload (e.g., a sudden spike in the number of tuples per second transported by a particular stream).

7.3.5   Data transport

A basic function of a SPS is to provide a communication substrate, or a transport layer, to applications such that they can ingest streaming data from external sources, stream data internally between their functional components, and finally output results to external consumers.

This substrate must transport tuples from a producer to a consumer, irrespective of whether these endpoints are sharing the same address space, as part of the same OS process, or are located on different hosts. Ideally, communicating within a host or across hosts should be transparent to the application developer. From this standpoint, the interfaces exposed by the data transport service include *configuration* as well as *send* and *receive* interfaces.

Basic inter-process communication mechanisms exist in most modern OSs. Thus, the data transport service is usually constructed on top of communication primitives made available by the host OS or, in some cases, on top of low-level communication middleware (e.g., MPI). In some cases it is also augmented by protocol stacks associated with additional network infrastructure (such as InfiniBand) and available via specific device drivers.

The data transport service is responsible for providing an abstraction that deals with low-level details required to ensure that the communication between producers and consumers is fast and reliable. It must be able to connect parties whose connections are statically described in the query or application flow graph. It must also ensure that these connections are fault-tolerant and continuously perform well in terms of throughput and latency.

The data transport service is also responsible for managing *dynamic* stream connections (in systems that support them), ensuring that, when new streams are advertised and/or new consumers are brought up, the appropriate protocol to establish a connection is followed. In this case, in addition to the services we described earlier, a mechanism to declare and update *stream properties* so they can be advertised, as well as a mechanism to declare and update *subscription predicates* is also provided.

7.3.6   Fault tolerance

SPSs are required to provide fault tolerance as a service because current OSs usually lack this as a native functionality. Fault tolerance services provide to applications as well as to the SPS itself a degree of resilience to failures in the computational infrastructure that includes hardware, software, and other middleware resources. This is achieved by continuously monitoring applications and the SPS's own components and by manually or automatically reacting to service degradation and outright failures.

Infrastructure fault tolerance services rely on the monitoring service (Section 7.3.4), but also require that applications themselves (or at least their *critical* segments[6]) make use of specially designed fault-tolerant mechanisms.

Application-level fault tolerance [18, 19, 20] is in some cases designed by employing programming language constructs [19]. These constructs can be used to designate portions of the application that should be replicated, operators that must be checkpointed, as well as constraints that must be satisfied to ensure continuous operation and high availability. As an example, this could include ensuring that two replicas of the same segment of the application flow graph (or query graph) representing the application are placed on different sets of hosts.

The fault tolerance services provide such infrastructure. For example, the support for replication requires both runtime coordination of the replicas (e.g., ensuring that only one is active) and proper placement of the replicated components (e.g., avoiding that different replicas are taken down by a single host failure). Similarly, checkpointing requires runtime management of the snapshots of the application state as well as an access interface to this state when a component recovers from failure.

Supporting end-to-end application failure recovery is a multi-step process. First, faults in applications and their PEs must be detected. Once a failure is reported, the fault tolerance service must determine whether this failure is fatal and, hence, leads to a non-recoverable failure of the whole application or whether it is non-fatal and normal operation can resume.

If the fault is deemed fatal, this information is relayed to the job manager and manual intervention is typically necessary if the application is to be restarted in the future. If, instead a fault is deemed non-fatal, the process of reacting to this event is initiated. This process may be as simple as restarting a failed segment of the application and bootstrapping it with accumulated state from a checkpointing repository. But it may

---

[6] Frequently, while an application must be fault-tolerant, not all of the components are equally critical in the sense that a failure in them disrupts the application to the point where it cannot recover [18]. In many cases, some components can be simply restarted cold or restarted from a checkpointed state. In other cases, a component has one or more replicas and a failure in one of them is not fatal as another replica becomes active.

also require a more complex orchestration if the segment to be restarted must first be placed elsewhere (e.g., moved to a different host) before being restarted.

7.3.7 Logging and error reporting

The SPS logging and error reporting service can log information regarding requests made by users, system and application components, as well as events produced as a result of state changes that occur in the hardware and software components used by the system. This service is used by several system administration, auditing, and debugging tasks.

The logging and error reporting service includes interfaces for log data entry, log data management, log curation, and data retrieval.

The log data entry interfaces are used to log events as well as to manage the logging process. Usually a log entry identifies when an event happened, its level of severity, where and what entity it affects as well as a description of what the nature of the logged event is. A *log entry* is stored in a (distributed) repository, which can later be queried and mined for offline data analysis. The logging process can be computationally expensive, possibly impacting the performance of the entity using the logging service. Thus, the verbosity and the level of logging can usually be controlled. For example, while *errors* are usually always logged, other information may only be logged upon request. Similarly, specific interfaces can be used to identify the sites where additional logging is active (e.g., runtime components as well as application segments) as well as to increase/decrease the amount of information to be logged when an event occurs.

The log data management interfaces provide subsequent access to log entries. In this way, users and automated tools can analyze the behavior of the application runtime environment and the individual applications for system administration purposes as well as for debugging, visualization and application understanding. The degree of sophistication in the querying mechanism varies across different SPSs. Naturally, distributed SPSs tend to have the most sophisticated interfaces and tooling, a consequence of a greater need for collecting, collating, and coalescing log entries generated by the different components and applications' PEs.

Finally, the log curation interfaces provide mechanisms to manually or autonomically expire older entries, ensuring that the log subsystem uses its storage efficiently [22].

7.3.8 Security and access control

In many cases, a SPS is deployed to support applications that are part of the *critical* data processing backbone of an enterprise. In this context, mechanisms for security must be in place to ensure appropriate access, logging, and auditing. Different SPSs include different levels of support for security services. Simpler centralized platforms rely purely on user authentication and access control mechanisms provided by the underlying OS. However, distributed stream processing platforms include complex services for security, authentication, and sophisticated entitlement-based access controls.

In this case, multiple mechanisms are usually available, including the ability to mutually authenticate[7] entities (e.g., where a runtime component might be attempting to use a service interface provided by another) as well as the ability to vet individual service requests between different components or entities [23].

7.3.9   Debugging

A SPS has a unique view of the interactions between the components of an application as well as of the interactions between them and the SPS services. This information can be used when an application is being debugged.

While a debugger's efficacy is often dictated by the interfaces and mechanisms for application execution control, the flexibility of these mechanisms depends on the existence of proper *hooks* or instrumentation points placed in the SPS and in the OS.

The debugging service in a SPS consists of the interfaces that are exposed to a debugger (and potentially to other tools) to indicate the occurrence of relevant execution-time events. These events, along with an application's execution context can be used to start specific debugging workflows, providing insight into its execution. In other words, the debugging service may include interfaces to activate and deactivate certain instrumentation points as well as the mechanisms to thread the execution through a debugger once an active instrumentation point is activated.

7.3.10   Visualization

A SPS provides an execution environment where multi-component applications can interact dynamically in non-trivial ways. These interactions might not be evident in a casual inspection of the source code describing the application. Additionally, the continuous and long-running nature of applications complicates the understanding, tuning, and evolution of its internal logic as well as the correction of implementation mistakes.

The visualization service in a SPS consists of internal infrastructure and interfaces that can be used to expose the *internals* of the system, as well as the state of an application. This information is used by tools to provide a cohesive and consistent view of the system and applications sharing the runtime environment. These interfaces allow active or passive retrieval of appropriately scoped information (e.g., the overall logical flow graph for a particular application), as desired. While these interfaces usually follow a *pull model* where a request for data must be issued, they can also support a *push model*, allowing the registering of callbacks that are asynchronously invoked when new data satisfying the request is available.

Most SPSs include visualization capabilities, allowing application, queries, performance metrics, and other artifacts to be rendered by a visual interface [24, 25]. Some systems, such as Streams, provide mechanisms for creating out-of-the-box visualizations of the data produced by an application (Section 6.2.4).

---

[7] Mutual authentication refers to the process whereby a client/user authenticates itself to a server and the server does the same with respect to the client/user.

7.4 Interaction with the system architecture

As described in Section 7.3, the architecture of a SPS comprises a large set of services, spanning a diverse set of management tasks that are not directly exposed to end users. Hence, the system requires mechanisms for user interaction including shells and GUI-based environments provided by OSs. Also, it must provide *system call*-like interfaces so applications can invoke system services.

Client-side tools come in different packages, from command-line utilities, to graphical tools and web-based tools. These tools include fully integrated management and development environments with functions for visualization, diagnostics, development, and debugging, as well as other types of more complex interactions.

A secondary way that a user is exposed to the system architecture is through *system calls* employed in an application. These calls are a subset of service interfaces made available to application developers. In general, they enable applications to closely integrate their internal processing with functions provided by the middleware.

Concrete examples of interacting with SPS services will be discussed in Chapter 8, where we will use InfoSphere Streams to solidify the conceptual discussion in this chapter.

7.5 Concluding remarks

This chapter provided an overview of the internal architecture of a conceptual SPS. It also sheds light on the existing services that an application can make direct use of, as well as the services such a platform can provide to application analysts and system administrators.

This information can help product managers, application designers, and developers understand what functions a stream processing platform must provide. Specifically, these decision-makers, when tasked with the evaluation of a SPS for acquisition and later deployment, can use the conceptual architectural building blocks described here as the basis to evaluate different technical alternatives.

References

[1] Meng S, Kashyap SR, Venkatramani C, Liu L. REMO: resource-aware application state monitoring for large-scale distributed systems. In: Proceedings of the International Conference on Distributed Computing Systems (ICDCS). Montreal, Canada; 2009. pp. 248–255.

[2] Nagar S, van Riel R, Franke H, Seetharaman C, Kashyap V, Zheng H. Improving Linux resource control using CKRM. In: Proceedings of the Linux Symposium. Ottawa, Canada; 2004. pp. 511–524.

[3] Globus. WS GRAM Documentation. The Globus Alliance; retrieved in November 2011. http://globus.org/toolkit/docs/3.2/gram/ws/index.html.

[4] Waldspurger CA. Memory resource management in VMware ESX server. In: Proceedings of the USENIX Symposium on Operating System Design and Implementation (OSDI). Boston, MA; 2002. pp. 181–194.

[5] Sullivan DG, Seltzer MI. Isolation with flexibility: a resource management framework for central servers. In: Proceedings of the USENIX Annual Technical Conference. San Diego, CA; 2000. pp. 337–350.

[6] Karczmarek M. Constrained and Phased Scheduling of Synchronous Data Flow Graphs for StreamIt Language [Masters Thesis]. Massachusetts Institute of Technology; 2002.

[7] Bouganim L, Fabret F, Mohan C, Valduriez P. Dynamic query scheduling in data integration systems. In: Proceedings of the IEEE International Conference on Data Engineering (ICDE). San Diego, CA; 2000. pp. 425–434.

[8] Graefe G. Query evaluation techniques for large databases. ACM Computing Surveys. 1993;25(2):73–169.

[9] Hellerstein J, Stonebraker M, editors. Readings in Database Systems. MIT Press; 2005.

[10] Kossman D. The state of the art in distributed query processing. ACM Computing Surveys. 2000 December;32(4):422–469.

[11] Abadi D, Ahmad Y, Balazinska M, Çetintemel U, Cherniack M, Hwang JH, et al. The design of the Borealis stream processing engine. In: Proceedings of the Innovative Data Systems Research Conference (CIDR). Asilomar, CA; 2005. pp. 277–289.

[12] Chandrasekaran S, Cooper O, Deshpande A, Franklin M, Hellerstein J, Krishnamurthy S, et al. TelegraphCQ: continuous dataflow processing for an uncertain world. In: Proceedings of the Innovative Data Systems Research Conference (CIDR). Asilomar, CA; 2003. pp. 269–280.

[13] Deshpande A, Ives ZG, Raman V. Adaptive query processing. Foundations and Trends in Databases. 2007;1(1).

[14] Wolf J, Khandekar R, Hildrum K, Parekh S, Rajan D, Wu KL, et al. COLA: optimizing stream processing applications via graph partitioning. In: Proceedings of the ACM/I-FIP/USENIX International Middleware Conference (Middleware). Urbana, IL; 2009. pp. 308–327.

[15] Wolf J, Bansal N, Hildrum K, Parekh S, Rajan D, Wagle R, et al. SODA: an optimizing scheduler for large-scale stream-based distributed computer systems. In: Proceedings of the ACM/IFIP/USENIX International Middleware Conference (Middleware). Leuven, Belgium; 2008. pp. 306–325.

[16] Saltz J, Crowley K, Mirchandaney R, Berryman H. Run-time scheduling and execution of loops on message passing machines. Journal of Parallel and Distributed Computing (JPDC). 1990;8(4):303–312.

[17] Murch R. Autonomic Computing. IBM Press; 2004.

[18] Jacques-Silva G, Gedik B, Andrade H, Wu KL. Fault-injection based assessment of partial fault tolerance in stream processing applications. In: Proceedings of the ACM International Conference on Distributed Event Based Systems (DEBS). New York, NY; 2011. pp. 231–242.

[19] Jacques-Silva G, Gedik B, Andrade H, Wu KL. Language-level checkpointing support for stream processing applications. In: Proceedings of the IEEE/IFIP International Conference on Dependable Systems and Networks (DSN). Lisbon, Portugal; 2009. pp. 145–154.

[20] Jacques-Silva G, Kalbarczyk Z, Gedik B, Andrade H, Wu KL, Iyer RK. Modeling stream processing applications for dependability evaluation. In: Proceedings of the IEEE/IFIP International Conference on Dependable Systems and Networks (DSN). Hong Kong, China; 2011. pp. 430–441.

[21] Sheu GW, Chang YS, Liang D, Yuan SM. A fault-tolerant object service on CORBA. In: Proceedings of the International Conference on Distributed Computing Systems (ICDCS). Baltimore, MA; 1997. pp. 393–400.

[22] Hildrum K, Douglis F, Wolf JL, Yu PS, Fleischer L, Katta A. Storage optimization for large-scale distributed stream processing systems. In: Proceedings of the IEEE International Conference on Parallel and Distributed Processing Systems (IPDPS). Long Beach, CA; 2007. pp. 1–8.

[23] Security Enhanced Linux; retrieved in November 2011. http://www.nsa.gov/research/selinux/.

[24] De Pauw W, Andrade H. Visualizing large-scale streaming applications. Information Visualization. 2009;8(2):87–106.

[25] Reyes JC. A Graph Editing Framework for the StreamIt Language [Masters Thesis]. Massachusetts Institute of Technology; 2004.

8 InfoSphere Streams architecture

8.1 Overview

In this chapter, we switch the focus from a conceptual description of the SPS architecture to the specifics of one such system, InfoSphere Streams. The concepts, entities, services, and interfaces described in Chapter 7 will now be made concrete by studying the engineering foundations of Streams. We start with a brief recount of Streams' research roots and historical context in Section 8.2. In Section 8.3 we discuss user interaction with Streams' application runtime environment.

We then describe the principal components of Streams in Section 8.4. We focus on how these components interact to form a cohesive runtime environment to support users and applications sharing a Streams *instance*. In the second half of this chapter we focus on services (Section 8.5), delving into the internals of Streams' architectural components, providing a broader discussion of their service APIs and their steady state runtime life cycles. We discuss Streams with a top-down description of its application runtime environment, starting with the overall architecture, followed by the specific services provided by the environment.

Finally, we discuss the facets of the architecture that are devoted to supporting application development and tuning.

8.2 Background and history

InfoSphere Streams can trace its roots to the System S middleware, which was developed between 2003 and 2009 at IBM Research [1, 2, 3, 4, 5, 6, 7, 8, 9, 10, 11, 12, 13]. The architectural foundations and programming language model in Streams are based on counterparts in System S. The design of System S benefited from substantial earlier and, sometimes, concurrent research in stream processing that also led to other systems, including DataCutter [14], Aurora [15], Borealis [16], TelegraphCQ [17], STREAM [18], and StreamIt [19].

Yet, System S and Streams are unique in many ways, providing an architecture for tackling large-scale and distributed applications, coupled with a programming language focused on the analytics and on the flow graph nature of SPAs. These design attributes provide a manageable, scalable, extensible, and industrial-strength environment that can seamlessly integrate with existing enterprise computing environments.

8.3 A user's perspective

As outlined in Section 3.4.2, Streams applications are developed using the SPL language and compiler. An application is created by composing its flow graph with the operators and streams representing its an analytical task. This is accomplished by using a standalone editor or the Streams Studio IDE (see Section 8.5.8). Once the application is implemented, it is built with the SPL compiler. The sc compiler generates a set of partitions (Section 5.3.1), or Processing Elements (PEs), each one hosting the operators that are part of the logical application data flow graph. The resulting set of PEs are connected by physical stream connections implemented, in most cases, by TCP/IP connections. As described in Section 5.3.2, intra-partition (or intra-PE) stream connections are implemented as function calls and, hence, require no external inter-process communication.

The compiler also produces a detailed runtime description of the application as a structured XML-based file referred to as an Application Description Language (ADL) file. The ADL file enumerates the PEs and their parameters: the binary executable, command-line parameters, scheduling restrictions, stream formats, and a description of the internal operator flow graph.

A user can start the compiled application by submitting its corresponding ADL description as a job to the application runtime, using the streamtool [20] command-line application (Section 8.4.4) or directly from Streams Studio (Section 8.4.4). Job submission employs the Streams' job management service (Section 8.5.1) and one of its service interfaces (i.e., submitJob) to accomplish this task.

This interface schedules and deploys the application, starting up its PEs. This operation is supported by the centralized Streams Application Manager (SAM) component, the set of distributed Host Controller (HC) components, as well as by the Scheduler (SCH) component (Section 8.4.2).

The job submission interface uses the scheduling service (Section 8.5.3) to efficiently place individual PEs on the runtime hosts, using the SCH's computeJobPlacement interface. It then initiates the physical deployment of the PEs using the HC's startPE interface. Starting distributed PEs requires coordination between the SAM and the HC instance residing on the host where the PE is to be started. Each PE runs in the context of a Processing Element Container (PEC) (Section 8.4.2). Once the job is submitted, it can be monitored, updated, or canceled, using a variety of SAM client interfaces, available via the streamtool application.

In addition to streamtool, a Streams system administrator can use other tools for managing the system. For example, the logical and physical state of the system may be inspected using a web-based interface, supported by the SWS component (Section 8.4.3). It is also possible to visually inspect a live application flow graph and the overall flow graph representing all applications sharing the runtime system using Streams Studio's *Instance Graph* view [2] (Section 8.5.8).

The Streams Resource Manager (SRM) and SAM components coordinate the monitoring and fault tolerance infrastructure provided by Streams (Section 8.5.5). It includes monitoring the health and load of the runtime hosts, the ability to add and remove hosts to/from the application runtime, and the ability to recover from partial failures in an application.

Finally, all of the user interactions with the application runtime environment as well as the interactions between components, are arbitrated by the Authentication and Authorization Service (AAS) component (Section 8.4.2), which ensures that the inter-component requests are pre-vetted, the parties are authenticated, and that appropriate audit logs are created (Sections 8.5.7 and 8.5.6).

All of the components mentioned in this section work in concert and are functionally grouped as part of a Streams *instance* (Section 8.4.1). Together, they expose all of the application as well as inter-component programming interfaces provided by Streams.

8.4   Components

In this section, we describe the InfoSphere Streams runtime as a set of modular components running on the hosts supporting the Stream Processing System (SPS). The design of the Streams runtime, its physical organization, and its logical organization are geared towards supporting the development of large-scale, scalable, and fault-tolerant applications.

The Streams application runtime environment includes a set of cooperating components running on the system's computational environment, typically a Linux cluster with multiple hosts. It includes centralized runtime components (Table 8.1) as well as distributed ones (Table 8.2).

The hosts used by Streams are separated into *management* and *application hosts*. Management hosts are primarily allocated to support the SPS components, and application hosts are primarily allocated to running an application's PEs. These roles are not exclusive, especially in small clusters.

Table 8.1 InfoSphere Streams centralized components.

Component name	Acronym	Executable name
Streams Application Manager	SAM	streams-sam
Streams Resource Manager	SRM	streams-srm
Scheduler	SCH	streams-sch
Name Service	NS	dnameserver
Authentication and Authorization Service	AAS	streams-aas
Streams Web Server	SWS	streams-sws

Table 8.2 InfoSphere Streams distributed components.

Component name	Acronym	Executable name
Host Controller	HC	streams-hc
Processing Element Container	PEC	streams-pec

Naming, creating, and customizing a *runtime instance* is accomplished using the `streamtool` command-line utility as follows:[1]

```
[user@streams01 ~]$ streamtool mkinstance -i myinstance
CDISC0040I The system is creating the myinstance@user instance.
CDISC0001I The myinstance@user instance was created.
```

We can confirm that this newly created runtime instance (`myinstance@user`) and a previously existing one (`streams@user`) are now available:

```
[user@streams01 ~]$ streamtool lsinstance
myinstance@user
streams@user
```

Note that the `streams@user` runtime instance is the *user* default. When this runtime instance is used the `-i` parameter in `streamtool` command invocations can be omitted. Once created, a runtime instance may be started as follows:

```
[user@streams01 ~]$ streamtool startinstance -i myinstance
CDISC0059I The system is starting the myinstance@user instance.
CDISC0078I The system is starting the runtime services on 1 hosts.
CDISC0056I The system is starting the distributed name service on the streams01.streamsland
host. The distributed name service has 1 partitions and 1 replications.
CDISC0057I The system is setting the NameServiceUrl property of the instance to
DN:streams01.streamsland:45118, which is the URL of the distributed name service that is
running.
CDISC0061I The system is starting in parallel the runtime services of 1 management hosts.
CDISC0003I The myinstance@user instance was started.
```

After this command, all of the runtime components (with the exception of the PEC) are bootstrapped on the appropriate hosts. The PEC is instantiated on demand after an application is submitted for execution. We can confirm that the runtime instance is operational as follows:

```
[user@streams01 ~]$ streamtool lshosts -i myinstance -l
Instance: myinstance@user
Host                    Services                 Tags
streams01.streamsland   hc,aas,nsr,sam,sch,srm,sws   build,execution
```

We can see that the host `streams01.streamsland` is running all of the runtime services and is tagged as both a `build` and `execution` host. A user can modify properties and configurations of the runtime instance (e.g., add or remove hosts) and manage it by employing `streamtool` and its other subcommands. The properties defining an instance can be queried as follows:

```
[user@streams01 ~]$ streamtool getproperty -i myinstance -a
AAS.ConfigFile=/home/user/.streams/instances/myinstance@user/config/security-config.xml
ConfigVersion=3.0
DNA.distributedNameServerPartitionServerCnt=0
DNA.distributedNameServerReplicationCnt=1
DNA.instanceStartedLock=user
DNA.instanceStartTime=2011-12-10T19:22:51-0500
    .
    .
    .
```

---

[1] Here we show the simplest forms of the commands, additional details may be obtained from the Streams official documentation [20]. The commands `streamtool man` and `streamtool help` can provide command-line syntax and options.

Finally, a runtime instance can be shutdown as follows:

```
[user@streams01 Vwap]$ streamtool stopinstance -i myinstance
CDISC0063I The system is stopping the runtime services of the myinstance@user instance.
CDISC0050I The system is stopping the hc service on the streams01.streamsland host.
CDISC0050I The system is stopping the sws service on the streams01.streamsland host.
CDISC0050I The system is stopping the sam service on the streams01.streamsland host.
CDISC0050I The system is stopping the sch service on the streams01.streamsland host.
CDISC0050I The system is stopping the srm service on the streams01.streamsland host.
CDISC0050I The system is stopping the aas service on the streams01.streamsland host.
CDISC0068I The system is stopping in parallel the runtime services of 1 hosts.
CDISC0054I The system is stopping in parallel the distributed name services of the following
1 hosts: streams01.streamsland
CDISC0055I The system is resetting the NameServiceUrl property of the instance because the
distributed name service is not running.
CDISC0004I The myinstance@user instance was stopped.
```

Next we describe the Streams runtime components as well as how they are organized as part of an instance.

8.4.1   Runtime instance

A Streams runtime instance defines an *isolated* environment that can be shared by multiple users. An instance is owned by and associated with the *instance owner*, i.e., the Linux `userid` used to create it.

Runtime instance isolation means that a Streams instance cannot directly interfere with activities and applications associated with other instances. Instance isolation is different from *physical* isolation, as multiple instances can share the same underlying computational resources. Note that this may result in indirect interference (e.g., performance degradation) due to contention for the shared resources.

A Streams runtime instance consists of a predefined configuration establishing the physical and logical layout of the middleware. This configuration indicates the set of hosts to use (from a possibly larger set belonging to a Linux cluster) as well as how the runtime components are to be laid out on them. It also defines how hosts are divided as management or application hosts (a host can also have both roles). Figure 8.1 depicts the set of Streams components (Section 8.4.2) deployed on a dual-purpose management and application host.

A runtime instance can be configured to support both small applications, requiring a single host, as well as large-scale applications, requiring a large cluster. Generally, in a production environment, a single Streams instance is setup to have exclusive use of the available computational resources. Such an instance can distribute its management components over more hosts, as shown in Figure 8.2. Multiple users may share this instance to deploy a set of applications that are inter-dependent and that collaborate in a larger analytical task. (Figure 8.3). In contrast, in a development environment, a larger number of small runtime instances (possibly one per developer or one per group of developers working on a team project), can be created to provide a level of isolation between developers.

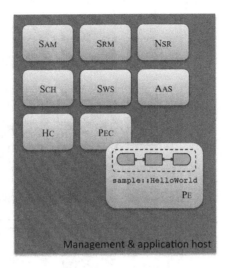

Figure 8.1 Single-host runtime instance with the Streams' runtime components.

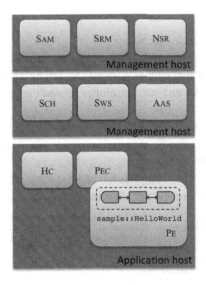

Figure 8.2 Multi-host runtime instance with the Streams' runtime components.

8.4.2   Instance components

A Streams runtime instance employs a set of components whose tasks range from managing the logical and physical state of the system to supporting the execution of applications. Among its management functions, an extensive set of services (Section 8.5) is provided. The components implementing these services are individually described next.

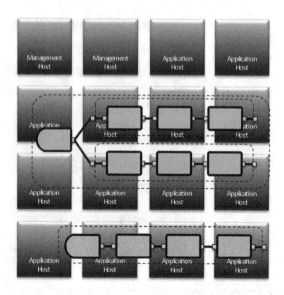

Figure 8.3 Cluster hosting a Streams' runtime instance with distributed applications.

Streams Application Manager

The Streams Application Manager (SAM) is the overseer of the logical system information related to applications currently hosted by a runtime instance. This includes the set of distinct states and state machines related to applications' jobs and each of the job's PEs.

Moreover, SAM also functions as the entry point for all job management services, many of which can be accessed using `streamtool`. As we will described shortly, a user can submit (`streamtool submitjob`) as well as cancel a job (`streamtool canceljob`). The complete list of SAM's interfaces is provided in Section 8.5.1.

In practical terms, a job is started as follows:

```
[user@streams01 Vwap]$ streamtool submitjob -i myinstance output/sample.Vwap.adl
CDISC0079I The system is Submitting 1 applications to the myinstance@user instance.
CDISC0080I Job ID 0 was submitted for the application that is stored at the
    following path:
output/sample.Vwap.adl.
CDISC0020I Submitted job IDs: 0
```

Once jobs have started, their state can be queried as follows:

```
[user@streams01 Vwap]$ streamtool lsjobs -i myinstance
Instance: myinstance@user
    Id  State    Healthy  User    Date                        Name
    0   Running  yes      user    2011-12-13T07:46:30-0500    sample::Vwap
```

In this case we can see that the application `sample::Vwap` was submitted to the system and is now `Running`. Similarly, the state for all PEs in the system can also be queried as follows:

```
[user@streams01 Vwap]$ streamtool lspes -i myinstance
Instance: myinstance@user
  Id State     RC Healthy Host       PID JobId JobName       Operators
   0 Running    - yes     streams01  8344     0 sample::Vwap  TradeQuote,...
```

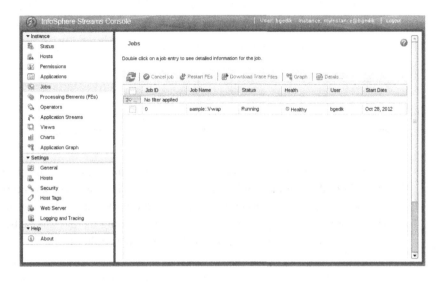

Figure 8.4 System management information displayed by Sws's Streams Console.

In this case we see that there is only a single PE currently Running and it is part of the job corresponding to the sample::Vwap application. SAM also supplies this information to administration and visualization tooling, including Sws and Streams Studio, as seen in Figure 8.4.

Once an application is no longer needed, a job can be terminated as follows:

```
[user@streams01 Vwap]$ streamtool canceljob -i myinstance -j 0
CDISC0021I Job ID 0 of the myinstance@user instance was cancelled.
```

In the course of carrying out these requests as well as other job management services, SAM usually interacts with the AAS and the SCH component. AAS is in charge of vetting operations to ensure that a user or a component (acting on behalf of a particular user) can indeed execute an operation. SCH is required to schedule the execution of an incoming job and to compute the placement for its PEs.

Once SAM obtains the proper AAS authorization, it interacts with one or more HC component instances to initiate (or cancel) the execution of a job's PEs.

All these components are described in greater detail later in this section. SAM and its interfaces are further described in Section 8.5.1.

Streams Resource Manager
The Streams Resource Manager (SRM) is the overseer of the physical state information related to the software and hardware components that make up a Streams instance. This includes monitoring the state of the individual components and hosts that form the computational infrastructure. At startup, SRM functions as the runtime bootstrapper, carrying out the initialization of all other components. In steady-state SRM switches its role to actively collecting and aggregating system-wide metrics on the health of hosts and of each of the individual runtime components.

Figure 8.5 State of the runtime instance hosts in the Streams Console running in a web browser.

The metric and state collection is carried out by direct interaction between SRM and each one of the runtime components. For example, information concerning each of the individual hosts is obtained by interacting with the HC instance managing that host. This information can be retrieved and visualized using the **Streams Console** as seen in Figure 8.5. This console can be manually launched as follows:

```
streamtool launch -i myinstance
```

Finally, the SRM also acts as a repository of performance metrics necessary for scheduling and system administration, making this information available via its service interface. SRM and its interfaces are further described in Section 8.5.2.

Scheduler

The scheduler (SCH) is responsible for computing the placement of PEs to be deployed on the Streams runtime environment. It is also responsible for continuously reassessing the overall placement of all PEs currently in the system and redistributing their workload when appropriate.[2]

As with the SAM and SRM, only one SCH component exists per runtime instance. SCH interacts with the SAM and SRM components primarily during a job submission workflow. Once required by SAM to compute the placement of a job, SCH retrieves the necessary runtime metrics from SRM, and generates a placement plan, which is then returned to SAM for execution. The SCH component and its interfaces are further described in Section 8.5.3.

Host Controller

Each Host Controller (HC) is responsible for *managing* an individual application host, ensuring that it is in good condition to support all of the PEs currently running on it. In a multi-host runtime instance there are multiple HC instances, one per application host.

The HC has two main roles. First, it acts as a SAM agent in carrying out the deployment of PEs on its host. Second, it acts as a SRM agent, continuously retrieving metrics to capture host-level workload metrics. They include overall host metrics (e.g., the overall load as computed by the OS) as well as PE metrics (e.g., the rate of incoming and outgoing tuples transported via stream connections whose arriving or departing endpoints are in HC's host). When an HC instance is tasked with deploying a PE, it starts

---

[2] The redistribution of the workload is not automatic, i.e., it must be requested by an administrator. This is done using the Streams Console (Section 8.4.3), where rebalancing recommendations for PE moves are shown [20].

up an instance of the PE container (described in the next section) that also has job and resource management functions.

The HC component and its interfaces are further described in Sections 8.5.1 and 8.5.2.

Processing Element Container

The Processing Element Container (PEC) envelopes a PE and provides interfaces used for servicing HC's job and resource management requests. The PE itself wraps a subset of the operators and streams from the application's flow graph (i.e., a partition).

Inside a PEC the PE manages the physical input and output ports that correspond to its internal operators' counterparts. Thus, a PE's input and output ports are endpoints for Streams data transport service, delivering incoming tuples to the PE's operators as well as delivering outgoing tuples produced by the PE's operators to external consumers (Section 8.5.4).

A PEC is started by the HC as a result of the job submission workflow. It then immediately installs its service interface, and initiates the PE it is supposed to oversee. The PE executable software is generated by the SPL compiler as a Dynamically Shared Object (DSO). Hence, the PEC loads the corresponding library and invokes its execution entry point, which then starts the operators in the PE.

Once the PEC is running, it can be notified of changes in inter-PE dynamic stream connections, as well as requests for termination as part of a job cancelation workflow. These notifications are carried out through requests to the PEC's service interface. Additionally, the PEC monitors and periodically collects performance metrics from its PE.

The PEC is further described in Sections 8.5.1 and 8.5.2.

Authentication and Authorization Service

The Authentication and Authorization Service (AAS) component provides *authorization* checks as well as inter-component *authentication*, vetting interactions between clients and servers. First, AAS authenticates the credentials provided by client components issuing service requests, before they can be processed. Second, AAS ensures that the authenticated client is authorized for the operation indicated by its request. The AAS component includes a repository of Access Control Lists (ACLs) associated with entities and objects using a Streams runtime instance. Thus, once a request has been properly authenticated, the AAS component verifies whether the *entity* is authorized to perform the requested operation on an *object* (e.g., can the user John cancel job 2011?). The AAS component also logs service requests, creating an audit trail that can be used to review the activities that have taken place in a Streams instance.

The AAS component is further described in Section 8.5.7.

8.4.3 Instance backbone

The components described so far have a *direct* role in managing and supporting the execution of an application. The Streams SPS also employs additional components whose roles do not clearly fall into the service categories outlined in Section 7.2. We refer to these component as the instance *backbone components*. The instance backbone is formed by two components that provide services to the SPS itself, rather than to applications.

Streams Web Server

The Streams Web Server (SWS) component provides web-based access to the Streams runtime system, proxying certain web-based requests to the internal service interfaces exposed by other runtime system components. The SWS component is a gateway, providing administrative and management capabilities from within a web browser, or from a web-enabled external application. It allows administrators and other types of users to securely connect to a Streams runtime instance without directly logging in to the cluster where the instance runs.

Each Streams runtime instance employs one SWS component. In regular operating mode, the SWS authenticates external users or applications (using the AAS service) and converts the web-based request into the corresponding service request for one of the runtime components.

The SWS web-based interfaces are defined using the Web Services Description Language (WSDL) [21], and clients communicate with the server exchanging messages over the Simple Object Access Protocol (SOAP) [22]. Hence, web-enabled applications like, for example, the Streams Console can be used to interact with multiple components (AAS, SAM, and SRM) of the runtime instance. The SWS is further described in Sections 8.5.1 and 8.5.2.

Name Service

The Name Service (NS) component provides a directory for storing references to objects. Each entry in the directory associates a *symbolic name* with an *object*. In the case of the Streams runtime, an object is usually the *location* of a component represented as a Common Object Request Broker Architecture (CORBA) Interoperable Object Reference (IOR). Thus, a component's location can be obtained by submitting a *lookup* request to the NS component using a human-readable symbolic name. The NS component translates this name to an IOR, which encapsulates the host and port where the component's service interface can be found. The issue of locating the NS component itself is solved by placing it in a well-known location and advertising it to the other runtime components as part of the bootstrapping process. Beyond lookup capabilities, the directory maintained by the NS component provides capabilities for registering new entries as well as for updating and removing existing entries (Figure 8.6).

During the bootstrapping process, each Streams component registers itself with the NS. When the runtime instance is operational, a component must first query the NS to locate other components, before submitting service requests. The NS component is also used to manage the data transport endpoints, which physically implement the stream connections (Section 8.5.4). This involves associating symbolic names that can be looked up with the specific attributes to connect a stream producer with a stream consumer (e.g., a host and its TCP socket port number). Furthermore, applications can

```
1    // Name Service inheritance definition
2    NAM: extends ApplicationLiveness
3
4    // lookup an entry
5    string lookupEntry(in name)
6
7    // list the directory entries matching a search pattern
8    EntryList listEntry(in pattern)
9
10   // register an entry, possibly overwriting an existing one
11   void registerEntry(in name, in value, in overwrite)
12
13   // unregister an entry
14   void unregisterEntry(in string name)
```

Figure 8.6 An excerpt of the NS service interfaces.

also make use of the NS component for storing and retrieving information relevant to their own naming needs.[3]

Finally, the NS component plays an important part in Streams' fault tolerance infrastructure (Section 8.5.5). After a failure, individual runtime components may be automatically relocated and restarted by the SRM. The restarted components re-register with the NS to update their directory entry. Other components that request service from a recently failed component may face transient Streams inter-component communication protocol errors and, as a result, will perform new NS lookup operations automatically to locate and resume the communication with the restarted component.

In general, multiple lookups may have to be attempted until the communication between the party requesting the service and the restarted component can be reestablished. As will be seen in Section 8.5.5, Streams inter-component communication protocol employs a fault-tolerant policy with timeouts and retry strategies in the face of transient failures, making such failures generally transparent to the involved parties.

8.4.4   Tooling

Several tools interact directly with the Streams runtime components, acting as clients and making use of their service interfaces. Some of these tools are designed to perform configuration and administration tasks, while others support application development, debugging, and performance tuning. These tools range from simple command-line tools to more sophisticated GUI- and web-based tools. We briefly describe them in the rest of this section.

Configuration and administration

As demonstrated by earlier examples, `streamtool` is a command-line utility that is used for performing instance management tasks. It also provides client-side access to many of the runtime component service interfaces.

---

[3] This support is limited to source and sink operators (e.g., `TCPSource` and `TCPSink`), which can register name-location pairs. These name-location entries can be browsed and searched using the `streamtool`'s `getnsentry` subcommand.

As seen in Section 8.4, it has subcommands to support instance creation and config-uration, for controlling the placement of service components on a subset of hosts, and for runtime instance bootstrapping.

It can also be used to access many of the application runtime environment services, including carrying out job management tasks.

Finally, `streamtool` exposes a great deal of status information about the hosts in the runtime instance. For example, the command below produces a snapthost of the instance state at a particular moment:

```
[user@streams01 Vwap]$ streamtool capturestate -i myinstance --select hosts=all
```

Its XML-formatted output can be interpreted and consumed by user-defined applica-tions and scripts. Here is a typical excerpt:

```
...
<host id="streams01.streamsland" state="running"
    schedulableState="schedulable " isMetricsStale="false">
    <metric name="nProcessors" lastChangeObserved="1323952301" userDefined="false">
        <metricValue xsi:type="streams:longType" value="1"/>
    </metric>
    <metric name="cpuSpeed" lastChangeObserved="1323952301" userDefined="false">
        <metricValue xsi:type="streams:longType" value="4247"/>
    </metric>
    <metric name="loadAverage" lastChangeObserved="1323952301" userDefined="false">
        <metricValue xsi:type="streams:longType" value="15"/>
    </metric>
    <service name="aas" state="running"/>
    <service name="hc" state="running"/>
    <service name="nsr" state="running"/>
    ...
</host>
```

As can be seen, it includes the status of a host's CPU, the status of the Streams components as well as other information that can be used for administration purposes.

Application development
Streams Studio is an IDE that supports application development and testing as well as monitoring, and visualization. Streams Studio (Figure 8.7) is an Eclipse plug-in [23] and provides many functions, ranging from syntax highlighting, syntax-aware comple-tion, and multiple code visualization perspectives, to on-the-fly application flow graph visualization and wizards for the development of operators and toolkits.

While most tasks and workflows supported by Streams Studio involve designing and coding applications in an offline mode (i.e., disconnected from a runtime instance), it can also connect to an online runtime instance using the job and resource management service interfaces and interact with the live system. In the online mode, several addi-tional functions are available, including launching, canceling, debugging, monitoring, and visualizing applications (Figure 8.8).

When connected to the live system, Streams Studio provides support for invoking sdb, the Streams debugger, and for capturing the internal processing logic of an operator as well as the stream traffic associated with its ports. Additionally, Streams Studio includes a live visualization of the overall logical and physical flow graph (*Instance Graph* view), with a variety of different perspectives and details.

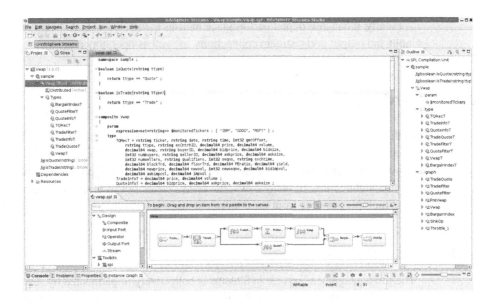

Figure 8.7 Streams Studio depicting an offline view of an application.

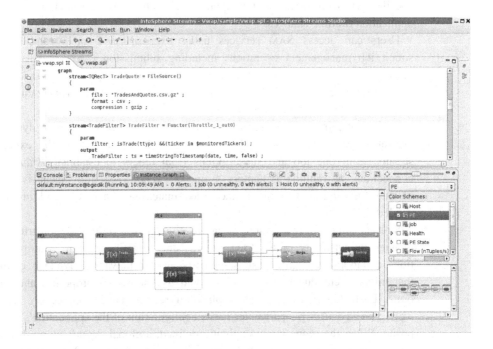

Figure 8.8 Streams Studio depicting a live graph for a running instance.

Finally, it can also depict performance metrics associated with different entities (e.g., input and output ports, stream connections, as well as operators and PEs).

8.5   Services

In this section, we describe the InfoSphere Streams runtime as a set of services roughly organized according to the conceptual architecture presented in Section 7.2. Note, however, implementations of SPSs, including Streams, do not precisely match that conceptual architecture. In many cases, the tasks carried out by some Streams components span multiple services from the conceptual architecture.

8.5.1   Job management

The job management service in Streams is collectively implemented by several components. The SAM component orchestrates most of the job management workflows and provides user-facing client interfaces. Requests sent to SAM (Figure 8.9) often trigger additional interactions with other components, including AAS, SCH, SRM, and individual HC and PEC instances. In these interactions, SAM acts both as a client, requesting specific services from the other components, as well as a server, when being notified of application runtime changes.

Users can access SAM services using `streamtool` or Streams Studio and solicit job management operations including submission (`submitJob`) and cancelation (`cancelJob`) as well as status reports (e.g., `getApplicationState`, `getSystemState`), and state snapshots used to generate the visualization perspectives rendered by Streams Studio.

In Streams, the job management service also needs to support dynamic collaboration and integration across jobs, especially in terms of the exporting of streams from one job and subscriptions from others to establish *dynamic* stream connections at runtime (Chapters 4 and 5). The SAM component is in charge of managing these dynamic connections, matching consumers and producers as needed.

Streams supports the notion of application *scope*, a logical construct used to manage the visibility of exported streams and to articulate how multiple applications collaborate. Scoping also safeguards against unwanted dynamic connections that might occur due to the uncoordinated use of stream properties and subscriptions. Each job belongs to an application scope that is defined by setting the `applicationScope` parameter in the configuration of the application's main composite. When the scope is not explicitly defined, a global *default* scope is associated with a job. These scope restrictions are enforced at runtime, with the exported stream being visible only within the application scope associated with that job, and subscriptions only considering exported streams originating from other jobs in the *same* scope.

```
1   // Streams Application Manager service definition
2   SAM: extends ApplicationLiveness PerfCounterService SrmNotificationService
3
4   // add a stream property associated with an output port
5   void addStreamProperty(in rpcTid, in oportid, in streamProperty, out
        samException)
6
7   // cancel a job
8   void cancelJob(in rpcTid, in session, in jobid, in cancelParams, out jobInfo)
9
10  // retrieve the application state
11  void getApplicationState(in session, in systemTopologyScoping, out
        systemTopology, out samException)
12
13  // retrieve a stream property associated with an output port
14  void getStreamProperty(in oportid, in streamPropertyName, out streamProperty,
        out samException)
15
16  // retrieve the stream subscription expression of an input port
17  void getSubscription(in iportid, out importedStreams, out samException)
18
19  // retrieve the state currently maintained by SAM
20  void getSystemState(in session, in systemStateScoping, out systemState, out
        samException)
21
22  // remove a stream property associated with an output port
23  void removeStreamProperty(in rpcTid, in oportid, in streamPropertyName, out
        samException)
24
25  // restart the execution of a PE
26  void restartPE(in rpcTid, in session, in peid, out peInfo, out samException)
27
28  // adjust the PE runtime state
29  void setPEState(in rpcTid, in peStateInfo, out samException)
30
31  // modify a stream property associated with an output port
32  void setStreamProperty(in rpcTid, in oportid, in streamProperty, out
        samException)
33
34  // set the stream subscription expression of an input port
35  void setSubscription(in rpcTid, in iportid, in importedStreams, out
        samException)
36
37  // halt the execution of single PE
38  void stopPE(in rpcTid, in session, in peid, out peInfo, out samException)
39
40  // submit an application for execution, resulting in a job
41  void submitJob(in rpcTid, in session, in applicationSet, in submitParams, out
        jobInfo, out samException)
```

Figure 8.9 An excerpt of the SAM service interfaces.

SAM *in action*

Jobs and PEs have well-defined life cycles, where different stages are marked as state transitions. A job can be in one of the following states: Failed, Canceling, and Running.

When a client starts up a job via the submitJob interface, the SAM component first creates an AAS session using either the createSessionByPwd or the createSessionByKey interface (Figure 8.19) and, subsequently, performs an authorization check using the checkPermissionSession interface.[4] If the user

---

[4] Other interfaces, for instance, the cancelJob, getApplicationState, getSystemState, restartPE, and stopPE also require a session (specified via the session parameter) so SAM can verify whether the caller holds the right to go through with the operation.

is not authorized to submit jobs, the job submission fails. Otherwise, SAM submits a scheduling request invoking the SCH's computeJobPlacement interface (Figure 8.15). If a scheduling plan cannot be obtained, the job is rejected and put in a Failed state. If a scheduling plan is available, it comes with specific host allocations for each of the PEs to be spawned by the application.

In this case, SAM submits requests to start individual PEs by invoking the HC's startPE interface (Figure 8.15). Finally, the HC instances spawn new PEC instances (one per PE created during the compilation of the application), and then invoke the PEC's startPE interface to start up a PE. Once this is done, the job moves into a Running state. From this point forward, each HC component instance monitors its local PEC instances and reports state changes to SAM (using SAM's setPEState interface) to update its internal management data structures.

A PE can be in one of the following states: Starting, Running, Restarting, Stopping, or Stopped. Once a PE has reached the Running state, it proceeds to create stream connections between operators it hosts and operators situated in other PEs. The physical connections corresponding to each stream are created (or updated) by the data transport service (Section 8.5.4), which also manages and maintains them throughout the PE's life cycle. Finally, the operators inside the PE start to process incoming tuples and to produce results as outbound tuples or sink-style output.

When an application makes use of dynamic composition, with stream subscriptions or exported streams, SAM performs additional work, carrying out a stream matching process to identify consumers for the streams being exported by the new application as well as producers for the subscriptions employed by operators in this new application. In either case, when a new pair of producer/consumer operators is identified, the PEC instance holding the consumer operator is notified using the PEC's routingInfoNotification interface (Figure 8.10), triggering the establishment of a new physical data transport connection.

SAM's dynamic matching process is performed continuously to account for changes in the application runtime environment's flow graph. For example, an operator can update its existing subscription (using SAM's setSubscription

```
1    // Processing Element Container service definition
2    PEC: extends ApplicationLiveness
3
4    // retrieve the metrics associated with the PE managed by this PEC
5    void getMetrics(out peMetricsInfo)
6
7    // retrieve this PEC's internal state
8    void getPECStateSnapshot(out pecStateSnapshot)
9
10   // notify the PEC about a routing change related to operators managed by this
          PEC's processing element
11   void routingInfoNotification(in rpcTid, in routingNotificationInfo, out
          rtException)
12
13   // start this PEC's processing element
14   void startPE(in pecId, in augmentedPEDescriptor, out string pecException)
```

Figure 8.10 An excerpt of the PEC service interfaces.

interface), or modify the properties associated with its exported streams (using `addStreamProperty`). Furthermore, as jobs arrive and depart the application runtime environment, different producer/consumer relationships are potentially established and/or severed.

Once an application is operational, the job management service is also responsible for managing the application's fault-tolerant infrastructure. Fault-tolerant operations are driven by application failures or as a result of user requests to stop, re-start, and move PEs. For example, after we launch a multi-PE application:

```
[user@streams01 Vwap]$ streamtool lspes -i myinstance
Instance: myinstance@user
  Id State     RC Healthy Host       PID JobId JobName       Operators
   1 Running    - yes     streams01 17271     1 sample::Vwap  TradeQuote
   2 Running    - yes     streams01 17279     1 sample::Vwap  TradeFilter
   3 Running    - yes     streams01 17287     1 sample::Vwap  QuoteFilter
   4 Running    - yes     streams01 17288     1 sample::Vwap  PreVwap
   5 Running    - yes     streams01 17289     1 sample::Vwap  Vwap
   6 Running    - yes     streams01 17290     1 sample::Vwap  BargainIndex
   7 Running    - yes     streams01 17291     1 sample::Vwap  SinkOp
```

any of its PEs can be stopped as follows:

```
[user@streams01 Vwap]$ streamtool stoppe -i myinstance 7
CDISC0022I The processing element ID 7 of the myinstance@user instance was stopped.
[user@streams01 Vwap]$ streamtool lspes -i myinstance
Instance: myinstance@user
  Id State     RC Healthy Host       PID JobId JobName       Operators
   1 Running    - yes     streams01 17271     1 sample::Vwap  TradeQuote
   2 Running    - yes     streams01 17279     1 sample::Vwap  TradeFilter
   3 Running    - yes     streams01 17287     1 sample::Vwap  QuoteFilter
   4 Running    - yes     streams01 17288     1 sample::Vwap  PreVwap
   5 Running    - yes     streams01 17289     1 sample::Vwap  Vwap
   6 Running    - yes     streams01 17290     1 sample::Vwap  BargainIndex
   7 Stopped    R no      streams01     0     1 sample::Vwap  SinkOp
```

In the example above, the PE with the `id` 7 was ordered to stop. The `RC` column from the command's output indicates the reason a PE reached that state: `R`, for a user request (as in the example); `A`, for automatic; `F`, for failure; `C`, for crash; `H`, for a host failure; and `V`, for a voluntary transition. *Restartable* PEs, which contain only operators declared *restartable* and whose maximum number of restarts have not yet been reached, can also be restarted as follows:

```
[user@streams01 Vwap]$ streamtool restartpe -i myinstance 7
CDISC0023I The processing element ID 7 of the myinstance@user instance was
    restarted.
```

Any attempt to restart a PE containing non-*restartable* operators will result in errors as shown:

```
[user@streams01 Vwap]$ streamtool restartpe -i myinstance 7
CDISR1001E The Streams Application Manager cannot restart the processing element ID 7. In the
application source code, the operators that the processing element executes are configured so
that they cannot be restarted.
CDISC5144E The processing element ID 7 cannot be restarted. See the previous error message.
```

While manual intervention of the type seen above is possible, it is often triggered automatically by the HC, when it detects unexpected termination during its monitoring cycle. The HC informs SAM by invoking its setPEState interface. SAM verifies whether the PE is restartable and proceeds to restart it by requiring the same HC to re-spawn a new PEC to host it. The PE is informed that it is being *restarted* and it can recover its pre-crash state by retrieving its previously checkpointed state (if any). This mechanism is further described in Section 8.5.5.

Severe failures, specifically where a compute host fails, require a complex set of job management operations. This situation is detected by SRM as part of its own monitoring cycle, which then re-starts the failed HC instance as well as informs SAM of the host failure using SAM's implementation of the hostChangeNotification interface. As a result, SAM proceeds to re-start the PEs hosting restartable portions of an application (if any) using a workflow similar to the one described earlier.

A user can request job cancelation by invoking SAM's cancelJob interface. SAM first verifies whether the user is authorized to perform this operation using AAS and then contacts the relevant HC instances (using the stopPE interface) to request the termination of the affected PEs. The PE then delivers a termination request to each of its operators via an asynchronous call. If for some reason, an operator instance does not terminate properly or any other abnormal situation causes the PEC to not terminate in an orderly fashion within an allotted amount of time, the PEC is forcefully terminated by HC using the Unix SIGKILL signal.

8.5.2   Resource management and monitoring

The resource management and monitoring services in Streams are collectively implemented by the SRM component, the set of distributed HC instances, and the set of PEC instances. The overall orchestration of the resource management and monitoring workflows is carried out by SRM, which also provides the service interfaces depicted in Figure 8.11. The monitoring information gathered by SRM is used for scheduling as well as for runtime and application recovery workflows. This information can be used for autonomic management as well as for human-driven system administration tasks required when performing performance tuning and debugging. In these cases, SRM's service interfaces are used to actively retrieve status and metrics and to display them via the visualization and monitoring tools (Streams Studio and streamtool). An example of the display of live runtime information to users is seen in Figure 8.12.

SRM *in action*

Resource management and monitoring services are implemented in separate, but intertwined threads of execution choreographed by SRM. It first starts the components that make up a runtime instance, using SRM's startService interface. After this the SRM continually collects and maintains the status of the system components. On a request-driven basis, it satisfies scheduling, system administration, and visualization inquiries through its getSCHMetricsData and getMetricsData interfaces.

The Streams monitoring infrastructure has four different functions. The first is to supervise *system health*. This type of supervision is performed by SRM and consists of

```
1    // Streams Resource Manager service definition
2    SRM: extends ApplicationLiveness
3
4    // add a host to a runtime instance
5    void addHost(in rpcTid, in HostConfig, out srmException)
6
7    // retrieve the state of the infrastructure supporting a runtime instance
8    void getInfrastructureState(out InfrastructureState, out srmException)
9
10   // retrieve the state of a runtime instance
11   void getInstanceState(out InstanceState, out srmException)
12
13   // retrieve the metrics associated with a runtime instance, filtered according
         to a particular a scope
14   void getMetricsData(in session, in instanceMetricsScoping, out instanceMetrics,
         out srmException)
15
16   // retrieve the metrics associated with a runtime instance for scheduling
         purposes, filtered according to a particular a scope
17   void getSCHMetricsData(in instanceMetricsScoping, out instanceMetrics, out
         srmException)
18
19   // quiesce a host, temporarily disabling it in a runtime instance
20   void quiesceHost(in rpcTid, in hostname, in forceKill, out srmException)
21
22   // remove a host, permanently detaching it from a runtime instance
23   void removeHost(in rpcTid, in hostname, in forceKill, out srmException)
24
25   // restart a component
26   void restartService(in rpcTid, in DaemonId, out srmException)
27
28   // start a component
29   void startService(in rpcTid, in DaemonId, out srmException)
30
31   // stop a runtime instance, globally stopping all of its runtime components
32   void stopInstance(in rpcTid, in forceKill, out srmException)
33
34   // stop a component
35   void stopService(in rpcTid, in DaemonId, in forceKill, out srmException)
36
37   // update the runtime metrics for a host
38   void updateMetricsData(in HostMetricInformation, out srmException)
39
40   // update the state of a host or runtime component
41   void updateLiveness(in rpcTid, in id, in alive, in isHost, out srmException)
42
43   // wake up a previously quiesced host, making it again available in the runtime
         instance
44   void wakeupHost(in rpcTid, in hostname, out srmException)
```

Figure 8.11 An excerpt of the SRM service interfaces.

probing all of the system components belonging to a runtime instance to ensure that they are active and responsive. The Streams management components implement a common *liveness service* interface, which provides a uniform mechanism for assessing whether a component is operational. This *inherited* interface, ApplicationLiveness, is seen, for example, in Figures 8.9, 8.11, and 8.13.

SRM performs a periodic sweep over the set of instance components and indirectly over the hosts that make up a runtime instance. Upon the detection of failures, it starts corrective, fault-tolerant measures, and coordinates with SAM to initate a corrective reaction in the job management infrastructure (Section 8.5.5). This coordination is enabled by the SAM implementation of SRM's notification service

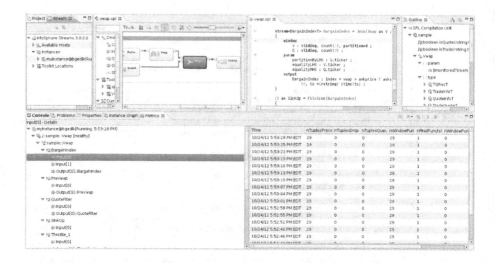

Figure 8.12 Live depiction of metrics for a running instance in Streams Studio's metric view pane.

```
1    // Host Controller service definition
2    HC: extends ApplicationLiveness
3
4    // retrieve the host metrics
5    void getMetrics(out hostMetricInfo)
6
7    // set the state of a processing element
8    void setPEState(in peId, in peRuntimeState, out hcException)
9
10   // start a processing element
11   void startPE(in augmentedPEDescriptor, out hcException)
12
13   // stop a processing element
14   void stopPE(in peId, out string hcException)
```

Figure 8.13 An excerpt of the HC service interfaces.

```
1    // SrmNotificationService - SRM's notification interface
2
3    // indicate a change in a host's state
4    void hostChangeNotification(in rpcTid, in HostState, out exceptionStr)
5
6    // indicate a change in a runtime instance
7    void instanceChangeNotification(in rpcTid, in instanceState, out exceptionStr)
```

Figure 8.14 The SRM notification interfaces.

SrmNotificationService (Figure 8.14) that includes interfaces for indicating changes in the runtime instance (instanceChangeNotification) as well as host state changes (hostChangeNotification).

The SRM can be queried about the overall instance state via its getInstanceState and getInfrastructureState interfaces.

SRM can also act on requests to preventively quiesce (i.e., temporarily disable) and later re-enable a host that must undergo maintenance. These operations use the `quiesceHost` and `wakeupHost` interfaces, respectively. They proceed to stop and restart components running on the host using the `stopService` and `restartService` interfaces, respectively.

The second monitoring function is to collect *system metrics*. The Streams management components implement a common *metric service* interface called `PerfCounterService` to describe and manage the specific metrics they wish to provide, relaying quantitative information about their internal load. The SRM uses this interface to retrieve metrics, and can be queried about the collected load metrics via its `getMetricsData` interface.

The third monitoring function is to supervise *application health*. This monitoring is completely distributed and carried out independently and asynchronously by the set of HC instances. Each HC monitors the PEC instances located on its host, verifying, periodically, that the corresponding PEs remain responsive. Failures are reported to SAM using its `setPEState`, triggering an application recovery workflow when necessary. SAM can be queried about the state of a particular application or about all applications via its `getSystemState` interface.

Finally, the fourth monitoring function consists of collecting *application metrics*. This type of supervision is also performed by the distributed HC instances. In this case, each HC retrieves the PE-related metrics from all the PEC instances (using the PEC's `getMetrics` interface) running on its host and aggregates them locally. The metrics reported by each PEC instance include the operator-level metrics, stored in performance counters, associated with the PE this PEC manages. Finally, the metrics gathered by an HC instance are periodically relayed to SRM using its `updateMetricsData` interface.

8.5.3 Scheduling

Scheduling in Streams is a combination of compile-time and online scheduling. The compile-time stage [11] leads to the creation of a multi-operator data flow graph grouped into application partitions (Section 5.3.2). The online scheduling workflow produces a suitable host placement for a new application and its partitions (PEs), given the current conditions of the system [10]. We now briefly describe these two steps.

As seen in Section 5.3.2, a SPL program can contain several placement constraints. These constraints are of two types: partition-level constraints and host-level constraints. Partition-level constraints are specified in SPL as `partitionColocation`, `partitionIsolation`, or `partitionExlocation`, as well as implied by partition ex-location related to `restartable` and `relocatable` fault-tolerant directives (defined later in this chapter) associated with an operator. Host-level constraints are specified as either `hostColocation`, `hostIsolation`, `hostExlocation` or implied by constraints related to host configurations.

At compile time, the first scheduling objective is to ensure that the constraints expressed in the SPL application source code are consistent. This task is performed by the *constraint checker*. If it does not detect any inconsistency, it computes the partitions that will be subsequently converted into PEs, and populates host-level constraints defined by the partitions. If the constraint checker detects an inconsistency, it outputs a *certificate of infeasibility*, which includes a subset of constraints explaining why a solution cannot be found. An application developer can then correct and refine the appliction's constraints and recompile it.

The constraint checker performs multiple steps to reach a feasibility/infeasibility outcome. First, it processes the `partitionColocation` constraints, forms candidate partitions and detects possible inconsistencies with the `partitionExlocation`, `partitionIsolation`, `hostExlocation` constraints, and the implied exlocation constraints due to the `restartable` and `relocation` settings associated with each operator. Second, the constraint checker processes the `hostColocation` constraints and detects possible inconsistencies with the host-implied settings as well as with the `hostIsolation` and `hostExlocation` constraints. Third, if the constraints are feasible, the constraint checker formulates the problem of assigning partitions to hosts as a *bipartite matching problem* [24]. This problem is solved using an iterative algorithm, and results in the partitions and host-level constraints to be used at runtime. These constraints set the stage for the online scheduling workflow that takes place when the application is finally submitted for execution.

Just as in the offline case, the online scheduling workflow makes use of a constraint checking algorithm for computing the placement of individual PEs. The outcome of this algorithm indicates whether the job can be scheduled given its constraints and the current load imposed by other applications that are sharing the runtime instance. If the constraints are deemed feasible, the placement for this job is computed. In this case, the scheduling algorithm uses load balancing as the optimization goal, assigning PEs to hosts with the most available remaining CPU resources.

The scheduling service in Streams is implemented by the SCH component. The resource availability information employed by the scheduling algorithm is aggregated by the SRM component from the set of distributed HC instances as discussed in the previous section. The SCH interfaces are depicted in Figure 8.15.

```
1   // Scheduler inheritance definition
2   SCH: extends ApplicationLiveness PerfCounterService
3
4   // compute the placement for a new job
5   void computeJobPlacement(in session, in jobSchedule, in
        disableHostLoadProtection, out peToHostAllocations, out schException)
6
7   // compute the placement for a single PE
8   void computePePlacement(in PEID, in jobSchedule, out peToHostAllocations, out
        schException)
9
10  // check wheter a PE can be placed on a given host
11  void computePePlacementHost(in PEID, in hostName, in jobSchedule, out
        peToHostAllocations, out schException)
12
13  // compute a set of PEs that could be restarted with new placements to improve
        load balance
14  void computeRestartPeRecommendations(in sessionId, in recommendationScope, in
        scopeId, in thresholdSpecification, out restartPeRecommendations, out
        schException)
```

Figure 8.15 An excerpt of the SCH service interfaces.

SCH *in action*

The online scheduling algorithm aims at balancing the load evenly across hosts. It does so by incrementally learning the computational cost of each PE, and using this information for both rebalancing the load as well as for improving scheduling decisions when a job is resubmitted for execution in the future. When a `computeJobPlacement` request is made to SCH by SAM, SCH assesses whether it is feasible to place it and produces a scheduling plan to be implemented by SAM. The list of PEs to be placed is traversed in decreasing order of computational cost and each PE is placed in the least loaded, *available*, host on which it can be scheduled. SCH obtains the current host load information by invoking SRM's `getSCHMetricsData` interface. When the computational costs for a PE is not available, the assignment follows a round-robin policy on the *available* hosts. In both cases, a host is deemed available if its load has not yet reached a threshold referred to as the host *protection level*, which ensures that some amount of slack is always available. This extra slack allows for moderate spikes in resource consumption by the PEs sharing the host. This is a *best effort* algorithm that produces good (not necessarily optimal) placements in practice.

Administrators may submit rebalancing requests to the scheduler at any time to improve the effective use of the computational resources. Usually, such requests are driven by observations of unbalanced load, when one host is more heavily loaded than others in the runtime instance. In general, rebalancing might require moving PEs from one host to another. Given other constraints associated with an application, the rebalancing operation itself may be infeasible or impractical.

An administrator can limit the scope of the rebalancing. Options range from *global* (including all PEs in all jobs owned by this administrator) to *local* (restricted to PEs running on a particular host or job). A rebalancing recommendation is usually predicated on the host load threshold setting associated with the runtime instance. Specifically, SCH determines the load on the host where a PE is currently placed versus possible alternative placements. It then provides a recommendation per PE, along with a category indicating the strength of this recommendation.

The following categories exist: *preferred*, the strongest recommendation grade; *high*, for PEs whose hosts are running above the specified load threshold and that *could* be moved to at least one candidate host currently below the load threshold; *directed*, for PEs on hosts that are running above the specified load threshold and whose other candidate hosts are also in the same situation, but movement may result in a marginal improvement; *low*, for PEs whose hosts are running below the specified load threshold and, hence, should not be considered for movement; and *none*, which is associated with PEs that should not be moved, either because their current hosts are running below the specified load threshold and no candidate hosts with lighter loads exist, or which are associated with a constraint stating that they must not be restarted or moved [20].

8.5.4   Data transport

The data transport service in Streams is the intrastructure that supports the stream connections, allowing operators that produce a stream to send tuples to operators that consume them. From an operator developer standpoint, the data transport provides a common and uniform interface, depicted by Figure 8.16. This interface is organized around two main objects, defined as classes implementing the pure abstract methods in the DataReceiver and DataSender classes. The data transport runtime workflows are extremely simple and designed to minimize any execution overhead.

On the receiving end (DataReceiver), the receiving endpoint (consumer) waits for the sending endpoints (producers) to connect (wait) before starting the execution cycle (run), which continues until shutdown is requested (shutdown). Every time a message is received, a callback is invoked to process the incoming message (onMessage). On the sending side, the sending endpoint establishes a connection with the receiving endpoints (connect) and sends messages corresponding to tuples, as they become available (write). When the sending endpoint supports dynamic connections, new receiving endpoints can be added (addConnection) and existing ones can

```
1   class DataReceiver
2   {
3     /// This class defines the callback used to process incoming messages
4     class Callback
5     {
6       ...
7       /// Process a message received by a Data Receiver
8       /// @param msg message data pointer
9       /// @param size message data size
10      virtual void onMessage(void* msg, uint32_t size) = 0;
11
12    };
13
14    ...
15    /// Run the receiver and process messages
16    virtual void run() = 0;
17
18    /// Ask the receiver to shutdown and stop processing messages
19    virtual void shutdown() = 0;
20
21    /// Wait for all the required senders to connect
22    virtual void wait() = 0;
23  };
24
25  class DataSender
26  {
27    /// Add a connection to a receiver
28    /// @param ns_label label of the receiver to lookup and connect to
29    virtual void addConnection(const std::string& ns_label) = 0;
30
31    /// Connect to the input port receivers
32    virtual void connect() = 0;
33
34    /// Remove a connection to a receiver
35    /// @param ns_label label of the receiver to remove connection to
36    virtual void removeConnection(const std::string& ns_label) = 0;
37
38    /// Shutdown
39    virtual void shutdown();
40
41    /// Send data to all the servers
42    /// @param data pointer to the data to send
43    /// @param size size of the data to send
44    virtual void write(const void* data, const uint32_t size) = 0;
45  };
```

Figure 8.16 An excerpt of the data transport service interfaces.

be removed (`removeConnection`) at runtime. This is an attractive approach because it preserves the uniformity of the interfaces, while the actual transport of tuples can occur over different types of communication infrastructure, depending on the physical layout of the application.

For example, operator instances placed in the same PE interact via direct function calls, whereas operator instances placed in different PEs can choose from different communication substrates (e.g., a socket-based substrate, or a proprietary high-performance communication library [7]). The high-level interface provided by the data transport service is homogeneous, regardless of the particular substrate that is in use. As a consequence, when implementing an operator (Section 5.2.3), a developer does not need to make any transport layer-specific provisions. Moreover, the same application can be instantiated in different physical data flow graph configurations (Section 5.3.1) and an operator's physical instance will employ the adequate transport substrate transparently.

The data transport service's internal architecture along with its homogeneous client-side interfaces helps applications to naturally scale up. Starting with the Streams compiler's ability to partition an application, all the way to the distributed processing support provided by the job management and scheduling services, the goal is to allow applications to handle larger workloads without requiring SPL source code modifications. For example, an application originally written to run on a laptop, to handle test workloads in a physical layout where its flow graph is fused in a single executable, can easily be scaled up to handle production-grade workloads by going through a recompilation and a new optimization cycle without the need for rewriting any code.

The offline compilation and re-partitioning steps along with the online scheduling and PE placement steps performed as part of the job submission workflow ensure that additional distributed computational resources can be used at runtime. In other words, when an application is re-partitioned and recompiled, a different transport layer substrate might be used to connect any two operators that were previously fused together and additional low-level code generated to allow the communication to occur over a network link.

Data transport layers

Multiple transport layers exist behind the data transport service interfaces. The simplest one, the *direct call* transport layer, relies on regular C++ function calls between operator instances within the same PE. Every tuple submission request via a `submit` call results in a call to the `process` interface associated with a downstream operator instance.

When a stream connection transcends a single PE, the default TCP transport layer uses regular socket-based interfaces [25]. The data transport endpoints in this case are a TCP client and server pair. This transport layer requires no other configuration beyond the communication endpoint labels automatically produced by the SPL compiler. These communication endpoint labels are managed by the data transport layer, which converts them to IP addresses and ports, which allows the stream producer (a TCP client) to locate the stream consumer (a TCP server).

The choice between the direct call transport layer and the TCP transport layer is made by the compiler and is completely transparent to the developer. The decision, made on a per operator instance basis, is a function of the application partitioning process carried out by the compiler (Section 5.3.1).

For applications with more stringent latency and/or throughput requirements, Streams also supports another distributed data transport layer referred to as the Low Latency Messaging (LLM) layer [26]. This layer has a large number of configuration options to tune buffering, thread usage, batching as well as the actual transport mechanism to be used (e.g., unicast or multicast). This advanced transport utilizes the IBM MQ Websphere Low Latency Messaging [26] middleware and can work with TCP/Ethernet, InfiniBand, or shared memory. The shared memory configuration can be used for establishing stream connections between PEs that are running on the same host.

Different types of physical connections can be used by different portions of an application and the specifics of how they are configured is discussed in Section 5.3.3.

Data transport in action

The communication endpoints employed by the distributed data transport layers have labels generated by the SPL compiler, when it builds the application. Different labels are associated with the server and client endpoints and are stored in the *augmented PE model instance* (Section 8.5.8). This descriptor is passed along to start up a PE when the PEC's startPE interface is invoked by HC.

The PE startup workflow also starts each of its operators. For an operator instance hosting a transport layer server (i.e., a stream consumer), the transport service binds the server to a TCP port and registers the label and the corresponding host and port with NS, using its registerEntry interface. In this way, before the data transport server becomes operational, the registration process ensures that its label (which is also known to its clients) can be located by a data transport client (via a NS lookupEntry operation).

Whenever an operator instance in a PE hosting a data transport client endpoint must establish or re-establish a stream connection, the PE submits a lookup request to NS, which returns information concerning the specific host and port to connect to. On a lookup failure, the operation is retried[5] until the client confirms that either a connection has been established, or a non-recoverable failure has been detected.

The data transport architecture, when relying on TCP connections, is shown in Figure 8.17. In this case, a stream is sent to two consumers, by two TCP clients, associated with the output port in the stream producer, to two TCP servers associated with different input ports belonging to two different PEs.

---

[5] The specific number of retries (on transient failures) and timeouts depends on the type of connection this information is associated with. For a static stream connection, there is an infinite retry policy in place. For a dynamic connection, there is a limited retry policy in place.

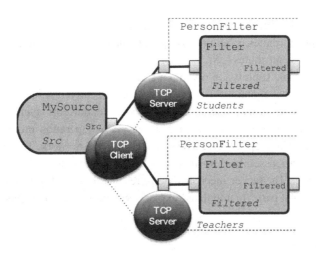

Figure 8.17 TCP data transport layer overlaid on a multi-PE application.

Data encoding

With the direct call transport layer the tuple object is sent directly from one operator instance to the next. In contrast, when a tuple is sent from one PE to another, the data transport service encodes the tuple content for efficient *tuple marshalling* and *unmarshalling*. These processes are required to, respectively, serialize and deserialize the tuple into and out of a stream of bytes, and can be computationally expensive.

In SPL, a tuple is created and manipulated using a specially built C++ class that contains appropriate *getters* and *setters*, and methods for marshalling, unmarshalling, encoding, and decoding the tuple content. The tuple manipulation classes are automatically produced by the SPL compiler as it identifies new tuple types when building an application.

Tuple marshalling converts the tuple content into a contiguous buffer representing a *packet* that can be directly manipulated by the data transport layer. Tuple unmarshalling is performed at the receiver side to convert the received packet back into a tuple. The *encoding* and *decoding* of tuples ensures that numeric attributes can be represented in an architecture-agnostic way inside a packet, by employing the appropriate endianness[6] [27]. The encoding also ensures that variable-size attributes in the tuple can encode their current length, along with the actual attribute value.

The data transport service has several mechanisms to mitigate the computational burden associated with marshalling, encoding, decoding, and unmarshalling steps. For instance, extremely high-performance applications manipulating tuples with fixed-size attributes can opt to use a tuple type with a façade that eliminates the copy overhead between the tuple object and the data packet buffer as well as the copy in the

---

[6] The encoding/decoding of numeric attributes is bypassed, if the cluster hosting the application is homogeneous.

opposite direction. The façade mechanism also allows *in place* manipulation of the tuple attributes, where no additional data copies are incurred, to set or update the value of these attributes [28].

A tuple submission operation starts with encoding and writing the tuple content to a buffer that contains the data packet. Subsequently, it proceeds by invoking the appropriate low-level *send* operation provided by the transport layer. Conversely, a tuple process operation first invokes the low-level *receive* operation provided by the data transport layer. This step retrieves a data packet from the network. The corresponding tuple object is then reassembled by linking it with the data stored in the buffer holding the data packet. The façade mechanism ensures that these underlying data packet buffers are, for the most part,[7] manipulated in place, whenever attributes must be set or read.

Data transport fault tolerance

The data transport service is resilient to transient failures. Specifically, whenever one of the data transport communication endpoints fails, the previously existing connection will eventually be automatically restored, once the failed party comes back online. This is part of the overall application fault tolerance strategy (Section 8.5.5).

With a distributed transport layer, transient failures may occur as a result of a host breakdown. When a failure occurs in an operator instance hosting a TCP server, the counterpart TCP client will periodically attempt to reconnect with the failed server until the stream tuple flow can be restored. This may require the client to perform a Ns lookup operation (using the label associated with this connection) to retrieve a possibly new location for the server (due to the restored server being bound to a different port or restarting on a new host), before attempting to reconnect.

If a failure occurs in an operator instance with a TCP client, an almost identical recovery process takes place. Once the failed operator instance is re-started, the client re-establishes connections with all servers corresponding to the stream consumers. This may also require multiple Ns lookup requests to locate the corresponding TCP servers.

The connection and reconnection processes rely on a data transport handshake mechanism. This mechanism ensures that when a client connects to a server, it can verify that the connected server is the one that has the correct label. If the server identity cannot be asserted, the connection/reconnection workflow is restarted. This is done to avoid cases when server ports (freed due to a crash) are reused by the OS for other services or operators.

Finally, an operator instance relying on the direct call transport layer doesn't generally suffer from a transient disconnection in the same sense as in other other transport layers. Communication failures when using this layer can only occur as a result of a crash in the PE hosting the co-located operator instances. Hence, both the producer and the consumer will be offline simultaneously. As a result, restoring the stream connection requires restarting the PE.

---

[7] Variable-size tuple attributes [20] (e.g., `rstrings`, `ustrings`, `lists`, `sets`, and `maps`) may prevent in-place manipulation when they shrink or grow as a result of updates to the tuple attributes.

8.5.5   Fault tolerance

The fault tolerance service in Streams spans the system software infrastructure, allowing the application runtime to survive host and management component crashes and to autonomically react to these failures. Additionally, application developers can use SPL language constructs to develop elaborate fault tolerance schemes for the application. In the following sections we will summarize these capabilities.

Runtime fault tolerance

The runtime components' fault tolerance mechanism relies on two architectural features. First, each component is built with a common infrastructure that allows it to store its internal state (as backup) onto an IBM DB2 Server. Second, service requests issued by a client to a particular runtime component use a fault-tolerant client/server communication protocol [29] that ensures resilience to transient communication failures.

The backup of a component's internal state (including internal tables and service queues) to a transactional database ensures that this state can be restored after failure. The database-persisted state includes a log of previously executed requests and the corresponding responses, indexed by a unique transaction identifier. This transaction identifier (rpcTid) is shown in the SAM's submitJob interface in Figure 8.9. It ensures that service requests are satisfied *at most once*, i.e., a request is either not satisfied (it does not exist in the component's internal log) or is satisfied only once (a second attempt will simply return the previously computed results, if any).

The fault-tolerant communication protocol considers whether a request is *idempotent* or not, assigns transactions identifiers that are tracked by the server component, and employs a retry logic that ensures that a *non-idempotent* request is always executed only once. Even if a component must be restarted after a failure, an earlier temporarily failed request, retried by the client, will be executed by either returning a previously computed result or by performing the operation from scratch.

These architectural features enable the relocation and restarting of Streams runtime components on a different host with minimal disturbance to the overall middleware and the applications that are running at any given moment.

Application-level fault tolerance

Each SPA has its own fault tolerance requirements [6, 30, 31], with different portions of the application flow graph being able to cope with failures in different ways. The SPL language coupled with certain runtime-provided support allows developers to mix and match different fault tolerance strategies to attain the desired level of application reliability. Three basic mechanisms are provided: operator restart, operator relocation, and operator checkpointing.

```
    ...
    () as SinkOp = FileSink(BargainIndex)
    {
        param
          file : "out";
          format : txt;
```

```
config
  restartable : true;
  relocatable : true;
}
...
```

In the preceding excerpt, two of these features are highlighted. The `FileSink` operator instance `SinkOp` has two configurations indicating that this operator instance is both `restartable` as well as `relocatable`. These features must be mapped to corresponding features associated with this operator's PE, making it restartable and/or relocatable. When the the SPL compiler fuses one or more operators (Section 5.3.2), it creates implicit partition ex-location constraints based on this operator's `restartable` and `relocatable` settings (Section 8.5.3) so other operators (which could be placed in its same partition) remain unaffected by them.

An operator instance can also be configured and implemented to periodically checkpoint its internal state, as shown below:

```
stream<ReliableOperatorT> ReliableStream = ReliableOperator(...) {
  ...
  config
    checkpoint : periodic(10);
    restartable : true;
}
```

The PE runtime performs periodic checkpointing using an operator-specific checkpoint method, at intervals defined by the `periodic` setting in the operator `config` (10 seconds in this example). On a PE restart, the PE runtime infrastructure retrieves the most recent checkpointed state to bootstrap the revived PE.

The system also supports non-periodic, operator-driven checkpoint operations. This behavior is activated by specifying the `operatorDriven` setting in the operator `config` clause. The operator developer is then responsible for providing the necessary functions for operator-driven checkpointing and state bootstrapping on restarts. This is discussed in more detail in Section 8.5.9.

8.5.6   Logging, tracing, and error reporting

The logging, tracing, and error reporting service in Streams allows its runtime components and applications to produce logs and traces documenting their operations. Logs contain information useful to system administrators and application managers. System administrators use logs to inspect the system behavior and diagnose abnormal conditions. Application managers use logs to monitor applications, their PEs, and operator instances [32]. Traces, on the other hand, are targeted at Streams product developers as well as application and toolkit developers. For Streams product personnel, the traces contain internal details of infrastructure events that may help diagnosing problems with the product. For application and toolkit personnel, the traces contain internal details of application and operator events that may help servicing the application and toolkits.

The Streams runtime component logs and traces are usually stored on the local file system as text files, with rotating overwrites of older files to cap the amount of used disk space.[8] For example, by default, individual component logs for a single host instance named `myinstance@user` are stored in `/tmp/streams.` `myinstance@user/logs`. The instance startup log is also stored and can be found in `/tmp/myinstance@user.boot.log`. In general, the location of the logs can be changed and remove log messages can be routed to the Unix *syslog* service.[9]

The logging and tracing levels are configurable for both runtime components as well as for applications and individual PEs.[10] The log and trace levels for infrastructure components can be changed at runtime as well.[11]

The log and trace levels for an application can be configured using the main composite operator `config` definition in an application's source code, using the `logLevel` and `tracing` configs. In addition, the log and trace levels can be specified as part of the job submission parameters.[12] Similarly, the log and trace levels for a PE can be configured using the operator `config` definition in SPL. Note that excessive log verbosity can result in noticeable performance degration at runtime, so the right log level for a production deployment must be chosen carefully.

By default, a runtime instance component's log and trace files are stored in the `/tmp` directory of the host where this component runs. When running a distributed runtime instance, it is possible to login on the different hosts and locally inspect the logs for each component. It is also possible to retrieve all logs and traces for an instance (including application logs) for local inspection, as follows:

```
[user@streams01 Vwap]$ streamtool getlog -i myinstance
CDISC0121I The system is collecting the log files of the following hosts: streams01.streamsland
CDISC0129I The operator that collects the log files for the streams01.streamsland host is finished.
CDISC0122I The log files were collected and archived to the /home/user/StreamsLogs.tgz file.
```

The logs are then collected and delivered compressed and `tarred`.[13] Individual logs may be inspected by untarring the archive:

```
[user@streams01 ~]$ tar xzvf StreamsLogs.tgz
config/
config/instance.properties
config/security-config.xml
streams01.streamsland/
streams01.streamsland/myinstance@user.boot.log
streams01.streamsland/streams.myinstance@user/
streams01.streamsland/streams.myinstance@user/logs/
streams01.streamsland/streams.myinstance@user/logs/hc.log
streams01.streamsland/streams.myinstance@user/logs/hc.out
streams01.streamsland/streams.myinstance@user/logs/hc.stdouterr
```

---

[8] Both the number of rotating log files to be used and their sizes are configurable during instance creation (using `streamtool`'s mkinstance subcommand with the `LogFileMaxSize` and `LogFileMaxFiles` parameters).

[9] These settings can also be configured using the `streamtool` command.

[10] When creating an instance using `streamtool mkinstance`, the parameter `-property LogLevel=<level>` can be used to set the default logging level for this instance, whereas the `-property InfrastructureTraceLevel=<level>` can be used to set the default tracing level. Tracing levels for individual components can also be set by component-specific parameters, e.g., `-property SAM.TraceLevel=<level>`.

[11] Using `streamtool`'s setproperty subcommand.

[12] Using `streamtool`'s submitjob subcommand.

[13] `tar` is a widely available archival tool.

```
streams01.streamsland/streams.myinstance@user/logs/srm.log
streams01.streamsland/streams.myinstance@user/logs/srm.out
streams01.streamsland/streams.myinstance@user/logs/srm.stdouterr
...
streams01.streamsland/streams.myinstance@user/jobs/
streams01.streamsland/streams.myinstance@user/jobs/0/
streams01.streamsland/streams.myinstance@user/jobs/0/pe4.out
streams01.streamsland/streams.myinstance@user/jobs/0/pec4.out
streams01.streamsland/streams.myinstance@user/jobs/0/pec.pe4.log
streams01.streamsland/streams.myinstance@user/jobs/0/pec.pe4.stdouterr
...
```

The runtime instance logs and traces are stored in the `logs` subdirectory as files with appropriate names (e.g., SRM component traces are stored in the `srm.out` file, logs in the `srm.log` file, and standard out/error messages in the `srm.stderrout` file). Application logs and traces are placed in the `jobs/<job identifier>` subdirectory in files corresponding to the different PEs that make up the application's job.

For instance, for job 0 corresponding to application `sample::Vwap`, the logs for PE 4 and its PEC are stored in `streams01.streamsland/streams.myinstance@user/jobs/0/pec.pe4.log`, the traces for PE 4 in the `pe4.out` file, the traces for its PEC in the `pec4.out` file, and finally the *standard out* and *standard error* for the PE and its PEC in the `pec.pe4.stdouterr` file.

The PE trace file contains the different ERROR (default), INFO, DEBUG, and TRACE trace entries. An example from a trace file for a `FileSource` operator's PE is shown below:

```
28 Oct 2012 15:17:30.099 [20190] ERROR #splapptrc,J[0],P[16],TradeQuote,spl_operator
M[TradeQuote.cpp:processOneFile:112] - CDISR5026E: The TradeAndQuotes.csv.gz input
file did not open. The error is: No such file or directory.
```

All log entries include a timestamp and its severity *level*, along with a source code location and a free-format text *message* to describe the nature of the event being logged. These entry attributes provide different elements that can be used for filtering as well as for searching through long log outputs.

Application and system logs as well as traces can also be retrieved and inspected using the Streams Studio's Streams Explorer view. An example of this is shown in Figure 8.18.

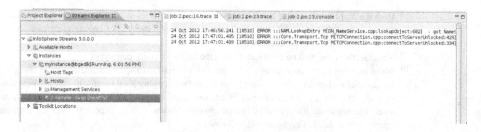

Figure 8.18 Streams Studio's Streams Explorer view. Logs and traces can be retrieved by right-clicking and selecting the appropriate element (e.g., instance, host, or job).

An operator developer can use client interfaces to Streams' logging and error reporting service to insert logging statements throughout the operator code, both in C++ as well as in SPL. An example tracing statement for a C++ operator is shown below:

```
SPLAPPTRC(L_INFO, "The value for x is '" << x << "'", "MODEL_COMPUTATION");
```

The arguments include the operator log level (L_ERROR, L_INFO, L_DEBUG, L_TRACE), the message with multiple expressions that are concatenated into a string using the << operator, and a set of comma-separated *debug aspects*, or tags indicating user-defined categories associated with the log entry. These tags can be used to filter specific log messages when inspecting the logs.

8.5.7   Security and access control

The security and access control service in Streams allows runtime components to authenticate users and individual interactions between components as well as to perform authorization checks. These operations are also logged, creating an audit trail that can later be used to re-trace events that took place in an instance. The security access control service is implemented by AAS. An excerpt of the AAS service interface is shown in Figure 8.19.

Authentication

A user must first be authenticated by the OS to be treated as a valid user of a Streams runtime instance. Users interacting with the system using streamtool can be authenticated using their regular Linux userid and password. A user interacting with the system through SWS must provide a user name and password through the web interface. The connection between a user's browser and SWS is protected by the Secure Sockets Layer (SSL) [34] traffic encryption mechanism. After AAS successfully authenticates a user, a *session* identifier is returned to the client. This identifier must be passed along with all the requests made to other Streams runtime components.

When using streamtool to interact with the Streams runtime, every new user interaction requires re-authentication. To simplify these interactions, AAS employs a public/private key management framework based on Secure SHell (SSH) [35].[14] When keys are available, clients can provide AAS with a signed proof of identity, which obviates the need for subsequently requesting a password from the user.

Authorization

Once a user is authenticated, the AAS *authorization* mechanism is used to verify that this user is allowed to perform the *action* or *actions* indicated by their request. This mechanism uses ACLs with permissions defined as a *function* of *objects* and *subjects*. Subjects are represented by a *principal*, which can be an individual user or a group of

---

[14] streamtool provides an interface to create public/private key pairs using streamtool's genkey subcommand. Private keys are stored in the user home directory and public keys are automatically copied to the .Streams/key directory in the streams instance owner's home directory.

```
1    // Authentication and Authorization Server service definition
2    SEC: extends ApplicationLiveness
3
4    // add an object associated with a session and a parent
5    Obj addObject(in rpcTid, in session, in objid, in parent)
6
7    // retrieve an object associated with a session
8    Obj getObject(in session, in objid)
9
10   // retrieve objects associated with a session
11   ObjectList listObjects(in session)
12
13   // delete an object associated with a session
14   void deleteObject(in rpcTid, in session, in objid)
15
16   // check the permission associated with a session for an object
17   boolean checkPermissionSession(in session, in obj, in mode)
18
19   // check the permission associated with a user for an object
20   boolean checkPermissionUser(in username, in obj, in mode)
21
22   // log an activity associated with a session for a particular operation
23   // on an object
24   void logActivity(in session, in obj, in op, in message)
25
26   // create a session using a password
27   string createSessionByPwd(in rpcTid, in userid, in password)
28
29   // create a session using a key
30   string createSessionByKey(in userid)
31
32   // complete an authentication challenge by submitting a response
33   string completeChallenge(in rpcTid, in response)
34
35   // obtain information related to a session
36   string getInfo(in session)
37
38   // obtain the user name associated with a session
39   string getUserName(in session)
40
41   // obtain the age of a session
42   unsigned long getSessionAge(in session)
43
44   // delete a session
45   void deleteSession(in rpcTid, in session)
46
47   // delete all sessions owned by the user of the session passed as a parameter
48   void deleteAllSession(in rpcTid, in session)
49
50   // stop the AAS component
51   void stop(in session)
```

Figure 8.19 An excerpt of the AAS service interfaces.

users. Objects are arbitrary labels representing resources or operations in the system and can include the runtime instance itself, its configuration and properties, its hosts, a collective set of jobs as well as specific jobs and logs.

Object hierarchy

The authorization mechanism relies on an object tree and the ACL associated with each object. Each object has two ACLs. The *effective* ACL, which controls access to the object, and the *default* ACL, which becomes the effective and default ACL of any child object that gets created. Each component (e.g., SAM and SRM) creating these objects (e.g., PEs, jobs, hosts, and metrics) is in charge of checking the adherence to this system-wide policy by interacting with AAS before executing an operation.

Access Control Lists

During authentication the list of groups a user belongs to is retrieved from the authentication backend (i.e., Pluggable Authentication Module (PAM) [33]- or Lightweight Directory Access Protocol (LDAP)-based [36]). The rights of a user, which are interpreted in context-specific ways by each service to perform an operation can be represented in the ACL as follows:

(a) *Add*: this right allows a user to add an object.
(b) *Delete*: this right allows a user to delete an object.
(c) *Own*: this right allows a user to change the permissions associated with an object.
(d) *Read*: this right allows a user to read (access) the object without changing its state or value.
(e) *Write*: this right allows a user to access the object and change its state or value.
(f) *Search*: this right allows a user to traverse a list of objects.

Instance-related objects have their ACLs initialized when the application runtime environment instance is created. The initialization is a function of whether the instance is created based on a *private* or *shared* security template.

The private security template, which is optimized for a single-user runtime instance, grants only to the instance owner the rights to control and manipulate the instance objects. The shared security template, on the other hand, is pre-configured for use by several users. Multiple system administrators (organized in an *administrator group*) and an optional set of users (organized in a *user group*) have rights to control and manipulate the instance objects. Both groups can be specified by defining specific runtime instance properties at its creation, as follows:

```
[user@streams01 ~]$ streamtool mkinstance --property AdminGroup=streamsadm
--property UsersGroup=streamsusr --template shared -i myinstance
CDISC0040I The system is creating the myinstance@user instance.
CDISC0001I The myinstance@user instance was created.
```

In the example, the instance is explicitly created to be a shared instance (`-template shared`) with the administrator group, defined by the `AdminGroup` property, and the user group, defined by the `UsersGroup` property. These groups correspond to the previously created `streamsadm` and `streamsusr` Linux groups, respectively.

A shared instance must have access to its users' public keys, which are stored in the `<instance-owner>/.streams/instances/<instance-id>/config/keys` directory. Hence, the users' public keys (created using the `streamtool genkey` command) must be transferred to the shared instance's key configuration directory. Once this is done, ACLs associated with the shared runtime instance can be manipulated. For example, an administrator can change the default permissions for the `jobs` runtime instance object to allow any jobs submitted by a specific user (e.g. `alice`) to export (i.e. *write*) data to *any* jobs submitted to this instance:

```
[user@streams01 ~]$ streamtool setacl -i myinstance default:user:alice+w jobs
CDISC0019I The access control list of the myinstance@user instance was
updated.
```

This change can be confirmed as follows:

```
[user@streams01 ~]$ streamtool getacl jobs -i myinstance@user
Instance: myinstance@user

# object: jobs
# parent: instance
# owner: nobody
# persistent: yes
group:user:--sa-o
default:user:alice:-w----
default:user:owner:rwsado
default:group:user:rwsado
```

Finer control is also possible. The following example shows how to change the access permissions to allow any jobs submitted by the user `alice` to export data to job 0 (assuming that job 0 is currently running on the system):

```
[user@streams01 ~]$ streamtool setacl -i myinstance default:user:alice+w job_0
CDISC0019I The access control list of the myinstance@user instance was
updated.
```

This change can also be confirmed using the `getacl streamtool` subcommand:

```
[user@streams01 ~]$ streamtool getacl job_0 -i myinstance@user
Instance: myinstance@user

# object: job_0
# parent: jobs
# owner: user
# persistent: no
user:owner:rwsado
user:alice:-w----
group:user:rwsado
default:user:owner:rwsado
default:user:alice:-w----
default:group:user:rwsado
```

This output states that, now, user `alice`'s jobs can export (or *write*) to job 0, which is known to AAS as the `job_0` object.

Audit logging

The *audit logging* mechanism captures audit log events generated by any of the Streams components as they perform specific operations on behalf of a user or in response to a user request. For instance, the AAS component itself logs events specific to authentication and authorization operations.

```
Thursday, January 12, 2012 1:08:38 PM|SUCCESS||system||log initialized|
Thursday, January 12, 2012 1:09:13 PM|SUCCESS|giop:tcp:[::ffff
    :10.6.24.114]:39688|user|instance|permission granted|--s---
Thursday, January 12, 2012 1:09:13 PM|SUCCESS|giop:tcp:[::ffff
    :10.6.24.114]:39688|user||new session (RSA key)|<<user,<user>>,e19190...
    e766a0,Jan 12, 2012 1:09:13 PM>
Thursday, January 12, 2012 1:09:32 PM|SUCCESS|giop:tcp:[::ffff
    :10.6.24.114]:39716|user|instance|permission granted|--s---
Thursday, January 12, 2012 1:09:32 PM|SUCCESS|giop:tcp:[::ffff
    :10.6.24.114]:39716|user||new session (RSA key)|<<user,<user>>,3c39cb
    ...51a382,Jan 12, 2012 1:09:32 PM>
Thursday, January 12, 2012 1:09:32 PM|SUCCESS|giop:tcp:[::ffff
    :10.6.24.114]:39719|system||session found|3c39cb...51a382
Thursday, January 12, 2012 1:09:32 PM|SUCCESS|giop:tcp:[::ffff
    :10.6.24.114]:39719|user|jobs|check permission|---a--
Thursday, January 12, 2012 1:09:32 PM|SUCCESS|giop:tcp:[::ffff
    :10.6.24.114]:39719|user|jobs|submitJob|new job is being submitted
Thursday, January 12, 2012 1:09:32 PM|SUCCESS|giop:tcp:[::ffff
    :10.6.24.114]:39719|system||session found|3c39cb...51a382
Thursday, January 12, 2012 1:09:32 PM|SUCCESS|giop:tcp:[::ffff
    :10.6.24.114]:39719|system||session found|3c39cb...51a382
Thursday, January 12, 2012 1:09:32 PM|SUCCESS|giop:tcp:[::ffff
    :10.6.24.114]:39719|user|jobs|permission granted|---a--
Thursday, January 12, 2012 1:09:32 PM|SUCCESS|giop:tcp:[::ffff
    :10.6.24.114]:39719|user|job_0|add object|job_0,jobs
Thursday, January 12, 2012 1:09:32 PM|SUCCESS|giop:tcp:[::ffff
    :10.6.24.114]:39719|user|jobs|registerJobWithAAS|job 0 is being submitted
```

Figure 8.20 An excerpt of the AAS audit log.

The parameter `-template audit` must be passed to `streamtool`'s `mkinstance` subcommand when the runtime instance is created to enable the audit log. By default, the audit log file (`audit.log`) is placed in the instance configuration subdirectory (`<instance-owner>/.streams/instances/<instance-id>/config`). An alternate location can be set by making use of the `AuditLogFile` property when creating the runtime instance. The file is readable and writable only by the instance owner, thus preserving the confidentiality of the entries. An example audit file is shown in Figure 8.20.

It shows some of the activities associated with an instance. From the establishment of a new user session (at 1:09:13pm) to the submission of a new job (at 1:09:32 pm) and the registration of a transient new AAS object associated with this job (`registerJobWithAAS`).

AAS *in action*

The AAS component provides a service interface (Figure 8.19) to handle authentication, object management, authorization, and logging requests. The workflows corresponding to these requests are described in more detail next.

The Streams components create and maintain hierarchical objects[15] in AAS by issuing object management requests. These requests take the form of calls to the

---

[15] The hierarchy of AAS objects is relevant because the rights held by an object are the ones passed on by its parent object. For instance, the rights associated with a PE's AAS object are inherited from the object corresponding to this PE's parent job.

`addObject` as well as to the `deleteObject` interfaces. For example, SAM creates an object for each job currently in the system. The owner of that object is the user submitting the job. The user identity is inferred from the session object previously created by a call to the `createSessionByKey`[16] or the `createSessionByPwd` interfaces.

AAS clients such as `streamtool` and SWS, pass user credentials to AAS to establish the user identity and to obtain a *session*. Subsequent requests to another Streams component requiring a session or the manipulation of a component's object are validated via a `checkPermissionSession` as well as a `checkPermissionUser` call to AAS.

AAS objects have a particularly important role when multiple users are sharing an instance. In this case, a session is used to verify whether the appropriate right to perform an operation is held by a user. For example, when a user submits a cancel job request, SAM submits an *authorization request* to AAS to verify whether the user has a *write* permission on the object corresponding to the job. If this is the case, SAM can proceed to perform the operation and terminate the job.

Finally, in addition to authorization operations, individual Streams components can submit a *logging request* (`logActivity` request) requiring AAS to record information about an operation for auditing purposes (Section 8.5.7).

8.5.8   Application development support

The application development support service in Streams provides a set of interfaces for the development, testing, and debugging of applications by interacting with a *development runtime instance*. This service is used directly from within Streams Studio. While there are specific service interfaces for development tasks, other previously defined client interfaces can also be accessed. For instance, an application can be submitted for execution, debugged, and monitored in a *development* runtime instance. In the rest of this section, we provide an overview of the application development and provisioning process and its interactions with the different runtime services.

Application development life cycle
The application development life cycle includes design, development, testing, and debugging of new applications, as well as maintenance of existing ones. In Streams, this process starts with writing or updating its source code and its *toolkit*s (see Chapter 3). The Streams Studio supports all these tasks, along with application visualization, and invocation of the SPL compiler and `sdb` debugger.

---

[16] When a call to `createSessionByKey` is made, a corresponding call to `completeChallenge` must be made to complete the creation of a session. This process includes the following steps. First, the client applies a transformation to the session string returned by the `createSessionByKey` call and uses its private key to encrypt the result. This result becomes the response parameter. Second, the server uses the client public key to decrypt the response and applies a reverse transformation to verify that it matches the session string.

Model-based code generation

The SPL compiler uses *model-based* code generation to create a representation of the application that can be understood by the runtime services. There are two types of objects that result from an application compilation: the PE binaries, in the form of DSOs, and the *application model instance* stored in the ADL file.

ADL is an XML-based language that describes the elements of an application. It is implemented as an XML Schema Definition (XSD) [37] schema (Figure 8.21).

An ADL file includes a description of the application's PEs and its operators, along with their properties, the format of the streaming data it manipulates, as well as submission-time directives to the runtime system. An example ADL file is shown in Figure 8.22.

An ADL file can be submitted to the Streams runtime to instantiate the application as a job. The Streams runtime processes the information in the ADL file to instantiate the application's PEs (Section 8.5.1) and also *augments* the model with runtime-related elements. For instance, logical processing host pools are *converted* to actual hosts

```
1    <xs:complexType name="pe">
2     <xs:sequence>
3      <xs:element name="operInstances"
4       type="operInstancesType"/>
5       ...
6     </xs:sequence>
7     <xs:attribute name="language"
8      use="required" type="languageType"/>
9     <xs:attribute name="restartable"
10     use="required" type="xs:boolean"/>
11    <xs:attribute name="relocatable"
12     use="required" type="xs:boolean"/>
13     ...
14   </xs:complexType>
```

Figure 8.21 Excerpt of the XSD-described application model. This model contains a list of PEs. The segment shown here defines a PE and its set of operator instances (lines 3–4) as well as this PE's properties, including the language in which it was implemented and its configuration regarding whether it is restartable and relocatable (lines 7–12).

```
1    <pe language="C++" relocatable="false"
2    restartable="false" ...>
3     <operInstances>
4      <operInstance class="Stats" name="Stats" ...>
5       <inputPorts> ... </inputPorts>
6       <outputPorts>
7        <outputPort isMutable="true" name="Stats"
8        tupleTypeIndex="0" ...>
9         <intraPeConnection iportIndex="0"
10        operInstanceIndex="1"/>
11       </outputPort>
12      </outputPorts>
13       ...
14     </operInstance>
15     ...
```

Figure 8.22 Excerpt of an XML-described application model instance, describing one of its PEs, with information about this PE's operator instances including its name and the name of its C++ implementation class (both seen in line 4), input and output ports (lines 5–12), including port mutability settings (line 7) and operator connections (lines 9–10).

```
1    <augmented:augmentedPE applicationScope="Default" id="0" jobId="0"
2     jobName="sample::VMStat" language="C++" launchCount="0" logLevel="error"
3     relocatable="false" restartable="false" user="user"...>
4     ...
5      <augmented:inputPorts>
6        <augmented:inputPort encoding="LE" id="1" peId="1" transport="TCP">
7          ...
8          <augmented:staticConnections>
9            <augmented:staticConnection oportId="0" peId="0"/>
10         </augmented:staticConnections>
11         ...
12       </augmented:inputPort>
13     </augmented:inputPorts>
14     ...
15   </augmented:augmentedPE>
```

Figure 8.23 Excerpt from an augmented application model instance showing an augmented PE model with its compile-time properties (line 3) as well as its runtime-enriched properties. For instance, its several ids, its launch count, and the user id of its owner (lines 1–3). Also shown are its ports and connections (lines 5–13) supplemented by their respective runtime ids.

and logical stream connections are *converted* to physical ones using the data transport service (Section 8.5.4).

SAM generates this augmented model instance (Figure 8.23), associating runtime identifiers and placement information to the descriptions extracted from the ADL file, and hands it to the HC components via their startPE interface (propagated to the PEC through its own startPE interface).

Once the corresponding Linux process starts, the PEC loads the DSOs with application code and the PE becomes operational.

The *model-based* description of a PE is also used to support visualization interfaces (Section 8.5.11) as well as other management tasks.

Standalone application

The SPL compiler can be configured to output a *standalone application* which places its flow graph in a single program, executable directly from the OS shell command line. A standalone application does not need the Streams application runtime environment and can simplify testing and debugging during application development, but it has limitations. For example, the program can only run on a single host, and features such as fault tolerance and dynamic connections are not supported.

Application provisioning

Application provisioning is the process of retrieving the executable components of an application from a staging location (e.g., a developer's workspace or an application asset repository) and placing them in a location where the application can be executed.

SPL applications are created using Streams Studio and compiled by the SPL compiler into binaries as well as ADL models. An application may also have additional dependencies (e.g., external libraries used by operators) and configuration data required to bootstrap its computation (e.g., calibration data). We use the term application *assets* to refer to the complete set of elements that form this application, with respect to what is necessary to run it on a Streams runtime instance.

Streams currently does not have a provisioning mechanism to transfer the application assets from a developer's environment to the production environment. Instead, it relies on a shared file system, for instance, Network File System (NFS) [38] or IBM's GPFS [39], as a distribution mechanism for them. Hence, the location where libraries, models, configuration data, as well as of the PE binaries are located must be accessible by all of the hosts that make up a Streams runtime instance.

8.5.9   Processing element

A PE is the smallest schedulable unit in Streams. The PE service interface provides an *execution context* for its operators. This interface is shown in Figure 8.24.

Once a PEC bootstraps a PE, it proceeds through its *startup process* that includes connecting all of its operator instances to their stream consumers and producers and initializing them. If the PE is restarting after a failure or relocation, this also includes any state recovery steps.

Two interfaces are important in this process. First, a PE tracks how many times it has been launched, by calling the getLaunchCount interface (Figure 8.24). A constantly failing PE will eventually no longer restart. Second, a PE retrieves its checkpointed state using the getCheckpointDirectory interface (Figure 8.24).

The operator service interface (Figure 8.25) provides access to both the PE service interface (via its getPE interface) and to its runtime execution context (via its getContext interface), depicted by Figure 8.26.

These interfaces include mechanisms for registering new threads needed by an operator (createThreads), for managing the inbound and outbound tuple flow (process and submit), and for accessing fault tolerance services (getCheckpoint, performSynchronousCheckpoint, and restoreCheckpoint).[17] They also include introspection mechanisms (e.g., getName, getCheckpointing Interval) and support for *system calls* (e.g., getMetrics) via the Operator Context object.

A PE may contain multiple threads of execution. There is one thread per PE input port, which is used for receiving tuples from the data transport and routing them to internal operator input ports. Each source operator within a PE also requires one additional thread, allowing it to process the incoming data flow from an external source and produce the corresponding outbound flow of tuples.

Individual operators may be configured to use threaded input ports, to allow the operator to manage the inbound tuple flow, dropping tuples according to a pre-established policy when congestion arises (Section 9.2.1.2). An operator input port using windowing (Section 4.3.6) with either a time-based trigger policy or a time-based eviction policy also requires an additional thread. Finally, an operator itself can start additional threads to support its processing logic. All these threads are registered and managed by the PE.

---

[17] The checkpointing methods getCheckpoint and restoreCheckpoint are implemented by the operator writer and use the services provided by additional methods for serializing an operator's state from its internal data structures and for deserializing the checkpointed state to repopulate data structures, respectively. These methods are automatically invoked by the runtime when a checkpointing or a PE restart operation takes place.

```
1   namespace SPL {
2     ...
3     class ProcessingElement {
4       /// Get the path to the data directory of the SPL application.
5       virtual const std::string & getDataDirectory() const = 0;
6
7       /// Get the path to the checkpointing directory of the PE
8       virtual const std::string & getCheckpointDirectory() const = 0;
9
10      /// Get the application scope for the job the PE is running in
11      virtual const std::string & getApplicationScope() const = 0;
12
13      /// Get the path to the SPL application directory
14      virtual const std::string & getApplicationDirectory() const = 0;
15
16      /// Get the path to the SPL application output directory
17      virtual const std::string & getOutputDirectory() const = 0;
18
19      /// Get the runtime identifier for this PE
20      virtual uint64_t getPEId() const = 0;
21
22      /// Get the runtime identifier for the job containing this PE
23      virtual uint64_t getJobId() const = 0;
24
25      /// Get the number of times this PE has been relaunched. For the first
26      /// execution, the value will be 0.
27      virtual uint32_t getRelaunchCount() const = 0;
28
29      /// Get the number of inputs for this PE
30      virtual uint32_t getNumberOfInputPorts() const = 0;
31
32      /// Get the number of outputs for this PE
33      virtual uint32_t getNumberOfOutputPorts() const = 0;
34
35      /// Check if the PE is running in standalone mode
36      virtual bool isStandalone() const = 0;
37
38      /// Check if the PE is running under optimized mode
39      virtual bool isOptimized() const = 0;
40
41      /// Check if the PE is running under fused mode
42      virtual bool isInFusedMode() const = 0;
43
44      /// Check if the PE is running under debug mode
45      virtual bool isInDebugMode() const = 0;
46      ...
47      /// Get the PE metrics
48      /// @return PE metrics
49      virtual PEMetrics & getMetrics() = 0;
50      ...
51    };
52  };
```

Figure 8.24 An excerpt of the SPL/Runtime/ProcessingElement.h include file, describing the PE interface (ProcessingElement).

When a PE receives a cancelation request from the PEC it triggers a *shutdown process*. This includes severing data transport connections and informing its operators about the impending termination. The operators proceed through their own shutdown steps, including checkpointing their state (if applicable) and terminating any threads started by them. The time allotted for shutdown is capped and operators are forcefully killed if they exceed this time. Finally, the PE terminates, allowing the PEC instance to shutdown.

```
1   namespace SPL {
2     class Operator {
3       ...
4       /// Get the processing element hosting this operator
5       ProcessingElement & getPE();
6
7       /// Get the context for this operator
8       OperatorContext & getContext() const;
9
10      /// Get the number of inputs for this operator
11      uint32_t getNumberOfInputPorts() const;
12
13      /// Get the number of outputs for this operator
14      uint32_t getNumberOfOutputPorts() const;
15
16      /// Get the operator input port
17      /// @param port input port index
18      OperatorInputPort & getInputPortAt(uint32_t port);
19
20      /// Get the operator output port
21      /// @param port output port index
22      OperatorOutputPort & getOutputPortAt(uint32_t port);
23
24      /// This function is called by the runtime when the operator receives a
25      /// tuple and the port on which the tuple is received is marked as
26      /// mutating in the operator model
27      /// @param tuple tuple value
28      /// @param port index for the input port, from which the tuple is
29      /// received
30      virtual void process(Tuple & tuple, uint32_t port) {}
31      ...
32      /// This function is called by the runtime once and only once for each
33      /// thread created, and runs the thread's main logic
34      /// @param idx thread index (starts from 0)
35      virtual void process(uint32_t idx) {}
36
37      /// Submit a tuple to a specified port
38      /// @param tuple tuple value
39      /// @param port port on which the tuple will be submitted
40      void submit(Tuple & tuple, uint32_t port);
41      ...
42      /// Create threads
43      /// @param numThreads number of threads to create
44      uint32_t createThreads(uint32_t numThreads);
45
46      /// Access a specific thread
47      /// @param index thread index
48      OperatorThread & getThread(uint32_t index);
49      ...
50      /// This function is called by the runtime when the operator state is to
51      /// be checkpointed
52      virtual void getCheckpoint(NetworkByteBuffer & opstate) {}
53
54      /// This function is called by the runtime when the operator state is
55      /// to be restored
56      /// @param opstate serialization buffer containing the operator state
57      virtual void restoreCheckpoint(NetworkByteBuffer & opstate) {}
58
59      /// Force a synchronous checkpoint
60      /// @param opstate serialization buffer containing the operator state
61      void performSynchronousCheckpoint(NetworkByteBuffer const & opstate);
62      ...
63    };
64  };
```

Figure 8.25 An excerpt of the SPL/Runtime/Operator/Operator.h include file, which provides access to the operator execution context (OperatorContext) as well as to the enveloping PE (ProcessingElement).

```
 1  namespace SPL {
 2     class OperatorContext {
 3        /// Get the name of the operator
 4        virtual std::string const & getName() const = 0;
 5
 6        /// Get the name of the C++ class corresponding to the operator
 7        virtual std::string const & getClassName() const = 0;
 8
 9        /// Get the PE hosting this operator
10        virtual ProcessingElement & getPE() = 0;
11
12        /// Get the index of the operator within the PE
13        virtual uint32_t const & getIndex() const = 0;
14
15        /// Get the operator metrics
16        virtual OperatorMetrics & getMetrics() = 0;
17
18        /// Check if this operator is running with profiling turned on
19        virtual bool isProfilingOn() const = 0;
20
21        /// Get the reservoir size used for profiling this operator
22        virtual uint32_t getProfilingBufferSize() const = 0;
23
24        /// Get the sampling rate used for profiling this operator
25        virtual double getProfilingSamplingRate() const = 0;
26
27        /// Check if operator checkpointing is on
28        virtual bool isCheckpointingOn() const = 0;
29
30        /// Get the checkpointing interval for the operator
31        virtual double getCheckpointingInterval() const = 0;
32        ...
33     };
34  };
```

Figure 8.26 An excerpt of the SPL/Runtime/Operator/OperatorContext.h include file, describing the operator execution context (OperatorContext).

PE and operator runtime

The PE also provides access to certain runtime services through *system calls*. These calls can be subdivided into three different types.

The first type is a system callback, or a *system notification*, which is invoked by a system management component and addressed to a particular PE with the internal mechanisms to handle it. For example, when a stream produced by one of the operators in a PE has a new subscriber, SAM issues a routing notification using the routingInfoNotification PEC interface (Figure 8.10). This notification causes the PE runtime to setup a new data transport connection for that new subscriber.

The second type is also a system callback referred to as an *operator system callback*. An operator system callback is made by the PE runtime to trigger different types of activity in an operator. In this way, an operator can be informed about the arrival of an inbound tuple by a call to its process interface, of checkpointing operations by calls to its getCheckpoint and restoreCheckpoint interfaces (Figure 8.25), as well as of many other events such as the readiness of its ports regarding their ability to process tuples, windowing signals (e.g., a window is full), and impending termination.

The third type of system calls is used by an operator to interact with a runtime component. For example, an operator can manipulate one of its input port's subscription by employing input port interfaces (Figure 8.27) to inspect (getSubscriptionExpression) as well as to change (setSubscription Expression) it when needed. A change in subscription is relayed to SAM via its setSubscription interface (Figure 8.9) and will cause the recomputation of inter-operator (and inter-PE) connections which will also result in routing notifications to the affected PEs, relayed to them by the PEC routingInfoNotification interface.

Similarly, an operator can change the properties associated with its exported streams by employing operator output port interfaces (Figure 8.28) to add (addStreamProperty), inspect (getStreamProperty), set (setStream Property), as well as remove (removeStreamProperty) properties. A call to some of these interfaces also triggers a similar recomputation of connections and routing notifications.

Some of these system calls, which are *query interfaces* on the operator and PE execution context objects, also provide support for adaptive processing that might include choosing, on-the-fly, a cheaper analytical algorithm when computational resources become scarce. Specifically, an operator can modify its internal processing logic based on the values of its own performance metrics, which can be retrieved using the

```
1   namespace SPL {
2      ...
3      class OperatorInputPort : public virtual OperatorPort {
4         /// Export type enumeration
5         enum ImportType
6         {
7            ByName,
8            BySubscription,
9            NotImported
10        };
11
12        /// Check if this is an imported port
13        virtual bool isImportedPort() const = 0;
14
15        /// Get the import type for the port
16        virtual ImportType getImportType() const = 0;
17
18        /// Set the subscription expression associated with this input port
19        /// @param subExpr the new subscription expression
20        virtual void setSubscriptionExpression(
21           SubscriptionExpression const & subExpr) = 0;
22
23        /// Return the subscription expression property associated with this
24        /// input port
25        virtual SubscriptionExpressionPtr getSubscriptionExpression() const = 0;
26
27        /// Check if this input port is connected to a PE input port
28        virtual bool isConnectedToAPEInputPort() const = 0;
29
30        /// Get the PE input port index for this operator input port
31        virtual uint32_t getPEInputPortIndex() const = 0;
32     };
33     ...
34  };
```

Figure 8.27 An excerpt of the SPL/Runtime/Operator/Port/OperatorInputPort.h include file, describing an operator's input port interface (OperatorInputPort).

```
1   namespace SPL {
2       ...
3       class OperatorOutputPort : public virtual OperatorPort {
4           /// Export type enumeration
5           enum ExportType
6           {
7               ByName,
8               ByProperties,
9               NotExported
10          };
11
12          /// Check if this is an exported port
13          virtual bool isExportedPort() const = 0;
14
15          /// Get the export type for the port
16          virtual ExportType getExportType() const = 0;
17          ...
18          /// Check if a property of this output port's stream exists
19          /// @param sp the property to be checked for existence
20          virtual bool hasStreamProperty(std::string const & sp) const = 0;
21
22          /// Get a property of this output port's stream
23          /// @param sp the property to be retreived ('out' parameter)
24          /// @param name the name of the stream property to be retrieved
25          virtual void getStreamProperty(StreamProperty & sp,
26              std::string const & name) const = 0;
27
28          /// Set a property of this output port's stream
29          /// @param sp the property to be set
30          virtual void setStreamProperty(StreamProperty const & sp) const = 0;
31
32          /// Add a property to this output port's stream
33          /// @param sp the property to be added
34          virtual void addStreamProperty(StreamProperty const & sp) const = 0;
35
36          /// Remove a property of an operator's output port's stream
37          virtual void removeStreamProperty(std::string const & sp) const = 0;
38
39          /// Check if this output port is connected to a PE output port
40          virtual bool isConnectedToAPEOutputPort() const = 0;
41
42          /// Get the PE output port index for this operator output port
43          virtual uint32_t getPEOutputPortIndex() const = 0;
44      };
45      ...
46  };
```

Figure 8.28 An excerpt of the `SPL/Runtime/Operator/Port/OperatorOutputPort.h` include file, describing an operator's input port interface (`OperatorOutputPort`).

`getMetrics` interface in the `OperatorContext` class, or on the values of its PE performance metrics, which can be retrieved using the `getMetrics` interface in the `ProcessingElement` class.

8.5.10   Debugging

The debugging service in **Streams** provides a set of interfaces that enable tools to introspect and examine the internal state of an application and control its execution. An application can be compiled using the `sc`'s `-g` option to indicate that the operator's debugging hooks are active and that the debugging wrapper should be used.

Once an application compiled in debugging mode is submitted for execution, the control of the execution of its PEs is handed off to the `sdb` debugger (Figure 8.29).

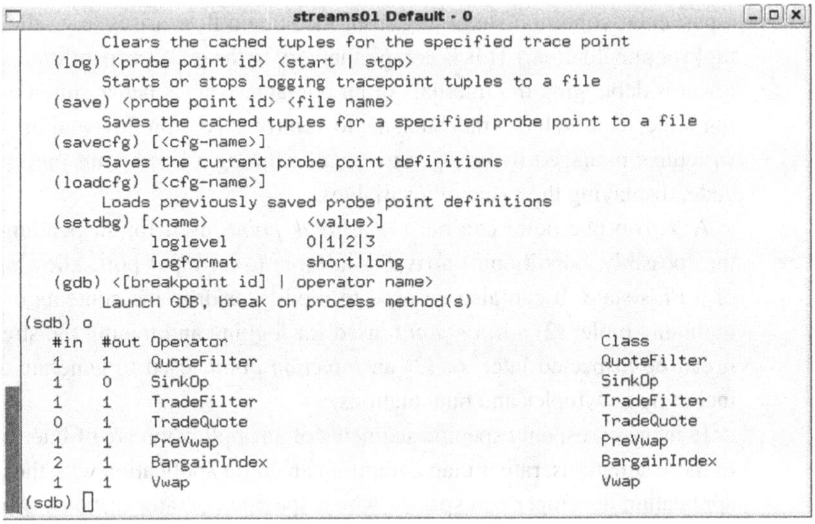

```
streams01 Default - 0
        Clears the cached tuples for the specified trace point
   (log) <probe point id> <start | stop>
        Starts or stops logging tracepoint tuples to a file
   (save) <probe point id> <file name>
        Saves the cached tuples for a specified probe point to a file
   (savecfg) [<cfg-name>]
        Saves the current probe point definitions
   (loadcfg) [<cfg-name>]
        Loads previously saved probe point definitions
   (setdbg) [<name>          <value>]
             loglevel        0|1|2|3
             logformat       short|long
   (gdb) <[breakpoint id] | operator name>
        Launch GDB, break on process method(s)
(sdb) o
   #in  #out Operator                                 Class
   1    1    QuoteFilter                              QuoteFilter
   1    0    SinkOp                                   SinkOp
   1    1    TradeFilter                              TradeFilter
   0    1    TradeQuote                               TradeQuote
   1    1    PreVwap                                  PreVwap
   2    1    BargainIndex                             BargainIndex
   1    1    Vwap                                     Vwap
(sdb) 
```

Figure 8.29 Streams' sdb debugger.

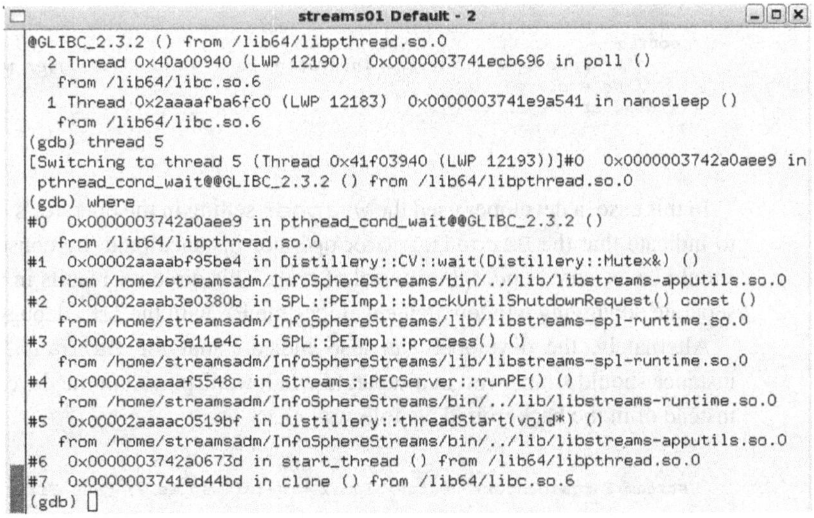

```
streams01 Default - 2
@GLIBC_2.3.2 () from /lib64/libpthread.so.0
  2 Thread 0x40a00940 (LWP 12190)  0x0000003741ecb696 in poll ()
   from /lib64/libc.so.6
  1 Thread 0x2aaaafba6fc0 (LWP 12183)  0x0000003741e9a541 in nanosleep ()
   from /lib64/libc.so.6
(gdb) thread 5
[Switching to thread 5 (Thread 0x41f03940 (LWP 12193))]#0  0x0000003742a0aee9 in
 pthread_cond_wait@@GLIBC_2.3.2 () from /lib64/libpthread.so.0
(gdb) where
#0  0x0000003742a0aee9 in pthread_cond_wait@@GLIBC_2.3.2 ()
   from /lib64/libpthread.so.0
#1  0x00002aaaabf95be4 in Distillery::CV::wait(Distillery::Mutex&) ()
   from /home/streamsadm/InfoSphereStreams/bin/../lib/libstreams-apputils.so.0
#2  0x00002aaab3e0380b in SPL::PEImpl::blockUntilShutdownRequest() const ()
   from /home/streamsadm/InfoSphereStreams/lib/libstreams-spl-runtime.so.0
#3  0x00002aaab3e11e4c in SPL::PEImpl::process() ()
   from /home/streamsadm/InfoSphereStreams/lib/libstreams-spl-runtime.so.0
#4  0x00002aaaaaf5548c in Streams::PECServer::runPE() ()
   from /home/streamsadm/InfoSphereStreams/bin/../lib/libstreams-runtime.so.0
#5  0x00002aaaac0519bf in Distillery::threadStart(void*) ()
   from /home/streamsadm/InfoSphereStreams/bin/../lib/libstreams-apputils.so.0
#6  0x0000003742a0673d in start_thread () from /lib64/libpthread.so.0
#7  0x0000003741ed44bd in clone () from /lib64/libc.so.6
(gdb) 
```

Figure 8.30 The gdb debugger launched during a sdb session.

The user can then inspect the application state as well as control its execution flow using sdb commands, including launching the Free Software Foundation (FSF) debugger, gdb [40], to examine (and act on) the state of C++ - implemented operators. An example is shown in Figure 8.30.

The sdb and gdb debuggers provide different debugging environments (Section 6.3) suitable for different tasks. sdb provides control over the streaming data flow, including the definition and management of *probe points*. A probe point is activated once a particular condition stemming from the stream flow arises (e.g., the arrival of a new tuple or punctuation.) This is complementary to the gdb's capabilities, which are geared towards debugging the internals of an operator. gdb is better suited to debug processing logic, as it includes mechanisms to control execution flow and inspect internal data structures managed by an operator (e.g., defining a breakpoint on a particular line of code, displaying the value of a variable).

A sdb probe point can be: (1) a *break point*, used for suspending a PE based on the (possibly, conditional) arrival of a tuple to an input port, allowing the inspection of a PE's state. It can also be used to modify or drop the contents of an incoming or outbound tuple; (2) a *trace point*, used for logging and tracing the stream data flow so it can be inspected later; or (3) an *injection point*, used to generate additional traffic including new tuples and punctuations.

In many cases, only specific segments of an application are of interest for debugging. In these situations, rather than compiling an entire application with the debug option, an application developer can specify which specific operator instance or instances should be wrapped in a particular way. For example:

```
...
stream<BargainIndexT> BargainIndex = Join(Vwap as V; QuoteFilter as Q)
{
...
config
    // wrapper configuration, indicating that the sdb debugger wrapper should
    // be employed
    wrapper : sdb;
}
...
```

In this case, a developer used the wrapper setting in the operator's config section to indicate that the BargainIndex operator instance (and, by consequence, its PE) should be executed under the control of sdb. This directive results in the creation of a separate debugging window that envelopes the PE with the sdb debugger.

Alternately, the developer can also indicate that the BargainIndex operator instance should simply run in a separate window, displayed in the developer's terminal, instead of in the background, as follows:

```
...
stream<BargainIndexT> BargainIndex = Join(Vwap as V; QuoteFilter as Q)
{
window
    V : partitioned, sliding, count(1);
    Q : sliding, count(0);
param
    partitionByLHS : V.ticker;
    equalityLHS : V.ticker;
    equalityRHS : Q.ticker;
output
    BargainIndex : index = vwap > askprice ? asksize * exp(vwap - askprice) :
        0d, ts = (rstring) ctime(ts);
config
    // wrapper configuration, indicating that the sdb debugger wrapper should
    // be employed
    wrapper : console;
}
...
```

In this case, the PE standard output, along with the messages printed by that operator, are displayed in the console window.

Beyond the `sdb` and `console`, other wrappers can also be specified. For instance, a developer can use the `gdb` wrapper (configured using the `gdb` setting) as well as the `valgrind` [41] wrapper (configured using the `valgrind` setting), for additional debugging support.

The `gdb` and `valgrind` wrapper `configs` instruct the runtime to hand off execution control of a PE to these tools. In contrast, when employing the `sdb` debugger, the operator instance and its surrounding PE code are *augmented* with hooks that allow `sdb` to exert control over the application.

All of these compile-time debug configurations are passed to the runtime as PE wrapper directives in the application model instance (Figure 8.22) stored in the ADL file produced by the compiler. The runtime acts on these directives when launching the application as part of the job submission workflow.

8.5.11   Visualization

The visualization service in Streams provides a set of interfaces that enable tools such as the Streams Studio as well as custom-built ones to retrieve, in real-time, the current state of the system. The available functions can, for example, fetch the entire flow graph representing the applications that currently share the runtime environment as well as performance counters associated with them. But they also include parametrization to manage the scope of the information to be retrieved, ranging from retrieving all of the information about a particular application to creating a snapshot of specific operators.

Streams also includes a visualization hub called *Instance Graph* live view (Section 6.2), which is part of the Streams Studio and provides the ability to examine and monitor the dynamic behavior of an application.

The visualization interfaces are implemented as functions provided by both the SAM as well as the SRM components. These interfaces include SAM's `getApplication State` as well as SRM's `getMetricsData` and `getSAMPEConnectionState`, which are used to retrieve the logical state of the data flow graph, the current values of an application's performance counters, and the state of the stream connections, respectively.

The client side of some of these interfaces is exposed via `streamtool`. For instance, the logical topology can be retrieved as follows:

```
[user@streams01 Vwap]$ streamtool getjobtopology -i myinstance -f topology.xml -j 0
```

As a result a file named `topology.xml` storing an XML representation for job 0's flow graph is produced. Portions of this file are shown in Figure 8.31 (describing the application's PEs), Figure 8.32 (describing the application's operators), and Figure 8.33 (describing the application's corresponding job).

The representation captured by the `sam:systemTopology` object provides a comprehensive snapshot of the runtime environment that can be translated into a visual

```
1    <sam:systemTopology>
2      <sam:scope>
3        <sam:jobs id="0"/>
4      </sam:scope>
5      <sam:applicationSets>
6        <sam:applicationSet name="sample::Vwap.appset" productVersion="2.0.0.1">
7          <description>sample::Vwap application set</description>
8          <application
9          applicationDirectory="/home/user/samples/spl/application/Vwap"
10         applicationScope="Default"
11         checkpointDirectory="/home/user/samples/spl/application/Vwap/data/ckpt"
12         dataDirectory="/home/user/samples/spl/application/Vwap/data"
13         logLevel="error" name="sample::Vwap"
14         outputDirectory="/home/user/samples/spl/application/Vwap/output"
15         version="1.0.0">
16           <libraries/>
17           <hostpools>
18             <hostpool index="0" membershipMode="shared" name="$default"/>
19           </hostpools>
20           <tupleTypes>
21             <tupleType index="0">
22               <attribute name="ticker" type="rstring"/>
23               ...
24             </tupleType>
25           </tupleTypes>
26           <pes>
27             <pe
28             class="Be...zu" index="0" language="C++" logLevel="error"
29             optimized="true" relocatable="false" restartable="false">
30             <envVars/>
31             <executable
32                 digest="tYaLuNKR4w52CCwbbdjy/pAf/Cg="
33                 dynamicLoad="true">
34               <wrapper>none</wrapper>
35               <executableUri>.../output/bin/Be...zu.dpe</executableUri>
36               <arguments/>
37               <dependencies/>
38             </executable>
39             ...
```

Figure 8.31 An excerpt from a system topology instance model object, including the set of applications in the system and the beginning of the description of the sample::Vwap application.

representation of the overall data flow graph (Figure 8.34), and its corresponding live view (Figure 8.8).

The systemTopologyScoping parameter in SAM's getApplicationState interface (Figure 8.9) enables the scoping of the actual information that should be returned to the visualization engine. This control parameter can be used to provide incremental and localized updates, thus reducing the amount of data exchanged between visualization tools and the runtime management components.

8.6   Concluding remarks

This chapter provided an overview of the internal architecture of Streams, including its history and research roots, its modular organization, and a description of its services based on the conceptual architectural description introduced in Chapter 7.

```
1    ...
2    <operInstances>
3     <operInstance
4      class="TradeQuote" definitionIndex="0" index="0" name="TradeQuote"
5      singleThreadedOnInputs="true" singleThreadedOnOutputs="true">
6      <resources>
7       <poolLocation poolIndex="0"/>
8      </resources>
9      <runtimeConstants>
10      <runtimeConstant
11       name="lit$0"
12       value="&quot;TradesAndQuotes.csv.gz&quot;"/>
13      </runtimeConstants>
14      <inputPorts/>
15      <outputPorts>
16       <outputPort index="0" isMutable="true" name="TradeQuote"
17        streamName="TradeQuote" tupleTypeIndex="0">
18        <intraPeConnection iportIndex="0" operInstanceIndex="1"/>
19        <intraPeConnection iportIndex="0" operInstanceIndex="2"/>
20       </outputPort>
21      </outputPorts>
22     </operInstance>
23     ...
24    </pe>
25   </pes>
26   <operDefinitions>
27    <operDefinition index="0" kind="spl.adapter::FileSource".>
28     ...
29   </operDefinitions>
```

Figure 8.32 An excerpt from a system topology instance model object, including the operator instances (`operInstances`) and operator definition (`operDefinitions`) belonging to the `sample::Vwap` application.

```
1    ...
2    <sam:jobSets>
3     <sam:jobSet>
4      <sam:job
5       applicationScope="Default"
6       appsetId="0"
7       id="0"
8       name="sample::Vwap"
9       state="INSTANTIATED"
10      submitTime="1323780390"
11      user="user">
12      <sam:pes>
13       <sam:pe host="192.168.13.135" id="0" index="0" isStateStale="false"
14        jobId="0" launchCount="1" reason="NONE" state="RUNNING">
15        <sam:health healthSummary="UP" isHealthy="true"
16          optionalConnectionsSummary="UP"
17          requiredConnectionsSummary="UP"/>
18        <sam:inputPorts/>
19        <sam:outputPorts/>
20       </sam:pe>
21      </sam:pes>
22      <sam:health isHealthy="true" peHealth="UP"/>
23     </sam:job>
24    </sam:jobSet>
25   </sam:jobSets>
26  </sam:systemTopology>
```

Figure 8.33 An excerpt of the system topology instance model object, including the runtime information (e.g., the job's id, name, state) corresponding to the `sample::Vwap` application.

Figure 8.34 The graphical representation of the runtime system topology, depicting the flow graph for the multi-operator `sample::Vwap` application.

This overview provides the reader with a hands-on understanding of the Streams runtime, its components, tools, and internal organization. Application designers and developers can use this information to ensure that applications utilize the underlying system effectively. Application analysts and system administrators can use this information to understand how an application can be monitored and tuned, as well as how the overall application runtime environment can be managed.

The architectural foundations in this chapter, coupled with the application development methods outlined in earlier chapters, provide a solid basis for the in-depth look into design principles and advanced stream analytics that are the target of the following chapters.

References

[1] Amini L, Andrade H, Bhagwan R, Eskesen F, King R, Selo P, *et al.* SPC: a distributed, scalable platform for data mining. In: Proceedings of the Workshop on Data Mining Standards, Services and Platforms (DM-SSP). Philadelphia, PA; 2006. pp. 27–37.

[2] De Pauw W, Andrade H. Visualizing large-scale streaming applications. Information Visualization. 2009;8(2):87–106.

[3] De Pauw W, Andrade H, Amini L. StreamSight: a visualization tool for large-scale streaming applications. In: Proceedings of the Symposium on Software Visualization (SoftVis). Herrsching am Ammersee, Germany; 2008. pp. 125–134.

[4] Gedik B, Andrade H, Frenkiel A, De Pauw W, Pfeifer M, Allen P, *et al.* Debugging tools and strategies for distributed stream processing applications. Software: Practice & Experience. 2009;39(16):1347–1376.

[5] Jacques-Silva G, Challenger J, Degenaro L, Giles J, Wagle R. Towards autonomic fault recovery in System S. In: Proceedings of the IEEE/ACM International Conference on Autonomic Computing (ICAC). Jacksonville, FL; 2007. p. 31.

[6] Jacques-Silva G, Gedik B, Andrade H, Wu KL. Language-level checkpointing support for stream processing applications. In: Proceedings of the IEEE/IFIP International Conference on Dependable Systems and Networks (DSN). Lisbon, Portugal; 2009. pp. 145–154.

[7] Park Y, King R, Nathan S, Most W, Andrade H. Evaluation of a high-volume, low-latency market data processing sytem implemented with IBM middleware. Software: Practice & Experience. 2012;42(1):37–56.

[8] Turaga D, Andrade H, Gedik B, Venkatramani C, Verscheure O, Harris JD, *et al.* Design principles for developing stream processing applications. Software: Practice & Experience. 2010;40(12):1073–1104.

[9] Wang H, Andrade H, Gedik B, Wu KL. A code generation approach for auto-vectorization in the SPADE compiler. In: Proceedings of the International Workshop on Languages and Compilers for Parallel Computing (LCPC). Newark, DE; 2009. pp. 383–390.

[10] Wolf J, Bansal N, Hildrum K, Parekh S, Rajan D, Wagle R, *et al.* SODA: an optimizing scheduler for large-scale stream-based distributed computer systems. In: Proceedings of the ACM/IFIP/USENIX International Middleware Conference (Middleware). Leuven, Belgium; 2008. pp. 306–325.

[11] Wolf J, Khandekar R, Hildrum K, Parekh S, Rajan D, Wu KL, *et al.* COLA: optimizing stream processing applications via graph partitioning. In: Proceedings of the ACM/IFIP/USENIX International Middleware Conference (Middleware). Urbana, IL; 2009. pp. 308–327.

[12] Wu KL, Yu PS, Gedik B, Hildrum KW, Aggarwal CC, Bouillet E, *et al.* Challenges and experience in prototyping a multi-modal stream analytic and monitoring application on System S. In: Proceedings of the International Conference on Very Large Databases (VLDB). Vienna, Austria; 2007. pp. 1185–1196.

[13] Zhang X, Andrade H, Gedik B, King R, Morar J, Nathan S, *et al.* Implementing a high-volume, low-latency market data processing system on commodity hardware using IBM middleware. In: Proceedings of the Workshop on High Performance Computational Finance (WHPCF). Portland, OR; 2009. article no. 7.

[14] Beynon M, Ferreira R, Kurc T, Sussman A, Saltz J. DataCutter: middleware for filtering very large scientific datasets on archival storage systems. In: Proceedings of the IEEE Symposium on Mass Storage Systems (MSS). College Park, MD; 2000. pp. 119–134.

[15] Balakrishnan H, Balazinska M, Carney D, Çetintemel U, Cherniack M, Convey C, *et al.* Retrospective on Aurora. Very Large Databases Journal (VLDBJ). 2004;13(4): 370–383.

[16] Abadi D, Ahmad Y, Balazinska M, Çetintemel U, Cherniack M, Hwang JH, *et al.* The design of the Borealis stream processing engine. In: Proceedings of the Innovative Data Systems Research Conference (CIDR). Asilomar, CA; 2005. pp. 277–289.

[17] Chandrasekaran S, Cooper O, Deshpande A, Franklin M, Hellerstein J, Hong W, *et al.* TelegraphCQ: continuous dataflow processing. In: Proceedings of the ACM International Conference on Management of Data (SIGMOD). San Diego, CA; 2003. pp. 329–338.

[18] Arasu A, Babcock B, Babu S, Datar M, Ito K, Motwani R, *et al.* STREAM: the Stanford stream data manager. IEEE Data Engineering Bulletin. 2003;26(1):665.

[19] Thies W, Karczmarek M, Amarasinghe S. StreamIt: a language for streaming applications. In: Proceedings of the International Conference on Compiler Construction (CC). Grenoble, France; 2002. pp. 179–196.

[20] IBM InfoSphere Streams Version 3.0 Information Center; retrieved in June 2011. `http://publib.boulder.ibm.com/infocenter/streams/v3r0/index.jsp`.

[21] Christensen E, Curbera F, Meredith G, Weerawarana S. Web Services Description Language (WSDL) 1.1. World Wide Web Consortium (W3C); 2001. `http://www.w3.org/TR/wsdl`.

[22] Gudgin M, Hadley M, Mendelsohn N, Moreau JJ, Nielsen HF, Karmarkar A, *et al.* SOAP Version 1.2 Part 1: Messaging Framework (Second Edition). World Wide Web Consortium (W3C); 2007. `http://www.w3.org/TR/soap12-part1/`.

[23] Clayberg E, Rubel D. Eclipse Plug-ins. 3rd edn. Addison Wesley; 2008.

[24] Cormen TH, Leiserson CE, Rivest RL. Introduction to Algorithms. MIT Press and McGraw Hill; 1990.

[25] Stevens WR. UNIX Network Programming: Networking APIs, Sockets and XTI (Volume 1). 2nd edn. Prentice Hall; 1998.

[26] IBM WebSphere MQ Low Latency Messaging; retrieved in September 2010. http://www-01.ibm.com/software/integration/wmq/llm/.

[27] Tanenbaum A, Wetherall D. Computer Networks. 5th edn. Prentice Hall; 2011.

[28] On-Demand Marshalling and De-Marshalling of Network Messages; 2008. Patent application filed as IBM Docket YOR920090029US1 in the United States.

[29] Wagle R, Andrade H, Hildrum K, Venkatramani C, Spicer M. Distributed middleware reliability and fault tolerance support in System S. In: Proceedings of the ACM International Conference on Distributed Event Based Systems (DEBS). New York, NY; 2011. pp. 335–346.

[30] Jacques-Silva G, Kalbarczyk Z, Gedik B, Andrade H, Wu KL, Iyer RK. Modeling stream processing applications for dependability evaluation. In: Proceedings of the IEEE/IFIP International Conference on Dependable Systems and Networks (DSN). Hong Kong, China; 2011. pp. 430–441.

[31] Jacques-Silva G, Gedik B, Andrade H, Wu KL. Fault-injection based assessment of partial fault tolerance in stream processing applications. In: Proceedings of the ACM International Conference on Distributed Event Based Systems (DEBS). New York, NY; 2011. pp. 231–242.

[32] De Pauw W, Letia M, Gedik B, Andrade H, Frenkiel A, Pfeifer M, et al. Visual debugging for stream processing applications. In: Proceedings of the International Conference on Runtime Verification (RV). St Julians, Malta; 2010. pp. 18–35.

[33] Geisshirt K. X/Open Single Sign-On Service (XSSO) – Pluggable Authentication. The Open Group; 1997. P702.

[34] Dierks T, Rescorla E. The Transport Layer Security (TLS) Protocol Version 1.2. The Internet Engineering Task Force (IETF); 2008. RFC 5246.

[35] Zeilenga K, Lonvick C. The Secure Shell (SSH) Protocol Architecture. The Internet Engineering Task Force (IETF); 2006. RFC 4251.

[36] Zeilenga K. Lightweight Directory Access Protocol (LDAP): Technical Specification Road Map. The Internet Engineering Task Force (IETF); 2006. RFC 4510.

[37] Fallside DC, Walmsley P. XML Schema Part 0: Primer – Second Edition. World Wide Web Consortium (W3C); 2004. http://www.w3.org/TR/xmlschema-0/.

[38] Sun Microsystems. NFS: Network File System Protocol Specification. The Internet Engineering Task Force (IETF); 1989. RFC 1094.

[39] IBM General Parallel File System; retrieved in November 2011. http://www-03.ibm.com/systems/software/gpfs/.

[40] Stallman RM, CygnusSupport. Debugging with GDB – The GNU Source-Level Debugger. Free Software Foundation; 1996.

[41] Valgrind Developers. Valgrind User Manual (Release 3.6.0 21). Apache Software Foundation; 2010.

Part IV

Application design and analytics

9 Design principles of stream processing applications

9.1 Overview

In the preceding chapters we described the stream processing programming model and the system architecture that supports it. In this chapter we will describe the principles of stream processing application design, and provide patterns to illustrate effective and efficient ways in which these principles can be put into practice.

We look at look at functional design patterns [1] and principles that describe effective ways to accomplish stream processing tasks, as well as non-functional ones [1] that address cross-cutting concerns such as scalability, performance, and fault tolerance.

This chapter is organized as follows. Section 9.2 describes functional design patterns and principles, covering the topics of edge adaptation, flow manipulation, and dynamic adaptation. Section 9.3 describes non-functional design patterns and principles, covering the topics of application composition, parallelization, optimization, and fault tolerance.

9.2 Functional design patterns and principles

We start by examining functional design patterns and principles, covering edge adaptation, flow manipulation, and dynamic adaptation.

9.2.1 Edge adaptation

SPAs consume data from external sources available in various different formats and accessible by employing different protocols. Similarly, results produced by streaming applications are often consumed by external systems in various formats and through different protocols. We term the process of interacting with external systems to receive and send data as *edge adaptation*. As an example, an operations monitoring application can consume log files produced by external systems (requiring edge adaptation for data ingest), detect anomalous conditions, and produce alarms that are inserted into a message queue for consumption by external applications (requiring edge adaptation for data egress).

The development of edge adapters requires an understanding of the nature of the interaction with a particular data source or sink (e.g., whether the data is pushed or pulled) and the nature of how the raw data is organized and made accessible (e.g., the data type and whether it is structured, unstructured, numeric, or categorical).

Many SPSs come with a set of built-in edge adapters supporting standard devices, interfaces, and data types. Their availability can greatly simplify application development. In Chapter 3 we looked at various edge adapters provided by Streams. Many SPSs also provide a set of well-defined interfaces that application developers can use to create new edge adapters.

This latter case is the one of greater interest to application and toolkit designers, as there are important considerations to make when implementing new edge adapter operators.

9.2.1.1 Principles of edge adaptation

Every edge adapter must examine certain characteristics of the external data source to determine the best architecture for ingesting data efficiently.

Push- versus *pull-based interactions*

A first consideration in the adapter design is the *push versus pull* nature of the interaction between the external source and the adapter. Before looking at this in detail, we first consider the way typical stream processing operators, that is non-source and non-sink operators, interact with their input and output.

Non-source operators generally employ a push-based protocol with respect to their interfacing with an upstream operator as well as with a downstream operator. In other words, inbound tuples are pushed into the input port of the consumer operator, which in turn produces outbound tuples that are pushed into its output port. In this arrangement, an operator handles the incoming traffic with an event-based interface, where specific handlers are invoked by the runtime on the arrival of tuples or punctuations (Section 8.5.9).

However, when it comes to connecting to an external source, a developer is at the will of the interface provided by that source. If an external source is providing a pull-based interface only, the source operator implementing the edge adaptation task must request the data explicitly and, subsequently, convert the pull-based interaction into a push-based one, *converting* the incoming data into tuples.

As an example of pull-based interactions, consider a file source operator, such as Streams' FileSource (Section 3.5.3). A file is a passive resource in the sense that one has to actively open it, read from it, and close it. Hence, this operator's implementation must rely on traditional file Input/Output (I/O) operations provided by the Operating System (OS).

In other words, the operator has to continually *pull* the data from the file. As it does this, the data is also progressively segmented, converted to tuples, and the resulting tuples are *pushed* to the downstream operators.

As an example of push-based interaction, consider an XML source operator that uses a Simple API for XML (SAX)-based parser.[1] In this case, the parser library provides a

---

[1] SAX [2] is an event-based, sequential access parser API for XML documents. For each element it encounters in the document it is parsing, it generates a corresponding event.

set of callback interfaces, which are hooked up to different parsing events. Therefore, the operator can register handlers with the library and these handlers are called when specific XML elements are encountered.

In other words, the parser library provides an event-based interface, where events of interest are *pushed* into the operator, which can then translate the data into tuples to be sent to the downstream consumers.

If both pull- and push-based interaction is possible with a particular external source, then the push-based approach is often more natural from the perspective of stream processing, as it resembles how a non-source operator typically operates.

Client/server relationship

Another consideration in the edge adapter's design is the client/server relationship between the edge adapter and the external source or sink. Client/server protocols are often used when inter-process communication is necessary to link an external source to a source operator, for example, via a socket or an RPC-based interface. The specific client and server relationship between these two parties impacts how a connection is established and, possibly, re-established as a result of failures.

Typically, an operator serving as a client initiates the connection to the external source. In this case, the operator is also responsible for detecting and re-establishing the connection upon disconnections. On the other hand, an operator acting as a server waits for and accepts connections from the external source, which, depending on how resilient the implementation is, can also re-establish the connection should a failure occur.

As an example of an operator that serves as a client, consider the Streams' TCPSink (Section 3.5.3), which can be configured by its role parameter to act as a client. At runtime, an instance of this operator will connect to an external TCP server at a specific IP address or host and also reconnect in the event of transient failures.

As an example of an operator that functions as a server, consider again the Streams' TCPSink whose role parameter can also assume the value server. With this configuration, a TCPSink operator instance will wait for and accept a connection from an external TCP client. In this case, the external client is responsible for re-establishing the connection after disconnections.

In general, if a SPA is designed to be long-lived, but the external system it is interacting with is short-lived or transient, it is preferable to make the edge adapter a server. In this way, the connection between the edge adapter operator and the external system will be established right away when the external system becomes available. If the edge adapter is instead configured as a client, it would have to periodically attempt to connect to a server, as the external source is transient and may come up at any time.

Conversely, if a streaming application is short-lived and the external system is long-lived, configuring the edge adapter as a client is the better choice.

Data source format and the communication protocol

One of the tasks a source edge adapter has to perform is to parse the incoming data from the external producer. Likewise, a sink edge adapter must also generate outgoing data in the format expected by an external consumer.

It is important to differentiate the incoming or outbound data format from the communication protocol employed to communicate with data producers or consumers. Having this separation usually enables greater reuse of the parsing and formatting code as well as of the communication protocol machinery, across different edge adapters.

As an example, consider the Comma-Separated Value (CSV) format. Data in this format can come from different sources, for example, from a file, from a TCP socket, from a UDP socket, or from a message queue interface. Decoupling the format from the protocol can enable the reuse of the format handling functionality by different protocols.

Many of the operators available in SPL's standard toolkit are implemented according to this principle, which has the additional benefit of ensuring consistent parsing and formatting behavior across operators.

Type mapping

Integrating a SPA with an external system, irrespective of whether it is a data producer or a data consumer system, introduces the issue of mapping the data types between the two likely dissimilar systems.

For instance, Data Base Management Systems (DBMSs) have a type system defined by SQL and often augmented with custom extensions.

Along the same lines, distributed inter-process communication mechanisms, such as CORBA [3], Google Protocol Buffers [4], or Apache Thrift [5], define their own type systems.

Finally, and not surprisingly, self-describing formats such as XML include mechanisms (e.g., an XSD definition) that enable the creation of custom vocabularies and, thus, the definition of application- or domain-specific types.

In general, different type systems may differ in terms of the support they offer for basic types, structured types, collection types, and nesting in terms of allowing attributes that are themselves defined as user-defined types. For instance, databases often support a *flat* type system, as it is a common practice to use *normalization* [6] and to employ multiple tables to represent nested structures. XML-based types, on the other hand, are inherently nested.

The first step in addressing type mapping issues is to consider the intersection between the types in the type systems provided by a particular SPS and the external system. Naturally, for values associated with corresponding or compatible types, transferring them from one system to another is trivial. In a second step, type conversions, decoding and re-encoding, or, in some cases, the dropping of non-representable values, are necessary tasks to translate data between the two systems.

In situations where data translation is required, it is important to make sure that both sources and sinks that interact with the same external system are consistent in how these operations are performed. Indeed, ensuring that what is written by a sink operator can be read by the corresponding source operator is a valuable testing feature when validating the implementation of the edge adapters.

As an example, consider both the ODBCSource and the ODBCAppend operators [7] from Streams' standard toolkit, respectively a source and a sink edge adapter.

```
1  <access_specification name="PersonRemainder">
2    <query query="SELECT id, fname, lname FROM personsrc" isolation_level="
        READ_UNCOMMITTED" replays="0" />
3    ...
4    <external_schema>
5      <attribute name="id" type="int32" />
6      <attribute name="fname" type="rstring" length="15" />
7      <attribute name="lname" type="rstring" length="20" />
8    </external_schema>
9  </access_specification>
```

Figure 9.1 A sample document showing an access specification for the SPL ODBC operators.

These operators allow a SPA to interact with relational databases. Specifically, the ODBCSource operator can retrieve database tuples and transform them into SPL tuples and the ODBCAppend operator can transform SPL tuples into database tuples that are subsequently appended to a relational table.

Both of these operators take an ODBC[2] access specification file as a parameter. This file specifies the mapping between SPL tuple attributes and the columns in a database table, where corresponding tuples are either retrieved from (ODBCSource) or stored into (ODBCAppend). The documentation for these operators specifies the type mappings they support.

Figure 9.1 shows an example access specification document. It provides the SQL query to be executed as well as the schema mapping for the database table. In particular, for each attribute in the table, the name of the attribute and the SPL type it corresponds to are specified.

When the type mapping strategy employed by an operator is not well designed, usability and performance problems might arise. On the usability side, a developer might be burdened with additional work required to fine-tune the mapping from an external type to a corresponding one in the SPS. As expected, such extra operations might also have a negative impact on performance.

Rightsizing the granularity of external interactions
As we have discussed, edge adaptation requires interaction with an outside system or with an external resource. These interactions might be computationally costly. In certain cases, edge adapters can become the performance *bottleneck* of an application, if the granularity of the interaction with the external entity is not properly chosen.

Batching read and write operations is often useful in reducing the overhead associated with these external interactions. In general, increasing the batch size (e.g., how many bytes are read/written from/to a file) trades off latency (or how long an operation takes) for increased throughput (or the volume of data that is read or written per unit of time). Nevertheless, finding the right balance between these two performance metrics is application-dependent. As a result, it is usually beneficial to design edge adapters that expose the batch size as an operator parameter.

---

[2] Open Data Base Connectivity (ODBC) [8] is a C-based application programming interface for interacting with a DBMS programatically.

As an example, consider Streams' FileSink (Section 3.5.3). It provides a flush parameter that can be used to specify how often tuples are to be flushed to a file. Similarly, Streams' InetSource operator [7] has a fetchInterval parameter that specifies how often to pull data from a web server.

Clearly, application developers who tune these parameters must make certain considerations. For example, an extremely small batch size in a source operator might cause decreased throughput as well as the inability to keep up with the rate of incoming data, which, in severe cases, can result in data loss. On the other hand, an extremely large batch size in a source operator usually increases the response time of the whole application, which might hinder its ability to produce a timely reaction to incoming events.

Considering back pressure in external interactions
A source edge adapter that is consuming data from an external source can create *back pressure* on the external source, if it cannot consume the incoming data at an equal or higher rate.

Back pressure can also be caused by the amount of *downstream* processing which propagates all the way to the source operator. In other words, if the downstream processing of any operator cannot keep up with the rate of ingest of an external source, the back pressure will eventually slow down all of the operators in that path, including the source operator and the external source (Section 6.3.4).

Note that, in certain cases, the topologically final operator in a flow graph, i.e., a sink edge adapter, might be the one responsible for the back pressure. This occurs, for example, when the external system connected to the sink is not able to consume the data produced by the application at the rate it is being produced.

As one practical example, consider the ingestion of a feed produced by a stock exchange where the data is provided via a UDP-based protocol [9]. If the application cannot keep up with the feed message rate, data loss will ensue due to the datagram-oriented nature of the UDP protocol [10]. In this example, the external protocol, UDP, naturally handles back pressure by dropping datagrams.

Similarly, consider a scenario where an application is processing a video feed transmitted by a TCP connection. If the application cannot keep up with the video frame rate, direct impact on the video source occurs in the form of a slowdown due to TCP's connection-oriented nature [10]. Eventually, the video source will notice that it is unable to sustain the frames per second it is designed to produce and will have to either reduce the video quality or drop frames.

Back pressure effects can occur even in the absence of an explicit network connection. Consider a system where log files are generated by its components and an application is used to post-process these files. If the log processing application cannot keep up with the rate that log entries are produced, the log files will pile up. In this scenario, if no mitigating measures are taken, the system may eventually run out of disk space and come to a halt.

While the ideal approach is to provision an application with enough resources to avoid back pressure and the resulting slowdown or data loss that might occur, it is sometimes not possible to do so, particularly, due to the dynamic nature of the workload imposed on such applications (Section 1.2). For example, unexpected transient spikes in the load or non-cooperative external systems can create situations where back pressure effects might still occur.

Yet, an application designer must consider whether the external system is prepared to handle unanticipated back pressure as well as put in place mitigation procedures should the problem arise at runtime.

For example, in some cases, tuples destined to an external system can be dropped, if the external sink is not keeping up with the rate of data. Alternatively, selectively skipping files or reducing the amount of analysis dedicated to each log entry in the log processing example can be very effective.

In other cases, more sophisticated methods must be employed to preserve the general statistical characteristics of the streams, even when individual tuples are dropped. These measures are collectively referred to as *load shedding* [11, 12, 13]. Its implementation usually relies on specifically designed operators [14, 15, 16] that selectively drop tuples according to different criteria that attempt to minimize the net effect on the accuracy and efficiency of a particular data analysis task.

9.2.1.2   **Patterns of edge adaptation**
We now look at concrete examples of edge adaptation using the SPL language.

Edge adaptation using files
SPAs are sometimes designed to retrieve and process data from files that are produced by an external system, to continuously analyze as well as to make inferences about the state of the external system.

As an example, a telecommunications switch may generate Call Detail Records (CDRs) as files. Similarly, a software instrumentation framework performing operational monitoring may emit log messages to a file. In situations like these, the files that are continuously updated must also be continually read by the SPAs that analyze them.

As is the case of any typical queuing system [17], as long as the time it takes for the SPA to process a log file is less than the inter-arrival time of new files, a no-loss continuous operation can be sustained.

Figure 9.2 provides sample code to illustrate this scenario. In summary, this application first ingests a continuous series of files from a source directory. After analyzing the input files, it generates a continuous series of output files that are then placed in a destination directory.

In this application, we assume that the external system that produces the log files places them in a directory named `"inDir"`. In this example, we make the assumption that the files appear in this directory atomically, once they are ready to be analyzed.[3] We also assume that the external system that consumes the output files generated by the streaming application retrieves them from the `"outDir"` directory.

---

[3] In general, this can be achieved by creating the file using a temporary name and, subsequently, performing an atomic file system *move* operation when it is ready to be processed.

```
1  composite FileDataAdaptation {
2    graph
3      stream<rstring file> Files = DirectoryScanner() {
4        param
5          directory: "inDir";
6          pattern: "input.*";
7          sortBy: date;
8        output
9          Files: file = FilePath();
10     }
11     stream<DataType> Data = FileSource(Files) {
12       param
13         format: csv;
14         deleteFile: true;
15     }
16     stream<ProcessedDataType> ProcessedData = DoProcess(Data) {}
17     () as Sink = FileSink(ProcessedData) {
18       param
19         file: "output.{id}";
20         format: csv;
21         closeMode: count;
22         tuplesPerFile: 10000;
23         moveFileToDirectory: "outDir";
24     }
25 }
```

Figure 9.2 File-based edge adaptation.

The FileDataAdaptation application uses a DirectoryScanner operator, to retrieve the file names from the "inDir" directory (configured as the operator instance's directory parameter), where it looks for files whose names match the "input.*" regular expression. The DirectoryScanner operator continuously scans this directory and produces a stream with file names. The order of file names seen in the Files stream is based on the file modification time associated with each file, with older files coming ahead of newer ones. This behavior is selected by using the setting date for the parameter sortBy in the operator configuration.

This Files stream is fed to a FileSource operator, which opens the file designated by the file attribute in the inbound tuples, reads its content, and converts each comma-separated line (as indicated by the csv setting in the operator's format parameter) into an outbound tuple, which is transported by the Data stream. Once the end of a file is reached, the FileSource operator removes the file from the source directory (a behavior selected by the deleteFile parameter). In this way, files that have already been processed are eliminated from the directory.

This application uses a composite operator called DoProcess, whose implementation is not shown, to perform the actual analytic task on the log entries. It also employs a FileSink operator to produce the results as a series of files. The sink operator generates a maximum of 10,000 tuples per file. Each file is named uniquely by incrementing the value of the {id} placeholder in the file name.

Specifically, the FileSink operator is configured with a tuple count-based closeMode, where the count threshold defined by the tuplesPerFile parameter is set to 10000.

Moreover, the file parameter is used to name the generated files with the id part replaced with a monotonically increasing integer representing the index of the file.

When a file is closed, it is subsequently moved to the "outDir" directory as indicated by the operator's moveFileToDirectory parameter. This operation is atomic, which means that the file appears in the destination directory only when it is complete.

Edge adaptation for XML data

XML is a popular data exchange format. As a result, many sources encode their data in this format, often following custom vocabularies defined either by an XML Schema Definition (XSD) [18] or by a Document Type Definition (DTD) [19]. Examples include web services [20], eXtensible HyperText Markup Language (XHTML) documents [21], Really Simple Syndication (RSS) feeds [22], as well as Atom feeds [23].

While XML is a flexible and expressive format, it usually isn't the most adequate choice for representing data for low-latency and high-performance processing due to its verbosity and text-based representation. As a result, many SPSs employ streamlined binary formats for representing data. Yet, in many situations XML sources must also be integrated with an application, making it necessary to convert XML-encoded data into tuples.

The ingestion of XML-encoded data can be performed by an edge adapter that extracts the relevant pieces of information from the XML data and converts them into tuple attributes.

In general, this conversion requires a mechanism for indicating what parts of the XML-encoded data are to be extracted and converted into tuples. One such mechanism is the XML Path Language (XPATH) [24], a query language with constructs for selecting nodes from an XML document, making it a natural choice for this task.

There are two common ways of processing XML data. One is to convert the XML document into a Document Object Model (DOM) [25], which is an in-memory tree representation of the document. Once the XML data is available in this format, parts of it can be extracted by running an XPATH query against the DOM representation. Nevertheless, this approach is generally not suitable for a stream processing, as the external source is continuously emitting new data.

The other approach is to have a streaming parser that processes the XML document and delivers events as it encounters specific XML elements and attributes in the document. One such engine is the SAX parser [2]. In most cases, a streaming parser can be used to efficiently evaluate certain types of XPATH queries on unbounded incoming XML streams.

As an example, consider the XMLDataAdaptation application depicted by Figure 9.3. In this example, a TCPSource operator is used to process inbound lines of text as indicated by the line setting for its format parameter. In this case, we assume that the inbound external data comes in as XML-encoded elements. This operator's output stream, Data, is connected to the XMLParse edge adapter operator, which is also available in the SPL standard toolkit.

The XMLParse operator's trigger parameter is specified to instruct the operator to look for customer elements in the XML stream, irrespective of the level in the XML hierarchy of the incoming data that they appear. The setting for this parameter, "/customer", is an XPATH expression.

```
1   composite XMLDataAdaptation {
2     graph
3       stream<rstring line> Data = TCPSource() {
4         param
5           role: client;
6           address: "my.xml.source.com";
7           port: 40120;
8           format: line;
9       }
10      stream<rstring name, list<tuple<rstring id, int32 cost>> transactions> Data
            = XMLParse(Data) {
11        param
12          trigger: "/customer";
13        output
14          Data: name = XPath("@name"),
15            transactions = XPathList("transaction",
16              {id=XPath("@id"), cost=(int32)XPath("@cost")});
17      }
18  }
```

Figure 9.3 XML-based edge adaptation.

This instance of the XMLParse operator indicates how a tuple is to be constructed from the data encoded in the XML customer elements. Specifically, the name outbound tuple attribute is assigned the value associated with the name XML attribute of each instance of the customer element. This transformation is specified via the XPath function and its parameter "@name".

In similar fashion, the transactions outbound tuple attribute, which is defined as a list of tuples, is populated with the values in the list of transaction XML elements associated with the customer XML element. This transformation is specified via the XPathList function, which makes use of the XPath function to specify the mappings for individual tuple attributes. In other words, for each transaction XML element, a tuple is extracted by assigning the id XML attribute to the id tuple attribute and the cost XML attribute to the cost tuple attribute, after casting it to the corresponding attribute type, int32 in this case.

As a concrete example, consider the following XML segment received by the source stream:

```
1   <customer name="Fred">
2     <transaction id="3" cost="300">
3       <order id="10123" cost="100"/>
4       <order id="10124" cost="200"/>
5     </transaction>
6     <transaction id="5" cost="600">
7       <order id="10568" cost="400"/>
8       <order id="10569" cost="200"/>
9     </transaction>
10  </customer>
```

The corresponding tuple based on the extraction specification given in Figure 9.3 as part of the XMLParse operator configuration, looks as follows:[4]

```
{name="Fred", transactions=[{id="3", cost=300}, {id="5", cost=600}]}
```

---

[4] The tuple (which is delimited by the { ... } notation) as depicted in this example consists of two attributes: name and transactions. The transactions attribute is a list (which is denoted by the [...] notation) and each element consists of two attributes: id and cost. The value for each simple type attribute follows the = sign.

Finally, this example also illustrates the principle of keeping the data source format independent of the communication protocol employed to talk to the external data source (Section 9.2.1.1). In this case, the TCPSource operator can easily be replaced with a different edge adapter without impacting the parsing performed by the XMLParse operator.

Edge adaptation for data on the Internet
SPAs may want to leverage Internet-based data sources as part of their analytics. For this purpose, a SPA must be able to fetch data using common Internet protocols such as HTTP and FTP.

In SPL, these protocols are available in the InetSource operator defined in the com.ibm.streams.inet namespace as part of the *inet toolkit*. The InetSource operator can fetch a document specified using a Uniform Resource Identifier (URI) from HTTP, HyperText Transfer Protocol Secure (HTTPS), and FTP servers, either as an one-off operation or on a periodic basis.

Figure 9.4 depicts a code excerpt where the InetSource operator is used to fetch weather data by making use of an HTTP-based operation on the United States' National Oceanic and Atmospheric Administration (NOAA) web server. The contents of those two webpages are periodically updated by this agency as it appends additional content to them.

The operator's URIList parameter is used to provide the set of pages to be retrieved. In this case, these pages are retrieved every 60 seconds as specified by the operator's fetchIntervalSeconds parameter.

When a web page is retrieved, its content is broken into lines and each line is placed in a single-attribute tuple, in this example, the outbound tuple's metarRecord attribute. Note that the operator can make use of the incrementalFetch parameter to avoid streaming old data, thus only text that is different from the previously retrieved content is streamed out. Finally, the operator can also be configured with the doNotStreamInitialFetch parameter to suppress the streaming out of the initial content of a web page.

Handling back pressure
As we discussed earlier, addressing the issue of back pressure is fundamental in designing, configuring, and employing both source and sink edge adapters. This is especially important in two cases. First, when it is not possible to exert control over the

```
1   stream <rstring metarRecord> Observations = InetSource () {
2     param
3       URIList:
4         ["http://weather.noaa.gov/pub/data/observations/metar/cycles/07Z.TXT",
5          "http://weather.noaa.gov/pub/data/observations/metar/cycles/08Z.TXT"];
6       fetchIntervalSeconds: 60u;
7       incrementalFetch: true;
8       doNotStreamInitialFetch: true;
9   }
```

Figure 9.4 Retrieving Internet-based data using the InetSource operator.

external data sources and sinks. Second, when an application is not sufficiently scalable (scalability techniques are described in Section 9.3.3) and/or not over-provisioned in terms of computational resources to handle spikes in the load.

As an example, we first consider the source code depicted by Figure 9.5, which illustrates how to be resilient to back pressure when interacting with an external source.

Inside the `InputBackPressure` composite operator, the edge adapter, an instance of the hypothetical `SourceAdapter` operator, is followed by a `Filter` operator (Section 3.5.1) named `PressureHandler`. The `PressureHandler` filter has a threaded input port. This port is configured with the *drop first* congestion policy and an associated queue size of $10,000$ tuples.

When the downstream processing is slow and causes back pressure, the queue associated with the threaded port will start filling up. When the queue is full, rather than propagating the back pressure further to the external source, the congestion policy will start dropping tuples. This process starts with the *older* tuples, i.e., the tuples that arrived first (using the `Sys.DropFirst` setting[5]), currently buffered in the threaded port's queue, thus protecting the external source from directly experiencing back pressure. Moreover, if the back pressure situation is transient, the queue buffer provides some amount of time for the rest of the application to catch up, sometimes without any data loss.

We now consider the code depicted by Figure 9.6, which illustrates how to be resilient to back pressure when interacting with an external data sink.

In this example, the sink edge adapter, an instance of the hypothetical `SinkAdapter`, is preceded by a `Filter` operator named `PressureHandler`. In similar fashion to the source edge adapter case, it is also configured with a threaded port.

```
1   composite InputBackPressure(output Data) {
2     graph
3       stream<T> Src = SourceAdapter() {}
4       (stream<T> Data) as PressureHandler = Filter(Src) {
5         config
6           threadedPort: queue(Src, Sys.DropFirst, 10000);
7       }
8   }
```

Figure 9.5 Handling back pressure in source edge adaptation.

```
1   composite OutputBackPressure(input Src) {
2     graph
3       (stream<T> Data) as PressureHandler = Filter(Src) {
4         config
5           threadedPort: queue(Src, Sys.DropFirst, 10000);
6       }
7       () as Out = SinkAdapter(Data) {}
8   }
```

Figure 9.6 Handling back pressure in sink edge adaptation.

---

[5] `Sys.DropLast` is another alternative and is used to drop the incoming tuples that came last.

The `Filter` operator's queue in this case protects the application from experiencing back pressure exerted by the external data sink application. In other words, when the external consumer application cannot handle the incoming data flow, it will initially cause queueing up in the `Filter` operator, followed by tuple dropping, if the consumer application cannot eventually catch up.

9.2.2 Flow manipulation

In several applications domain-specific analytic flows contain, or are connected to each other via, operators that manipulate these flows in some way.

These flow manipulation tasks include data reduction to decrease the volume of data in a stream, partitioning to demultiplex it, flow splitting to distribute the processing so it can be tackled in parallel downstream, enrichment to supplement the inbound data with additional information, and, finally, correlation to create relationships between tuples that come from different streams. In this section, we describe each of these tasks as well as the basic principles and associated patterns governing them.

9.2.2.1 Principles of flow manipulation

Every application must decide on how to handle the incoming data flow, focusing on ways to perform its analytic task efficiently.

Data volume reduction
As previously discussed, in many domains, applications must process large volumes of data with low latency to keep up with the live nature of the data they consume. An effective way to achieve higher throughput and, thus, to be able to scale up to process high data rates, is to perform data reduction.

Data reduction is accomplished by applying one or more of these methods:

- *Filtering* is the operation that discards data that is not relevant to the downstream processing. Filtering can be stateless or stateful. A stateless filter makes a decision to discard a tuple solely based on its contents. As an example, in a log monitoring application, stateless filtering can be applied to remove tuples that do not correspond to the particular log event type we are interested in. On the other hand, a stateful filter makes a decision to discard a tuple based not only on the current tuple contents, but also based on the state accumulated by the filter, which is a function of the earlier tuples it has seen. For instance, in a sensor network application, we could use a stateful filter to remove tuples if a reading is within a small margin of the last reported value.
- *Sampling* is the operation that selects a representative subset of the tuples from a stream (Section 10.5). This is usually achieved through a statistical process. This process can be as simple as random sampling, which is based on a fixed probability p of selecting any given tuple and $1 - p$ of dropping it. Alternatively, it can be implemented by a more advanced process, such as *reservoir sampling* [26]. This is an algorithm that restricts the number of samples, while ensuring that these samples remain true to the statistical characteristics of the whole population.

- *Aggregation* is the operation that gathers multiple individual tuples into a single one that somehow represents a summary of the group using, for example, descriptive statistics like the *mean* or the *median* of a particular tuple attribute. Aggregations can be used to reduce the level of detail present in a stream to match the needs of an application. For instance, a network monitoring application may need data usage information on a per minute basis and thus may aggregate per second information produced by a data source to generate the necessary summaries.
- *Projection* is the operation that removes or replaces one or more of the tuple attributes from a stream. Unlike the other forms of data reduction we have seen so far, a projection does not change the number of tuples produced, but potentially reduces the size of each tuple by removing or replacing unnecessary attributes.
- *Quantization* is the operation that maps values from a larger domain to a smaller domain of discrete values (Section 10.7). A common case involves mapping floating point numbers to integers with a smaller range. For instance, an image processing application can quantize an image, mapping each 32-bit color pixel to an 8-bit shade of gray pixel. Similar to a projection operation, quantization does not necessarily reduce the number of tuples, but instead shrinks the size of each tuple.

These data reduction techniques can be applied individually as well as combined, depending on the specific needs of an application. For instance, a stream can be *sampled* before tuples are *aggregated*. Similarly, a stream can be *filtered* to remove non-relevant image frames before *quantization* is applied to bring the image resolution down to the necessary level required by a hypothetical video processing application. Specific online algorithms for data reduction are discussed in detail in Chapter 10.

Partitioning multiplexed flows

Many applications must process multiple *logical* substreams that originate from a single *physical* stream. As an example, consider a stream carrying data from the stock market. All tuples sharing the same value for a ticker symbol attribute, i.e., the attribute that identifies a specific stock symbol, can be considered as an individual substream. In other words, all of the tuples bearing the IBM ticker symbol[6] (associated with the IBM Corporation) form a substream.

Similarly, consider a physical stream that transports network protocol packets in an Intrusion Detection System (IDS). In this case, all of the tuples that share a common value for the IP address attribute can be deemed as a separate logical substream.

Frequently, the number of unique substreams multiplexed in a stream is either not known at application development time (e.g., a set of pairwise Short Message System (SMS) conversations), or it is dynamic and changes periodically (e.g., a new stock symbol may be added or a company might be delisted), or it is too large to enumerate explicitly (e.g., IP addresses). Yet, it is often the case that the analytical processing carried out by an application has to be performed on each substream independently.

---

[6] The stock symbols employed here correspond to the ticker symbols used by the New York Stock Exchange (NYSE).

A typical example of an application processing streaming data from the stock market is the computation of the Volume Weighted Average Price (VWAP) for a company's stock (Section 4.3.1). In this case, the VWAP value, i.e., a weighted moving average of a particular stock's price, is computed for each logical stream, independently of the others, and is periodically emitted.

In general, the first step necessary to process multiplexed stream flows is to identify one or more attributes in a tuple's schema as the *partitioning attributes*. Each combination of values for these attributes represents a different logical substream. Naturally, in the VWAP computing example, the partitioning attribute is the stock symbol attribute.

Once the partitioning attributes are defined, the original physical stream as well as the derived ones produced by other operators in the application can be processed in a multiplexed fashion. This approach avoids incurring the additional cost of physically separating each logical substream as well as deploying additional instances of the subgraph of operators (one for each substream). Naturally, for this to occur all of the other operators must internally support multiplexed processing, i.e., segmenting the internal state and processing logic according to the partitioning attributes.

A common technique for implementing an operator that supports multiplexed processing is to partition its internal state using a hash table [27]. In this way, for each tuple received, the partitioning attributes are used to locate and retrieve the relevant portion of the state in the hash table, if available, or alternatively to create and initialize it. Once this portion of the state is available, the operator can manipulate it as part of its internal processing and, subsequently, update it. Thus, a single operator instance can act as multiple *virtual* ones operating independently on disjoint parts of the internal state.

In the SPL standard toolkit, there are several such operators. For example, the `Aggregate`, `Join`, `Barrier`, and `Sort` operators support multiplexed processing. This capability is available when these operators' instances are configured with the `partitionBy` parameter, which indicates the attributes to be used to logically demultiplex the inbound stream.

Streaming data enrichment

In many applications it is necessary to enrich the incoming data. Generally, this task is performed using auxiliary, mostly static information kept in persistent storage such as external relational databases, text files, or in other systems and applications accessible through remote procedure calls.

As an example, consider a telecommunications application designed to process phone call records. In this case, a stream transporting phone CDRs is formed by a sequence of tuples where each one contains attributes storing the identification numbers representing the two customers involved in a phone call. In this scenario, the customer profiles are usually kept in a separate relational database. Therefore, if our hypothetical SPA requires customer information (e.g., the customer's name, address, or his calling history) to carry out its analytical task, the original CDR stream can also be *enriched* with the information from the customer database.

Enrichment is typically implemented by performing a lookup operation based on information found in the tuple being processed (e.g., the customer identifier in our example) to retrieve the relevant data from the external dataset.

Naturally, enrichment imposes an additional computational cost that can slow down a SPA and act as a source of back pressure (Section 9.2.1.1). Therefore, it is important to evaluate the impact of the enrichment workflow on the speed at which the data can flow through the application and to devise strategies to mitigate any possible slowdown. Different strategies might be used depending on the characteristics of the enrichment as well as of the external data to be used.

When the static data set is small enough to fit in the memory of a single host, it can be indexed and loaded once, typically during the initialization of the SPA using this data. In this case, the resulting in-memory lookup table might generally allow the enrichment operation to be performed at streaming speeds. This approach is called cache-based enrichment.

When the static data set is too large to fit in a single host's main memory, the cache-based enrichment can be extended to employ multiple hosts. This is accomplished by making use of a hash-based partitioning strategy similar to the approach discussed for handling multiplexed flows discussed in Section 9.2.2.1. Referring back to the phone call record processing application, non-overlapping subsets of the customer data would be kept on different hosts and the incoming CDR stream would be physically partitioned based on the same criteria used to split the customer data. This approach amounts to parallelizing the application and is further discussed in Section 9.3.2.

When the enrichment dataset can be periodically updated and, hence, is not static, cache-based enrichment can still be applied by either periodically or on-demand reconstructing and reloading the lookup table.

Finally, in some cases, the cache-based enrichment approach cannot be used and other techniques must be considered. A reasonably simple yet effective tactic, consists of reducing the cost of lookups by keeping the connection open with the external data source and performing the lookups in batches. This approach usually amortizes the fixed costs of connecting and disconnecting to and from the external data source (e.g., opening/closing a file, creating and scheduling a query plan in a relational database).

Streaming data scoring

As we discussed extensively, an application implements data in-motion analytics on potentially unbounded data, yet as one such application runs, its internal state must be bounded and is frequently kept in main memory. Generally the latest data is stored in windows, capturing the characteristics of the current and most recent events. Nevertheless, many SPAs require historic knowledge, usually in the form of a *model* distilled from older data using data mining techniques, to carry out their own analytic task in a process referred to as *scoring* (Section 10.2).

Scoring a data stream is similar in nature to enriching it. Even so, rather than directly retrieving additional information from other data sources to *augment* the current tuple, the tuple is *scored* against a model derived from stored data. This process might also yield new attributes or updates to existing ones.

As is the case with an enrichment operation, a scoring operation can also substantially benefit from an in-memory cache-based approach. In other words, the data mining models required for scoring should be preloaded, whenever possible, and updated, periodically or on-demand, as appropriate.

As an example, consider a hypothetical *sentiment analysis* application that processes a stream of *tweets* produced by the Twitter [28] social networking service. The application can use streaming text analytics to extract product and associated sentiment information (e.g., "my new Massive Dynamics smart phone is·perfect!") from the input data.

In this application, one of the possible analytical goals is to report reasons and frequencies for negative and positive sentiment associated with the products being talked about. This sort of scoring usually requires the association of comments to a particular product as well as of certain textual characterizations to a positive or negative sentiment.

This example presents another dimension related to the use of scoring models. As this application runs, it may eventually detect that the current models used for analysis is no longer yielding scores for certain tweets. This situation usually arises when the data used to build the current incarnation of the scoring models is no longer representative.

In general, applications making use of them will rebuild these models as part of their continuous life cycle, perpetually evaluating, evolving, and updating them. In Chapter 10, the analytical side of these techniques is further discussed and the Streams' data mining toolkit, with several operators that employ Predictive Model Markup Language (PMML) (Section 1.3.3) scoring models, is introduced.

Flow splitting

A widely applicable flow manipulation principle is to split a processing flow, essentially a subgraph of the application, into multiple flows for independent downstream processing. When this is possible, it creates the opportunity to employ different modes of parallel processing on the streaming data to potentially scale up the application so it can handle increased data volumes. Stream splitting comes in various forms.

A flow can be *replicated* by splitting it into two or more identical flows. This is often useful to perform different tasks in parallel on the replicated flows. As an example, in a hypothetical video surveillance application a video frame stream can be replicated into two, where one of the flows is tasked with *face detection*, whereas the other is tasked with *object detection*. As can be inferred from this example, this type of splitting is used to leverage task parallelism (Section 9.3.2) in an application.

A flow can be *vertically distributed*[7] by splitting it into two or more flows that each contain a subset of the tuple attributes. As an example, a video stream can be split into an audio stream and an image stream.

---

[7] We use the term *vertically* because a stream schema, seen as a line-by-line list of attribute types and names, can be vertically split into two or more sub-schemas, or sub-lists of attributes.

A flow can also *horizontally distributed*[8] by splitting it into two or more flows that each contain a subset of the tuples from the source stream.

There are two basic ways of implementing a horizontal split. An application might use a *non-content-based* flow splitting mechanism, which does not take into consideration the contents of the tuples to decide how to route each one of them. For instance, in the *sentiment analysis* application introduced above, the stream carrying the tweets can be split in round-robin fashion to allow multiple instances of the text analytics subgraph to operate simultaneously on different tuples.

Alternatively, an application might employ a *content-based* flow splitting mechanism, which does inspect the contents of the tuples to decide how to route each one of them. For instance, in the log processing application discussed earlier in this section, the inbound log stream might contain multiple types of log formats, which can be split into multiple outbound streams so that different parsers can analyze the log entries.

Naturally, each of these different splitting approaches impact the application performance and can be used to aid in the scaling up of an application. These topics are discussed in more detail in Section 9.3.2.

Cross stream correlation

Many applications bring together multiple streams as they seek to *correlate* data extracted from different sources. Such correlation forms the basis of multi-modal analytics, where different types of data (e.g., from raw text to structured records, from audio to video) as well as different types of analysis techniques (from descriptive statistics and signal processing to data mining) are put together with a common analytical goal.

Consider, for example, an application that integrates a stock exchange market feed with data extracted from news and blog posts to augment the pricing and risk models for stocks and options. In this hypothetical case, news and blog items must first be processed to identify which companies and market sectors they relate to (if any), before they can be correlated with specific trades that occurred in the past or are occurring live.

The fundamental building blocks supporting a correlation task include the following operations:

- *Union* is the operation performed on two or more streams where tuples from these streams are interleaved into a single one with a common schema (which must be a subset of all the input stream schemas). The tuples are projected out according to the outbound stream schema and are emitted in the order they were received.
- *Barrier* is the operation performed on two or more streams to match corresponding tuples from each stream based on the arrival order of their tuples. This is often necessary to recombine flows that were created as a result of a splitting operation, upstream in the flow graph.
- *Merge* is the operation performed on two or more streams to combine streams that were sorted on a given attribute into a single stream whose tuples are also ordered on the same attribute.

---

[8] We use the term *horizontally* because a stream, seen as a left to right sequence of tuples, can be horizontally split into two or more substreams, containing, typically, different subsets of tuples.

- *Join* is the operation used to match tuples across windows maintained over two or more streams, based on a match condition that specifies when tuples should be combined.

Naturally, more sophisticated forms of correlation can also be performed, sometimes following simpler and computationally cheaper operations like those we just described.

Note that, in many cases, the streams being correlated must be *aligned*. In other words, there must exist a mechanism, explicit or otherwise, that allows the correlation algorithm to relate a tuple from one stream to a tuple in the other stream.

For instance, if we are performing a barrier operation between two streams, the relative tuple arrival order determines how the correlation occurs. More specifically, the first tuple of the first stream is correlated with the first tuple of the second stream, and so on. In this case, the alignment between the two streams is *implicit*.

Analogously, if we are performing a join operation between two streams, the operation's match condition, a boolean expression, defines a property on the attribute values from the tuples (possibly, buffered in windows) coming from each stream to determine whether the tuples should be correlated. In this case, the alignment between the two streams is *explicit*.

Naturally, a consideration for any correlation operation that requires the streams to be *aligned* is the amount of memory required to establish such alignment when live data is being processed.

Consider a merge operation. In this case, in order to emit a tuple the merge algorithm has to establish that the tuple's ordering attribute value is smaller than or equal to that of any other tuple that is currently buffered, or that can be received in the future, from any of the incoming streams being correlated.

Clearly, if for some reason, one of the streams is lagging behind the others, the operator performing the merge might end up having to buffer a large number of tuples from the other streams.

For streams where the ordering attribute is a wall clock timestamp, closely tracking the actual passage of time, such lags are often bounded, resulting from the transmission delay between the data source and the application. Thus, the amount of buffering necessary cannot grow arbitrarily.

On the other hand, if we are testing an application using the same data, but rather than using live streams, the data is being played back from files residing on disk, the buffering behavior might change considerably. This is particularly true if the stream playback does not respect the original inter-arrival time of tuples. As a result, streams might be running at very different paces, imposing more extreme buffering demands on the merge operator.

Temporal pattern matching

Detecting patterns across time is a common task in stream processing. For instance, consider a stream with stock market transactions. Each tuple in this stream contains, among other attributes, a stock symbol and its price. Assume that an application is interested in sequences of tuples that share the same stock symbol. Furthermore, assume that the sequence of interest must include an *M*-shaped movement in the stock price within a certain time frame. In other words, the sequence must include increases in the price, followed by a sequence of decreases that do not take the price below its initial price, followed by another sequence of increases, and, finally, a sequence of decreases that take the price below its initial price. Finding such an *M*-shaped trend requires locating a *temporal pattern* in a stream's sequence.

This problem is similar to performing regular expression matches over strings. The main difference is that, in regular expression matching the operation consists of matching characters, whereas in temporal pattern matching the operation consists of matching *events*.

These events correspond to predicates defined over the stream. Specifically, when the movement in a sequence of tuple attribute values matches one or more of these predicates, the corresponding events can be teased out. Pattern matching is then carried out over these events. For instance, an event called *drop in price* can be defined as the value of the current tuple being lower than its last value. In this case, a tuple that matches this condition will represent the *drop in price* event.

A temporal pattern defined over a stream can use this event as part of its definition. A temporal pattern definition involves common pattern matching operations, such as applying the Kleene star operator, indicating zero or more appearance of an event; detecting optional events, which may or may not be present; detecting disjunctive events, where at least one is present; and detecting event groups, where a sequence of events is treated as one.

A temporal pattern matching predicate defines the events over which the actual matching is performed and can, typically, reference two kinds of values. The first is the value of an attribute present in the current tuple. In our earlier example, the current price of a particular stock is such a value. The second is the value of an aggregation function defined on a tuple attribute over a sequence of tuples. Examples of such functions include the summation of a sequence of attibute values (Sum), the average value of a sequence (Average), the minimum and maximum value in a sequence (Min and Max), the first and last value in a sequence (First and Last), among many others. In our earlier example, the *last* price for a stock can be computed by the Last aggregation function.

9.2.2.2   Patterns of flow manipulation

In Chapter 3 we introduced various relational and utility operators available in SPL's standard toolkit. Later, we also discussed how user-defined processing can be added to a Custom operator. As part of the discussion in this section, we will use these features and operators, often in conjunction, to illustrate specific flow manipulation techniques.

Time-based sampling

Time-based sampling is a simple technique used to reduce the data volume of an incoming stream. We will look at two possible implementations of this technique in SPL.

In the first implementation, *inline time-based sampling*, a per-tuple decision is made on whether the current tuple should be forwarded downstream or not. This decision is

```
1   composite InlineTimeBasedSampler(output Sampled; input Source) {
2     param
3       expression<float64> $period;
4     graph
5       stream<Source> Sampled = Custom(Source) {
6         logic
7           state:
8             mutable float64 nextTime = 0.0;
9           onTuple Source: {
10            float64 time = getTimestamp();
11            if (time>=nextTime) {
12              submit(Source, Sampled);
13              while (nextTime<=time)
14                nextTime += $period;
15            }
16          }
17      }
18  }
```

Figure 9.7 Inline time-based sampling.

based on segmenting the continuous time line into discrete chunks with some *period* duration.

Every time a tuple arrives, the algorithm checks whether this arrival has happened in a period in which no tuples have been emitted. If that is the case, the tuple is forwarded downstream, otherwise it is discarded. Naturally, when the inbound stream has an intermittent flow and no tuples arrive in a given period, the inline sampler will not produce any tuples during that time interval.

Figure 9.7 depicts the source code for the InlineTimeBasedSample composite operator. It uses a Custom operator to perform the sampling. The next time a tuple should be sent out is kept as a state variable (nextTime). If the current time, obtained through the getTimestamp function in fractional seconds, is equal to or greater than the next time a tuple should be sent out, then the tuple that just arrived is sent out. The period parameter of the composite operator is used to update the nextTime a tuple should be sent out after each submission.

The second kind of time-based sampling, *fixed-interval time-based sampling*, produces tuples at fixed intervals, *always* outputting the most recently received tuple. To preserve this property, it can emit the same tuple in successive periods of time, if no new tuples arrive during a period of time. Figure 9.8 provides the code for this algorithm, which is encapsulated in the FixedIntervalTimeBasedSampler composite operator.

In this case, a Custom operator is used to cache the last received tuple as a state variable (lastTuple). Differently from our previous example, this Custom operator also has a second input port, which periodically receives tuples from a Beacon operator, which set the sampling rate.

The duration of the sampling period is defined by the period parameter in the composite operator. With this organization, when the Custom operator receives an incoming tuple on its second stream (Beat), it emits the last cached tuple, received from its first inbound stream (Source), on the Sampled stream.

Several other types of sampling techniques exist. We will discuss some of them in the context of streaming analytics in Section 10.5.

```
1   composite FixedIntervalTimeBasedSampler(output Sampled; input
2   Source) {
3     param
4       expression<float64> $period;
5     graph
6       stream<int32 a> Beat = Beacon() {
7         param
8           period: $period;
9       }
10      stream<Source> Sampled = Custom(Source; Beat) {
11        logic
12          state: {
13            mutable boolean cached = false;
14            mutable tuple<Source> lastTuple = {};
15          }
16          onTuple Source: {
17            lastTuple = Source;
18            cached = true;
19          }
20          onTuple Beat: {
21            if (cached)
22              submit(lastTuple, Sampled);
23          }
24      }
25  }
```

Figure 9.8 Fixed-interval time-based sampling.

Nested aggregation

Applications often compute aggregations over time-based windows. Such aggregations help observe and monitor variations in metrics of interest computed over the values of individual tuple attributes at different time scales.

An effective way of implementing such aggregations is to use *nested aggregations*, where the results of a finer-grain aggregation are used to compute the results for a coarser-grain one, progressively increasing the size of the time-based window.

Figure 9.9 depicts sample SPL code used to implement a nested aggregation algorithm. The NestedAggregation composite operator processes one inbound stream and outputs three streams, one that contains per-second aggregations (PerSecond), one that contains per-minute aggregations (PerMinute), and another one that contains per-hour aggregations (PerHour).

The per-second aggregation is computed using a time-based tumbling window. The subsequent aggregations use count-based tumbling windows to create aggregations at larger scales. For instance, the per-minute aggregations are updated every 60 seconds, whereas the per-hour aggregations are updated every 60 minutes.

Note that this particular mechanism of nesting aggregations only works when the aggregation function is *associative* as is the case for the Max function used in this example. Functions such as Sum and Min also fall into this category, but others, such as average, do not.

Multiplexed flow partitioning

We now look at how multiplexed flow partitioning can be performed using SPL operators. We consider an application where the goal is to compute the total amount of revenue for a business as well as trending information (i.e., has the revenue gone

```
1   composite NestedAggregation(output PerSecond, PerMinute, PerHour;
2     input Source)
3   {
4     graph
5       stream<Source> PerSecond = Aggregate(Source) {
6         window
7           Source: tumbling, time(1);
8         output
9           Temp: value = Max(value);
10      }
11      stream<Source> PerMinute = Aggregate(PerSecond) {
12        window
13          PerSecond: tumbling, count(60);
14        output
15          PerMinute: value = Max(value);
16      }
17      stream<Source> PerHour = Aggregate(PerMinute) {
18        window
19          PerMinute: tumbling, count(60);
20        output
21          PerHour: value = Max(value);
22      }
23  }
```

Figure 9.9 Nested aggregation.

```
1   type CompanyRevenue = tuple<rstring sector, rstring company,
2                               float64 revenue, int32 year>;
3   type SectorRevenue = tuple<rstring sector, float64 revenue, int32 year>;
4   type SectorRevenueWithTrend = tuple<SectorRevenue, tuple<boolean up>>;
5   composite RevenueBySector(
6     output stream<SectorRevenueWithTrend> SectorRevenuesWithTrend;
7     input stream<CompanyRevenue> CompanyRevenues) {
8     graph
9       stream<SectorRevenue> SectorRevenues = Aggregate(CompanyRevenues) {
10        window
11          Revenues: tumbling, delta(year, 0), partitioned;
12        param
13          partitionBy: sector;
14        output
15          SectorRevenues: revenue = Sum(revenue);
16      }
17      stream<SectorRevenueWithTrend> SectorRevenuesWithTrend as O
18          = Custom(SectorRevenues as I) {
19        logic
20          state:
21            mutable map<rstring, float64> lastRevenue = {};
22          onTuple I: {
23            tuple<O> otuple = {};
24            assignFrom(otuple, I);
25            if (sector in lastRevenue)
26              otuple.up = (lastRevenue[sector]<revenue);
27            else
28              otuple.up = false;
29            lastRevenue[sector] = revenue;
30            submit(otuple, O);
31          }
32      }
33  }
```

Figure 9.10 Multiplexed flow partitioning.

up/down since the last time period?) for each business sector, using per-company revenue information as our source.

Figure 9.10 depicts the code for the RevenueBySector composite operator implementing this logic. This operator takes as input the CompanyRevenues stream, where each tuple stores time-stamped revenue information for a company and outputs the `SectorRevenuesWithTrend` stream, whose tuples contain per-sector revenue aggregation and trending information.

In this example, `sector`, the tuple attribute used to indicate the business sector a company belongs to, is the *partitioning attribute*. This attribute is explicitly used as the `partitionBy` attribute for the partitioned `Aggregate` operator that computes the total annual revenues per sector by adding up (`Sum`) the individual revenues for each company. The aggregation is performed using a partitioned window, during a period of one year, using a delta-based trigger policy (`delta`).

This example also illustrates how a new stateful operator supporting multiplexed partitioning can be built. Following the annual partitioned aggregation results, we employ a `Custom` operator to compute trends on the individual per-sector annual revenues, by checking if the revenue in the sector has gone up since the last time the aggregation was performed. The `lastRevenue` map, associating a sector to its revenue, is built and updated to perform this assessment. The boolean trend result stored in the `up` attribute augments each of the original incoming tuples, indicating whether the revenue has increased.

Note that both operators used in this example operate on individual implicit substreams, one per sector, independently of the others. While the standard toolkit `Aggregate` operator is instructed to do so via its `partitionBy` setting, the user-defined `Custom` operator uses a map *state variable*, whose key is the partitioning attribute, to achieve the same goal, explicitly.

Cache-based enrichment

As discussed in Section 9.2.2.1, cache-based enrichment is a technique to efficiently augment the streaming data by performing lookups on quasi-static data structures that fit in memory.

The code implementing the `CustomerNameLookup` composite operator depicted in Figure 9.11 illustrates how this technique can be implemented with a `Custom` operator.

In this example, the application goal is to enrich the incoming tuples that contain the identifier attribute `id` with additional information associated with the customer corresponding to this identifier, specifically his or her name. This additional data is retrieved from a file and includes the mappings between `id`s and `name`s.

The `CustomerNameLookup` operator employs a `FileSource` operator to read the "`CustomerInfo.txt`" file, producing the `CustomerInfo` stream.

The `CustomerNameLookup` operator also employs a `Custom` operator, which consumes the `CustomerInfo` stream, storing the mapping between a customer identifier and his/her name in the `idToName` state variable. The composite operator's incoming stream to be enriched, `Source`, is also connected to this `Custom` operator by a different input port.

Every time a tuple is received by the input port connected to the `Source` stream, the customer name is looked up in the `idToName` map using the tuple's `id` attribute as the key and the resulting name is assigned to the `name` attribute in the outgoing tuple.

```
1    composite CustomerNameLookup(output Enriched; input Source) {
2      graph
3        stream<rstring name, uint64 id> CustomerInfo = FileSource() {
4          param
5            file: "CustomerInfo.txt";
6        }
7        stream<Source, tuple<rstring name>> Enriched = Custom(Source; CustomerInfo){
8          logic
9            state:
10             mutable map<uint64, rstring> idToName = {};
11           onTuple Source: {
12             tuple<Enriched> outTuple = {};
13             assignFrom(outTuple, Source);
14             if(Source.id in idToName)
15               outTuple.name = idToName[Source.id];
16             else
17               outTuple.name = "N/A";
18             submit(outTuple, Enriched);
19           }
20           onTuple CustomerInfo: {
21             idToName[CustomerInfo.id] = CustomerInfo.name;
22           }
23       }
24   }
```

Figure 9.11 Cache-based enrichment.

In practice, the Source stream should be delayed during startup to wait for the lookup table to be fully built. We will discuss this very mechanism in Section 9.2.3.2.

Note that streaming in the enrichment data, as we did in this example, has two main benefits. First, the source of this data can be easily changed from a file to a different source such as a database or a live external source. Second, this data can be updated as needed, for example, when information about new customers becomes available.

Declarative pattern matching

As seen earlier in this section, pattern matching, particularly temporal pattern matching, is a common operation in SPAs.

Temporal pattern matching over a stream can be implemented by maintaining a record representing the sequence of values seen so far and thus tracking it with respect to the pattern being sought. However, implementing pattern matching in this way can be cumbersome and usually requires implementing a custom-made state machine to keep track of events and transitions between their occurrences.

Alternatively, one can make use of a general purpose state machine where a pattern matching operation is described declaratively. In the rest of this section, we will look at one such mechanism.

In this declarative approach, the developer simply specifies the list of events as well as the pattern that specifies the order in which such events must happen.

Let's reconsider the problem of finding M-shaped stock price movements we described earlier in this chapter. Figure 9.12 shows the code for the FindMShape composite operator, which processes a Quotes stream as input and generates a Matches stream as output. The former contains a sequence of tuples, each with a stock trade transaction that includes the individual price for a share, whereas the latter contains tuples with attributes representing the stock symbol and the sequence of price changes forming an M-shape, i.e., the sought pattern.

```
1    composite FindMShape(
2       output stream<rstring ticker, list<float> prices> Matches;
3       input stream<rstring ticker, float price> Quotes) {
4     graph
5       stream<string ticker, list<float> prices> Matches = MatchRegex(Quotes) {
6        param
7          partitionBy: ticker;
8          pattern: ". up+ down+ up+ down* under";
9          predicates: {
10           up = price>First(price) && price>Last(price),
11           down = price>=First(price) && price<Last(price),
12           under = price<First(price) && price<Last(price)
13         }
14       output
15         Matches: prices = Collect(price);
16     }
17  }
```

Figure 9.12 Declarative temporal pattern matching.

MatchRegex is an operator in the SPL CEP toolkit (in namespace com.ibm.
streams.cep) that performs stream pattern matching using a declarative syntax. In
Figure 9.12, the partitionBy attribute specifies that pattern matching will be performed independently for each stock symbol. The predicates parameter specifies
the events the application is interested in.

The up event is triggered when the share price associated with the current tuple is
greater than that of the last tuple as well as that of the first tuple in the sequence. Here,
the Last and the First aggregation functions are used. The down event is triggered
when the price in the current tuple is smaller than that of the last one, yet greater than
or equal to that of the first tuple in the sequence. The under event is triggered when
the price in the current tuple is smaller than that of the last tuple as well as the first
tuple in the sequence.

The pattern parameter specifies the temporal pattern as a function of the events
defined by the set of predicates. In this example, the application is interested in:

- any tuple (indicated by .), forming the starting point of the M shape;
- followed by a sequence of tuples creating up events (indicated by up+), forming the
  first line of the M shape;
- followed by a sequence of tuples creating down events (indicated by down+),
  forming the second line of the M shape;
- followed by a sequence of tuples again creating up events (indicated by up+),
  forming the third line of the M shape;
- followed by a potentially empty sequence of tuples creating down events and the
  last tuple of the sequence creating an under event, where the final price for a share
  is below the initial one in the sequence (indicated by down* under), forming the
  fourth and final line of the M shape.

The MatchRegex operator's output clause specifies the outgoing tuple to be
generated when a pattern is found. Each such tuple includes the collection of price
movements in the sequence represented by the prices list attribute associated with
the (implicit) ticker attribute representing the stock symbol.

The semantics of the `MatchRegex` operator is such that, when a pattern is encountered, the operator generates an outgoing tuple. The pattern matching process restarts with the next incoming tuple and, therefore, locates non-overlapping sequences.

9.2.3 Dynamic adaptation

In many application domains, streaming data volumes are uneven. Both the tuple rate and the tuple inter-arrival times fluctuate as a result of *external* conditions (e.g., increased number of transactions during a sale, increased market transactions as a result of quarterly financial results).

Moreover, *internal* conditions such as the computational effort and the storage and networking requirements of an application, as well as their availability can also fluctuate over time.

Therefore, the ability to dynamically adapt to both external and internal changes is important to applications, particularly due to their continuous and long-running characteristics.

In Chapter 4, we discussed various constructs that form the basis for engineering dynamic adaptation techniques, including feedback loops, dynamic connections, and incremental application composition. In this section, we will make use of these constructs and look at the principles, programming patterns, and techniques used to implement applications that are adaptive.

9.2.3.1 Principles of dynamic adaptation

Every application must include strategies and mechanisms to be resilient and adaptive to changes in the workload and in resource availability.

Autonomic application feedback

The primary mechanism enabling operators to work together in an application is the stream connection, which is responsible for transferring *data* from a producing operator to a consuming one.

Nevertheless, in some cases, operators may want to exchange *control* directives, allowing an operator to transmit information that can alter the behavior of another. Stream connections transporting control directives are no different than connections transporting data.

A stream connection in the form of a feedback loop in an application's data flow graph enables a downstream segment of the application to logically modify the processing performed as part of an upstream segment. This is done based on the findings of the downstream segment, where more refined analysis of the data is performed.

As an example, consider a hypothetical application used to monitor and control a manufacturing plant. This type of application might employ online model learning techniques (Chapter 11) to spot and anticipate potential problems in the production line.

In general, the upstream, *model builder* segment of this application might include one or more operators charged with building the predictive model for the application's analytical task. Because this type of application must be adaptive, the model must evolve and learn from the data that continuously flows into it. For this reason, this application might also include a downstream, *model checker* segment where an operator examines whether the predictions of the model are accurate.

The results of these predictions can be used to further improve the model. Therefore, a feedback loop in the form of a stream connection between the model checker and the model builder is usually put in place to continuously convey this information.

Typically, the stream connection supporting the feedback loop is directed to an upstream operator's *control port*, a special kind of input port tasked with receiving tuples carrying control directives. A control port's processing logic does not result in the generation of tuples, but instead the control directives it receives are used to update the internal state of the operator.

Note that employing a feedback loop for reasons other than establishing a control connection of the sort we discussed can be problematic, potentially leading to deadlocks in the application and, hence, should be avoided.

Dynamic application configuration

Adaptivity in an application can also result from an external interaction with a system administrator or an external system designed to tweak the application's internal logic by altering its configuration at runtime, usually by dynamically modifying the configuration of its operators. This capability is often useful to match the processing logic of an operator to the changing analytical needs of this application.

Consider a hypothetical cybersecurity application used to filter tuples transporting IP traffic from a set of blacklisted IP addresses. In this scenario, the use of a dynamically configurable IP address list allows the application to change its behavior, focusing on new threats immediately.

A possible way to dynamically configure an operator is to design it with a *control port*. This input port can be connected to a source operator linked to a *control console*, perhaps as part of an external system. Internally, the processing logic associated with the control port implements mechanisms to adjust the operator's internal logic as a result of the directives it receives.

In the cybersecurity application, for example, a new IP address can be manually added to the blacklist by a system administrator using the external control console. Or, it can be added automatically, by an offline batch process that mines historical network traffic data and sends updates via the control part. Once the filter operator receives a new IP address in this port, it reacts by updating its internal data structure used to keep track of blacklisted addresses and by immediately starting to screen the network traffic based on the new set of addresses in the list.

Dynamic application composition

In Section 4.2 we looked at dynamic composition mechanisms. In that discussion, we showed how an input port's subscription expression and an output port's stream properties could be used to establish dynamic connections between applications.

These mechanisms can be used for many tasks including the discovery of new data sources and sinks, which might become available/unavailable at any point in time, and the incremental deployment of multi-component applications. These same mechanisms can also be employed to implement *hot swap* procedures for replacing a component in need of fixes or improvements. In this way, it becomes possible to minimize the downtime during an application's maintenance cycle as well as to support self-evolving applications whose data flow graph evolves autonomically as a function of the workload it processes.

Naturally, the effective use of dynamic composition hinges on applications being designed to be modular, with input and output ports that will serve as integration points across component boundaries. This design must also consider the impact of establishing dynamic connections at runtime, including the increased processing and networking loads as well as the potential for back pressure on the data producing components. Indeed, an implementation using these features should always include mechanisms to minimize or mitigate these effects.

Finally, it is important to emphasize that dynamic application composition is an active area of research, especially along the dimension of autonomic planning and orchestration [29, 30].

9.2.3.2   Patterns of dynamic adaptation

As discussed, employing dynamic adaptation is essential to SPAs with their unique combination of continuous processing and evolving analytical needs.

To support these requirements, multiple programming features in the SPL language can be used to implement the patterns that form the basis of an adaptive application. We will explore them next.

Feedback loops for application signaling

The continuous nature of streaming data as well as the difficulties of controlling the incoming data flow and *aligning* streams for processing, create numerous programming challenges when a developer is implementing an application.

One of these flow control problems relates to how to delay the processing in certain segments of the data flow graph until another segment's processing is completed. We saw an instance of this problem in Section 9.2.2.2 (Figure 9.11), as part of the cache-based enrichment example. In that application, a delay was imposed on the stream to be annotated until the base data used for providing these annotations became available.

In general, problems like this can be resolved by having the data processing segment that sits further downstream notify the upstream segment via a feedback loop when the *bootstrapping* task is complete. We will demonstrate how this can be achieved by extending the implementation of the `CustomerNameLookup` composite operator we discussed earlier. The code and its corresponding flow graph are depicted by Figures 9.13 and 9.14, respectively.

There are two important improvements in this version. First, the standard toolkit `Switch` operator is used to block the `Source` stream until the `idToName` map is fully constructed. The `Switch` operator has two input ports. The first is a port that receives the stream to be blocked/unblocked on demand. The second is a control port, that receives tuples indicating whether the data stream should flow, if it is currently

```
1   composite CustomerNameLookup(output Enriched; input Source) {
2     graph
3       stream<Source> ToEnrich = Switch(Source; Control) {
4         param
5           initialStatus: false;
6           status: Control.on;
7       }
8       stream<rstring name, uint64 id> Info = FileSource() {
9         param
10          file: "CustomerInfo.txt";
11      }
12      (stream<Source, tuple<rstring name>> Enriched;
13       stream<bool on> Control) = Custom(ToEnrich; Info) {
14        logic
15          state:
16            mutable map<uint64, rstring> idToName = {};
17          onTuple ToEnrich: {
18            tuple<Enriched> outTuple = {};
19            assignFrom(outTuple, ToEnrich);
20            if(ToEnrich.id in idToName)
21              outTuple.name = idToName[ToEnrich.id];
22            else
23              outTuple.name = "N/A";
24            submit(outTuple, Enriched);
25          }
26          onTuple Info: {
27            idToName[Info.id] = Info.name;
28          }
29          onPunct Info: {
30            if (currentPunct()==Sys.WindowMarker)
31              submit({on=true}, Control);
32          }
33      }
34  }
```

Figure 9.13 Using feedback to improve the implementation of the CustomerNameLookup operator.

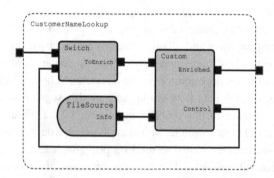

Figure 9.14 Flow graph corresponding to the CustomerNameLookup operator implementation.

stopped, or stopped, if it is currently flowing. The status parameter of the Switch operator controls which attribute in the control stream will be used to adjust the flow status.

In this particular example, the Switch operator is configured to stop the Source stream on initialization, as the value of this operator instance's initialStatus parameter is set to false. It is also configured to allow the Source stream to start flowing when an inbound tuple arriving from the `Control` stream has its `on` attribute set to `true`.

Second, the `Custom` operator, used to create the `idToName` map and to perform the enrichment operations, is modified with the inclusion of an additional output port. This output port produces a control tuple that turns on the `Switch` operator. This outbound control tuple is produced as a result of receiving a window punctuation marker, which is handled via the `onPunct` clause associated with the operator's second input port. The punctuation is received and processed after this input port has received all the tuples used to populate the `idToName` map.

In other words, the `onPunct` clause is used to detect the punctuation marker signaling that the `idToName` map is fully populated. This marker is generated by the `FileSource` operator when it finishes reading the entire file. Once the map is built, an outbound tuple is sent on the `Control` stream to announce that the data from the `Source` stream can now be processed.

Note that the `Control` stream is connected to the control input port of the `Switch` operator, forming a feedback loop in the application's flow graph.

Dynamic external application configuration
Dynamic external application configuration comes in many forms. An interesting example is the modification of a filter condition at runtime.

Figure 9.15 depicts one such case. The `BlackListFilter` composite operator has a single input port. It filters the incoming stream based on the set of IP addresses that have been blacklisted, producing an outgoing stream with the filtered tuples.

This composite operator makes use of the standard toolkit's `DynamicFilter` operator to screen inbound tuples based on whether their `ip` attribute value appears in the blacklisted set.

```
1   composite BlacklistFilter(output ScreenedIPTraffic; input IPTraffic) {
2     graph
3       stream<rstring ip> InitialIPs = FileSource() {
4         param
5           file: "initialBlackList.txt";
6       }
7       stream<rstring ip> AddIPs = FileSource() {
8         param
9           file: "addIpToBlackList.txt";
10          hotFile: true;
11      }
12      stream<rstring ip> RemoveIPs = FileSource() {
13        param
14          file: "removeIpFromBlackList.txt";
15          hotFile: true;
16      }
17      stream<IPTraffic> ScreenedIPTraffic = DynamicFilter(IPTraffic;
18          InitialIPs, AddIPs; RemoveIPs) {
19        param
20          key: IPTraffic.ip;
21          addKey: AddIPs.ip;
22          removeKey: RemoveIPs.ip;
23      }
24  }
```

Figure 9.15 Dynamic external application configuration.

The DynamicFilter operator has two control ports, one to add and one to remove IP addresses from the blacklist, using the addKey and removeKey parameters, respectively. The filtering itself is done by checking whether the key attribute (specified via the key parameter) in the incoming tuples on the IPTraffic stream can be found in the operator's internal key map.[9]

The BlackListFilter composite operator employs a FileSource operator to stream the initial list of blacklisted IP addresses via the InitialIPs stream. Two other FileSource operators are used to create streams that contain the IP addresses that are dynamically inserted in or removed from the blacklist, which are connected to the second and third input ports of the DynamicFilter operator, respectively.

Note that the FileSource operators producing the AddIPs and RemoveIPs streams employ *hot files*, which can pick up the incremental updates to be streamed out. In this way, to insert a new IP address into the blacklist, an external user can simply append a new line to the "addIpToBlackList.txt" file. In this example, there is a simplifying assumption that the whole application has access to a common file system.

Dynamic connection re-routing

An important form of dynamic adaptation occurs when a segment of an application consuming data switches to a different data provider. Usually a data consumer initiates such a change as a reaction to its evolving interest, often resulting from analysis it performed on the data it received earlier.

The code in Figure 9.16 shows how this type of change can be implemented by re-routing the incoming stream traffic using dynamic composition (Section 4.2). The corresponding, fully expanded flow graph is depicted by Figure 9.17.

The application has two components, the exporting one represented by the Exporter composite operator and the importing one represented by the Importer composite.

The exporting component (Figure 9.18) produces and exports two streams. One of the them is exported with the name="A" property (line 7) and the other is exported with the name="B" property (line 11).

The importing component (Figure 9.19) starts up by subscribing to one of the exported streams, initially setting its subscription expression to name=="A" (line 5).

```
1   composite DynImportMain {
2     graph
3       () as E = Exporter() {}
4       () as I = Importer() {}
5   }
```

Figure 9.16 Dynamic connection re-routing using the Exporter (Figure 9.18) and the Importer (Figure 9.19) composite operators.

---

[9] Note that the DynamicFilter operator filters tuples based on the keys that can be found in the blacklist. If the opposite filtering condition is needed, the outflow from the operator's optional second output port can be used instead.

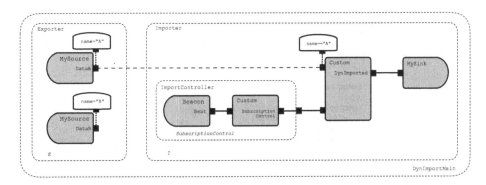

Figure 9.17 Fully expanded flow graph corresponding to the `DynImportMain` operator (Figure 9.16).

```
1    composite Exporter {
2      graph
3        stream<int32 id, rstring name> DataA = MySource() {}
4        stream<int32 id, rstring name> DataB = MySource() {}
5        () as ExporterA = Export(DataA) {
6          param
7            properties: { name = "A" };
8        }
9        () as ExporterB = Export(DataB) {
10         param
11           properties: { name = "B" };
12       }
13   }
```

Figure 9.18 The `Exporter` operator used by the `DynImportMain` application (Figure 9.16).

```
1    composite Importer {
2      graph
3        stream<int32 id, rstring name> Imported = Import() {
4          param
5            subscription: name=="A";
6        }
7        stream<Imported> DynImported = Custom(Imported; SubscriptionControl) {
8          logic
9            onTuple Imported:
10             submit(Imported, DynImported);
11           onTuple SubscriptionControl:
12             setInputPortImportSubscription(subscription, 0u);
13       }
14       stream<rstring subscription> SubscriptionControl = ImportController() {}
15       () as Sink = MySink(DynImported) {}
16   }
```

Figure 9.19 The `Importer` composite operator used by the `DynImportMain` application (Figure 9.16).

Periodically, it changes its subscription expression (using the `setInputPortImport Subscription` function[10] in line 12), alternating between `name=="A"` and `name=="B"`, thus switching the actual `Imported` stream it consumes.

---

[10] The `setInputPortImportSubscription` function's first parameter is the updated subscription expression. Its second parameter is the index of the input port for which the subscription expression is being updated.

```
1   composite ImportController(output SubscriptionControl) {
2     graph
3       stream<int8 dummy> Beat = Beacon() {
4         param
5           iterations: 1u;
6       }
7       stream<rstring subscription> SubscriptionControl = Custom(Beat) {
8         logic
9           onTuple Beat: {
10            while(!isShutdown()) {
11              block(1.0);
12              submit({subscription="name==\"A\""}, SubscriptionControl);
13              block(1.0);
14              submit({subscription="name==\"B\""}, SubscriptionControl);
15            }
16          }
17      }
18  }
```

Figure 9.20 The `ImportController` operator used by the `Importer` operator (Figure 9.19).

```
1   composite DynExportMain {
2     graph
3       () as E = Exporter() {}
4       () as I = Importer() {}
5   }
```

Figure 9.21 Dynamic connection re-routing using the `Exporter` (Figure 9.23) and `Importer` (Figure 9.24) operators.

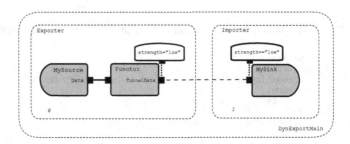

Figure 9.22 Fully expanded flow graph for the `DynExportMain` operator (Figure 9.21).

The `ImportController` composite operator (Figure 9.20), used by the importing component, creates a new subscription expression every second (lines 12 and 14), using a `Beacon` operator to initiate this process and a `Custom` operator to rotate between the two alternative subscription expressions. The `Subscription Control` stream created by this `Custom` operator is ultimately output by the `ImportController`.

As with dynamically updating the stream subscription expressions, the stream properties can also be modified dynamically. Figure 9.21 shows a sample application that illustrates this type of change, which is implemented by re-routing the outgoing stream traffic using dynamic composition. Figure 9.22 shows this application's fully expanded flow graph.

```
1   composite Exporter {
2     graph
3       // assume 'signal' is in the range [0,1]
4       stream<float64 signal> Data = MySource() {}
5       stream<Data> TunnelData = Functor(Data) {
6         logic
7           state: mutable boolean low = true;
8         onTuple Data: {
9           if (value<0.5) {
10            if (!low) {
11              low = true;
12              setOutputPortExportProperties({strength="low"}, 0u);
13            }
14          } else {
15            if(low) {
16              low = false;
17              setOutputPortExportProperties({strength="high"}, 0u);
18            }
19          }
20        }
21      }
22      () as Sink = Export(TunnelData) {
23        param
24          properties: {strength="low"};
25      }
26  }
```

Figure 9.23 The Exporter operator used by the DynExportMain operator (Figure 9.21).

The application has two components, the exporting one represented by the Exporter composite operator and the importing one represented by the Importer composite.

The exporting component (Figure 9.23) produces a stream called Data that contains a single float64 value called signal. We assume that this value lies in the range [0, 1]. The exporting component uses a Functor operator to adjust the stream property called strength used for exporting the Data stream, based on the current value of the signal value. In particular, when the last received value is below 0.5, the stream property strength is set to "low" (if not already set to that) or to "high" (if not already set to that). This is done by the setOutputPortExportProperties function.[11] Naturally, in a real application the changes in the signal strength are not arbitrary and will be driven by some external event.

The importing component (Figure 9.24) simply subscribes to exported streams whose properties satisfy the condition strength=="low". The net result is that, when the content of the exported stream starts containing low signal values, the exported stream's strength property will be set to "low" by the exporting component and the importer component will start receiving that stream. Otherwise, the exporting component will set the value of the strength property to "high" and, thus, the importing component will stop receiving the stream.

---

[11] The setOutputPortExportProperties function's first parameter is the set of properties of the exported stream. Its second parameter is the index of the output port for which the properties are being set.

```
1   composite Importer {
2     graph
3       stream<float64 signal> Data = Import() {
4         param
5           subscription: strength=="low";
6       }
7       () as Sink = MySink(Data) {}
8   }
```

Figure 9.24 The `Importer` operator used by the `DynExportMain` operator (Figure 9.21).

9.3   Non-functional principles and design patterns

In this section, we look at non-functional principles and design patterns, covering the topics of application design and composition, parallelization, performance optimization, and fault tolerance.

9.3.1   Application design and composition

Application composition, i.e., the design of a multi-component application where each component might itself be another application, is fundamental to large-scale stream processing.

Such a design is based on hierarchically decomposing a large *conceptual* application into modular components and designing how these modules can be integrated based on their exposed interfaces.

In the rest of this discussion, we will use the Streams nomenclature, where we refer to the conceptual application as a *solution* and, to its application components or modules, simply as *applications*. Hence, a solution is a set of applications.

Considering the hierarchical nature of a solution, at the highest level, a developer must first decide how to decompose it into applications. Going down the hierarchy, for each application in the solution, the next decision is how to decompose it into a set of composite operators. This is a recursive decision as each composite operator might itself employ other composite operators in a nested structure.

A *terminal* composite operator only employs other primitive operators and it is not further decomposable. Therefore, the primitive operators sit at the lowest level of the solution hierarchy.

In the rest of this section, we will look at the principles involved in making such decomposition decisions an iterative part of the application development process.

9.3.1.1   Principles of application design and composition

Modular and component-based design have been pillars of programming for many decades. Most modern programming languages include mechanisms for creating and integrating components. This notion serves the dual purpose of breaking a large problem into smaller and more manageable pieces as well as of making these pieces reusable across different problems.

The crux of the problem, however, lies in how to design each module, finding the ideal granularity for each of them. In this section, we discuss some of the principles behind the problem of application decomposition, focusing primarily on how to better engineer modular applications.

Primitive operator design

A primitive operator is a non-decomposable building block used for creating an application and is represented as a node in an application's data flow graph.

As seen in Section 5.2.3, a primitive operator can be developed in a general purpose programming language using an event-based interface or, in SPL, using a `Custom` operator to envelope the desired processing logic.

The basic decision in designing a new primitive operator consists of determining what chunk of a large computational problem should be carved out and implemented as an *operator*.

Ideally, a primitive operator should perform *one* well-defined task. When this principle is violated, it usually leads to architectural and computational shortcomings.

First, the web of interactions between tasks results in increased complexity for the operator internal design and in its parameterization. These issues, in turn, make the operator more difficult to use, particularly when only one portion of the functionality is desired.

As an example, consider the tasks of scanning the contents of a directory and streaming out the data stored in the files in that directory. One approach is to create a monolithic primitive operator that performs both tasks at once. This approach has certain disadvantages.

First, if the way this operator scans a directory (e.g., by ordering the entries alphabetically) does not correspond to a particular developer's needs, the operator must be ruled out, even if the data streaming function matches his or her requirements.

Naturally, an alternative and more effective design consists of creating two primitive operators, decoupling the two tasks. Indeed, this is the approach taken by SPL's standard toolkit. It provides both the `DirectoryScanner` operator, which can sweep the contents of a directory in different ways, as well as the `FileSource` operator, which can read different types of files and stream them out.

With this arrangement, if an application has specific requirements that cannot be fulfilled by one of these two operators, a developer can simply create a replacement for the missing part and make use of the other one.

Second, by its very nature, a multi-task *primitive* operator cannot be further decomposed. On the other hand, multiple simpler primitive operators can be combined and fused into a single partition by the SPL compiler, as discussed in Section 5.3.2. Therefore, an operator designed for a single task empowers the compiler with an additional degree of freedom as it tries to match the application physical layout to the computational infrastructure that will host the application. Naturally, these two primitive operators can also be used together as part of a composite operator, whose physical deployment can also be optimized by the SPL compiler.

Third, a monolithic primitive operator encompassing multiple tasks makes it difficult to take advantage of parallelism opportunities that might exist in the application flow graph (Section 3.2), hindering scaling up efforts. This can happen because individual tasks that by themselves are, for example, data parallel might not be so when grouped together by a multi-task operator.

Additionally, when operators are designed to perform a single task, they are simpler, and can be easily organized in an execution pipeline, with computationally balanced stages, which can facilitate this type of parallelization of an application. The same argument applies to employing a task parallel organization, as will be discussed in Section 9.3.2.

Nevertheless, there are cases where it might be preferable to perform closely related tasks within a single primitive operator. This usually happens with tightly coupled, performance critical tasks, where, for instance, they must both access a common data structure or share another non-splittable resource. An example of this situation occurs in certain types of finance engineering applications, where network protocol decoding, content parsing, and data splitting must be tightly integrated to cope with the high influx of messages from a stock exchange [9].

In general, however, using multiple single-task operators at the composition-level and relying on partition co-location directives (Section 5.3.2) to achieve better performance, should be the first choice.

Composite operator design
In general, the development of a large program in a general purpose programming language such as C++ or Java consists of breaking the processing logic into classes and functions. The class methods and functions are usually themselves broken down into a set of smaller methods and functions when they become either too complex or too large.

Similarly, the design of a composite operator in a stream processing language such as SPL should follow an analogous modularization process and be broken down into a set of smaller composite operators as it becomes either too complex or too large.

In a composite operator, it is the size of the data flow subgraph implied by its specification that determines whether it has become too large. Naturally, the larger the subgraph, the more difficult it is for a developer to grasp what the operator does. This is usually an indication that the composite operator's processing can be split into additional composite operators via nested composition.

The input and output interfaces exposed by a composite operator resulting from this refinement process should be carefully designed as this operator and others generated by the same process can be reused elsewhere.

In fact, if similar data flow subgraphs appear in different sites in a larger processing graph, reusing a composite operator to factor out the common processing in these subgraphs improves the design of the application in several ways. First, code readability is enhanced as it becomes clear that there is a well-defined, common operation taking place. Second, code quality and productivity are improved as the same code is written, debugged, and tested *once*. Third, code maintenance is easier and, as the reused code gets improved, all the sites that use it reap the benefits.

In the SPL language, modular design techniques also benefit from the availability of higher-order composite operators (Section 4.2.3). Such composites allow the reuse of a *structurally* similar subgraph where *operational* differences might exist in the operators employed by the composite's data flow subgraph.

In other words, a composite operator's data flow subgraph, where one of the nodes might make use of different operators, can be implemented as a higher-order composite. This operator can later be instantiated in different ways by employing different parameters, which define the actual operators to be used. This capability makes it possible to reduce code bloat by reusing structural patterns present in an application's data flow graph.

Application organization

As seen in Section 4.2.2, dynamic composition is an important engineering feature available to application developers to design adaptive applications.

Dynamic composition can also be used to engineer a *solution* that comprises multiple SPL applications that, in conjunction, act as a single application. This approach can be advantageous with respect to the alternative of having the solution implemented as a single SPL application.

First, when a solution is engineered with the dynamic composition of multiple applications, each one can be built, maintained, and deployed independently.

From a project management standpoint, this organization enables the simultaneous development and testing of each component application by different software developers from a large team.

From a runtime deployment standpoint, this organization enables the incremental launching of the solution, potentially speeding up the overall process, as different component applications are independently and, potentially, simultaneously submitted for execution.

From the standpoint of maintenance tasks, this organization can potentially reduce the overall solution's down time as only the relevant component applications have to be taken offline. In this case, however, benefits are only seen if the remaining components can still temporarily operate while the maintenance cycle is ongoing.

Second, a multi-component solution exposes another type of reuse. Each of the component applications can be used, individually or by other multi-component solutions. Furthermore, if a specific component application cannot be directly reused, it might be possible to modify it by employing configurable submission-time parameters[12] to make it suitable to other uses. This approach is particularly attractive because the different instantiations of such a component application do not require recompilation. They share the same underlying implementation, differing only in how they are configured at the time of deployment.

---

[12] Submission time parameters are used to customize an application at the time of job submission. In SPL, they are specified using the `-P` parameter when invoking `streamtool` during job submission, and their values can be accessed from within a program by the `getSubmissionTimeValue` function.

Finally, different application components in a solution may have different fault tolerance requirements and, hence, each one of them can use an individually appropriate mechanism to attain resilience.

As seen in Section 8.5.4, a stream connection created by dynamic composition, unlike a stream connection created by static composition, is transient by nature and, hence, creates a *loose* coupling between a data producer and a data consumer. In other words, in a static stream connection when the consumer becomes unavailable, perhaps due to a temporary failure, the stream producer must wait until the connection is re-established, a behavior that in some cases might be undesirable. Conversely, in a dynamic stream connection the producer can continue to stream out data to other consumers, even if one of them becomes unavailable. Hence, depending on the fault tolerance semantics of the solution (such as its sensitiveness to data loss), a more appropriate choice between static and dynamic composition can be made.

9.3.2   Parallelization

The data flow graph nature of an application makes it simple to logically express the various forms of parallelization that can be present in it. Likewise, it also simplifies the ways in which the execution runtime can exploit such parallelism.

Three forms of parallelism are particularly useful in the context of stream processing:

- *Pipeline parallelism* is a form of parallelism where sequential stages of a computation execute concurrently for different data items. As an example, consider how a modern CPU processes instructions [31]. Each one is executed in various stages, including fetch, decoding, execution, and memory access. While these stages must be performed sequentially, under certain conditions pipeline parallelism can be applied so that, while one instruction is in a particular stage, a subsequent instruction can simultaneously be in an earlier stage. This kind of parallelism manifests itself naturally when an application is seen through the lenses of its data flow graph representation.
- *Data parallelism* is a form of parallelism where the same computation takes place concurrently on *different* data items [31]. As an example, consider again how a modern CPU executes a Single Instruction Multiple Data (SIMD) instruction. Essentially, it performs the same computation on multiple values at the same time, for instance, when multiplying four pairs of 32-bit integers by executing four concurrent multiplications. Nevertheless, this technique requires a good understanding of the nature of the processing, for example, whether the operators in the segment to be parallelized are stateful as well as whether the outgoing tuples from a parallel segment must be emitted in a specific order.
- *Task parallelism* is a form of parallelism where independent processing stages of a larger computation are executed concurrently on the same or distinct data items. As an example, consider again how a modern *hyperthreaded* Central Processing Unit (CPU) [32] can execute different instructions from different threads, concurrently. As with pipeline parallelism, this kind of parallelism can be exploited naturally in an application where a stream that is fanned out to multiple downstream operators has

its tuples independently processed by each of them. More sophisticated uses are also common, where, for example, different tasks (or PEs as is the case in Streams) are placed on different CPU cores, or on different hosts, and execute concurrently.

Figure 9.25 depicts simple examples of these three forms of parallelism in the context of a SPA. In the figure, the operators that are executing concurrently are marked with stripes. The labels A, B, and C mark the operator instances that collectively parallelize a segment of an application's data flow graph.

The first form of parallelism shown in Figure 9.25 is pipeline parallelism. The execution time line on the right side of the diagram shows how incoming tuples 1–6, received from a source operator, are processed as time progresses. It can be seen that while tuple n is being processed by operator C, tuples $n - 1$ and $n - 2$ are also being processed by operators B and A, respectively. In this *three*-stage pipeline, up to *three* concurrent operators might be processing tuples at any time.

The second form of parallelism shown in Figure 9.25 is data parallelism. In this scenario, the operator immediately downstream from the source is used to split a stream into three outbound ones, distributing tuples to three operators. Effectively, different tuples are concurrently processed by independent instances of the A operator. For instance, as the time line suggests, tuples 1–3 are processed by each of the different instances of the A operator, simultaneously. Depending on whether the subsequent processing, downstream from the split point, must respect the original order that the tuples were emitted by the source, the outbound tuples from the parallel segment may need to be merged back and the order restored before they are further processed. The merging and reordering process is not transparent, hence, if it is necessary, an additional operator tasked with this job must also be included in the application.

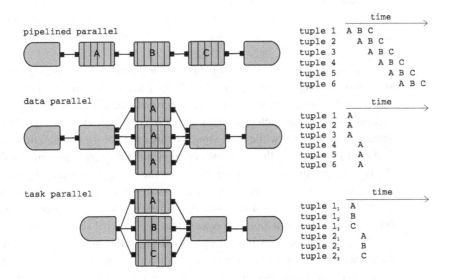

Figure 9.25 Pipeline, data, and task parallel computation in stream processing.

The third form of parallelism shown in Figure 9.25 is task parallelism. In this scenario, the fan-out in the stream produced by the source operator is used to send the same stream to three different downstream operators. As a result, three different tasks encapsulated by operators A, B, and C are performed, concurrently, on different copies of the same tuple. In this case, the operator immediately downstream from the parallelization segment is a `Barrier` operator, which implements a *barrier* synchronization point. In this way, the three instances of same logical tuple are brought together before being sent out for further processing.

It is important to note that the overall speed up that can be achieved through parallelization is always limited by the sequential segments of an application [33]. This is captured by *Amdahl's law* [34]: if we use P to represent the fraction of an application that can be parallelized and N to be the number of parallel processors, the speed up that can be achieved by the parallel execution over the sequential execution is given by $\frac{1}{(1-P)+P/N}$. Note that the speed up is bounded by $1/(1-P)$ no matter how many processors are available.

9.3.2.1   Principles of parallelization

In this section, we discuss some of the parallelization principles that can be applied to stream processing, focusing mainly on the types of parallelism we just discussed.

Pipeline parallelism

For an application to take advantage of pipeline parallelism, its computation has to be expressed as a sequence of stages, where each stage is an operator. While this is a necessary condition and most SPAs can be expressed in this way, this alone is not sufficient to extract performance gains.

The effective use of pipeline parallelism is predicated on increasing the *utilization* of limited computational resources by breaking down a large computation into multiple stages where there is *balance* between the amount of work performed by each stage.

Simply adding an additional stage may not be effective, for example, if that additional stage has to compete for resources with the earlier ones. Furthermore, the series of operators that form a pipeline must have computationally balanced stages. Balance in this case is defined as each pipeline stage having roughly the same throughput. This will minimize their idling, while increasing the utilization of the computational resources.

As a result, devising the *right* level of application decomposition into multiple stages is critical. Specifically, if certain stages of the computation are implemented using instances of a `Custom` operator, or if new primitive operators are going to be designed, or, perhaps, a combination of both approaches is going to be used, deciding how to break down the computation into *equitable* computational stages is fundamental.

To accomplish these goals, each operator, irrespective of how it is implemented, must have a well-defined purpose, where its computation is not overly complex. Indeed, also in line with the arguments that support modularity and reuse (Section 9.3.1), pipeline parallelism works best with *modular* operators.

The steps we outlined above are iterative. A developer must assess, possibly refine the design, and reassess both how the computational task is broken into pipeline stages as well as the complete makeup of the pipeline. Naturally, to iterate over and refine possible design alternatives, a developer must have access to profiling information about the application performance.

Specifically, to figure out how to improve the overall *utilization* of the computational resources, the percentage of the CPU cycles consumed by a series of sequentially executing operators must be monitored as part of a profiling run. Naturally, the pipeline design can be tweaked and the utilization increased by either adding additional stages if no fundamental problems already exist, or by improving the existing stages if imbalances exist, or, often, by a combination of these two measures.

On the other hand, to figure out how to *balance* each pipeline stage, the size of the input port queue (or any other indicator of utilization) for each operator in the pipeline must be inspected. In general, in the presence of imbalance, excessive data queued at a particular stage is an indication that this stage is a bottleneck and that the pipeline is not at its peak efficiency. In this case, the pipeline design can be tweaked by optimizing the bottleneck stage, for instance, by improving its internal algorithm, by splitting it into multiple stages to divide the work more equitably, or by parallelizing the stage via other types of parallelization (e.g., data parallelism). Again, the developer may need to iterate over multiple alternatives.

Data parallelism

Data parallelism involves splitting a stream across multiple channels, leading to a set of replicated segments of the data flow graph in place of the original one. Each segment consists of the same set of operators now processing a subset of the incoming tuples in parallel, and, in the last processing stage, their results are merged back into a single stream.

The splitting and merging parts of this flow can take different forms, depending on whether the operators involved are stateful or not, as well as on whether the downstream processing expects the original order of the tuples to be maintained. The type of data parallelism, *stateless*, *partitioned-stateful*, or *stateful*, dictates how these tasks are performed.

A *stateless data parallelism* configuration exists when all of the operators in the segment of the data flow graph to be improved are stateless. As a result, if that segment is replicated multiple times, any of the replicated segments can process any one of the incoming tuples. Naturally, the splitting of the incoming data and their assignment to a replicated parallel segment can be performed arbitrarily, since any parallel segment can process any of the tuples.

This type of data parallelism occurs, for instance, when a face detection algorithm operates on an incoming set of images, one at a time. In this example, data parallelism can be used to accelerate the application of the face detection algorithm (perhaps implemented as a set of operators) by simply splitting the incoming stream and assigning each image to a different instance of the algorithm running in parallel with the others. The collective outgoing flow is simply sent downstream. Depending on the nature of the downstream processing, re-ordering of the outbound tuples may be required to bring them back into their original order.

A *partitioned-stateful data parallelism* configuration exists when some of the operators in the segment of the data flow graph are *partitioned stateful* (see Section 4.3.1) and others are stateless. In this case, after splitting the stream to be routed to one of the parallel replicas of that segment, the tuple routing must be performed in a *consistent* manner, such that all tuples with the same values for the partitioning attribute are always sent to the same segment replica.

This type of data parallelism occurs, for instance, in the earlier processing stages of several finance engineering applications. Consider the computation of the VWAP of a particular stock. Recall that we introduced this computation as part of the partitioning of multiplexed flows in Section 9.2.2.1.

In this case, the routing of tuples to different replicas of the VWAP computation must ensure that each replica consistently gets only tuples associated with the ticker symbols it is configured to handle. For example, assuming the incoming stream is partitioned based on disjoint ranges of ticker symbols, the replica assigned to handle symbols "C" (i.e., Citigroup Inc) to "F" (i.e., Ford Motor Company) should never receive "IBM" (i.e., IBM Corporation) stock market transactions.

Alternatively, when there is a series of data parallel segments in an application's data flow graph, where each segment is partitioned stateful, but the partitioning attribute is different for each segment, the application can still be parallelized by employing *shuffling* to properly route the traffic.

As an example, consider a stream whose tuples contain log entries from an operational monitoring application (similar to the example in Section 9.2.1.2), where each log entry contains, among other elements, a `site` and a `mission` attribute. Consider also that we have a processing segment that contains a set of operators performing multiplexed processing on a per `site` basis, followed by a processing segment that contains a set of operators performing multiplexed processing on a per `mission` basis.

In this case, each of these two segments can be *independently* parallelized as shown by the top diagram in Figure 9.26, where three-way parallelism is applied on the first segment and two-way parallelism on the second one. Yet, this configuration has a potential bottleneck in that all of the streams coming out of the first parallel segment must be merged and then repartitioned to be fed to the second parallel segment. In essence, the middle segment is sequential and can potentially slowdown the whole application.

Shuffling can be applied to remove this bottleneck. This is shown in the lower diagram in Figure 9.26. In this configuration, rather than merging back all the streams from the first parallel segment and then perform a repartitioning, we employ a `Split` operator at the end of each instance of the first parallel segment. This split operation, which also occurs in parallel, employs the partitioning attribute required by the second segment. Each parallel instance of the second segment starts with a `Merge` operator, receiving one stream from each one of the upstream `Split` instances, thereby performing the merge also in parallel.

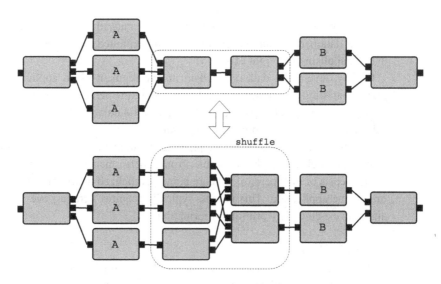

Figure 9.26 Using shuffles to improve the performance of series of partitioned stateful data parallel sections partitioned on different keys.

A *stateful data parallelism* configuration exists when some of the operators in the segment of the data flow to be improved are stateful. In this case, data parallelism cannot be leveraged unless a developer takes into consideration application-specific properties, which might make it possible to preserve its overall correctness in a parallel configuration. Usually, special merge operators along with synchronization techniques are required to achieve this objective.

All forms of data parallelism are subject to *ordering requirements*, as parts of an application downstream of a data parallel segment may need to receive the resulting tuples in an order logically corresponding to the original one.

In such a situation, after the parallel segment, a merge operation must take place to restore the original order in the data flow. In general, merging in this case must rely on additional attributes, such as sequence numbers, that might not have existed originally in the sequential implementation, but must be added as part of the parallelization design.

More complex configurations where ordering must be preserved, but not as strictly as on an individual tuple basis, are also possible. For example, in some cases the downstream operators do not require total ordering, as long as ordering is preserved at window boundaries.

For example, when a commutative aggregation operation such as the *average* or the *summation* over the values of a tuple attribute for a collection of tuples in a window takes places, the individual ordering of these tuples within the window is irrelevant. Yet, despite the fact that these incoming tuples may all be processed in parallel in a non-deterministic order, all of the tuples must be present in the window at the time that the aggregation is computed.

In summary, a developer must first understand where the processing bottleneck is in the flow graph, as is the case with any performance optimization endeavor. Once this information is available, it is necessary to study this particular site to determine whether data parallelism opportunities can be exploited as well as the specific *form* of data parallelism that might exist.

Clearly, leveraging data parallelism in any segment of an application's data flow graph consisting of stateless and partitioned stateful operators is usually a simple engineering task. On the other hand, employing stateful data parallelism can be complex as substantial parts of an application might have to be re-designed.

As discussed in the context of pipeline parallelism, whether exploiting any form of data parallelism will be profitable, from a performance standpoint, depends on the relative costs of the operators that are being parallelized as well as on the overall structure of the application. This process typically requires iterating over a set of possible alternatives, guided by profiling information.

Task parallelism

In stream processing, task parallelism is closely aligned with the modularity of an application. In a well-engineered application, the analytical task is usually decomposed into a set of logical tasks to simplify both its design and its implementation process. In many situations, the corresponding physical layout associated with a modular application will be naturally task parallel because modules will be able to operate independently from others. Yet, this type of "accidental" task parallelization can, in most cases, be further improved based, again, on an iterative analysis of the application's performance.

As part of this refinement, the individual tasks to be executed in parallel should take approximately the same amount of time on a per unit of work basis. In other words, assuming that the outcomes of each parallel task (a unit of work) will eventually have to be brought together at a synchronization point, ideally all tasks should output their result roughly at the same time to minimize idling.

In general, due to this need to periodically synchronize the tasks, the overall throughput of a task parallel flow is limited by the throughput achieved by the task that takes the largest amount of time. Clearly, when task imbalance is present, the faster tasks will eventually be slowed down, generally as a result of back pressure, which reduces the overall utilization of the available computational resources. In extreme cases, the presence of significant imbalances may render the task parallel decomposition ineffective when compared to a sequential configuration.

As is the case with pipeline and data parallelism, application profiling can aid a developer to figure out whether imbalance is impacting the effectiveness of task parallelism.

In particular, an increasingly longer queue associated with an operator's input port (where this operator is part of a parallel task) is usually an indication that imbalance has become a problem in an application. Similarly, high CPU utilization associated with a task contrasted with other tasks running on CPUs that are partially idle can also be an indication of imbalance if all of them are, a priori, determined to be CPU-bound tasks.

9.3.2.2   **Patterns of parallelization**

As discussed earlier, exploiting parallelism is one of the primary mechanisms available to a developer to speed up an application as well as to scale it up so it can handle larger workloads.

A parallel application differs from a sequential one in how it uses the computational resources where it runs. Most modern OSs provide two abstractions for exploiting parallelism: threads and processes.

A *thread* is the smallest processing unit an OS can schedule and run. A *process*, on the other hand, is also an OS scheduling unit and corresponds to an instance of a program. A process can host multiple threads, which share the same address space, and run concurrently.

A parallel SPA can make use of either construct as well as combine them in a distributed configuration, where an application spans multiple hosts. In Streams, both of these constructs are exposed and can be used in different ways.

Threads can be used directly, when an operator's implementation creates them and, indirectly, by employing *threaded* input ports in an operator's instance (Section 8.5.9).

Processes, on the other hand, are indirectly exposed. As seen in Section 8.4.2, a Streams' PE corresponds to a process and each one hosts an application partition. Partitions can be directly specified by an application developer and the SPL compiler outputs them when the application is built.

Thus, an application that employs one or more multi-threaded operators, or threaded input ports, and consists of several partitions, is making use of multiple OS threads and processes. Yet, the question that remains is how to employ these constructs to improve an application's performance.

Clearly, the effective utilization of parallelization techniques requires an understanding of an application's internal processing logic, how it was decomposed, the computational environment where this application runs, and the workload it is subjected to. Such understanding is necessary because, in most cases, parallelizing an application requires experimenting with different physical configurations and design alternatives, guided by profiling results extracted from runtime traces.

As a starting point, we will explore the parallelization techniques available in SPL: the use of threaded input ports, the use of data partitioning, as well as the use of multiple application partitions.

In each case, we will also showcase a different type of parallelism, noting that the relationship between the specific technique and the type of parallelism shown is not exclusive. In other words, the techniques discussed in the rest of this section can also be used to express other types of parallelism.

Pipeline parallelism using threaded ports

Consider an application, FaceDetector, designed to perform face recognition in a two-stage process. To start, the incoming Images stream is cleaned up and the amount of noise in the individual images is reduced by the NoiseReducer operator and, subsequently, the FaceDetector operator performs a face recognition task on these

```
1   composite FaceDetector {
2     graph
3       stream<Image image> Images = MyImageSource() {}
4       stream<Image image, Image imageRN> RNImages = NoiseReducer(Images) {
5         output
6           RNImages: imageRN = ReduceNoise(image);
7         config
8           threadedPort: queue(Images, Sys.Wait);
9       }
10      stream<Image image, list<Rect> faces> Faces = FaceDetector(RNImages) {
11        output
12          Faces: faces = Detect(imageRN);
13        config
14          threadedPort: queue(RNImages, Sys.Wait);
15      }
16      () as Sink = MyImageSink(Faces) {}
17    config
18      placement: partitionColocation("all");
19  }
```

Figure 9.27 Using threaded ports to expose pipeline parallelism.

```
1   composite FaceDetector {
2     graph
3       stream<Image image> Images = MyImageSource() {}
4       stream<Image image, Image imageRN> RNImages = NoiseReducer(Images) {
5         output
6           RNImages: imageRN = ReduceNoise(image);
7         config
8           placement: partitionExlocation("pipeline-parallel");
9       }
10      stream<Image image, list<Rect> faces> Faces = FaceDetector(RNImages) {
11        output
12          Faces: faces = Detect(imageRN);
13        config
14          placement: partitionExlocation("pipeline-parallel");
15      }
16      () as Sink = MyImageSink(Faces) {}
17  }
```

Figure 9.28 Using multiple partitions to expose pipeline parallelism.

improved images, generating the list of people it was able to identify in each image.
Figure 9.27 depicts this application, implemented as a parallel pipeline.

Note that in this example, all of the operators are fused in a single application partition, using the `partitionColocation` configuration, resulting in only one OS
process. Hence, there is no multi-process parallelism in this case. On the other hand,
both the `NoiseReducer` and the `FaceDetector` operators employ threaded input
ports and, as a result, both of them can concurrently execute their tasks as the resulting
PE makes use of multiple threads of execution.

This arrangement is an example of pipeline parallelism as each of the computational stages might be, simultaneously, operating on different images. Naturally, other parallel designs are also possible, including, for example, the use of
`partitionExlocation` constraints to place the two operators on different PEs
(Figure 9.28), which can optionally be executed on different hosts. Clearly, finding the
best performing configuration requires further experimentation.

```
1   composite VWAP {
2     graph
3       stream<StockTicker> Tickers = StockExchangeFeed() {}
4       (stream<StockTicker> TickersL;
5        stream<StockTicker> TickersR) = Split(Tickers) {
6         param
7           index: toCharacterCode(symbol[0]) < toCharacterCode("K") ? 0: 1;
8       }
9       stream<StockTicker, tuple<float64 vwap>> VwapL = Vwap(TickersL) {
10        window
11          TickersL: sliding, count(100), count(1);
12        param
13          partitionBy: symbol;
14        config
15          threadedPort(TickersL, Sys.Wait);
16      }
17      stream<StockTicker, tuple<float64 vwap>> VwapR = Vwap(TickersR) {
18        window
19          TickersR: sliding, count(100), count(1);
20        param
21          partitionBy: symbol;
22        config
23          threadedPort(TickersL, Sys.Wait);
24      }
25      stream<StockTicker, tuple<float64 vwap>> Results = Union(VwapL; VwapR) {}
26    config
27      placement: partitionColocation("all");
28  }
```

Figure 9.29 Using data partitioning to expose data parallelism.

Data parallelism using flow partitioning

Let us revisit a variation of the VWAP application, introduced in Section 4.3.1. This application computes the VWAP of stock prices traded in an electronic exchange and is depicted by Figure 9.29.

In this case, a source operator (StockExchangeFeed) is used to read and stream out the trading transactions. A Split operator is then used to divide the original stream into two subflows depending on whether the stock ticker symbol associated with a tuple is lexicographically below or above "K" (i.e., the Kellogg Company in the New York Stock Exchange). In other words, the Split operator ensures that transactions with the same ticker symbol are always routed to the same outbound stream.

The VWAP computation carried out by the Vwap operator[13] is a stateful operation performed on a per-partition basis, where the partitioning attribute is the ticker symbol. This configuration ensures that the operator's internal state is split and segregated.

Two instances of the Vwap operator are used to process the two streams output by the Split operator. As a final step, the output produced by the two Vwap operators is merged using a Union operator (available in the SPL standard toolkit) on a first-come, first-served basis.

To expose the partitioned-stateful data parallelism in this application, threaded input ports are used by the two instances of the Vwap operator. This configuration ensures that their two incoming streams are processed in parallel. In this case, we chose to fuse

---

[13] In this case, the Vwap operator is assumed to be a primitive operator for simplicity. As seen in Section 4.3.1, an Aggregate and a Functor operator can be used to implement an equivalent computation.

all operators into a single application partition using the `partitionColocation` directive.

Nevertheless, while this configuration will make the application multi-threaded, in some cases, a better alternative might turn out to be the use of multiple application partitions in a multi-process configuration.

Task parallelism using threaded ports

Now we consider the `CustomerTrendAnalysis` application, a hypothetical application designed to analyze Twitter feeds and extract customer sentiment about products, from their tweets. Figure 9.30 depicts this application's implementation in SPL.

In this scenario, the `TweeterFeed` source operator is used to ingest the Tweeter stream. Subsequently, for each tweet, stored in a tuple whose only attribute is a chunk of unstructured text, two independent tasks must be performed.

First, the *sentiment* implied in the tweet must be extracted using the `SentimentDetector` operator. This operator can employ a combination of several text analytics techniques, implemented by the `DetectSentiment` function. This function's results consist of a list of strings representing the customer's sentiment and is output as a tuple.

Second, the product mentioned in the `tweet` is detected using the `Product Detector` operator that also uses text analytics, here encapsulated in the `DetectProduct` function.

Observe that both of these analytic tasks can be carried out independently, an indication that there is latent task parallelism in this application. Once tweets are analyzed by these two tasks, the outgoing streams from each operator are brought together at a synchronization point implemented via a `Barrier` operator. This operator produces tuples whose contents include the original `tweet` as well as the `sentiments` and `products` that have been detected.

```
1   composite CustomerTrendAnalysis {
2     graph
3       stream<rstring tweet> Tweets = TweeterFeed() {}
4       stream<rstring tweet, list<rstring> sentiments>
5         Sentiments = SentimentDetector(Tweets) {
6       output
7         Sentiments: sentiments = DetectSentiment(tweet);
8       config
9         threadedPort(Tweets, Sys.Wait);
10      }
11      stream<rstring tweet, list<rstring> products> Products =
12        ProductDetector(Tweets) {
13      output
14        Products: products = DetectProduct(tweet);
15      config
16        threadedPort(Tweets, Sys.Wait);
17      }
18      stream<Sentiments,Products> Results = Barrier(Sentiments; Products) {}
19    config
20      placement: partitionColocation("all");
21  }
```

Figure 9.30 Unlocking task parallelism.

As was the case in prior examples, all of the operators in this application are fused into a single partition, resulting in one PE and, thereby, one OS process. Again, to unlock the untapped task parallelism in the sequential version of this application, threaded ports are used by the `SentimentDetector` and the `ProductDetector` operators, allowing them to run as two independent threads of execution.

Once again profiling this application in its runtime environment with its actual workload might reveal that further parallelization improvements might be possible. These improvements might come, perhaps, by further decomposing the analytic tasks performed internally by the `SentimentDetector` and the `ProductDetector` operators, by replicating them, by employing multiple processes in the form of multiple and, potentially, distributed application partitions, or by a combination of all these techniques.

9.3.3   Performance optimization

In general, optimizing distributed applications involves experimentation as well as an understanding of algorithms and system issues that might affect their performance. A diverse set of techniques and approaches, including the parallelization techniques we described in the previous section, are part of the tools available to developers.

In this section we discuss design principles and implementation patterns that can aid in performance optimization tasks. We focus primarily on increasing an application's throughput, but we also discuss latency when the throughput-driven optimizations adversely impact it.

Performance optimization might be time-consuming and result in reconfiguring an application's data flow graph, by changing and improving the implementation of certain operators, choosing or tuning their analytics, as well as modifying how an application is physically deployed. Yet, developing performance optimization skills in the context of stream processing is important, as several continuous processing applications must process vast amount of data in a timely manner (Section 1.2).

9.3.3.1   Principles of optimization

Certain application design principles, if applied in the earlier stages of application development, can establish strong foundations for a scalable system. We discuss some of these principles next.

Early data volume reduction
The computational costs of a SPA can be significantly decreased by filtering out data identified as *irrelevant* as early as possible as well as by reducing the volume through other means.

Identifying data as irrelevant earlier in the processing usually requires employing dynamic adaptation techniques of the sort described in Section 9.2.3.1. On the other hand, reducing an application's data volume can usually be accomplished through sampling, filtering, quantization, projection, and different types of aggregation as seen in Section 9.2.2.1.

Figure 9.31 Operator reordering: performing a data reduction as early as possible.

A more subtle form of data reduction consists of reordering the processing in a data flow graph to account for the computational costs and selectivity[14] of different operators. More concretely, consider a simple configuration with two streaming operators, A and B.

Assume that operator A generates the stream consumed by B. Furthermore assume that operator A has a per-tuple computational cost c_A and a selectivity s_A, whereas operator B has a per-tuple cost c_B and selectivity s_B. The computational cost of executing this flow is given by $c_A + s_A \cdot c_B$. Now assume that it is possible to reorder the operators such that the results from operator B are sent to operator A. In this setup, the new computational cost of the flow is given by $c_B + s_B \cdot c_A$.

In general, executing the computationally cheaper operator (i.e., the one with the lower cost c) and/or the more selective operator (i.e., the one with the lower s) earlier reduces the overall cost of the segment. Figure 9.31 illustrates this principle.

Naturally, it may not always be possible to move the more selective or lower cost operator as this rearrangement might violate the application semantics. For example, when the downstream operator relies on tuple attributes that are computed by the upstream one, this is clearly the case.

There has been substantial research on this topic, especially when it comes to relational operators [35] as most relational databases automatically exploit these types of optimizations. Currently, making use of such of optimizations in a Streams application is a manual process.

Redundancy elimination
The removal of redundant segments from an application's data flow graph can free up computational resources and, potentially, result in performance improvements.

A common structural pattern in SPAs is the replication of a flow to expose task parallelism (Section 9.3.2), i.e., processing the same tuples via different sets of operators.

In some cases, there is a common subset of operators used by the replicated flows, performing the same work multiple times. Figure 9.32 shows a redundancy elimination transformation, where the superfluous operators, in this case, two instances of the A operator, are moved before the split. Here, the split could be a fully replicating split (all tuples are sent to all segments) or partially replicating split (some tuples are sent to all segments).

The type of transformation shown in Figure 9.32 is preferred under sequential execution, since the total amount of work is reduced. Naturally, the more computationally costly the replicated operator is, the higher are the savings realized by this optimization.

---

[14] An operator's selectivity is defined as the number of tuples produced per tuples consumed; e.g., a 0.1 or 10% selectivity means that the operator emits 1 tuple for every 10 consumed.

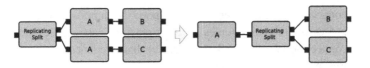

Figure 9.32 Redundancy elimination by sharing common processing.

If the part of the flow graph before the split and the two subflows after the split are each being run by a different thread of execution, then this transformation is only beneficial when the post-split subflows are application bottlenecks. If the pre-split flow is the bottleneck, then eliminating the redundancy and moving the operator to the pre-split flow will reduce the overall throughput.

Note that this optimization is safe when the redundant operators are stateless. If the replicated operators are stateful, the optimization is only safe when the split operator is fully replicating the flow, so that each subflow sees the same content as the original stream before the split.

Operator fusion

Different transport layers can be used to implement a stream connection between operators, depending on how the consumer and producer operators are placed with respect to the application's PEs (Section 5.3.3). From a performance standpoint, each type of transport layer incurs different types of I/O and computational overheads.

For instance, when the operators at the two ends of a stream connection are running in the context of PEs placed on different hosts, there is a non-negligible transport cost incurred for transferring tuples across this connection via the network. On the other hand, if these two operators are fused inside a single PE in the same host, the direct call transport layer (Section 8.5.4) is used, bypassing most of the costs associated with networking. SPL implements operator fusion via partition placement directives, as we have seen in Section 5.3.2.

Operator fusion can be effective in improving the application's performance when the per-tuple processing cost of the operators being fused is *low* compared to the cost of transferring the tuples across the stream connection. This is something that must be assessed by profiling the application.

However, since operator fusion is implemented as a function call, it eliminates the potential for pipeline parallelism that otherwise exists in the multi-Processing Element (PE) configuration. Naturally, one can modify the application further to explicitly make it multi-threaded. This can be done either by modifying one of the operator's internal implementation so it employs multiple threads, or by appropriately making use of threaded ports in the operator's configuration.

In most cases, fusing operators is semantically safe as it changes only the physical layout of a portion of the application. Nevertheless, when an application makes use of a feedback loop in its data flow graph, fusing operators might result in an infinite loop. This might occur in cases where all the stream connections involved in the loop are connected to *tuple generating* input ports. Such ports generate new tuples as a result of processing incoming tuples. In most practical cases, a feedback loop will include one connection (the feedback connection) that is connected to a *control* input port. Control input ports only modify the internal state of the operator as a result of processing incoming tuples, but do not cause the operator to emit new tuples.

Tuple batching

Processing a group of tuples in every iteration of an operator's internal algorithm or, more simply, employing tuple batching, is an effective technique for adjusting the trade-off between latency and throughput.

Batching tuples can increase an algorithm's throughput (i.e., the number of results the algorithm produces per unit of time) at the expense of higher latency (i.e., the delay involved in processing a particular tuple).

Figure 9.33 illustrates the effect of tuple batching in a SPA. Note that a *batcher* operator is used to group together multiple incoming tuples into a single outbound tuple containing a collection of the original tuples grouped in a list. In this case, the operator downstream from the *batcher* must be modified to work with a batch of tuples, rather than a single tuple at a time.

Many algorithms can be implemented more efficiently when the incoming data arrives as a batch. Some computational savings occur because various *warm-up* and initialization tasks performed by these algorithms are performed only once per batch as opposed to once per tuple. Moreover, an algorithm that involves executing large amounts of code might also benefit from improved instruction cache locality when data is processed in batch. Finally, an algorithm that involves querying a large amount of state, internal or external, can amortize this cost if a single or fewer queries can be issued on behalf of the current batch of incoming tuples.

In general, tuple batching is semantically safe and can be applied as long as a version of the operator, downstream from where batches are created, can be made to work with them. Nevertheless, to ensure that batching can indeed improve an application's performance, it is often necessary to profile the application. Profiling in this case must include a study on how the size of the batch affects the application's throughput and, more importantly, whether the per tuple latency remains within the allowed limits for the application.

Load balancing

As seen in Section 9.2.2, some SPAs make use of flow partitioning to distribute their workload, usually as a means to expose data or task parallelism. When such an approach is used, ensuring that the load is distributed evenly across the different subflows is usually a source of performance improvements, as threads and processes use the available

Figure 9.33 Batched processing: trading higher latency for higher throughput.

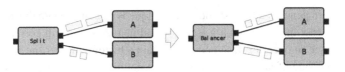

Figure 9.34 Load balancing by distributing the workload evenly.

Figure 9.35 Load shedding by gracefully degrading under overload.

CPU resources in a more balanced fashion and minimize idling. This is generally known as *load balancing* and is shown in Figure 9.34.

Naturally, optimizing an application by seeking an improved load balancing is more effective when there is inherent imbalance in the workload. Sources of imbalance include skews in the processing costs associated with different tuples, as well as diverging computational costs across different split flows. Again, application profiling is usually necessary to verify whether such conditions arise as well as to suggest what kind of changes might address the imbalance.

In general, application changes driven by attempting to improve load balancing are semantically safe when consistent partitioning conditions are preserved.

Load shedding

Load shedding describes the techniques used by an operator to reduce the amount of computational resources, such as CPU cycles and memory, it uses to more efficiently process high data rate streams.

These techniques are often employed by applications that face soaring workloads on a constant or transient basis. In general, load shedding has two goals. First, they attempt to sidestep sustained increases in memory utilization, e.g., the sizes of internal data structures and buffers an operator might manage. Second, they attempt to keep the operator's latency low and throughput high, by limiting the amount of work an operator will perform per unit of time (effectively relieving back pressure). As a generic example, Figure 9.35 illustrates load shedding in action in the form of a standalone *shedder* operator.

Load shedding techniques cover a wide spectrum of sophistication. On one end of this spectrum, simply dropping incoming tuples based on an application-specific policy can provide immediate relief under transient overload. If more intelligence is needed in this process, different data reduction techniques (Section 9.2.2.1), such as sampling, can be tapped to preserve certain statistical properties of the incoming flow while reducing the operator load (Section 10.5). On the other extreme of the spectrum, developers can also make use of adaptive techniques that employ a combination of strategies. They include reducing the incoming flow and continuously changing the operator's internal algorithm based on an assessment of the degree of operational stress.

In general, a load-shedding technique is effective when the dropped tuples minimally impact the application's accuracy and its results are still correct. More concretely, consider an operator designed to maintain a histogram over the values of a particular tuple attribute of an incoming stream. By definition a histogram is a statistical summary. If an operator *carefully* drops half of the incoming tuples, the accuracy of the histogram will most certainly not drop by half, instead it should face a rather small degradation in quality with respect to the histogram computed with the totality of the incoming flow (Chapter 10).

Unlike most of the other forms of application optimization discussed so far, load shedding will alter the outcome of an operator and, hence, the results produced by a SPA. Yet, a well designed load shedding technique aims at minimizing the impact on the *accuracy* of the results, while achieving the desired level of resource utilization reduction. In other words, a good load shedding technique provides graceful degradation under operational overloading.

9.3.3.2   Patterns of optimization

Making use of techniques to improve the performance of an application requires understanding its specific performance needs as a function of throughput and latency, the characteristics of its workload, the environment where the application runs, as well as its tolerance to potential decreases in the accuracy of the results produced.

The SPL language includes certain constructs to simplify the implementation of these optimization-driven patterns. From volume reduction, to batching, as well as load balancing and load shedding strategies, the language makes available ways to express and combine these strategies. As we will see, when it comes to optimizing an application, certain implementation patterns emerge, and they are the focus of this section.

Tuple batching

Tuple batching can be effective in increasing an operator's throughput. As discussed, when this strategy is used, the primary concern is to limit the potential latency increase that this operator might face.

The simplest way to cap the increase in latency is to limit the size of the batches to be created, either by restricting the maximum number of tuples to be batched (referred to as tuple batching *by count*), or the amount of time an operator waits to accumulate additional tuples (referred to as tuple batching *by time*).

Tuple batching by count works well as long as the flow rate of the incoming stream is mostly steady. An intermittent flow causes the batching interval and, hence, the operator's latency to fluctuate as a function of the tuple inter-arrival intervals. Hence, tuple batching by time, when dealing with an intermittent tuple flow, provides better control over the operator's latency.

In some cases, both techniques can be combined. If the maximum number of tuples for a batch is not reached within a fixed time frame, a partial batch is released after that amount of time has elapsed. The application depicted in Figure 9.36 illustrates this strategy.

```
1   composite Batcher(output Out; input In) {
2     param
3       expression<float64> $maxTimeToWaitForABatch;
4       expression<uint32> $maxNumberOfTuplesToBatch;
5     graph
6       stream<int8 dummy> Beat = Beacon() {
7         param
8           period: 1.0;
9       }
10      stream<list<In> batch> Out = Custom(In; Beat) {
11        logic
12          state: {
13            mutable tuple<list<In> batch> batchedTuple = {};
14            float64 startTime = 0.0;
15          }
16          onTuple In: {
17            if (size(batchedTuple.batch)==0)
18              startTime = getTimestamp();
19            insertM(batchedTuple.batch, In);
20            if (size(batchedTuple.batch)==$maxNumberOfTuplesToBatch) {
21              submit(batchedTuple, Out);
22              clearM(batchedTuple.batch);
23            }
24          }
25          onTuple Beat: {
26            float64 time = getTimestamp();
27            if ((size(batchedTuple.batch)>0) &&
28                (time-startTime>=$maxTimeToWaitForABatch)) {
29              submit(batchedTuple, Out);
30              clearM(batchedTuple.batch);
31            }
32          }
33      }
34  }
```

Figure 9.36 Tuple batching.

In this case, the `Batcher` composite operator takes two parameters, one specifying the maximum number of tuples in a batch (`$maxNumberOfTuplesToBatch`) and, the other specifying the maximum time, in seconds, to wait for a complete batch (`$maxTimeToWaitForABatch`).

The `Batcher` composite employs a `Custom` operator with two input ports, one for receiving the incoming stream (`In`) whose tuples should be batched and another to receive a regular 1-second timeout signal (`Beat`). Having the ability to react to the arrival of tuples from either port (using two `onTuple` directives) is what makes this operator able to implement both types of batching.

The `Custom` operator accumulates the incoming tuples in a state variable (`batchedTuple`), using SPL's `insertM` built-in function. It then emits this tuple holding the batch when its maximum size[15] is reached. Alternatively, it might also emit the tuple holding the batch as a result of the maximum batching time having elapsed. This is accomplished by employing the operator's second input port that receives a periodic beat. Once a batch is streamed out, the operator's internal state is reset by clearing the `batchedTuple` using SPL's `clearM` built-in function, which resets the batch-holding `list`.

---

[15] The size of the batch is computed using SPL's built-in `size` function on the `batch` attribute of the `batchedTuple` state variable. This attribute has a `list` type and holds the current batch of incoming tuples.

Load balancing

Load balancing can be implemented in many different ways depending on the needs of an application. Regardless of how it is done, a well-designed load balancing strategy hinges on splitting the incoming flow of tuples in a *balanced* manner. In this way the different subflows can be independently processed and the threads, in the case of multi-threaded load balancing, or processes, in the case of multi-processing load balancing, perform roughly the same amount of work.

The SPL standard toolkit provides an operator called `ThreadedSplit` that can perform this task for scenarios when *any* of the downstream data flow subgraphs can handle *any* of the incoming tuples. This operator manages a fixed-size buffer for each of its output ports (as many as the number of split subflows) and uses separate threads to stream out the accumulated tuples independently.

The operator provides workload balancing based on the status of the individual buffers. Specifically, when one of the downstream subgraphs is busy processing a tuple that triggers a more computationally intensive process, the `ThreadedSplit` buffer where this tuple originated will likely fill up. As a result, a future incoming tuple that arrives at the `ThreadedSplit` operator will be stored in one of the buffers that still has space and, thus, will get processed by one of less busy downstream subgraphs.

The application implemented by the `BalanceLoad` composite operator depicted by Figure 9.37 shows how the `ThreadedSplit` operator can be used in practice. The instance of this operator used in the `BalanceLoad` operator splits the incoming `Flow` stream into two subflows: the `SubFlow1` and the `SubFlow2` streams. In this case, each of the `ThreadedSplit` output ports has a buffer with the capacity to hold a single outgoing tuple, as defined by the operator instance's `bufferSize` setting. In this way, an incoming tuple will be routed to any output port whose buffer is currently empty. If neither buffer has space, back pressure propagates upstream.

Adaptive load shedding

Employing load shedding is only relevant when an application is facing overload. Therefore, there are two pieces to consider when engineering such a solution. First, there is the *detection* mechanism by which an application proactively senses that overloading is about to ensue. And, second, there is the *reaction* mechanism, or the load shedding algorithm itself.

One way to detect the onset of overloading is to employ a fixed-size buffer to store incoming tuples and use the buffer occupancy as a proxy for the intensity of the load facing the application. The `LoadShedder` composite operator sketched in Figure 9.38

```
1   composite BalanceLoad {
2     graph stream<Data> Flow = MySource() {}
3     (stream<Data> SubFlow1, stream<Data> SubFlow2) = ThreadedSplit(Flow) {
4       param
5         bufferSize: 1u;
6     }
7     stream<Result> Results1 = Process(SubFlow1) {}
8     stream<Result> Results2 = Process(SubFlow2) {}
9   }
```

Figure 9.37 Load balancing using the standard toolkit `ThreadedSplit` operator.

```
1   composite LoadShedder (input In) {
2     graph
3       () as Worked = Custom(In) {
4         logic
5           onTuple In: {
6             mutable int64 nQueued = 0;
7             getInputPortMetricValue(0u, Sys.nTuplesQueued, nQueued);
8             if(nQueued>9001)
9               ignoreWork();
10            else if(nQueued>5001)
11              performWorkCheaply(In);
12            else
13              performWork(In);
14          }
15        config
16          threadedPort: queue(In, Sys.Wait, 1000);
17      }
18  }
```

Figure 9.38 Adaptive load shedding.

demonstrates this technique. This composite makes use of a `Custom` operator whose input port is associated with a fixed-size `queue`, used to buffer the incoming tuples.[16]

Whenever the `Custom` operator receives an incoming tuple, it uses the built-in `getInputPortMetricValue`[17] function to check the number of tuples currently queued on its input port. If the queue's buffer occupancy is below 500, the tuple is processed normally, using the `performWork` function. If the buffer occupancy is above 500, but less than 900, then a function with lower computational costs, `performWorkCheaply`, is employed to process this incoming tuple. If the buffer occupancy is above 900, then this incoming tuple might be *ignored*, i.e., the specific action is whatever the `ignoreWork` function does, including, potentially, dropping it altogether.

Note that this example assumes that any application using this composite operator is amenable to graceful degradation, where quality can be traded off for cheaper processing, either by using a less accurate processing function to reduce the computational cost or by dropping tuples to avoid the processing cost entirely.

9.3.4 Fault tolerance

The long-running nature of SPAs creates the potential for accumulating substantial internal state over time. This state, in the form of internal data structures, statistical models, and other artifacts, is a significant factor in determining the fault tolerance requirements associated with any continuously running application.

Depending on the nature of the analytic task implemented by an application, the state it maintains can be highly *transient* or relatively *stationary*. For instance, keeping the last ten stock transactions to compute an average price for a particular stock showcases a

---

[16] The specific configuration for this queue can accommodate up to 1000 tuples and, once the buffer is full, the operator blocks as indicated by the `Sys.Wait` directive.

[17] This function is used to retrieve metrics associated with input ports of an operator. Its first parameter is used to specify the input port index. The second one is used to specify the metric of interest. And the third one, which is modified as a result of the call, is used to store the metric value.

highly transient state. Naturally, there is an assumption here that most stocks are highly liquid and traded frequently. Furthermore, stock prices for the most part do not exhibit extremely large fluctuations in prices in a very small amount of time. In other words, a short-lasting disruption in this type of processing is not particularly concerning, as the state can be rebuilt from scratch without any major processing hiccups.

On the flip side, consider an application that builds an online graph of edges and nodes over SMS messages, where this graph captures the communication patterns between mobile phone users. Assume that in a particular application's context, an instance of this graph is maintained over a one week window. In this scenario, the state represented by this graph changes slowly. In other words, even a short-lasting disruption that causes this state to be lost is potentially problematic for this application, as its state cannot be rebuilt easily from future incoming SMS messages. In general, the state maintained over larger windows is more stationary, whereas state maintained over smaller windows is more transient.

Another important reliability facet of a SPA is how critical and time-sensitive its analytic task is, particularly, if it faces an operational disruption. In general, a consumer of the results produced by a continuous application expects that these results arrive in a timely manner. Some applications can be considered highly *latency sensitive*, as results that are produced late can be deemed worthless.

For instance, in many algorithmic trading applications [36], if a decision to buy or sell a particular financial instrument, after analyzing the current market conditions, is made too late, it usually will not be acted upon, either because the opportunity is gone or because another market competitor already took advantage of it. In other types of applications, long latency might be totally unacceptable as it might result in loss of life or equipment. In either case, failures in the application's critical infrastructure will result in increases in latency.

Finally, yet another dimension concerning an application's reliability relates to the fact that not every segment of an application's flow graph as well as the state it keeps is equally important.

From this point forward, we will refer to an application that cannot tolerate any loss of data or state as a *loss intolerant* application and one that can as a *loss tolerant* application.

Hence, devising a fault tolerance strategy for an application must take into consideration the nature and importance of the different components of its state, as well as the costs, computational and monetary, to put in place resilience measures to guard against failures.

In general, for a loss intolerant and/or latency sensitive application, maintaining the timeliness and operational continuity requires the implementation of a *robust* fault tolerance strategy aiming at providing high availability from the onset. On the other hand, for a loss tolerant application, the analysis must include an evaluation of how much tolerance it has to data loss. In this case, a *partial* fault tolerance strategy aiming at mitigating risks might be sufficient.

In the rest of this section, we look at different ways of implementing fault tolerance strategies in Streams, focusing primarily on the nature of an application's internal state.

9.3.4.1   Principles of fault tolerance

An application failure might stem from an error in the application itself, a software bug, a failure in the SPS middleware, or a failure in the computational infrastructure that supports the application and the SPS, e.g. the hosts, the networking infrastructure, or the storage system.

When devising a fault tolerance strategy for an application, failures in the application logic itself are of less interest, as they typically require fixing the application as part of its maintenance cycle. Failures in the middleware will also require software modifications that are typically outside the control of application developers and administrators.

Hence, when discussing fault tolerance, we focus on the third type of failures, i.e., failures in the computational infrastructure that are typically *transient* in nature. In other words, a host failure (or other similar types of equipment failure) might lead to that host being temporarily shutdown. Despite that, we assume that the overall runtime environment can be *quickly* reconfigured with a replacement host or that it will simply remain generally functional, even if working with a reduced processing capability. In either case, the applications or their parts that used to run in the disrupted host will themselves face a disruption, which requires some form of reaction, manual or automated, to restore their normal operation. As discussed in Section 8.5.5, the Streams application runtime is fault tolerant and provides several mechanisms to support the development of autonomic and fault-tolerant applications. Yet, the specific mechanisms used by an application are determined by its implementation and operational requirements.

The principles behind any fault tolerance strategy, when applied to stream processing, include making use of three basic mechanisms: cold restart, checkpointing/restart, and replication. We discuss them next, followed by an examination of a partial fault tolerance strategy.

Cold restart

The simplest way to react to an application failure stemming from an infrastructure failure is to restart the whole application or the segment affected by the failure. Naturally this approach, a *cold restart*, is only feasible if the application's internal state is *transient* and can rapidly be rebuilt from new incoming data.

In this case, a cold restart will take place once the SPS middleware has had a chance to detect the failure and has reconfigured itself. In Streams, both the runtime and the application can be autonomically restarted.

Consider a hypothetical operational monitoring application that processes system logs to compute per-minute averages on resource usage by other applications, and generates alerts upon the detection of abnormal levels.

This application can rebuild its entire state upon restart. Assuming that it is acceptable for the monitoring system to be out of service for a small amount of time, when facing an unlikely failure event, a simple cold restart scheme suffices as the fault tolerance strategy for this application.

Figure 9.39 shows a sample code fragment where *all* partitions within an application are configured to be `restartable` and `relocatable`. As it can be seen, this is

```
1  composite Main {
2    graph
3      ...
4    config
5      restartable : true;
6      relocatable : true;
7  }
```

Figure 9.39 Marking all partitions as restartable and relocatable for performing cold restart upon failures.

achieved by making use of settings at the level of an application's *main* composite, indicating that all of the application's PEs can be restarted and moved to a different host in the event of a failure.

Checkpointing

Certain applications, while loss tolerant, might want to minimize the disruption due to transient failures. An application can run for long periods, potentially uninterrupted, until a failure occurs. During this period its internal state might be continuously building up at a low pace. Checkpointing can be used by this application, where the state is mostly stationary, thus protecting it against losing all of the state it has already accumulated.

Checkpointing an application's internal state periodically or, possibly *on demand*, consists of taking a snapshot (e.g., by writing its state to a disk) of the internal data structures managed by the application. In this way, a *recent* version of the state is always available after a failure. Thus, when an application or some of its components must be restarted, it is no longer a cold restart, but rather the application's processing can resume from its most recent snapshot.

A checkpointing operation can often be performed incrementally [37], thus reducing the computational overhead it introduces and, perhaps, enabling more frequent checkpoint operations. Naturally, more frequent checkpointing reduces the likelihood that a substantial amount of the accumulated state will be lost.

Nevertheless relying on checkpoints does not imply that no data is lost. Moreover, incoming streaming traffic will not cease while the checkpoint is being restored after a failure, and the application is still being brought back online.

Yet, checkpointing can be very effective. As an example, consider a weather tracking application that builds a model of the daily precipitation levels on a per-month basis with the goal of making predictions for the near future. Without checkpointing, the loss of the model will result in the application not being able to make *accurate* predictions for a while. Conversely, if periodic snapshots of the model are available, a failure will impact analytic accuracy only minimally as recent snapshots likely hold enough information to still yield very accurate predictions.

In Section 8.5.5, we have illustrated how periodic checkpointing can be configured for individual operators using SPL. However, this often requires support from each operator's implementation. In particular, they must provide the code for serializing the state to be checkpointed, as well as the code for deserializing the state to be restored after restarts [7, 37].

Replication

For loss intolerant applications, fault tolerance is usually provided via the use of *replication*. Replication consists of having multiple replicas of the whole application, or of its portions that cannot face any data loss. In this configuration, the replica that is being used at a particular time is the *active* replica, whereas all the other ones are *backup* replicas.

Along with the replicas, such a configuration must also include a mechanism for detecting failures and for selecting between the data output of the multiple replicas. Designing this mechanism along with the one to activate an alternate replica can be challenging. Moreover, a replication configuration by its very nature employs more computational resources and is more complicated to manage.

The form of replication often employed by SPAs is called *active replication*, where multiple copies of the same application (or of the fault-tolerant segment) run concurrently. At any given time, only the results from one of them is actively used. When the active replica fails, one of the backup replicas is designated as the new active replica, keeping the application operationally intact.

In a typical deployment, replicas must be placed on different hosts to make sure that a single host failure does not impact more than one replica at the same time. A consideration in this design is that a failed replica cannot enter back into the pool of backup replicas until it has somehow *synchronized* its state with the other replicas. Indeed, the physical layout of an application making use of replication as well as the synchronization issues that may arise at runtime can be daunting, and go beyond the scope of our abridged discussion.

Nevertheless, let us consider a simpler issue. Figure 9.40 depicts a code excerpt for a scenario where arbitration between the use of one of the two replicas of an application's `CriticalComponent` is necessary.

For the purpose of this example, we assume that two instances of the `CriticalComponent` application are running simultaneously, each one configured with a different value for the `$replicaNumber` parameter, specifically, 1 or 2. The output produced by each instance of this component is carried by an exported stream whose `replicaNumber` property value is associated with the specific instance that produces it.

Note also that the `CriticalComponent` operator is configured with a `hostIsolation` placement configuration and, as a result, the different replicas (held by different application partitions and, correspondingly, different PEs) are placed on different hosts.

In this example, we assume that a single instance of the `ResultCombiner` composite operator is deployed by an application. This operator is responsible for receiving the result streams from the two active replicas of the `CriticalComponent` composite operator and for selecting the `Result` stream from one of them to use.

The actual task of arbitrating between the replicas is carried out by the `Arbitrator` composite operator, which, based on the health of the replicas (a task implemented by the `HealthMonitor` operator), selects between the possibly two available `Result` streams. The replica currently active is deemed the *master replica*.

```
 1  composite CriticalComponent {
 2    param
 3      expression<int32> $replicaNumber = getSubmissionTimeValue("replicaNumber");
 4    graph
 5      stream<ResultType> Result = Process() {}
 6      () as Exported = Export(Result) {
 7        param
 8          properties = { name = "criticalComponentOutput",
 9              replicaNumber = $replicaNumber };
10      }
11    config
12      placement: hostIsolation;
13  }
14
15  composite Arbitrator(output Data;
16                       input Data1, Data2, stream<int32 sel> Selector) {
17    stream<ResultType> Data = Custom(Data1; Data2; Selector) {
18      logic
19        state: mutable int32 activeSrc = 1;
20        onTuple Data1: {
21          if (activeSrc==1)
22            submit(Data1, Data);
23        }
24        onTuple Data2: {
25          if (activeSrc==2)
26            submit(Data2, Data);
27        }
28        onTuple Selector: {
29          activeSrc = sel;
30        }
31    }
32  }
33
34  composite ResultCombiner {
35    graph
36      // Result1 - results from replica 1
37      stream<ResultType> Result1 = Import() {
38        param
39          subscription: appName == "criticalComponentOutput" && replicaNumber == 1;
40      }
41      // Result2 - results from replica 2
42      stream<ResultType> Result2 = Import() { // from replica 2
43        param
44          subscription: appName == "criticalComponentOutput" && replicaNumber == 2;
45      }
46      // Monitors the well being of replicas 1 and 2, elects which
47      // CriticalComponent replica is the master one
48      stream<int32 sel> Control = HealthMonitor() {}
49      // Outputs the Result stream from the master replica
50      stream<ResultType> Result = Arbitrator(Result1; Result2; Control) {}
51      () as Exported = Export(Result) {
52        param
53          properties = { name = "activeCriticalComponent" };
54      }
55  }
```

Figure 9.40 Arbitration for active replication.

The HealthMonitor operator can be implemented as a primitive operator using Streams job management APIs (Section 8.5.1) to periodically check the status of the PEs hosting each of the two replicas. When a failure of the (current) master replica is detected by the HealthMonitor, an output tuple in the Control stream is produced to indicate a request to replace it. This tuple is used by the Arbitrator operator to actually switch the flow, activating the output of the alternate instance of the CriticalComponent.

Finally, the selected `Result` stream, with the associated name property with the value `activeCriticalComponent`, is then exported for use by other applications.

Partial fault tolerance

As hinted at earlier in this section, when it comes to a distributed SPA's fault tolerance strategy, not every segment of its data flow graph is equally important. Tolerance to sporadic data loss is, in many cases, acceptable for portions of the application as long as the amount of error can be bounded, and the accuracy of the results can be properly assessed and preserved.

Other application segments, however, cannot tolerate any failures as they may contain critical stationary states or produce latency sensitive results. Hence, an important aspect in designing a complex application is to rely on *partial* fault tolerance constructs.

Furthermore, employing fault-tolerant mechanisms increases the computational as well as the capital costs associated with an application. Hence, a developer must take into account where and how fault-tolerant capabilities should be provided [38]. In other words, an application should employ fault tolerance strategies only in segments that can actually benefit from them given the operational failure risks it faces [39].

9.4   Concluding remarks

In this chapter we examined several design principles for building SPAs. We also presented multiple design patterns, illustrated with SPL code segments, showcasing the application of these principles in practice.

We presented functional design patterns and principles in Section 9.2, covering a wide range of issues that often come up during the development of production-grade SPAs.

We also presented non-functional design patterns and principles in Section 9.3. They aim at improving the effectiveness of SPAs in the presence of requirements, such as maintainability, scalability, performance, and fault tolerance.

References

[1] Wiegers KE. Software Requirements. Microsoft Press; 2003.
[2] Brownell D. SAX2. O'Reilly Media; 2002.
[3] The Object Management Group (OMG), Corba; retrieved in September 2010. `http://www.corba.org/`.
[4] Protocol Buffers – Google's data interchange format; retrieved in August 2011. `http://code.google.com/p/protobuf/`.
[5] Apache Thrift; retrieved in August 2011. `http://thrift.apache.org/`.
[6] Elmasri R, Navathe S. Fundamentals of Database Systems. Addison Wesley; 2000.

[7] IBM InfoSphere Streams Version 3.0 Information Center; retrieved in June 2011. `http://publib.boulder.ibm.com/infocenter/streams/v3r0/index.jsp`.

[8] ISO. Information Technology – Database Languages – SQL – Part 3: Call-Level Interface (SQL/CLI). International Organization for Standardization (ISO); 2008. ISO/IEC 9075-3.

[9] Park Y, King R, Nathan S, Most W, Andrade H. Evaluation of a high-volume, low-latency market data processing sytem implemented with IBM middleware. Software: Practice & Experience. 2012;42(1):37–56.

[10] Tanenbaum A, Wetherall D. Computer Networks. 5th edn. Prentice Hall; 2011.

[11] Babcock B, Datar M, Motwani R. Load shedding in data stream systems. In: Aggarwal C, editor. Data Streams: Models and Algorithms. Springer; 2007. pp. 127–146.

[12] Tatbul N, Çetintemel U, Zdonik SB, Cherniack M, Stonebraker M. Load shedding in a data stream manager. In: Proceedings of the International Conference on Very Large Databases (VLDB). Berlin, Germany; 2003. pp. 309–320.

[13] Tatbul N, Çetintemel U, Zdonik SB. Staying FIT: efficient load shedding techniques for distributed stream processing. In: Proceedings of the International Conference on Very Large Databases (VLDB). Vienna, Austria; 2007. pp. 159–170.

[14] Chi Y, Yu PS, Wang H, Muntz RR. LoadStar: a load shedding scheme for classifying data streams. In: Proceedings of the SIAM Conference on Data Mining (SDM). Newport Beach, CA; 2005. pp. 346–357.

[15] Gedik B, Wu KL, Yu PS. Efficient construction of compact source filters for adaptive load shedding in data stream processing. In: Proceedings of the IEEE International Conference on Data Engineering (ICDE). Cancun, Mexico; 2008. pp. 396–405.

[16] Gedik B, Wu KL, Yu PS, Liu L. GrubJoin: an adaptive, multi-way, windowed stream join with time correlation-aware CPU load shedding. IEEE Transactions on Data and Knowledge Engineering (TKDE). 2007;19(10):1363–1380.

[17] Molloy M. Fundamentals of Performance Modeling. Prentice Hall; 1998.

[18] Fallside DC, Walmsley P. XML Schema Part 0: Primer – Second Edition. World Wide Web Consortium (W3C); 2004. `http://www.w3.org/TR/xmlschema-0/`.

[19] ISO. Information Processing – Text and Office Systems – Standard Generalized Markup Language (SGML). International Organization for Standardization (ISO); 1986. ISO 8879.

[20] Booth D, Haas H, McCabe F, Newcomer E, Champion M, Ferris C, et al. Web Services Architecture – W3C Working Group Note. World Wide Web Consortium (W3C); 2004. `http://www.w3.org/TR/ws-arch/`.

[21] Pemberton S. XHTML 1.0 The Extensible HyperText Markup Language (Second Edition). World Wide Web Consortium (W3C); 2002. `http://www.w3.org/TR/xhtml1/`.

[22] Cadenhead R, RSS Board. RSS 2.0 Specification. RSS Advisory Board; 2009. `http://www.rssboard.org/rss-specification`.

[23] Nottingham M, Sayre R. The Atom Syndication Format. The Internet Engineering Task Force (IETF); 2005. RFC 4287.

[24] Clark J, DeRose S. XML Path Language (XPath) Version 1.0. World Wide Web Consortium (W3C); 1999. `http://www.w3.org/TR/xpath/`.

[25] Hégaret PL. Document Object Model (DOM). World Wide Web Consortium (W3C); 2008. `http://www.w3.org/DOM/`.

[26] Vitter JS. Random sampling with a reservoir. ACM Transactions on Mathematical Software (TOMS). 1985;11(1):37–57.

[27] Cormen TH, Leiserson CE, Rivest RL. Introduction to Algorithms. MIT Press and McGraw Hill; 1990.

[28] Twitter; retrieved in March 2011. http://www.twitter.com/.

[29] Bouillet E, Feblowitz M, Feng H, Ranganathan A, Riabov A, Udrea O, *et al.* MARIO: middleware for assembly and deployment of multi-platform flow-based applications. In: Proceedings of the ACM/IFIP/USENIX International Middleware Conference (Middleware). Urbana, IL; 2009. p. 26.

[30] Jacques-Silva G, Gedik B, Wagle R, Wu KL, Kumar V. Building user-defined runtime adaptation routines for stream processing applications. Proceedings of the VLDB Endowment. 2012;5(12):1826–1837.

[31] Hennessy JL, Patterson DA. Computer Architecture: A Quantitative Approach. 2nd edn. Morgan Kaufmann; 1996.

[32] Marr DT, Binns F, Hill DL, Hinton G, Koufaty DA, Miller AJ, *et al.* Hyper-threading technology architecture and microarchitecture. Intel Technology Journal. 2002;6(1):4–15.

[33] Andrade H, Gedik B, Wu KL, Yu PS. Processing high data rate streams in System S. Journal of Parallel and Distributed Computing (JPDC). 2011;71(2):145–156.

[34] Amdahl G. Validity of the single processor approach to achieving large-scale computing capabilities. In: Proceedings of the American Federation of Information Processing Societies Conference (AFIPS). Anaheim, CA; 1967. pp. 483–485.

[35] Molina HG, Ullman JD, Widom J. Database Systems: The Complete Book. Prentice Hall; 2008.

[36] Zhang X, Andrade H, Gedik B, King R, Morar J, Nathan S, *et al.* Implementing a high-volume, low-latency market data processing system on commodity hardware using IBM middleware. In: Proceedings of the Workshop on High Performance Computational Finance (WHPCF). Portland, OR; 2009. article no. 7.

[37] Jacques-Silva G, Gedik B, Andrade H, Wu KL. Language-level checkpointing support for stream processing applications. In: Proceedings of the IEEE/IFIP International Conference on Dependable Systems and Networks (DSN). Lisbon, Portugal; 2009. pp. 145–154.

[38] Jacques-Silva G, Gedik B, Andrade H, Wu KL. Fault-injection based assessment of partial fault tolerance in stream processing applications. In: Proceedings of the ACM International Conference on Distributed Event Based Systems (DEBS). New York, NY; 2011. pp. 231–242.

[39] Jacques-Silva G, Kalbarczyk Z, Gedik B, Andrade H, Wu KL, Iyer RK. Modeling stream processing applications for dependability evaluation. In: Proceedings of the IEEE/IFIP International Conference on Dependable Systems and Networks (DSN). Hong Kong, China; 2011. pp. 430–441.

10 Stream analytics: data pre-processing and transformation

10.1 Overview

Continuous operation and data analysis are two of the distinguishing features of the stream processing paradigm. Arguably, the way in which SPAs employ *analytics* is what makes them invaluable to many businesses and scientific organizations.

In earlier chapters, we discussed how to architect and build a SPA to perform its analytical task efficiently. Yet, we haven't yet addressed any of the algorithmic issues surrounding the implementation of the analytical task itself.

Now that the stage is set and we have the knowledge and tools for engineering a high-performance SPA, we switch the focus to studying how existing stream processing and mining algorithms work, and how new ones can be designed. In the next two chapters, we will examine techniques drawn from data mining, machine learning, statistics, and other fields and show how they have been adapted and evolved to perform in the context of stream processing.

This chapter is organized as follows. The following two sections provide a conceptual introduction of the mining process in terms of its five broad steps: data acquisition, pre-processing, transformation, as well as modeling and evaluation (Section 10.2), followed by a description of the mathematical notation to be used when discussing specific algorithms (Section 10.3).

Since many issues associated with data acquisition were discussed in Section 9.2.1, in this chapter, we focus on the second and third steps, specifically on the techniques for data pre-processing and transformation. We discuss the third and fourth steps of modeling and evaluation in Chapter 11.

In the rest of this chapter, we describe algorithms for computing summaries and descriptive statistics on streaming data (Section 10.4), for sampling (Section 10.5), for computing sketches (Section 10.6), as well as techniques for quantization (Section 10.7), dimensionality reduction (Section 10.8), and for implementing data transforms (Section 10.9).

10.2 The mining process

The data mining process for knowledge discovery is commonly partitioned into five stages:

(a) *Data acquisition* is the stage that covers the various operations involved in collecting data from external data sources. In the stream processing setting, it includes decisions on how to connect to a data source, how to stream data from this source to an application while managing delays and losses, how to scale up, and how to ensure that this process is resilient and fault-tolerant.

(b) *Data pre-processing* is the stage that prepares the data for further analysis. Typical pre-processing operations include cleaning to filter out noisy data elements, interpolation to cope with missing values, normalization to handle heterogeneous sources, temporal alignment and formatting, and data reduction methods to deal with high-dimensional or high-volume data.

(c) *Data transformation* is the stage, sometimes referred to as feature extraction and selection. It includes operations for representing the data appropriately and selecting specific features from this representation. It also includes summarization to either provide a synopsis of the data or to compress it, such that interesting pieces of information can be located easily.

(d) *Modeling* is the stage, also referred to as *mining*, that includes knowledge discovery algorithms and other methods used for identifying interesting patterns in the data, for identifying similarity and groupings, for partitioning the data into well defined classes, for fitting different mathematical functions to the data, for identifying common dependencies and correlations, as well as for identifying abnormal or anomalous data.

(e) *Evaluation* is the stage that includes operations for the use or evaluation of a mining model and for the interpretation of the results produced by the modeling process.

Note that these stages are not completely disjoint from each other, and there are techniques that perform functions common to different stages. Additionally, there are also interactions between the techniques used in the different stages that can affect end-to-end results in various ways. For instance, the choice of data pre-processing and transformation techniques impacts the choice of the techniques for the modeling step. Hence, in the end-to-end design of a stream mining application the selection of techniques must be made while considering these interactions, along with the data characteristics, and the analysis of other requirements (e.g., fault tolerance and performance).

In the rest of this and also in the next chapter, we discuss these processing stages in greater detail.

We start by describing techniques for data pre-processing and transformation. These techniques are broken down into the following six categories:

(a) *Descriptive statistics*, which are techniques focused on extracting simple, quantitative, and visual statistics from a dataset. Often, in the context of streaming data, these statistics must be computed in a continuous manner as tuples of a stream flow into an application.

(b) *Sampling*, which focuses on reducing the volume of the incoming data by retaining only an appropriate subset for analysis. In the context of streaming data it may also involve discarding some tuples based on specific criteria.

(c) *Sketching*, which aims at building in-memory data structures that contain compact synopses of the properties of the streaming data such that specific queries can be answered, in some cases approximately.

(d) *Quantization*, which focuses on reducing the fidelity of individual data samples to remove noise as well as to lower computational and memory costs of other downstream stages of a data analysis task.

(e) *Dimensionality reduction*, which lowers the number of attributes in each data item (i.e., attributes of a tuple) to decrease the data volume and/or to improve the accuracy of the models built on the data.

(f) *Transforms*, which convert data items or tuples and their attributes from one domain to another, such that the resulting data is better suited for further analysis.

For each of these classes of techniques, we formally define the problem, provide a summary of the well-known algorithms, and discuss extensions required to apply them in a stream processing scenario. We also include the implementation of one specific illustrative technique for each class, and briefly review the research in the area.

10.3   Notation

As defined in Chapter 2, a data *tuple* is the fundamental data item embedded in a data *stream*. Each tuple in a data stream is typically associated with a certain time step (or timestamp), either in terms of an arrival time, a creation time, a time-to-live threshold, or, simply, a sequence number. Additionally, each tuple includes a collection of named *attributes*[1] with different individual types and values. Hence, we formalize a tuple as follows:

$$A(t) = \{a_0(t), \ldots, a_{N-1}(t)\}$$

where t corresponds to a time index, and $a_i(t)$ corresponds to the i-th (out of N) attribute of a tuple. In addition, we are interested in a special class of tuples with all scalar numeric attributes that may then be treated as an N-dimensional vector. We represent this as:

$$A(t) \equiv \mathbf{x}(t) = \begin{bmatrix} x_0(t) \\ \vdots \\ x_{N-1}(t) \end{bmatrix}$$

where the elements of the vector correspond to the scalar attributes of a tuple, specifically, $x_i(t) = a_i(t)$ for $0 \leq i < N$. These numeric data tuples are commonly encountered in many SPAs due to the nature of streaming data sources (e.g., sensor networks often produce numeric values). Additionally, even if the data source is non-numeric, there are often techniques used to convert categorical values to numeric ones (for instance, using a bag-of-words approach [1] to translate text to numbers), that allow for a greater range of analytical processing.

---

[1] In the rest of this chapter, we use the terms *attribute, dimension,* and *feature* interchangeably.

A stream S is the potentially infinite sequence of such data tuples.

$$S = A(t_0), A(t_1), A(t_2), \ldots$$

with $t_j < t_{j+1}$. A regular stream is one where the timestamps of the tuples are equally spaced, specifically, $t_{i+1} - t_i = t_{j+1} - t_j$ for all i, j.

These definitions for tuples and their attributes are useful as they will allow us to formalize the description of different operations that are part of the stream mining process.

In the equations used in this and the next chapter, we use $\cdot$ to represent scalar multiplication (including for multiplying a matrix with a scalar), $\times$ for matrix-matrix multiplication, $\bullet$ for inner product on vectors, ^{T} for transpose on matrices, and . for attribute access.

10.4 Descriptive statistics

Descriptive statistics [2] can be used to extract information from the tuples in a stream such that its statistical properties can be characterized or summarized, and the *quality* of the data it transports can be assessed.

Descriptive statistics can also be computed across multiple streams to capture correlations and relationships among them. An example is depicted in Figure 10.1. In this case, an operator ingests the incoming stream, creating an in-memory structure holding descriptive statistics of different types, which may later be queried and used by an application.

Measures of centrality
Several well-known properties of a series of data items can be captured and represented statistically, including multiple measures of centrality, which are used as a proxy for the average properties of the data. These measures include the moments of the data, which are simple to compute. Given a window with Q one-dimensional numeric values,

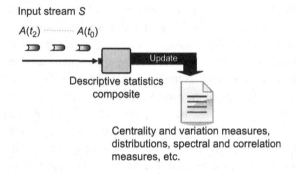

Figure 10.1 Quantitatively characterizing a stream with descriptive statistics.

$x(t_0), \ldots, x(t_{Q-1})$, for a particular tuple attribute x, the i-th sample moment m_i can be computed as follows:

$$m_i = \frac{1}{Q} \sum_{j=0}^{Q-1} \left(x(t_j) \right)^i$$

The *first* sample moment[2] is the sample mean, the *second* sample moment can be used to derive the sample variance, and the *third* sample moment can be used to derive the sample skew.

Other statistical properties include the *median*, which indicates the value that lies in the middle when all the samples are sorted in ascending or descending order; the *mode*, which indicates the most frequently occurring value; measures of variation such as the *standard deviation* and *variance*; measures of extreme limits such as the *maximum* and the *minimum* values; as well as indications of range of values.

Histograms

In an analytical application, it may also become necessary to characterize the function that describes the distribution associated with the values of a tuple attribute. These functions are called Probability Density Functions (PDFs) for tuple attributes taking continuous numeric values, and Probability Mass Functions (PMFs) for tuple attributes taking either discrete numeric or categorical values.

In practice, the density of the data can be estimated from the tuple attribute values by computing a *histogram*, essentially, a function that captures counts of the number of values that fall into each of a disjoint set of *bins*.

The histogram bins are used to discretize the space of possible values a tuple attribute can take, such that the counts can be computed and maintained with appropriate trade-offs between accuracy and efficiency. For instance, consider a numeric tuple attribute that can take real values in the range $[0, 100)$. In this case, it is possible to create a histogram $\mathcal{H}_1$ with bins $B_i = [(i-1) \cdot 10, i \cdot 10)$, where $1 \leq i \leq 10$. Similarly, it is also possible to create a different histogram $\mathcal{H}_2$ with bins $B_j = [(j-1) \cdot 20, i \cdot 20)$, where $1 \leq i \leq 5$. In this case, the first histogram represents the statistical distribution of the data with more accuracy at the expense of requiring more storage space than the second one.

The bins used by a histogram can also have different sizes, as is the case for *equi-depth histograms*, where the goal is to make the counts for the values in each bin roughly the same. Equi-depth histograms can be advantageous in various types of data processing, particularly when providing approximate answers to selectivity queries [3]. Other extensions such as the v-optimal histogram [4], which minimizes the cumulative weighted variance of the bins, have also been proposed.

A histogram can also be created for categorical tuple attributes. In this case, each bin can either correspond to an individual category or to groups of them. For instance, if a tuple attribute represents the country of origin of a credit card transaction, individual bins can be created to count the transactions broken down by individual countries, or

---

[2] In the rest of this chapter we use the term moments to represent the sample moments.

by subsets of countries. Naturally, if bins are used to represent transaction counts per continent of the world, the summarization held in the histogram will occur at a much coarser granularity than if bins represented individual countries. In general, it is often hard to determine the optimal number of bins as well as their size ahead of time.

Estimating other measures from histograms
Histograms have been used extensively because of their conceptual simplicity as well as ease of visualization and interpretation. Furthermore, the information kept in a histogram can be used to quickly estimate several types of statistical measures, including the moments of the data, its median, its mode, as well as its quantiles [5].

Interestingly, quantiles have a strong relationship with the equi-depth histogram. In this type of histogram the values must be partitioned in equal-sized sets (in terms of their counts) and quantiles can be thought of representing bin boundaries. For instance, if Q numeric attribute values are received, sorted, and partitioned into q equal-sized subsets (with Q/q values each), then the k-th q-quantile ($0 < k < q$) represents the data value such that there are k subsets with data values below it and $q - k$ subsets with data values above it.

In terms of probability, the k-th q-quantile represents the value v such that the probability of data values being smaller than v is given by k/q. For instance, when $q = 2$, the first 2-quantile is the median of the values. Numerous other similar techniques to compute quantiles exist [5].

Besides distribution characteristics and measures of centrality and variance, there are other descriptive statistics of interest. For instance, periodicity measures that capture the properties of a numeric time series [6], correlation and cross covariance measures that capture relationships among values associated with different steps of a time series (that is the different values an attribute takes as represented by the sequence of tuples), as well as relationships between values across different streams [2].

Furthermore, several information-theoretic measures such as entropy, entropy rate, Kolmogorov complexity, information gain, and bottleneck, as well as measures of density similarity such as Kullback–Leibler divergence can also be computed for a sequence of numeric values [7].

Continuous data
In SPAs, descriptive statistics may be computed since the beginning of a stream or over different time horizons using different types of windows. Some of the descriptive statistics measures such as the mean and the standard deviation can be computed incrementally from the data as it arrives. Others, for instance, quantiles cannot, and are computationally and memory-wise more demanding. Computing such statistics may require storing all of the values pertaining to the time horizon of interest, potentially even the whole stream.

In these cases, several streaming algorithms making use of accuracy-computational cost tradeoffs have been proposed [8]. These algorithms can be very useful depending on the type of data analysis performed by an application. Indeed, algorithms exist for computing frequency moments, for identifying frequently occurring values (or, heavy-hitters), for counting distinct elements (to create histograms and to compute quantiles), and for computing entropies.

Discussion

The choice of which set of descriptive statistics to compute and maintain, as well as of the method to perform such computations in the context of a SPA, depend heavily on its analytical goals as well as on its computational and storage requirements.

More importantly, care must be taken to ensure that the computational cost versus approximation guarantees provided by the selected method are considered. This may be hard to model, and may require extensive empirical experimentation with real data.

Furthermore, because descriptive statistics computation is often the first step in a sequence of data mining operations, care must also be taken to understand the impact of using approximations, in lieu of actual measures, on the end-to-end application results as those approximations are used and combined in downstream processing steps.

10.4.1   Illustrative technique: BasicCounting

In this section, we consider the BasicCounting algorithm, which is used to determine approximate counts of binary values from a stream, using a sliding window [9]. This algorithm is a basic step in maintaining a (continuous) histogram, in computing aggregates such as averages and sums, as well as in estimating the distinct values seen in the window.

To describe the algorithm let's consider a stream with single-attribute tuples whose values are numeric and can be either 0 or 1. Note that a non-numeric binary categorical attribute can be handled as well by a straightforward mapping to either 0 or 1. Consider also that the goal is to determine the count of 0s and 1s over the most recent W tuples.

An obvious solution to this problem requires storing the last W tuples in a W-tuple sliding window with a 1-tuple slide and using this information to compute the number of 0s and 1s incrementally. Clearly, this solution requires storing $O(W)$ values. For large value of W, this approach may become too expensive in its use of memory. In contrast, the BasicCounting algorithm ensures that by using $O\left(\frac{1}{\epsilon} \log^2 W\right)$ bits of memory the count estimate will lie within a factor $1 \pm \epsilon$ of its actual value.

The data structure

To understand the operations performed by the algorithm, let's consider a few concepts exemplified in Figure 10.2. For each input tuple $x(t_i)$, two indices are used, its *arrival* index, as well as its *recency* index, representing the difference between a tuple's arrival index and the index of the most recent tuple, plus 1.

In this way, any recency index is at least 1 and, because this is a sliding window algorithm, the only tuples that matter are those whose recency index are smaller than or equal to W. Note also that the recency indices of existing tuples change every time a new tuple arrives.

The algorithm uses a specially designed data structure to capture the temporal counts in an efficient manner, such that both the accuracy and memory consumption guarantees hold. This data structure consists of a set of variable-size temporal bins (different from a regular histogram bin discussed earlier) $\mathcal{T}$, where the $T_j \in \mathcal{T}$ bin has three attributes: its size $T_j.C$, its arrival index $T_j.a$, and its recency index $T_j.r$.

The size $T_j.C$ represents the number of successive, but not necessarily contiguous, 1s the temporal bin has captured. The recency index $T_j.r$ represents the smallest recency index of all 1s captured in the bin. And the arrival index $T_j.a$ represents the largest arrival index of all 1s in the bin. While both the recency index and the arrival index were used in our description of the data structure, these values are clearly redundant and only one must be stored. In practice, it is better to keep the arrival index as it does not require updating when a new tuple arrives. Not keeping the recency index means that we also need to keep track of the last received tuple's arrival index to compute recencies.

To understand how the data is maintained we refer back to Figure 10.2. Temporal bin T_0 with $T_0.C = 1$, $T_0.r = 2$, and $T_0.a = 17$ captures information about the most recent 1 value in the stream as its recency index has a value of 1. Similarly, a temporal bin with $T_1.C = 2$, $T_1.r = 6$, and $T_1.a = 13$ captures information about two successive 1s, with the most recent of them having a recency index of 6. Hence, this temporal bin captures information about the 1s occurring at arrival index 11 and 13. Clearly, a bin T_j with $T_j.r > W$ does not need to be stored, as all the successive 1s it represents are older than W and, thus, outside of the window.

At a certain point in time, the data structure will consist of m non-overlapping temporal bins $T_0, \ldots, T_{m-1}$, with bins being numbered in increasing order of their recency. Hence, T_0 is the bin with the smallest recency index and bin T_{m-1} is the one with the largest recency index. Therefore, T_0 has information on recently seen tuples, whereas T_{m-1} has information on tuples seen in the more distant past.

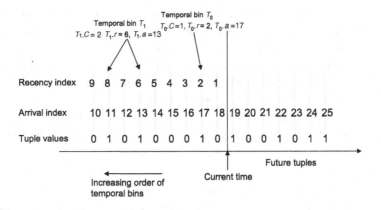

Figure 10.2 The BasicCounting algorithm notation.

Intuitively, this temporal data structure elegantly represents the occurrence of multiple successive 1s within a single bin, effectively compressing this information. Multiple such temporal bins are kept so it becomes possible to incrementally expire and discard information as new tuples arrive, without having to modify all of the stored values.

In essence, the BasicCounting algorithm boils down to a recipe to create, update, and expire these bins such that the desired compression and accuracy performance can be achieved.

The algorithmic steps

When a new tuple arrives, the recency index for each stored bin T_j is incremented by one. If a bin's recency index exceeds W, this bin can be discarded. Indeed, with the way the bins are ordered, only the last bin T_{m-1} might be discarded on the newest tuple arrival. Hence, if $T_{m-1}.r > W$, the bin is excluded and its storage reclaimed.

If the newly arrived tuple carries a 1 value, a new temporal bin is created. Considering the bin ordering as a function of its recency index, the new bin is inserted at the beginning of the list of bins, becoming the T_0 bin, reflecting the fact that it has information about the most recent 1 value. The numbers for the other bins in the list are adjusted accordingly.

To keep the number of bins from growing too large, potentially reaching W, additional constraints are imposed, which causes existing bins to be merged. First, a more recent bin is always smaller or equal in size to older ones, that is $T_{j-1}.C \leq T_j.C$ for $1 \leq j < m$. Second, more than a certain threshold number of bins with the same size are not allowed. This threshold is derived from the algorithm's ϵ error tolerance. In this way, there are at most $k + 1$ bins with size 1, and at most $\frac{k}{2} + 1$ bins allowed with the same size for bins with size greater than 1, where $k = \left\lceil \frac{1}{\epsilon} \right\rceil$.

If the insertion of a new bin causes the size constraint to be violated, the two oldest bins of size 1 are merged into a bin of size 2. The result of this operation might then trigger a violation of the constraint on the number of bins with size 2 and, therefore, additional merges can potentially be triggered until they are no longer required. As a result, the value for $T_j.C$ is always a power of 2.

For example, assume a scenario where $k = 2$ ($\epsilon = 0.5$) and only 1s are received. In this case, the bin counts, without expiration, evolve as shown in Table 10.1. As seen, merges for bins of size 1 happen at arrival index $i = 4$, $i = 6$, and $i = 8$, because the insertion operations at these time steps result in $4 > k + 1 = 3$ bins of size 1. Similarly, the merge for bins of size 2 happens at arrival index $i = 8$, because the merge for bins of size 1 results in $3 > \frac{k}{2} + 1 = 2$ bins of size 2.

Understanding the algorithm's accuracy

The merge policy and ϵ are intimately related. Considering the expiration policy, only bin T_{m-1} can include information about tuples older than W, even if $T_{m-1}.r < W$. Hence, any error in the estimate of non-zero values within the last W tuples comes from

Table 10.1 Evolution of temporal bin size due to insertions and merges.

i	$x(t_i)$	m	$T_0.C$	$T_1.C$	$T_2.C$	$T_3.C$	$T_4.C$	$T_5.C$	Operation
1	1	1	1	–	–	–	–	–	Insert
2	1	2	1	1	–	–	·	–	Insert
3	1	3	1	1	1	–	–	–	Insert
4	1	4	1	1	1	1	–	–	Insert
4	1	3	1	1	2	–	–	–	Merge
5	1	4	1	1	1	2	–	–	Insert
6	1	5	1	1	1	1	2	–	Insert
6	1	4	1	1	2	2	–	–	Merge
7	1	5	1	1	1	2	2	–	Insert
8	1	6	1	1	1	1	2	2	Insert
8	1	5	1	1	2	2	2	–	Merge
8	1	4	1	1	2	4	–	–	Merge

the older unexpired values represented in this bin. The count of non-zero values within the last W tuples is estimated from these bins as follows:

$$\hat{C} = \sum_{j=0}^{m-2} T_j.C + \frac{T_{m-1}.C}{2}$$

This computation makes the assumption that roughly half of the size captured in bin T_{m-1} is due to expired, older tuples. Hence the expected error in the count of 1s is $\frac{T_{m-1}.C}{2}$. Now, since all bins besides T_{m-1} contain information only about unexpired tuples and bin T_{m-1} has at least one unexpired tuple (as its recency is smaller than or equal to W), the smallest estimate of the number of 1s in the last W tuples is $1 + \sum_{j=0}^{m-2} T_j.C$. Hence the expected largest possible relative error err is

$$err \leq \frac{T_{m-1}.C}{2 \cdot \left(1 + \sum_{j=0}^{m-2} T_j.C\right)}$$

Given our choice for k, we can ensure that $err \leq \frac{1}{k}$, and hence $err \leq \epsilon$, because $k = \left\lceil \frac{1}{\epsilon} \right\rceil$.

Data structure management
The pseudo-code in Algorithm 1 demonstrates how the updating and maintenance of the bins is performed. The algorithm keeps track of the arrival indices $T.a$ and maintains the total for all of the bin counts (C), incrementally. The last bin is excluded to avoid iterating over all bins every time a tuple is received.

When a tuple is received, the algorithm increments the current timestamp and checks whether the last bin has expired by comparing the difference between its timestamp and the current tuple index to the window size W. If the last bin has expired, it is removed from the list and C is adjusted appropriately.

If the newly arrived value is 1, the algorithm goes through a number of additional steps. First, a new bin is created, its count is set to 1, and its timestamp is set to the

---

Algorithm 1: BasicCounting.

Param : ϵ, relative error threshold
Param : W, number of most recent tuples to consider

$k \leftarrow \left\lceil \frac{1}{\epsilon} \right\rceil$ ▷ k will be used to bound the successive # of bins with the same size
$\mathcal{T} \leftarrow []$ ▷ The ordered list of bins, initially empty
$i \leftarrow 0$ ▷ Current tuple index
$C \leftarrow 0$ ▷ Total of all bin counts except the last
while *not end of the stream* **do**
  read tuple $x(t)$ ▷ Get the next binary tuple in the stream
  $i \leftarrow i + 1$ ▷ Increment the current tuple index
  $m \leftarrow |\mathcal{T}|$ ▷ Current number of bins
  if $m > 0 \wedge i - T_{m-1}.a \geq W$ **then**   ▷ The last bin has expired
   $\mathcal{T} \leftarrow \mathcal{T} - T_{m-1}$ ▷ Remove from the list of bins
   $m \leftarrow m - 1$ ▷ Decrement the number of bins
   if $m > 0$ **then**   ▷ More bins remain
    $C \leftarrow C - T_{m-1}.C$ ▷ Update the total count
  if $x(t) = 1$ **then**   ▷ We have a 1 in the current tuple
   $T \leftarrow \text{createNewBin}()$ ▷ Create a new bin
   $T.C \leftarrow 1$ ▷ Set the count of 1s to 1
   $T.a \leftarrow i$ ▷ Set the timestamp to i
   $\mathcal{T} \leftarrow [T, \mathcal{T}]$ ▷ Add the new bin to the list of bins
   $m \leftarrow m + 1$ ▷ Increment the number of bins
   if $m > 1$ **then**   ▷ There are other bins
    $C \leftarrow C + T.C$ ▷ Update the total count
   ▷ Search for possible merges through the bins
   $c \leftarrow 1$ ▷ Count of 1s for the current bin group, initially 1
   $l \leftarrow k + 1$ ▷ Number of repetitions allowed for the current bin group
   for $j \leftarrow 0$ *to* $m - 1$ **do**   ▷ For each bin
    if $T_j.C = c$ **then**   ▷ A new bin for the same group
     if $l = 0$ **then**   ▷ We are above the repetition limit
      $T_j.C \leftarrow T_j.C + T_{j-1}.C$ ▷ Add count from previous bin
      $T_j.a \leftarrow T_{j-1}.a$ ▷ Set the arrival index from previous bin
      if $j = m - 1$ **then**   ▷ If this is the oldest bin
       $C \leftarrow C - T_{j-1}.C$ ▷ Update the total count
      $\mathcal{T} \leftarrow \mathcal{T} - T_{j-1}$ ▷ Remove the previous bin that was merged in
      $c \leftarrow T_j.C$ ▷ Update the 1 count for the current bin group
      $l \leftarrow k/2 + 1$ ▷ Update the number of repetitions allowed
      $m \leftarrow m - 1, j \leftarrow j - 1$ ▷ Update the # of bins, and the index
     $l \leftarrow l - 1$ ▷ One repetition less remaining
    else   ▷ This is a new group but no merges took place for the last one
     break   ▷ No more merges, safe to break
  if *user wants count estimate* **then**
   $L \leftarrow T_{m-1}.C$ ▷ The count for the last bin
   output $\hat{C} = C + L/2$ ▷ The last bin is counted as half

---

current timestamp. This new bin is then added to the list of bins. If this new bin is not the only bin in the list, then C is also updated. Second, the list of bins is screened for merges.

The bins are processed from the most recent to the oldest. Bins with the same count are treated as a group. For each group, l, the number of repetitions allowed,[3] is used as

---

[3] The value for l is a pre-computed value [9] based on the value of k we discussed earlier.

the algorithm scans a group, decrementing it as it progresses. If l reaches 0 and there are other remaining bins in the group, a merge operation will be required.

A merge is carried out by removing the previous bin, adding its count to the current bin, and setting the current bin's timestamp to that of the previous bin. The value C is also adjusted if this is the last bin. If a group of bins is fully scanned without a merge, the process stops, since the remaining bins cannot change.

As the process repeats for each new tuple that arrives, if and when a request is made for the count estimate, the algorithm returns the value $C + L/2$, where L is the count of the oldest bin and C is the count for the other remaining bins.

10.4.2   Advanced reading

The BasicCounting algorithm has been extended to consider non-binary attribute values, where an estimate of the sum of non-zero values with attributes taking integer values in a range $[0, R]$ can be computed [9]. Other extensions [8] consider the problem of estimating histograms, as well as the value for a set of functions such as the *sum* over a sliding window. The key modification to the algorithm is the change in the size of the temporal bins to reflect the value of the function to be estimated. This approach is analogous to the BasicCounting temporal bin definition, where its size represents the count of non-zeros captured by the bin. Note that interpolation may have to be performed on the last bin, which may include information from tuples that are now too old to get an appropriate estimate for the function of interest.

There have also been several other research efforts to estimate variance, median, and quantiles [10, 11] as well as to estimate different types of histograms on streaming data [12, 13].

Algorithms to model descriptive statistics for temporal characteristics such as auto-correlation of stationary, seasonal, and non-seasonal time series have been developed using the Auto Regressive Integrated Moving Average (ARIMA) models [14]. Several other linear and non-linear time series characterization models have also been proposed specifically to track and to provide time series descriptive statistics for forecasting. They include the Kalman filter [15], the Wiener filter [15], as well as models for capturing periodicity and seasonality such as the Holts–Winters additive and multiplicative models [16], and the Kernel Smoothing methods [17].

10.5   Sampling

Sampling methods are used to reduce the data stream by selecting some tuples (and discarding others) based on one or more criteria. An illustration of sampling is shown in Figure 10.3.

As is shown, the sampling operator discards incoming tuples $A(t_2)$ and $A(t_3)$ while retaining the rest. Hence, in the output stream $\hat{S}$, we have $\hat{A}(\hat{t}_0) = A(t_0)$, $\hat{A}(\hat{t}_1) = A(t_1)$, and $\hat{A}(\hat{t}_2) = A(t_4)$.

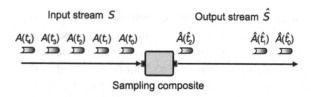

Figure 10.3 The sampling process.

Sampling techniques have been studied extensively in different areas of research including statistics, signal processing, approximation theory, and, more recently, stream processing and mining. A large number of sampling techniques were primarily derived to support the needs of Analog-to-Digital (A/D) conversion (i.e., converting an analog, continuous signal to a discrete digital signal) by deciding which samples in time to retain and which samples to discard. Similarly, many techniques were also created to extract representative samples of large populations for a census or marketing survey.

In the stream processing context, several of these algorithms have been adapted to construct concise summaries of the data and to facilitate data reduction. Sampling strategies may be primarily partitioned into the following three classes:

- *Systematic sampling*: These strategies involve selecting items (e.g., tuples) using a constant interval between the selections. *Uniform* sampling, where we sample every k-th tuple, falls into this category. In some cases systematic sampling involves starting with a random seed to pick the first sample randomly, but then the rest of the samples are chosen in constant intervals. In the most general setting, systematic sampling can be defined with an appropriate trigger condition that sets off at constant intervals (e.g., wall clock time or tuple count) and then one or more tuples can be retained at each trigger time. In many signal processing applications, systematic sampling for numeric tuples may be coupled with some pre-processing, such as digital filtering [18].
- *Random sampling*: These strategies involve selecting items or tuples randomly using a certain probability distribution. The probability distribution may be determined a priori or may be modified during the process of sampling, to allow dynamic modifications of the scheme. These schemes often result in temporally non-uniformly spaced samples. There is a large amount of research in terms of determining the appropriate probability distribution to use and its impact on the result of the sampling. Several techniques [19] for *hierarchical* and *quota sampling* are based on random sampling. In the former, entities are sampled hierarchically, (e.g., by sampling network packets by subnets before sampling packets within a single subnet), and in the latter the sampling is designed such that it results in a fixed number of samples.
- *Data-driven sampling*: These strategies use the value of the data item being sampled to determine whether to retain or discard it [18]. This is unlike the systematic and random sampling techniques that focus only on the timestamp and or the sequence

number attribute associated with a tuple. A simple example of data-driven sampling for numeric tuples is linear interpolation-driven sampling, where a data tuple is retained if it cannot be predicted (within a certain error bound) using a linear combination of previously retained tuples. Several non-linear extensions of these techniques have been proposed for numeric data, and include cubic and spline interpolations. For numeric data, the error bounds may be defined in terms of the residue error, i.e., the error incurred because the tuple is discarded. Data-driven sampling techniques for non-numeric tuples can also be defined by determining the conditions on the attributes to retain/discard a tuple.

Sampling strategies have several properties that make them suitable for use in stream processing. In many cases they are simple and efficient to implement. Several systematic, random, and data-driven sampling strategies can be implemented using a simple `Filter` operator in SPL. Additionally, operators for systematic and random sampling are reusable across different types of data streams and data types due to their independence from the actual data content.

Sampling strategies are also mature and have been studied in multiple settings in terms of their impact on several different types of error metrics. For instance, there are several strong theoretical results in signal processing (e.g., Nyquist theorem) as well as in statistics that bound the impact of sampling in terms of mean squared and other errors (for numeric data), as well as *selection bias* or random sampling error, both of which can lead to a retention of tuples that are not representative of the entire stream. In many cases, it is possible to bound the effectiveness of a variety of applications with sampling methods.

Sampling strategies also retain the inter-attribute correlations across the sampled tuples. Consider a correlation between attribute $a_i(t)$ and attribute $a_j(t)$ of a tuple, such as $a_i(t)$ decreases when $a_j(t)$ increases. Sampling retains this correlation, as it does not change the attributes themselves.

However, there are some disadvantages to sampling. In general, sampling is lossy and it may not be possible to reconstruct the original stream perfectly. Additionally, a poorly chosen sampling strategy may result in different types of biases and errors in the resulting data stream that may impact the performance of the application.

It is critical that the impact of sampling be well understood before deployment, and that the sampling strategy be dynamically changed based on the needs of the application and the characteristics of the data.

When using sampling in stream processing, it is also necessary to consider that the totality of the data points (i.e., the entire *population*) is not available to be sampled in advance. Rather, one must dynamically maintain samples over the entire course of the computation.

Sampling can be ineffective when the incoming stream data contains rare events of interest that need to be preserved. In this case, the sampling strategy must be data-driven and may require complex analysis to determine if a particular event is of interest. There has been some work on providing estimates for counts of such rare events [20], but this is still an active area of research. Finally, it is often "too easy" to devise a complex sampling strategy whose impact and properties are hard to analyze. This is especially true for data-driven approaches, where care needs to be taken to deploy a well-understood and tested strategy.

10.5.1 Illustrative technique: reservoir sampling

We now describe an example algorithm to illustrate how a sampling strategy can be implemented in practice. Reservoir sampling [21] is a quota-based random sampling approach popular in stream processing that was originally proposed in the context of one-pass access of data from magnetic storage devices. This algorithm is designed to extract a fixed number of tuples as a summary of a data stream of finite (but a priori unknown) length. The sampling is performed dynamically as more tuples of the stream are received. The scheme is called reservoir sampling because it maintains a *reservoir* of sampled tuples that are released when all the tuples for the stream are received.

Consider the case where we want to retain q tuples from the original data stream S with finite length Q (which is not known a priori). After sampling, the resulting stream $\hat{S}$ contains only tuples $\hat{A}(\hat{t}_0), \hat{A}(\hat{t}_1), \ldots, \hat{A}(\hat{t}_{q-1})$. We want to determine which of the tuples in the original stream to retain. The reservoir sampling technique allows us to obtain an *unbiased* set of q tuples to represent the stream.

The reservoir sampling algorithm is initialized by simply adding the first q tuples, $A(t_0), \ldots, A(t_{q-1})$, from stream S to the reservoir. Subsequently, when tuple $A(t_i)$ is received, it is added to the reservoir with probability $q/(i+1)$. When a tuple is chosen to be added, it randomly replaces a tuple already in the reservoir, ensuring that we do not exceed the reservoir capacity q. The replacement is performed by randomly sampling from among the previously retained tuples with uniform probability.

While reservoir sampling was originally designed to maintain a reservoir of q samples over the entire (finite) population, the same idea works with streaming data, where the sequence is, at least conceptually, infinite. Hence, an intuitive approach to using reservoir sampling on data streams is to apply it on tumbling windows on an incoming stream, where the size of the tumbling window may be changed dynamically. For each window, our goal is to retain q samples. This is achieved quite simply by the pseudo-code in Algorithm 2.

In this pseudo-code, *rand* represents a random number generator that produces random numbers uniformly distributed in the range $[0, 1)$. For each tumbling window (the end of which is indicated by a punctuation), we run reservoir sampling to store q samples, out of the Q original samples. Note that the number of original samples may be different in each window. The variable i represents the index of the tuple in the current window.

As shown, when the reservoir has space, the current tuple is inserted into its next free location. Otherwise, a random number p is generated in the range $[0..i]$ and the sample at index p in the reservoir is replaced with the current tuple $A(t)$ if and only if the generated random number is smaller than q (i.e., an event with probability $q/(i+1)$). It has been shown that this replacement strategy results in sampling the incoming tuples uniformly with a probability $\min(1, q/Q)$, despite the fact that Q is not known a priori.

Algorithm 2: Reservoir sampling algorithm.

Param : q, the reservoir size
while *not end of the stream* **do**
  $i \leftarrow 0$ ▷ Tuple index in window
  $R, T \leftarrow \emptyset$ ▷ Tuples and timestamps
  while *not end of tumbling window* **do**
    read tuple $A(t)$ ▷ Get the next tuple
    if $i < q$ **then**   ▷ Reservoir is not full
      append $A(t)$ to R ▷ Add to tuple list
      append t to T ▷ Add to timestamp list
    else   ▷ Reservoir is full
      $p \leftarrow \lfloor rand() * (i + 1) \rfloor$
      if $p < q$ **then**   ▷ Do replacement
        delete $R(p), T(p)$ ▷ Delete the replaced tuple, timestamp
        append $A(t)$ to R ▷ Add the new tuple
        append t to T ▷ Add its timestamp
    $i \leftarrow i + 1$
  $Q \leftarrow i$ ▷ # of tuples in the window
  for $i \leftarrow 0$ *to* $min(q, Q) - 1$ **do**   ▷ For each tuple in reservoir
    $\hat{t} \leftarrow T(i)$ ▷ Set the output timestamp
    $\hat{A}(\hat{t}) \leftarrow R(i)$ ▷ Set the output tuple
    emit tuple $\hat{A}(\hat{t})$

This property is important as the algorithm produces a stream summary that is uniformly random across each window of the original stream.

Reservoir sampling is used in many different real-world SPAs and SPSs, including by the Streams performance monitoring tooling for profiling SPL applications (Section 8.5.2).

Finally, it is important to note that reservoir sampling has also been extended to cases when sampling occurs over sliding windows [22].

10.5.2   Advanced reading

There are several other advanced sampling approaches developed by different communities. Recent advances in information theory have led to the development of novel *compressive sampling* techniques [23]. They go past traditional wisdom in data acquisition and assert that one can recover certain signals using far fewer samples than the Nyquist rate. Compressive sampling techniques are used in several sensor network data acquisition applications, approximate stream processing, and image and video compression.

Several data-driven sampling approaches for numeric data using forecasting and tracking techniques, such as particle filters [24] as well as cubic and spline interpolation [25], are also used in several domains.

There is a large amount of research on the use of random sampling techniques for *approximate query processing* [26, 27]. Several extensions to the reservoir sampling approach have been proposed to improve its effectiveness in cases where the data stream characteristics change significantly over time. Different bias functions may be used to modify (e.g., decrease) the probability of retaining older samples [28] and to allow sampling from various lengths of the stream's history. While the design of reservoir sampling algorithms with arbitrary bias functions may be difficult, certain bias functions such as the exponential bias [28] have been shown to allow for simple replacement strategies.

Finally, random sampling strategies have been used for load shedding in distributed SPAs [29]. In similar fashion, other random sampling techniques such as *concise sampling* [30] have also been extended for use in SPAs [8].

10.6   Sketches

Sketches [31] are in-memory data structures that contain compact, often lossy, synopses of the streaming data. These synopses are designed to capture temporal properties of the stream such that specific queries (e.g., finding the most frequently occurring items in a stream) can be answered, in some cases approximately, from only the information they hold.

In general, a sketch provides a compact, *fixed size* summary of the data, independent of the size of the incoming data. Hence, employing them in an application often results in the use of less memory and computational cycles. The actual size of a sketch can usually be controlled by adjusting certain input parameters and is a function of the tolerable approximation error and the likelihood of violating an error bound. Both of these aspects are usually part of an application's set of requirements.

Sketches are popular in stream processing because of their efficiency and flexibility, especially in terms of supporting *approximate query processing*. Each type of sketch is designed to support a specific type of query. The (approximate) answer for this query can then be computed from it, sometimes with the help of additional processing. In SPAs, as new tuples arrive, the sketch must be incrementally updated to maintain its capability to provide accurate query results. This is depicted by Figure 10.4.

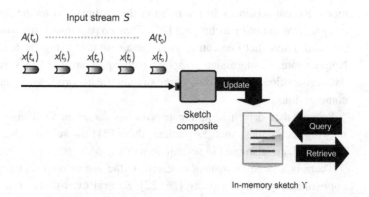

Figure 10.4 The sketch construction and querying process.

As can be seen, the tuples' numeric attributes $\mathbf{x}(t)$ are used by a composite operator to incrementally update the in-memory synopsis Υ. This synopsis is then available to be used by other components of the application to answer certain types of queries, as described later in this section.

Sketches that are of interest to SPAs are those that can be built, updated, and queried efficiently, allowing the application to keep up with a high rate of data ingest. Hence, we focus on *linear sketches* where each update is handled in the same way, irrespective of the history of prior updates. These updates are the result of the arrival of new tuples or of the deletion of existing ones as an older tuple is evicted.

A linear sketch is computed as linear transformations of the data. Hence, if the input is a numeric vector $\mathbf{y}$, a sketch may be computed as $\Upsilon = \Lambda \times \mathbf{y}$, where Λ is a transform matrix (Section 10.9).

A sketch can be computed across multiple tuples in time. Thus, if we have Q tuples in a window, $\mathbf{y}$ is a $Q \times 1$ vector that consists of values from one attribute across multiple tuples (e.g., the k-th attribute from each tuple). Naturally, if the incoming tuples contain N numeric attributes, it is possible to compute N such different sketches.

Updating a linear sketch is straightforward as the effect on Υ of a change, such as insertion or deletion of a value in $\mathbf{y}$, can be computed incrementally using just the transform matrix and the updated row value by either subtracting or adding a vector to the previously computed sketch. Indeed, in the worst case, the update time is linear in the size of the sketch (size of Υ) [31], which makes sketches well suited for dealing with real-time updates.

Nevertheless, as Q increases, it becomes impractical to maintain the transform matrix Λ. Hence, linear sketches are usually implemented using a transform requiring a much smaller amount of information, often employing hash functions [31]. Space saving approaches similar to this are also used by linear transform techniques (e.g., the Haar transform described in Section 10.9).

The most popular linear sketches are designed to answer two broad classes of queries: *frequency-based queries* and *distinct value-based queries*.

A *frequency-based query* is usually posed to retrieve the properties of the frequency distribution of a dataset, including the frequency of specific values, the set of heavy hitter values (i.e., the values that occur very frequently), as well as specific quantiles such as the median (Section 10.4). As an aside, while one might be inclined to use the same types of sketches to also compute statistical attributes of a distribution such as its mean and some of its moments, these values can be computed *exactly* using incremental algorithms, without sketches.

Sketches designed to answer frequency-based queries can also be used to estimate the size of the result of a join operation between different streams as well as the size of self-join operations. Likewise, the size of the output of range queries and histogram representations of the streaming data can also be computed from these specially built sketches.

Some common examples of sketches belonging to this class are the Count sketch [32], the Count-Min sketch [33], and the Alon, Matias, and Szegedy (AMS) sketch [34].

The Count and Count-Min sketches can be used to answer queries about the frequency of occurrence of different values in a dataset (a frequency vector, for short) and can also be used to identify its *heavy hitters*.

The AMS sketch is designed to answer queries related to the squared sum of the frequency vector, i.e., the squared sum of the frequency of occurrences for different values. This particular computation has been shown to be useful in estimating the size of the output of self-join operations.

A *distinct value-based query* is usually posed to identify attributes related to the cardinality of a value, including the number of distinct values in a dataset.

Sketches designed to support these types of queries can be used to estimate the cardinality of a given attribute or combination of attributes, as well as the cardinality of various operations such as set *union* and *difference*.

Some common examples of sketches that belong to this class include the Flajolet–Martin sketch [35], the Gibbons–Tirthapura sketch [36], and the Bar-Yossef, Jayram, Kumar, Sivakumar, Trevisan (BJKST) sketch [37].

The Flajolet–Martin sketch, one of the earliest of this type, is used to efficiently estimate the number of distinct values in a stream, in Online Analytical Processing (OLAP) applications.

Discussion

Many types of sketches can be implemented in SPL using a Custom operator. In most cases, one such operator will have to be stateful because it must maintain the data structure associated with the sketch. Moreover, since the state held by this operator must be made available to other components, the overall infrastructure around the sketch usually requires the use of other operators and additional processing logic.

Indeed, in many applications, multiple sketches must be maintained to capture the different properties of the data that are relevant to the analytical task. Hence, the overall design, including post-processing steps on results extracted from the sketches, as well as accuracy-cost-compute tradeoffs must be carefully evaluated to ensure it matches an application's needs.

10.6.1   Illustrative technique: Count-Min sketch

As mentioned earlier, the Count-Min sketch is a linear sketch designed to support frequency-based queries. It is called Count-Min because it counts groups of items and takes the minimum of various counts to produce an estimate.

The Count-Min sketch can also be used to compute approximations to different summarization functions such as point, range, and inner products, as well as to address other problems common in stream processing such as determining specific quantiles.

Consider that the incoming tuples $x(t)$ include a numeric attribute that takes one of M values. Note that M can be very large, such as the list of all 64-bit integers. Without loss of generality, assume that these values lie in the set $\{0, \ldots, M - 1\}$.[4]

---

[4] If the stream tuples take one of M values these can be directly mapped onto the range $0, \ldots, M - 1$ by simply creating an array with all the values and using the index to represent the value.

The Υ Count-Min sketch will store a concise summary of the data in a $d \times w$ matrix with depth d (number of rows) and width w (number of columns). These parameters are used in the construction of the sketch as follows.

The Count-Min sketch employs a set H of d *pairwise independent* hash functions h_j with $0 \leq j < d$, each of which maps the tuples uniformly onto one of w values from the $\{0, \ldots, w - 1\}$ set.

Any set H of hash functions, each mapping from range M to range w, is said to be pairwise independent, if, for all $x_1 \neq x_2$ in range M, and all y_1 and y_2 in range w, the following property holds:

$$\Pr_{h \in H} \left[h(x_1) = y_1 \wedge h(x_2) = y_2 \right] = \frac{1}{w^2}.$$

Note that these hash functions implement a linear transform that corresponds to creating linear projections of the input vector using appropriately chosen random vectors [33].

Konheim [38] outlines a process by which such set can be constructed. First, a prime number p that lies in the range $M < p < 2 \cdot M$ is selected. The existence of such number is guaranteed by the Bertrand–Chebyshev theorem [39]. Second, two numbers $a_j \neq 0$ and b_j in the range $\{0, \ldots, p - 1\}$ are randomly selected. Thus, the h_j hash functions can be defined as follows:

$$h_j : x \rightarrow ((a_j \cdot x + b_j) \bmod p) \bmod w$$

where *mod* is the modulus operator. It can be shown that hash functions designed in this manner are pairwise independent.

Once the hash functions are selected, each input tuple $\mathbf{x}(t)$ is mapped to an index $0 \leq m < M$ and, then, the d pairwise independent hash functions are applied to it, resulting in d indices whose values are in the range $\{0, \ldots, w - 1\}$.

Finally, the sketch is updated:

$$\Upsilon(j, h_j(m)) = \Upsilon(j, h_j(m)) + 1 \ \ \text{for } 0 \leq j \leq d - 1$$

In its most general configuration, the Count-Min sketch allows the input value $\mathbf{x}(t)$ to be associated with a count parameter $c(t)$, indicating how the frequency for this value should be updated. This count can be positive (more tuples with this value have arrived) or negative (tuples with this value have been removed). In this case (Figure 10.5), the sketch update is carried out as follows:

$$\Upsilon(j, h_j(m)) = \Upsilon(j, h_j(m)) + c(t) \ \ \text{for } 0 \leq j \leq d - 1$$

As discussed, this sketch can be used to estimate $\hat{f}(i)$, the frequency of occurrence of a value i $(0 \leq i < M)$ in the stream, as follows:

$$\hat{f}(i) = \min_{0 \leq j < d} \Upsilon(j, h_j(i))$$

The accuracy of this estimate depends on the choice of the parameters d and w. If the chosen values are $w = \lceil 2/\epsilon \rceil$ and $d = \lceil \log(1/\delta) \rceil$, the error in the estimate, $|\hat{f}(i) - f(i)|$,

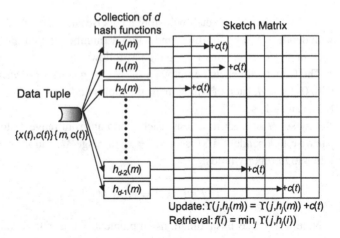

Update: $\Upsilon(j, h_j(m)) = \Upsilon(j, h_j(m)) + c(t)$
Retrieval: $f(i) = \min_j \Upsilon(j, h_j(i))$

Figure 10.5 Count-Min sketch for point query retrieval.

Algorithm 3: Count-Min sketch.

Param : M, number of distinct values that the input tuples can take
Param : ϵ, controls the tolerable error in the frequency estimate
Param : δ, controls the probability with which the estimate has tolerable error
$d \leftarrow \lceil \log(1/\delta) \rceil$ ▷ Determine the number of hash functions
$w \leftarrow \lceil 2/\epsilon \rceil$ ▷ Determine the size of the hash result range
$\Upsilon \leftarrow \mathbf{0}_{d \times w}$ ▷ Initialize the sketch to all zeros
for $j \leftarrow 0$ *to* $d - 1$ **do**   ▷ Generate d pairwise independent hash functions
    $h_j \leftarrow GenerateHash(M, w)$
while *not* end of the stream **do**
    Read tuple $\mathbf{x}(t), c(t)$ ▷ Read the current tuple
    $m \leftarrow Map(\mathbf{x}(t))$ ▷ Map the tuple to a value in range M
    for $j \leftarrow 0$ *to* $d - 1$ **do**   ▷ For each hash function
        $\Upsilon(j, h_j(m)) \leftarrow \Upsilon(j, h_j(m)) + c(t)$ ▷ Update the hashed cell of the sketch

---

is less than $\epsilon \cdot F$ with probability $1 - \delta$, where F is the sum of the actual frequencies:
$F = \sum_{i=0}^{M-1} f(i)$.

The intuition behind this estimate is as follows.[5] Since $w < M$, the hash operations employed during the sketch construction might result in collisions where multiple input index values (m) may be mapped to the same cell. Hence, the cells in the Υ sketch will likely contain an overestimate of the frequency for an individual value. Because d hash functions are employed, it is likely that different values will collide with a particular value of interest for each of these functions in different rows. Hence, taking the smallest estimate of the frequency across all the rows is likely to provide the closest estimate to the true frequency.

Algorithm 3 describes the steps used to construct the sketch. It requires M, d and w as input, alongside the error bound parameters ϵ and δ.

---

[5] We refer the interested reader to the detailed statistical analysis used in the original paper [33], where this analysis is carried out more formally.

Algorithm 4: Function: *GenerateHash*.

Param : M, number of distinct values that the input can take
Param : w, number of distinct values that the output can take
$p \leftarrow M$ ▷ To hold a prime number greater than M
for $p = M + 1$ *to* $2 \cdot M - 1$ **do**   ▷ A prime exists (Bertrand–Chebyshev theorem)
   if *IsPrime*(p) **then break**;   ▷ Miller-Rabin primality test can be used
$a \leftarrow 1 + random() \cdot (p - 1)$ ▷ Generate a random number in range $[1, p)$
$b \leftarrow random() \cdot p$ ▷ Generate a random number in range $[0, p)$
return h s.t. $h(x) = ((a \cdot x + b) \bmod p) \bmod w$ ▷ Return p, a, and b that define the function h

---

The algorithm's first step is to generate d-pairwise independent hash functions using the *GenerateHash* function (Algorithm 4). Subsequently, for each incoming tuple, it maps the vector $\mathbf{x}(t)$ to an index m, and then computes the hash functions to update the Υ sketch. Υ is kept in memory and is available for other applications to query and retrieve value frequency estimates.

10.6.2   Advanced reading

The method used by the Count-Min sketch can also be naturally extended to other problems such as computing the second-order moments [33] with flexible tradeoffs between storage (in-memory), computation cost, and accuracy of the answer. The flexibility of sketch-based methods has been exploited to build different types of summaries on streaming data. Several summaries for approximate query processing, sampling [40], histograms [41], and wavelets [42] can be extracted from sketch information. Finally, while we have focused on random vector-based sketch transforms, there has also been work on deterministic sketch functions [43] for advanced types of queries.

10.7   Quantization

Quantization is the process of reducing the *fidelity* of individual data samples using either scalar, vector, uniform, or non-uniform techniques [44]. A very simple and commonly used example of quantization is *rounding*, where a floating point numeric value is replaced with its nearest integer.

An illustration of the quantization process is shown in Figure 10.6. This process is usually applied to a numeric attribute in a tuple. As can be seen, while b bits are used to represent each tuple (or each input vector) in the S stream, the resulting $\hat{S}$ stream, modified with a quantization technique, requires $\hat{b} \leq b$ bits to represent each tuple. Thus, the actual numeric values for a tuple attribute are transformed, but the process does not change the number of tuples in a stream.

The quantization process consists of applying a *quantizer* function γ to the input data to create the quantized output, i.e., $\hat{\mathbf{x}}(t_i) = \gamma(\mathbf{x}(t_i))$. In practice, the quantizer function needs to be derived dynamically based on the data characteristics to minimize the quantization error, as expressed by the difference between the original and the quantized

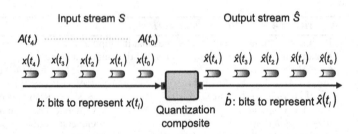

Figure 10.6 The quantization process.

vectors. The quantization error[6] is usually assessed in terms of metrics such as the absolute error (i.e., the $L1$ norm of the vector difference between the original and the quantized values) or the squared error (i.e., the $L2$ norm of the vector difference between the original and the quantized values).

A large number of quantization techniques were initially devised to support data compression, as the means to reduce the volume of data when handling audio, video, and images. Quantization techniques have also been considered extensively in signal processing as well as in approximation theory research.

For similar reasons, quantization techniques are also very well suited for stream processing, as they can often be used to reduce the complexity of processing as well as to discard irrelevant information. For instance, it is generally cheaper computationally to manipulate integers as opposed to floating point numbers. Along the same lines, quantization techniques can also be used to reduce the amount of memory as well as of offline data storage required by an application. Once a floating point number is converted into an index through quantization, the value can be stored using a small range integer to save space.

Indeed, the savings in computational cost and memory footprint can be large, especially when complex mining operations are performed on streaming data.

Quantization techniques may be primarily partitioned into two classes *scalar quantization* and *vector quantization*.

Scalar quantization
Scalar quantization is used when only a single attribute in a tuple is to be transformed. A very simple example of a scalar quantizer is the *uniform scalar quantizer* that maps input x into output $\gamma(x)$ using a fixed step size Δ as follows:

$$\gamma(x) = \text{sgn}(x) \cdot \Delta \cdot \left\lfloor \frac{|x|}{\Delta} + \frac{1}{2} \right\rfloor$$

where $\lfloor \cdot \rfloor$ is the *floor* operation, and sgn is the *sign* function.

When $\Delta = 1$ this is a regular *rounding* operation, otherwise the uniform quantizer maps the input into multiples of Δ. In general, instead of using $\gamma(x)$, we can choose to

---

[6] The error properties of quantizers and its impact have been studied by various researchers [44, 45].

use only the quantization index k as a replacement for the original value. This index is defined as follows:

$$k = \text{sgn}(x) \cdot \left\lfloor \frac{|x|}{\Delta} + \frac{1}{2} \right\rfloor$$

Using this method, the reconstruction of the quantized value can be made by computing $\gamma(x) = k \cdot \Delta$.

Clearly, using k requires fewer bits than the original value, which may be a floating point number. Hence, this method allows the use of integer numbers when carrying out certain (but not all) operations on the data, as opposed to floating point operations, thereby potentially reducing the complexity of subsequent data mining steps.

Note that when using small step sizes Δ, much smaller than the variance in the data, the average squared error between the input and output values may be approximated as $\frac{\Delta^2}{12}$ [44].

Besides uniform quantizers, several non-uniform quantizer functions have been designed for scalar data. In general, a B bit scalar quantizer specifies $2^B + 1$ boundary values $v_0, \ldots, v_{2^B}$ and 2^B reconstruction levels $r_1, \ldots, r_{2^B}$, with:

$$\gamma(x) = r_i \text{ if } v_{i-1} \leq x < v_i$$

These boundaries and levels for the quantizers are designed for specific statistical properties of the data, and to minimize different types of error metrics. Later in this section, we will discuss one such technique, Moment Preserving Quantization (MPQ), in greater detail as an illustrative example.

Vector quantization

For vectors with more than a single dimension, a possible approach for quantization is to use one independent scalar quantizer per dimension (i.e., one per tuple attribute). However, this approach does not exploit the correlations between the different elements in the vector (or between the attributes whose values are being transformed within a tuple).

Alternatively, another approach is to use vector quantizers [46, 47]. Their design is analogous to clustering, as it also involves partitioning a multi-dimensional space into regions (Voronoi cells [47]) and assigning reconstruction levels as the centroids of these regions.[7] Indeed, one of the commonly used vector quantization techniques is the Linde–Buzo–Gray (LBG) algorithm [48], which has similarities to the k-means clustering algorithm (Section 11.4).

In this case, once a quantizer function is designed, the quantized value corresponding to the input vector is the value of its closest centroid. The index of this centroid can be used as the quantized value to save space. For reconstruction (i.e., for computing an approximation to the original vector), the centroid corresponding to the index is used. This is called nearest-neighbor reconstruction. Clearly, the application of vector quantization to new data is often quite simple, requiring only the computation of multi-dimensional distances.

---

[7] This is a generalization of the scalar quantizer concepts of boundaries and reconstruction levels.

Discussion

Quantization is used to reduce the fidelity of the representation of values prior to additional processing. The manipulation of lower-fidelity values usually results in reduced computational complexity as well as in decreased memory and disk usage. Indeed, under certain conditions, it has been shown that a binary quantizer, with only two reconstruction values, may be sufficient for the needs of mining applications [49].

Additionally, when incoming data is transformed through quantization prior to its analysis, the interpretation of the results may be *easier* or more understandable. Naturally, when the data used as input to a mining algorithm is limited to a small finite set of values, the mining results are also expressed in terms of this smaller set, simplifying the task of identifying the properties that impact the mining performance and the results themselves.

The application of a given quantizer function is usually simple to implement. In most cases, it requires a small number of comparisons and distance computations. In SPL, this can usually be coded using a Functor or a Custom operator.

Quantization has limitations. First, the techniques can only be applied to tuples with numeric attributes. Second, quantization is inherently a lossy function and, as a consequence, the original numeric values cannot be recovered exactly from the quantized data. Third, some algorithms implementing quantizer functions can be computationally expensive. This is the case, for instance, when the LBG algorithm is used on vectors with a large number of dimensions. Fourth, the design of a quantizer function can be complex, particularly when dealing with continuously flowing data, as its characteristics may not be fully known a priori.

Considering our focus on stream processing, it is important to note that this last difficulty can be circumvented with one of two approaches: by making use of micro-clustering [50] (Section 11.4) and designing quantizers that operate incrementally; or by modifying the quantizer function periodically, and managing the tradeoffs between delay, accuracy, and complexity of the quantizer.

As is the case with any data transformation technique, it is critical that the impact of quantization be well understood and that it matches the analytical needs of an application, as well as the characteristics of the data to be processed.

10.7.1   Illustrative techniques: binary clipping and moment preserving quantization

We now consider two simple scalar quantization techniques: binary clipping and MPQ [51]. Both techniques are in the family of shape-preserving quantizers.

Binary clipping was originally proposed by Bagnall *et al.* [49]. When using this technique, the streaming data is processed in windows and data samples are converted to 0 or 1 based on whether they lie above or below the mean value baseline in that window. Note that converting the incoming stream to a binary stream provides significant space and computational advantages [49].

While binary clipping can capture simple trends from the original stream, it is only a crude approximation to the data. Yet, it has interesting theoretical underpinnings and has been employed for speeding up the execution of different stream mining algorithms.

MPQ, on the other hand, retains the properties of the original sequence of values in terms of trends, peaks, and dips, by preserving the moments of the data. In other words, moment preservation during quantization means that the moments of the quantized output data are the same as the moments of the input data. This ensures that some properties of the "shape" of the original signal are also retained.

Specifically, given a window with Q one-dimensional numeric tuples $x(t_0), \ldots, x(t_{Q-1})$, the i-th sample moment m_i is defined as follows:

$$m_i = \frac{1}{Q} \sum_{j=0}^{Q-1} \left(x(t_j)\right)^i$$

The first sample moment is the sample mean, the second sample moment captures the spread (if the mean is removed, this captures the variance), while the third sample moment captures the skew in the data samples.

The key idea behind MPQ, depicted by Figure 10.7, can be described as follows. Given the Q tuples $x(t_j)$, where $0 \leq j < Q$, with sample mean μ and sample variance σ^2, we construct a 1-bit quantizer with two reconstruction levels as follows:

$$\hat{x}(t_j) = \begin{cases} \mu + \sigma \sqrt{\frac{Q - Q_g}{Q_g}} & x(t_j) \geq \tau \\ \mu - \sigma \sqrt{\frac{Q_g}{Q - Q_g}} & x(t_j) < \tau \end{cases} \tag{10.1}$$

where Q_g is the number of input values that have magnitude greater than or equal to τ, a threshold chosen a priori such that there is at least one tuple attribute value mapping to each one of the reconstruction levels.

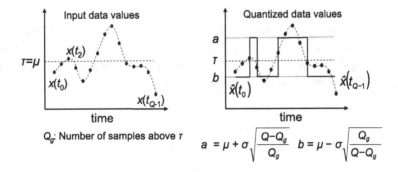

Figure 10.7 Moment preserving quantization.

Algorithm 5: Moment preserving quantization.

while *not end of the stream* **do**	
$\quad i \leftarrow 0$	▷ Tuple index in window
$\quad R, T \leftarrow \emptyset$	▷ Tuples and timestamps
$\quad \mu, \sigma, \tau \leftarrow 0$	▷ Mean, std. dev, and threshold
$\quad$ **while** *not end of tumbling window* **do**	
$\qquad$ read tuple $A(t)$	▷ Get the next tuple
$\qquad R(i) \leftarrow A(t)$	▷ Store the tuple
$\qquad T(i) \leftarrow t$	▷ Store its timestamp
$\qquad \mu \leftarrow \mu + R(i)$	▷ Update the sum
$\qquad \sigma \leftarrow \sigma + (R(i))^2$	▷ Update the squared sum
$\qquad i \leftarrow i + 1$	▷ Update the count
$\quad Q \leftarrow i$	▷ # of tuples in the window
$\quad Q_g \leftarrow 0$	▷ # of tuples above τ
$\quad \mu \leftarrow \mu / Q$	▷ Compute the mean
$\quad \sigma \leftarrow \sqrt{\sigma/Q - (\mu)^2}$	▷ Compute the std. deviation
$\quad$ **for** $i \leftarrow 0$ *to* $Q - 1$ **do**	▷ For each item in window
$\qquad$ **if** $R(i) \geq \mu$ **then**	▷ If in upper level
$\qquad\quad Q_g \leftarrow Q_g + 1$	▷ Increment the upper level count
$\quad a \leftarrow \mu + \sigma \sqrt{\dfrac{Q-Q_g}{Q_g}}$	▷ Upper level value
$\quad b \leftarrow \mu - \sigma \sqrt{\dfrac{Q_g}{Q-Q_g}}$	▷ Lower level value
$\quad \tau \leftarrow \mu$	▷ τ is set to mean
$\quad$ **for** $i \leftarrow 0$ *to* $Q - 1$ **do**	▷ For each item in window
$\qquad \hat{t} \leftarrow T(i)$	▷ Set the output timestamp
$\qquad$ **if** $R(i) \geq \tau$ **then**	▷ If in upper level
$\qquad\quad \hat{x}(\hat{t}) \leftarrow a$	▷ Quantize to upper level value
$\qquad$ **else**	▷ If in lower level
$\qquad\quad \hat{x}(\hat{t}) \leftarrow b$	▷ Quantize to lower level value
$\qquad$ submit tuple $\hat{x}(\hat{t})$	

It is straightforward to show that the resulting quantizer guarantees the preservation of the first two moments of this set of data values, independent of the threshold τ.[8] Moreover, by carefully selecting τ, for example, by making it equal to μ, we can also use this 1-bit quantizer to retain the first three moments of the original signal.

As seen in Algorithm 5, the MPQ algorithm is performed on tumbling windows holding the streaming data. The algorithm includes both the quantizer design step and the application of the quantization function step.

---

[8] As an illustration, after employing the quantizer, there are Q_g outputs with value $\mu + \sigma\sqrt{\dfrac{Q-Q_g}{Q_g}}$ and $Q - Q_g$ outputs with value $\mu - \sigma\sqrt{\dfrac{Q_g}{Q-Q_g}}$. Hence their resulting mean is $\dfrac{Q_g}{Q} \cdot \left(\mu + \sigma\sqrt{\dfrac{Q-Q_g}{Q_g}}\right) + \dfrac{Q-Q_g}{Q} \cdot \mu - \sigma\sqrt{\dfrac{Q_g}{Q-Q_g}}$, which is the same as μ, and independent of τ.

However, because the quantizer design step is non-incremental, the algorithm stores the tuples in the R window buffer and this data is used to compute the sample mean μ, sample standard deviation σ, number of samples above the mean Q_g, and total number of samples Q. Note that the multiple passes over the data in the window result in a delay. Subsequently, the algorithm submits the quantized values one by one for downstream processing.

As an alternative implementation, we may choose to omit the quantizer design step (as, in some cases, it can be done offline) and use predefined reconstruction levels and boundaries to quantize the data samples. In this case, there is no additional delay between receiving a value and emitting its quantized representation. Additionally, instead of using the quantized values $\hat{x}(t)$ themselves, the algorithm may simply emit the corresponding binary indices, with the reconstruction levels sent once per window.

This algorithm also works with more than two reconstruction levels by tweaking its quantization boundaries [52]. In general, when using a B-bit quantizer, it is possible to retain the first $2 \cdot B + 1$ moments of the signal. The MPQ 1-bit quantizer provides much better fidelity than the binary clipping quantizer. It has been used successfully in stream mining applications in domains such as financial services as well as automotive data processing [52].

Finally, while MPQ is a scalar quantization technique, in cases when the input consists of vectors whose attributes are not correlated with each other,[9] this scalar quantization may be applied in parallel for each dimension of the input vector.

10.7.2 Advanced reading

There has been a significant amount of research on designing scalar and vector quantization techniques to support data mining and analytical applications.

As previously mentioned, quantization is used extensively in data compression, so it is natural to consider its use for reducing streaming rates between different parts of a distributed application. Indeed, there is a large body of research on rate-distortion-complexity optimization [53] in video compression and transmission that can be adapted for dynamically changing the quantization strategy employed by an application. In other words, these techniques can be used to exploit tradeoffs between data volume and rates, the amount of distortion in mining results introduced by quantization, as well as memory and algorithmic computational complexity.

Recent research in scalar quantization techniques has been focused on *compressed sensing* [54, 55] applications, which perform mining tasks on data collected from distributed sensor networks. Vector quantization research has focused on the development of streaming and incremental clustering methods [56] as the use of these techniques in continuous data processing applications is becoming more common.

---

[9] Having non-correlated attributes may be a property of the original data or the data may have been pre-processed using a decorrelating and dimensionality reducing transform such as Principal Component Analysis (PCA) (Section 10.8).

10.8   Dimensionality reduction

Dimensionality reduction methods are used to reduce the data volume of a stream by decreasing the number of tuple attributes by transforming them based on one or more criteria.

A dimensionality reduction method is often applied to tuples with numeric attributes. These methods do not change the number of tuples in a stream, thus, for each incoming tuple, there is still a corresponding outbound one. An illustration of a dimensionality reduction operation is shown in Figure 10.8.

As seen in the figure, the incoming stream consists of tuples $\mathbf{x}(t_i)$ with N numeric attributes, while the outgoing stream consists of tuples $\hat{\mathbf{x}}(\hat{t}_i)$ each with $\hat{N}$ attributes, where $\hat{N} \leq N$.

Dimensionality reduction is used extensively in many machine learning and data mining techniques as it decreases the computational complexity associated with modeling, as well as aids in removing noisy or irrelevant attributes. Dimensionality reduction techniques may be partitioned into two broad types: *feature selection* and *feature extraction*.

Feature selection

Feature selection [57] is the process of identifying a subset of (unmodified) dimensions to retain based on different criteria, mostly related to enabling the building of robust data mining models.

Consider an example where tuples contain numeric information about a company's customers such as `age`, `income`, and `idnumber`. Consider also that an application is trying to determine if a particular employee is a good candidate to receive a certain advertisement. In most cases the `idnumber` attribute is not likely to be predictive for determining an individual's taste and consumption preferences, hence we can likely discard that attribute and retain only the `age` and `salary` attributes.

The choice of the optimal features to retain for a particular mining problem may involve an exhaustive search across all subsets of available attributes. Clearly, this can sometimes be infeasible. In such cases, practical approaches have been identified and include techniques such as *feature ranking* and *subset selection*.

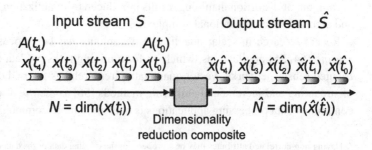

Figure 10.8 The dimensionality reduction process.

Most feature ranking techniques (also referred to as filtering techniques) [58] grade individual features based on a computed score that measures a feature's relevance or its degree of redundancy. The metrics used to compute this score include correlation and mutual information measures [59]. Afterwards, these techniques select the *top-k* features to use, according to the feature ranking formed by each of the features' scores.

Most subset selection techniques (also referred to as wrapper techniques) [60] perform an iterative greedy search (e.g., greedy hill climbing [61]) or other type of search (e.g., genetic algorithms) across subsets of features to determine a good subset to use. These search strategies evaluate each subset of features against the data using a black box learning strategy. A combination of feature search and learning posed as a joint optimization problem (called embedded search) [57] has also been used in some cases for subset selection.

Feature extraction

Feature extraction is the process of transforming an input set of attributes into a smaller set of features or dimensions.

Consider for instance a case where tuples contain numeric measurements from the Electrocardiogram (ECG) (or Elektrokardiogramm (EKG)) signal of a patient in a hospital. For an application to determine whether a patient has a particular condition, the raw ECG signal is first transformed into a compact cepstral representation [62][10] using a combination of a Discrete Fourier Transform (DFT) and a Discrete Cosine Transform (DCT). This step reduces the dimensions of the input signal using linear transforms, while also removing noise and converting the original data into a more intuitive representation for analysis.

During a feature extraction operation the original attributes are transformed into new attributes using *linear* and *non-linear* transforms.

Linear transforms [63] such as DCT, DFT, the multiple variations of the wavelet transform [63], and Linear Predictive Coding (LPC) have been used extensively for dimensionality reduction, especially when the input attributes are related to unstructured data such as audio and video.

Nevertheless, among all linear transforms that convert an input N dimensional vector $\mathbf{x}$ into a $\hat{N} < N$ dimensional vector $\hat{\mathbf{x}}$, the PCA [64], also known as the Karhunen–Loève Transform (KLT), provides the smallest approximation error, i.e., $\|\mathbf{x} - \hat{\mathbf{x}}\|^2$ on average.

Yet, the PCA technique requires estimation of second order statistics and the computation of correlations (specifically, the covariance matrix) from the data, making it difficult to use in a stream processing setting. These difficulties notwithstanding, iterative PCA approximations for continuous processing have been proposed [65] and we discuss one such approach later in this section.

Non-linear transforms include the Semi-Definite Embedding (SDE) algorithm [66], kernel methods [67], as well as Manifold Learning Algorithms (MLAs) [66]. These

---

[10] A *cepstrum* is the result of applying a Fourier transform to the logarithm of the estimated spectrum of a signal.

methods are more complex than linear transform methods and usually difficult to implement in a stream processing setting.

Discussion

Dimensionality reduction techniques are particularly useful in stream processing. First, the algorithmic complexity of several mining algorithms is often greater than linear, usually quadratic in the number of dimensions of the input feature vector (i.e., in the number tuple attributes). Therefore, reducing the number of input dimensions before mining can lead to a substantial reduction in total computational cost. Second, dimensionality reduction can remove irrelevant or noisy attributes and, as a byproduct, improve the quality of the mining models. Consequently, these models will become more robust and generalizable.

On the one hand, the use of selection-based techniques improves the degree to which a model can be understood and interpreted. In other words, the results of a mining operation can be explained in terms of the retained dimensions, allowing domain experts (who are not mining experts) to better evaluate and interpret the models.

On the other hand, extraction-based techniques, making use of well-known transforms such as DCT, DFT, or wavelet transforms, are very simple to implement. In SPL, they can be coded using matrix-vector multiplication in a `Functor` or in a `Custom` operator. Additionally, several of these techniques, such as PCA, result in the creation of *linearly uncorrelated* attributes, i.e., the correlation coefficient between attributes $\hat{x}_i$ and $\hat{x}_j$ is zero for $i \neq j$. This is advantageous because it can lead to reduced cost in data modeling and mining complexity.

Naturally, dimensionality reduction also has limitations. First, the techniques can only be applied to tuples with numeric attributes. Second, some dimensionality reduction techniques can be computationally expensive. For instance, the search-based selection techniques cannot be implemented to work incrementally, as they often require multiple passes over the data. Third, the use of extraction-based dimensionality reduction techniques can, in certain cases, reduce the degree to which results can be interpreted due to the use of transformed as opposed to original attributes.

Some of these limitations can be addressed with heuristics. For instance, non-numeric attributes can be mapped to numeric values in some cases. Similarly, non-incremental dimensionality reduction techniques may be executed periodically or offline. For instance, when using a search-based selection technique, we may use newly identified feature subsets for fresh incoming tuples. Indeed, the use of this approach exposes a tradeoff between computational complexity (how frequently the step is executed) and accuracy (how good the selected subset of features is at any given point in time). These types of tradeoffs will be discussed in greater detail in subsequent sections.

Finally, as is the case with any data analysis technique, it is critical that the impact of dimensionality reduction be well understood and that it matches the analytical needs of an application, as well as the characteristics of the data to be processed.

10.8.1   Illustrative technique: SPIRIT

We now describe an algorithm to illustrate how dimensionality reduction can be implemented in practice. We focus on the Streaming Pattern dIscoveRy in multIple Timeseries (SPIRIT) [68] algorithm, which is a transform-based feature extraction technique. This algorithm has been used in several applications ranging from forecasting and anomaly detection, to pattern detection on streaming data.

Consider a stream with tuples $\mathbf{x}(t)$, each tuple with N attributes, $x_0(t), \ldots, x_{N-1}(t)$. SPIRIT transforms this sequence into *hidden variables* $\hat{\mathbf{x}}(t)$ with $\hat{N} \leq N$ attributes, i.e., $\hat{x}_0(t), \ldots, \hat{x}_{\hat{N}-1}(t)$, at each point in time t. These hidden variables represent the projection of the input vector $\mathbf{x}(t)$ onto $\hat{N}$ *participation weight* vectors, where each weight vector $\mathbf{y}_d$ is of size N, i.e., $y_{d,0}, \ldots, y_{d,N-1}$, with $0 \leq d < \hat{N}$. Mathematically, this may be written in terms of the individual incoming tuple attributes as:

$$\hat{x}_d(t) = y_{d,0} \cdot x_0(t) + \cdots + y_{d,N-1} \cdot x_{N-1}(t), \, 0 \leq d < \hat{N}$$

Considering that $\mathbf{x}(t)$ and $\mathbf{y}_d$ are vectors ($N \times 1$), we can also express this operation in matrix notation:

$$\hat{x}_d(t) = \mathbf{y}_d^T \times \mathbf{x}(t)$$

Intuitively, the participation weight vectors $\mathbf{y}_d$ are analogous to basis vectors of the reduced $\hat{N}$ dimensional space, and the operation we are performing consists of projecting the original vector $\mathbf{x}(t)$ into this space. These basis vectors capture time-varying properties of the incoming stream and the correlations between the multiple attributes in each tuple. The hidden variables $\hat{x}_d$ represent the contribution of a particular basis vector to the original vector. Hence, an approximation to the original vector $\tilde{\mathbf{x}}(t)$ may be constructed using the basis vectors $\mathbf{y}_d$ and the hidden variables $\hat{x}_d$ as follows:

$$\tilde{\mathbf{x}}_d(t) = \sum_{d=0}^{\hat{N}-1} \hat{x}_d(t) \cdot \mathbf{y}_d$$

Thus, these (fewer) hidden variables can be used to approximately recover the original signal. This approximation is usually accurate, if the incoming stream does not exhibit significant changes and random variations over time.

Let us look at the specific steps performed by the SPIRIT algorithm (Algorithm 6). It requires three input values: $\hat{N}$, the number of output dimensions; λ, a *forgetting factor*, which allows the algorithm to focus on recent temporal properties from the incoming stream; and ϵ, a small positive value to prevent numerical instability.

The number of output dimensions $\hat{N}$ hat can be selected over time based on the properties of the data stream, with more output attributes required when the stream has a greater amount of variation. As for λ, it has been shown [68] that a value in the range [0.96, 0.98] is usually a good choice when the numeric values in the stream do not change significantly over time.

Algorithm 6: SPIRIT dimensionality reduction.

Param : N, # of tuples in the input tuple
Param : $\hat{N}$, # of dimensions in the reduced tuple
Param : ϵ, a small positive value to prevent numerical instability
Param : λ, the forgetting factor
for $d \leftarrow 0$ *to* $\hat{N} - 1$ **do**   ▷ For each reduced dimension
    $\Delta_d \leftarrow \epsilon$ ▷ Perform initialization
    $\mathbf{y}_d \leftarrow \mathbf{0}_N$
    $y_{d,d} \leftarrow 1$
while *not end of the stream* **do**
    $\mathbf{z} \leftarrow$ read tuple $\mathbf{x}(t)$ ▷ Get the next tuple
    for $d \leftarrow 0$ *to* $\hat{N} - 1$ **do**   ▷ For each reduced dimension
        $\hat{z}_d \leftarrow \mathbf{y}_d^T \times \mathbf{z}$ ▷ Project $\mathbf{z}$ onto $\mathbf{y}_d$'s direction
        $\Delta_d \leftarrow \lambda \cdot \Delta_d + (\hat{z}_d)^2$ ▷ Update the energy of $\mathbf{y}_d$'s direction
        $\mathbf{e} \leftarrow \mathbf{z} - \hat{z}_d \cdot \mathbf{y}_d$ ▷ Find the approximation error using $\hat{z}_d$
        $\mathbf{y}_d \leftarrow \mathbf{y}_d + (\hat{z}_d/\Delta_d) \cdot \mathbf{e}$ ▷ Realign $\mathbf{y}_d$ using the computed error $\mathbf{e}$
        $\mathbf{z} \leftarrow \mathbf{z} - \hat{z}_d \cdot \mathbf{y}_d$ ▷ Remove the contribution of $\mathbf{y}_d$'s direction from $\mathbf{z}$
    submit $\hat{\mathbf{z}}$ as tuple $\hat{\mathbf{x}}(\hat{t})$ with $\hat{t} = t$

The algorithm starts by updating each individual output attribute $\hat{z}_d$ as the inner product between the corresponding participation weight vector $\mathbf{y}_d$ and the input vector $\mathbf{z}$. Subsequently, the algorithm uses the residue in the projection to update the participation weight vector itself, using the Δ parameter for normalization.

The process iterates over all of the output dimensions at each time interval, ensuring that the contribution of the d-th participation weight vector is removed before updating the $d + 1$-th vector.

Intuitively, the steps taken by the algorithm may be described as follows. It first computes $\hat{z}_d$, a projection of the current vector $\mathbf{z}$ onto the d-th participation weight vector (basis vector) $\mathbf{y}_d$. In other words, this is the amount that $\mathbf{z}$ aligns with the direction pointed by $\mathbf{y}_d$. Next, the algorithm updates the corresponding Δ_d to capture (roughly) the amount of energy norm in the tuple along that direction.

The residue $\mathbf{e}$ represents how much of the input vector is not along the direction of $\mathbf{y}_d$. Next, the algorithm updates $\mathbf{y}_d$ with $\mathbf{e}$ normalized by the energy Δ_d along that direction. In simpler terms, if a particular direction $\mathbf{y}_d$ captures a large amount of the norm of the data stream vectors, it should not be perturbed by a large value.

The algorithm then repeats this process after removing the contribution of the updated $\mathbf{y}_d$ to the current vector $\mathbf{z}$. By performing the updates in this way, the vectors $\mathbf{y}_d$ are kept orthogonal to each other.

This technique has similarities to the PCA technique, where the participation weight vectors $\mathbf{y}_d$ may be viewed as *principal directions* of the $\hat{N}$ dimensional space. Likewise, the $\hat{x}_d$ ($\hat{z}_d$ in the algorithm) values represent the principal components along these principal directions.

Due to these similarities, the results of the SPIRIT algorithm can, under some assumptions [68], produce close approximations to the PCA technique. Yet, the SPIRIT algorithm can be implemented as a sequence of incremental steps, making it suitable for dimensionality reduction operations in SPAs.

Furthermore, the SPIRIT algorithm includes an adaptive mechanism to determine the transform dynamically and incrementally. Indeed, SPIRIT can be used even when the number of output dimensions $\hat{N}$ changes dynamically over time based on the measured variability in the incoming tuples [68].

10.8.2 Advanced reading

Dimensionality reduction is an active area of research, a result of increasingly more sophisticated demands stemming from machine learning and mining algorithms operating on high-dimensional streaming data. Masaeli *et al.* [69] provide a good comparison of many of the available methods.

Recently developed selection-based techniques [70, 71] have focused on feature ranking (or filter) methods due to their scalability. Several non-linear techniques have also been developed for extraction-based dimensionality reduction [72, 73].

The use of dimensionality reduction techniques in stream processing has also been examined by multiple researchers. Lin *et al.* [74] have used wavelet transform-based techniques, while Zhu and Shasha [75] have used DFT-based techniques for dimensionality reduction and clustering of large multi-dimensional data streams. Guha *et al.* [76] have shown the use of dimensionality reduction using Singular Value Decomposition (SVD) in determining correlations between synchronous and asynchronous data streams. Finally, the MUlti-SequenCe LEast Squares (MUSCLES) [77] method uses dimensionality reduction to identify lag correlations in data streams and also to forecast future values. Other research in this space is focused on the use of distributed dimensionality reduction [78, 79] and on scaling up to streams with very large number of dimensions.

10.9 Transforms

A transform method is used to convert tuples or tuple attributes from one domain[11] to another. Transforms may be applied to both numeric and non-numeric data tuples.

Transform methods have been used in different areas of research including signal processing, physics, data mining, and, more recently, data warehousing as well as stream processing.

Given the general nature of possible data transforms, many of the methods described in this section, including the ones for sampling (Section 10.5), quantization (Section 10.7), dimensionality reduction (Section 10.8), and sketches (Section 10.6) may be viewed as special classes of data transforms.

Many transforms were designed specifically to draw out certain properties of the data as well as to construct concise summaries. For instance, the DFT (Section 10.8) was designed to capture the frequency spectrum of the input data, and is used extensively in many forms of data processing. Similarly, the KLT was designed specifically for decorrelating [64] input values and for data reduction.

---

[11] A variable's domain designates the set of values that it can take.

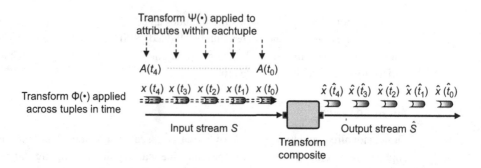

Figure 10.9 The transform process.

When a transform is applied across multiple tuples, we label it a *temporal* transform. A transform may also be qualified as *linear* or *non-linear*, *lossless* or *lossy*, and may modify the number of outbound tuples as well as the values of their attributes. Combinations along these variations are possible as well.

An illustration of the transform process is shown in Figure 10.9. As can be seen, the Φ transform is applied across different tuples, while the Ψ transform is applied across attribute values in each tuple. The dashed arrows in the figure indicate the direction in which the transform is applied, across tuples (Φ), or within a tuple (Ψ). The distinction between these across-tuple (temporal) and within-tuple transforms is discussed in more detail later, when we describe a specific transform method. As just mentioned, there are primarily two classes of transforms, which we will examine in detail next.

Linear transforms
Linear (as well as non-linear) transforms are exclusively applied to numeric data[12] such as the numeric attributes in a tuple. Given two input vectors y_1 and y_2, a linear transform Φ exhibits two properties. First:

$$\Phi(y_1 + y_2) = \Phi(y_1) + \Phi(y_2)$$

And, second, given a scalar α and a vector y, the following equality also holds:

$$\Phi(\alpha \cdot y) = \alpha \cdot \Phi(y)$$

In many cases, a linear transform is defined using a transform matrix Λ. The transform process is then represented as the multiplication of a matrix by a vector as follows:

$$\Phi(y) = \Lambda \times y$$

For a transform applied to the attributes of a tuple, the input vector y consists of the tuple attribute values themselves, that is $x(t)$. In other words, we can define an $N \times N$ tuple transform and set the input to the transform such that $y = x(t)$, as seen in Figure 10.10.

---

[12] Sometimes a prior transformation can be used to convert non-numeric data to a meaningful numeric representation.

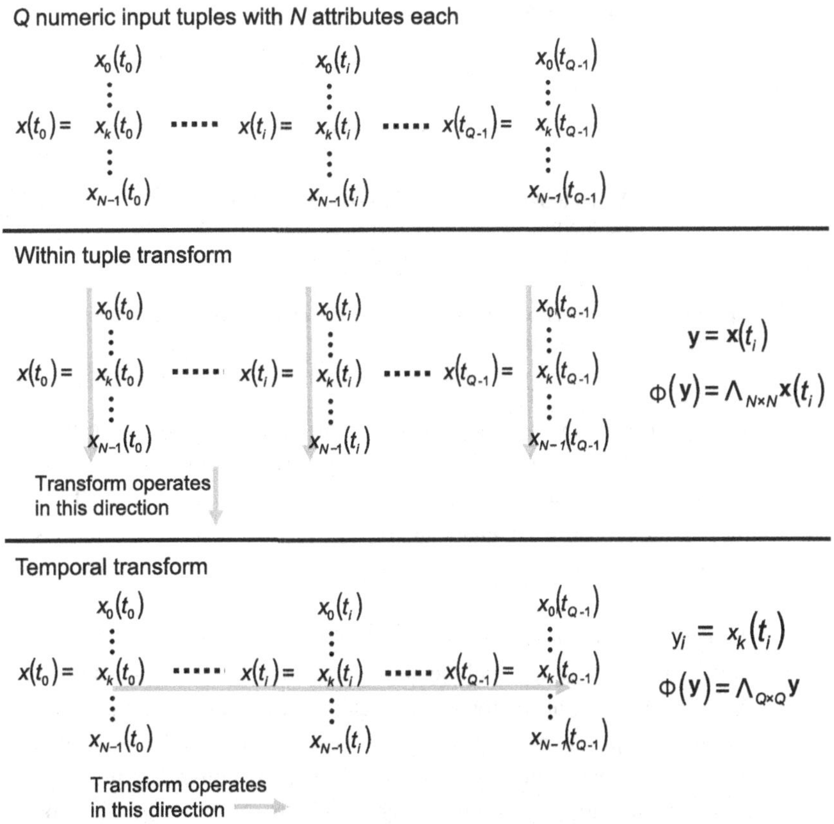

Q numeric input tuples with N attributes each

$$x(t_0) = \begin{matrix} x_0(t_0) \\ \vdots \\ x_k(t_0) \\ \vdots \\ x_{N-1}(t_0) \end{matrix} \quad \cdots \quad x(t_i) = \begin{matrix} x_0(t_i) \\ \vdots \\ x_k(t_i) \\ \vdots \\ x_{N-1}(t_i) \end{matrix} \quad \cdots \quad x(t_{Q-1}) = \begin{matrix} x_0(t_{Q-1}) \\ \vdots \\ x_k(t_{Q-1}) \\ \vdots \\ x_{N-1}(t_{Q-1}) \end{matrix}$$

Within tuple transform

$$x(t_0) = \begin{vmatrix} x_0(t_0) \\ \vdots \\ x_k(t_0) \\ \vdots \\ x_{N-1}(t_0) \end{vmatrix} \quad \cdots \quad x(t_i) = \begin{vmatrix} x_0(t_i) \\ \vdots \\ x_k(t_i) \\ \vdots \\ x_{N-1}(t_i) \end{vmatrix} \quad \cdots \quad x(t_{Q-1}) = \begin{vmatrix} x_0(t_{Q-1}) \\ \vdots \\ x_k(t_{Q-1}) \\ \vdots \\ x_{N-1}(t_{Q-1}) \end{vmatrix}$$

$$y = x(t_i)$$
$$\phi(y) = \Lambda_{N \times N} x(t_i)$$

Transform operates
in this direction

Temporal transform

$$x(t_0) = \begin{matrix} x_0(t_0) \\ \vdots \\ x_k(t_0) \\ \vdots \\ x_{N-1}(t_0) \end{matrix} \quad \cdots \quad x(t_i) = \begin{matrix} x_0(t_i) \\ \vdots \\ x_k(t_i) \\ \vdots \\ x_{N-1}(t_i) \end{matrix} \quad \cdots \quad x(t_{Q-1}) = \begin{matrix} x_0(t_{Q-1}) \\ \vdots \\ x_k(t_{Q-1}) \\ \vdots \\ x_{N-1}(t_{Q-1}) \end{matrix}$$

$$y_i = x_k(t_i)$$
$$\phi(y) = \Lambda_{Q \times Q} y$$

Transform operates
in this direction

Figure 10.10 Linear transforms applied within a tuple and across tuples as a temporal transform.

The definition of a temporal transform is a little more complex. Assuming each tuple, represented as an input vector $x(t)$, has N attributes and Q of these vectors $x(t_0), \ldots, x(t_{Q-1})$ are stored in a tumbling window, a temporal transform may be applied attribute by attribute across the Q tuples. Hence, for the k-th attribute, we can define:

$$y_i = x_k(t_i), \ 0 \le i < Q.$$

In this case, the input vector to the transform is Q-dimensional and it can be applied N times. In other words, we can define a $Q \times Q$ temporal transform, take each attribute from the incoming tuples and collect them into a $Q \times 1$ vector y, which is then set as the input to the transform. In this case, we can perform an independent transform per attribute of an incoming tuple. This is shown in Figure 10.10.

In the most general case, a transform can be applied both across the attributes within a tuple as well as across the tuples in a window. This collection of tuples can be seen as a single $N \times Q$ matrix X, which will have a 2-D linear transform applied to it:

$$\Lambda_1 \times X \times \Lambda_2^T$$

2-D Transform

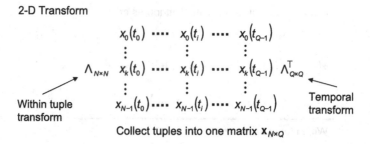

Collect tuples into one matrix $\mathbf{x}_{N \times Q}$

Figure 10.11 Multi-attribute temporal linear transform applied to multiple tuple attributes belonging to multiple tuples.

where Λ_1 and Λ_2 are two transform matrices, operating along the columns (within a tuple) and along the rows (temporal) respectively. This is shown in Figure 10.11.

The Discrete Fourier Transform (DFT), the Discrete Cosine Transform (DCT), and the Karhunen–Loève Transform (KLT) (Section 10.8) are examples of popular linear transforms.

Non-linear transforms
Non-linear transforms include numerical transforms where non-linear functions are applied to an input vector to generate the transformed output. Non-linear functions can include polynomial, logarithmic, exponential, or other such functions. Non-linear transforms have been studied in the context of non-linear regression as well as part of different numeric methods such as Gauss–Newton, gradient descent, and in the Levenberg–Marquardt [80] algorithm.

Mediation transforms
In terms of how transforms are used, one category is particularly important: *mediation transforms*. They are transforms designed specifically for *mediation* and edge adaptation tasks (Section 9.2.1), where data originating from one subsystem must be transformed so it can be consumed by another subsystem.

Mediation transforms are common in data warehouse applications, where they might include format conversion, normalization, and standardization of incoming data.

Mediation transforms may sometimes be specified as rules that use either a regular expression or a procedural method to define matching and replacement operations. There are also several languages that are well suited for specifying such transforms, including Perl [81] and AWK [82] for text transformations, as well as the eXtensible Stylesheet Language Transformations (XSLT) [83] for coding XML data transformations.

Finally, certain mediation transforms are also implemented with a sequence of relational operations, as is often the case in Extract/Transform/Load (ETL) processes used in databases and data warehouses. Note that, when it comes to the nature of these transforms, ETL transforms have several common characteristics with equivalent ones employed by SPAs.

Discussion

Transforms are usually required to make the original data usable for stream mining as well as for other types of continuous processing, where the original format is not immediately adequate. There are also different considerations to take into account when selecting the appropriate transform methods for an application. For instance, an application designer must understand the tradeoffs between the additional computational cost for implementing a transform versus the benefits resulting from having a more adequate representation of the original data.

Linear transforms are often very simple to implement, in some cases requiring just a single SPL Custom operator. Also, by using a combination of windowed operators, it is often possible to implement many types of temporal transforms.

Several linear transforms used for decorrelating a signal (e.g., DCT and PCA) provide good energy compaction,[13] resulting in dimensionality reduction. Moreover, these types of transforms also have two additional functions, which are important for certain types of applications. First, they simplify the task of identifying certain signal characteristics such as its frequency spectrum. And, second, they provide a multi-resolution representation of the data as is the case for many types of wavelet transforms.

As discussed earlier, while non-linear transforms are not as extensively used, in some cases they can provide better performance than linear transforms at the expense of possibly more computational complexity.

10.9.1   Illustrative technique: the Haar transform

One of the more commonly used wavelet transforms, the Haar transform, can be used for constructing a synopsis of numeric values. In general, wavelet transforms have two desirable properties, making them particularly suitable for processing streaming as well as temporal data.

First, they provide a multi-resolution way to represent the data, allowing the original data to be reconstructed at different resolutions. Second, they provide good energy compaction for data such as audio and images. Indeed, wavelet transforms are used by most image compression techniques, including by the JPEG2000 standard [84].

Let's first examine a simple example to demonstrate how the Haar transform works. Consider a 4-dimensional input vector y holding, for instance, four numeric attributes of a tuple.[14] In this case, the Haar transform can be applied at $\log_2 4 = 2$ levels (0 and 1).

---

[13] The *energy* of a vector refers to its $L2$ norm, the sum of the squared values held by it. Energy compaction occurs when the energy of a vector is concentrated in a small number of dimensions. In this case, only a small number of dimensions have large absolute values whereas others hold small ones. This property is useful for compression because the small values can usually be discarded without significantly impacting the energy of the resulting vector.

[14] In the rest of this subsection, our notation uses an input vector y that is constructed from the incoming tuples. The vector is populated differently depending on whether the operation is a *temporal* Haar transform or not. In the non-temporal case, one or more attributes of a single tuple are being transformed and the values for these attributes are placed in y. In the temporal case, the temporal Haar transform is operating across tuples and the values for the attributes from these multiple tuples are placed in y.

At level 0, the Haar transform takes pairs of the input values and transforms them into an average and a difference. Therefore:

$$l_i = \frac{y_{2 \cdot i} + y_{2 \cdot i+1}}{2}, \quad i = 0, 1$$

and

$$h_i = \frac{y_{2 \cdot i} - y_{2 \cdot i+1}}{2}, \quad i = 0, 1$$

where l_0 and l_1 are the average, or low-pass coefficients, and h_0 and h_1 are the difference, or high-pass coefficients of the input vector. Note that if the input vector stores four reasonably homogeneous values, such as four consecutive values from a time series, it is expected that the difference coefficients are small in magnitude. Intuitively, it can be seen that a coarse approximation of the input vector can be obtained by using just the low-pass coefficients:

$$\tilde{y}_j = l_{\lfloor j/2 \rfloor}, \quad j = 0, \ldots, 3$$

The Haar transform is then applied recursively at level 1 on the two low-pass coefficients l_0 and l_1 to create another pair of averages and differences:

$$ll_0 = \frac{l_0 + l_1}{2}$$

and

$$lh_0 = \frac{l_0 - l_1}{2}$$

The resulting ll_0 coefficient is the average of the four input values, while the lh_0 coefficient is the difference between the average of the first two values and the average of the next two values. Again, for homogeneous input vectors, it is expected that lh_0 will have a smaller magnitude compared to ll_0, a fact related to the energy compaction property of this transform. Thus, the two-level transform results in four output coefficients, ll_0, lh_0, h_0, and h_1, with most of the energy concentrated in the ll_0 coefficient.

Note that the ll_0 coefficient is a very coarse approximation of the input values, equivalent to replacing all four input values with their average. This approximation demonstrates the multi-resolution property of this transform as well as its potential for data compression. In other words, it is possible to discard up to three of the four resulting coefficients and retain a reasonable approximation of the input vector.

If we generalize the steps above, the Haar transform can be applied by making use of the following transform matrix:

$$\hat{y} = \begin{bmatrix} 1/4 & 1/4 & 1/4 & 1/4 \\ 1/4 & 1/4 & -1/4 & -1/4 \\ 1/2 & -1/2 & 0 & 0 \\ 0 & 0 & 1/2 & -1/2 \end{bmatrix} \times y$$

For ease of description, assume that Q, the number of dimensions of the input vector, is a power of 2 (i.e., $Q = 2^K$).[15] The Haar wavelet decomposition defines one coefficient that is the average of all values of the input vector along with 2^{k-1} *difference* coefficients at level $K - k$, where $1 \leq k \leq K$.

Each of these 2^{k-1} difference coefficients corresponds to a contiguous portion of the vector of length $Q/2^{k-1}$. The i-th of these 2^{k-1} coefficients (with $0 \leq i < 2^{k-1}$) corresponds to the segment in the series starting from position $i \cdot \frac{Q}{2^{k-1}}$ to position $(i + 1) \cdot \frac{Q}{2^{k-1}} - 1$. For this particular segment, the difference coefficient may be defined as half of the difference between the average of the first half of the segment and the average of the second half of the segment. These coefficients may be represented as a tree of $\log_2 Q$ depth, the *error tree*, which corresponds to the hierarchical representation of the entire set of values.

These coefficients are used in lieu of the original data by emitting them, or a subset of them, for downstream processing. The order in which they are transmitted (or used) is, however, crucial. The average coefficient is sent first, followed by the difference coefficients in decreasing order of their level k. In other words, the single difference coefficient at level $K - 1$ is sent, then the two difference coefficients at level $K - 2$, all the way to the 2^{K-1} difference coefficients at level 0.

This order is important because it makes it possible to progressively refine the approximation of the input vector as more coefficients are received, as opposed to waiting for all of the coefficients to become available. The order is also important when the input vector is to be compressed, as compression is achieved by simply discarding the last few coefficients.

Since using smaller values of k corresponds to geometrically reducing segment sizes, one can easily recover the basic trends at different granularity levels. For example, each difference coefficient is half the quantity by which the first half of its corresponding segment is larger than the second half of the same segment.

Based on this definition of the Haar transform, it is easy to compute the output coefficients by employing a sequence of averaging and subtraction operations, in batch, or by implementing a window-based algorithm to process streaming data. If Q is fixed, it is possible to construct the Haar matrix upfront and carry out matrix-vector operations to compute the coefficients.

The Haar transform is not energy-preserving, thus the sum of the squares of the resulting coefficients is not the same as the sum of the squares of the input values for each dimension of the vector. To counter this effect, the transform's individual coefficient calculation is often normalized with a $\sqrt{2}$ factor (instead of the 2 factor seen earlier in this section), resulting in the $\frac{y_i \pm y_j}{\sqrt{2}}$ term.

Algorithm 7 formalizes these steps and can be used to apply the Haar transform to groups of tuples from a data stream by making use of a tumbling window. In this case, we first collect the Q tuples (each of which is an N dimensional vector) in a window into a $N \times Q$ matrix $\mathbf{X}$. In our description, we consider a stream with regular time intervals

[15] This assumption does not affect the generality of the following steps, as it is always possible to decompose a series into its segments, each of which has a length that is a power of 2. Alternatively, the input vector can be padded or augmented with 0s such that its size is a power of 2.

Algorithm 7: Temporal Haar transform.

$m \leftarrow 0$ ▷ Tuple index in stream
while *not end of the stream* **do**
    $i \leftarrow 0$ ▷ Tuple index in window
    $\mathbf{X} \leftarrow []$ ▷ Matrix to keep all the values in the window
    while *not end of tumbling window* **do**
       read tuple $\mathbf{x}(m \cdot \delta)$ ▷ Read the next tuple in the regularly sampled stream
       $\mathbf{X} \leftarrow \begin{bmatrix} \mathbf{X} & \mathbf{x}(m \cdot \delta) \end{bmatrix}$ ▷ Add to the current window matrix
       $i \leftarrow i + 1$ ▷ Update the tuple index for window
       $m \leftarrow m + 1$ ▷ Update the tuple index for stream
    $Q \leftarrow i$ ▷ Dimension of the temporal transform (also window size)
    $\Lambda \leftarrow ConstructHaarMatrix(Q)$ ▷ The Haar matrix
    $\hat{\mathbf{X}} \leftarrow \mathbf{X} \times \Lambda^T$ ▷ Perform the transform
    for $i \leftarrow 0$ **to** $Q - 1$ **do**   ▷ For each transformed tuple
       $\hat{t} \leftarrow (m - Q + i) \cdot \delta$ ▷ The output timestamp
       $\hat{\mathbf{x}}(\hat{t}) = \hat{\mathbf{X}}(:, i)$ ▷ The i-th transformed tuple, i.e., the i-th column in the matrix
       emit tuple $\hat{\mathbf{x}}(\hat{t})$ ▷ Emit the output result

---

Algorithm 8: Function: *ConstructHaarMatrix*.

Param : d, number of input vector dimensions
$K \leftarrow \log_2 d$ ▷ Number of levels in the wavelet
$\Lambda \leftarrow \mathbf{0}_{d \times d}$ ▷ Initialize Haar matrix to zero
$\Lambda(0,:) \leftarrow 2^{-K} \cdot \mathbf{1}_{d \times 1}^T$ ▷ First row takes an average of the series
$n \leftarrow 0$ ▷ Current row index of the Haar matrix
for $k \leftarrow 1$ **to** K **do**   ▷ For each level in the error tree
    for $j \leftarrow 0$ **to** $2^{k-1} - 1$ **do**   ▷ For each coefficient in the level
       $n \leftarrow n + 1$ ▷ Update the row index
       $a \leftarrow j \cdot d / 2^{k-1}$ ▷ Index of first position in the relevant segment
       $b \leftarrow (j + 1) \cdot d / 2^{k-1}$ ▷ Index of last position in the relevant segment
       for $m \leftarrow a$ **to** $b - 1$ **do**   ▷ For each position in the segment
          if $m < (a + b)/2$ **then**   ▷ If in the first half
             $\Lambda(n, m) \leftarrow 2^{-(K-k+1)}$ ▷ Positive contribution
          else   ▷ If in the second half
             $\Lambda(n, m) \leftarrow -2^{-(K-k+1)}$ ▷ Negative contribution
return Λ

---

between tuples, i.e. $t_i - t_{i-1} = \delta, \forall i$. Using the earlier definition of the linear transform, we then perform a temporal transform across the columns of this matrix multiplying it by a $Q \times Q$ transposed transform matrix Λ^T.

Note that this matrix multiplication step is equivalent to collecting the individual attribute values across the tuples in time into a $Q \times 1$ vector $\mathbf{y}$ and multiplying it by a $Q \times Q$ transform matrix Λ. This matrix is constructed on the fly using the function *ConstructHaarMatrix* described by Algorithm 8.

The use of the Haar matrix in Algorithm 7 requires $O(Q^2)$ space, where Q is the number of tuples in the window. The Haar transform can also be computed using a recursive formulation using only $O(Q)$ space. Indeed, this is the formulation employed by the alternative implementation described by Algorithm 9.

Algorithm 9: Function: *ComputeHaarTransform*

Param : **y**, the input vector
Param : d, the dimension of the input vector
if $d = 1$ **then**   ▷ The base case
  | **return y**   ▷ Return the single remaining item
$A \leftarrow []$ ▷ The list of averages
$C \leftarrow []$ ▷ The list of coefficients
for $i = 0$ *to* $d/2$ **do**   ▷ For each pair of consecutive entries in vector
  | $A \leftarrow [A \ (\mathbf{y}(2 \cdot i) + \mathbf{y}(2 \cdot i + 1))/2]$ ▷ Append to the averages
  | $C \leftarrow [C \ (\mathbf{y}(2 \cdot i) - \mathbf{y}(2 \cdot i + 1))/2]$ ▷ Append to the coefficients
return $[ComputeHaarTransform(A, d/2) \ C]$ ▷ Recursive call

As can be seen, at each level, the transform computes the pair-wise averages and difference for consecutive dimensions of the input vector and recursively passes these averages to the next level of the transform.

10.9.2   Advanced reading

Wavelet as well as other multi-resolution transforms have been extensively researched in the past [63]. Of more direct interest to stream processing is the research into single pass algorithms for wavelet transforms [85], which can be used in SPAs in a straightforward manner.

10.10   Concluding remarks

In this chapter we provided a brief overview of the mining process, and examined several algorithms and techniques for the data pre-processing and transformation stages of a mining task. Our discussion included descriptions of techniques for computing descriptive statistics, for sampling, for generating sketches, for quantization, for dimensionality reduction, as well as for applying transforms to streaming data.

In the next chapter, we conclude the discussion on streaming analytics by describing techniques for data modeling and evaluation, including classification, clustering, regression, frequent pattern and association rule mining, as well as anomaly detection.

References

[1] Joachims T. Learning to Classify Text Using Support Vector Machines: Methods, Theory and Algorithms. Kluwer Academic Publishers; 2002.

[2] Papoulis A, editor. Probability, Random Variables, and Stochastic Processes. McGraw Hill; 1991.

[3] Ioannidis YE. The history of histograms (abridged). In: Proceedings of the International Conference on Very Large Databases (VLDB). Berlin, Germany; 2003. pp. 19–30.

[4] Jagadish H, Koudas N, Muthukrishnan S, Poosala V, Sevcik K, Suel T. Optimal histograms with quality guarantees. In: Proceedings of the International Conference on Very Large Databases (VLDB). New York, NY; 1998. pp. 275–286.

[5] Hyndman R, Fan Y. Sample quantiles in statistical packages. The American Statistician. 1996;50(4):361–365.

[6] Devore J. Probability and Statistics for Engineering and the Sciences. Brooks/Cole Publishing Company; 1995.

[7] Cover TM, Thomas JA. Elements of Information Theory. John Wiley & Sons, Inc and Interscience; 2006.

[8] Aggarwal C, editor. Data Streams: Models and Algorithms. Springer; 2007.

[9] Datar M, Gionis A, Indyk P, Motwani R. Maintaining stream statistics over sliding windows. SIAM Journal on Computing. 2002;31(6):1794–1813.

[10] Babcock B, Datar M, Motwani R, O'Callaghan L. Maintaining variance and k-medians over data stream windows. In: Proceedings of the ACM Symposium on Principles of Database Systems (PODS). San Diego, CA; 2003. pp. 234–243.

[11] Arasu A, Manku G. Approximate counts and quantiles over sliding windows. In: Proceedings of the ACM Symposium on Principles of Database Systems (PODS). Paris, France; 2004. pp. 286–296.

[12] Chaudhuri RM, Narasayya V. Random sampling for histogram construction. How much is enough? In: Proceedings of the ACM International Conference on Management of Data (SIGMOD). Seattle, WA; 1998. pp. 436–447.

[13] Gibbons YM, Poosala V. Fast incremental maintenance of approximate histograms. In: Proceedings of the International Conference on Very Large Databases (VLDB). Athens, Greece; 1997. pp. 466–475.

[14] Percival D, Walden A. Spectral Analysis for Physical Applications. Cambridge University Press; 1993.

[15] Brown R, Hwang P. Introduction to Random Signals and Applied Kalman Filtering. John Wiley & Sons, Inc.; 1996.

[16] Goodwin P. The Holt–Winters approach to exponential smoothing: 50 years old and going strong. Foresight: The International Journal of Applied Forecasting. 2010;19: 30–34.

[17] Wand MP, Jones MC. Kernel Smoothing. Chapman & Hall and CRC Press; 1995.

[18] Pharr M, Humphreys G. Physically Based Rendering: From Theory to Implementation. Morgan Kaufmann; 2010.

[19] Ardilly P, Tillé Y. Sampling Methods. Springer; 2006.

[20] Ross S. Simulation. 4th edn. Academic Press; 2006.

[21] Vitter JS. Random sampling with a reservoir. ACM Transactions on Mathematical Software (TOMS). 1985;11(1):37–57.

[22] Babcock B, Datar M, Motwani R. Sampling from a moving window over streaming data. In: Proceedings of the ACM/SIAM Symposium on Discrete Algorithms (SODA). San Francisco, CA; 2002. pp. 633–634.

[23] Candès E, Wakin M. An introduction to compressive sampling. IEEE Signal Processing Magazine. 2008;25(2):21–30.

[24] Doucet A, Freitas N, Gordon N. Sequential Monte Carlo Methods in Practice. Springer; 2001.

[25] Thévenaz P, Blu T, Unser M. Interpolation revisited. IEEE Transactions on Medical Imaging. 2000;19(7):739–758.

[26] Babcock B, Chaudhuri S, Das G. Dynamic sample selection for approximate query processing. In: Proceedings of the ACM International Conference on Management of Data (SIGMOD). San Diego, CA; 2003. pp. 539–550.

[27] Chaudhuri S, Das G, Narasayya V. Optimized stratified sampling for approximate query processing. ACM Transactions on Data Base Systems (TODS). 2007;32(2).

[28] Aggarwal C. On biased reservoir sampling in the presence of stream evolution. In: Proceedings of the International Conference on Very Large Databases (VLDB). Seoul, Korea; 2006. pp. 607–618.

[29] Babcock B, Datar M, Motwani R. Load shedding for aggregation queries over data streams. In: Proceedings of the IEEE International Conference on Data Engineering (ICDE). Boston, MA; 2004. pp. 350–361.

[30] Gibbons P, Matias Y. New sampling-based summary statistics for improving approximate query answers. In: Proceedings of the ACM International Conference on Management of Data (SIGMOD). Seattle, WA; 1998. pp. 331–342.

[31] Cormode G, Garofalakis M, Haas P, Jermaine C. Synopses for Massive Data: Samples, Histograms, Wavelets, Sketches. Now Publishing: Foundations and Trends in Databases Series; 2011.

[32] Charikar M, Chen K, Farach-Colton M. Finding frequent items in data streams. In: Proceeding of the International Colloquium on Automata, Languages and Programming (ICALP). Malaga, Spain; 2002. pp. 693–703.

[33] Cormode G, Muthukrishnan S. An improved data stream summary: the count-min sketch and its applications. Journal of Algorithms. 2005;55(1):58–75.

[34] Alon N, Matias Y, Szegedy M. The space complexity of approximating the frequency moments. In: Proceedings of the ACM Symposium on the Theory of Computing (STOC). Philadelphia, PA; 1996. pp. 20–29.

[35] Flajolet P, Martin GN. Probabilistic counting algorithms for database applications. Journal of Computer and System Sciences. 1985;31(2):182–209.

[36] Gibbons P, Tirthapura S. Estimating simple functions on the union of data streams. In: Proceedings of Symposium on Parallel Algorithms and Architectures (SPAA). Crete, Greece; 2001. pp. 281–291.

[37] Bar-Yossef Z, Jayram T, Kumar R, Sivakumar D, Trevisan L. Counting distinct elements in a data stream. In: Proceedings of the International Workshop on Randomization and Approximation Techniques (RANDOM). Crete, Greece; 2002. pp. 1–10.

[38] Konheim AG. Hashing in Computer Science: Fifty Years of Slicing and Dicing. John Wiley & Sons, Inc.; 2010.

[39] Aigner M, Ziegler G. Proofs from the Book. Springer; 2000.

[40] Monemizadeh M, Woodruff DP. 1-pass relative-error lp-sampling with applications. In: Proceedings of the ACM/SIAM Symposium on Discrete Algorithms (SODA). Austin, TX; 2010. pp. 1143–1160.

[41] Gilbert A, Guha S, Indyk P, Kotidis Y, Muthukrishnan S, Strauss M. Fast, small-space algorithms for approximate histogram maintenance. In: Proceedings of the ACM Symposium on the Theory of Computing (STOC). Montreal, Canada; 2002. pp. 389–398.

[42] Gilbert A, Kotidis Y, Muthukrishnan S, Strauss M. Surfing wavelets on streams: one-pass summaries for approximate aggregate queries. In: Proceedings of the International Conference on Very Large Databases (VLDB). Rome, Italy; 2001. pp. 79–88.

[43] Ganguly S, Majumder A. CR-precis: a deterministic summary structure for update data streams. In: Proceedings of the International Symposium on Combinatorics (ESCAPE). Bucharest, Romania; 2007. pp. 48–59.

[44] Sayood K. Introduction to Data Compression. Morgan Kaufmann; 2005.

[45] Tseng IH, Verscheure O, Turaga DS, Chaudhari UV. Quantization for adapted GMM-based speaker verification. In: Proceedings of the International Conference on Acoustics, Speech, and Signal Processing (ICASSP). Toulouse, France; 2006. pp. 653–656.

[46] Xu R, Wunsch D. Clustering. Series on Computational Intelligence. John Wiley & Sons, Inc and IEEE; 2009.

[47] Gersho A, Gray RM. Vector Quantization and Signal Compression. Kluwer Academic Publishers; 1991.

[48] Linde Y, Buzo A, Gray R. An algorithm for vector quantizer design. IEEE Transactions on Communications. 1980;28:84–95.

[49] Bagnall AJ, Ratanamahatana CA, Keogh EJ, Lonardi S, Janacek GJ. A bit level representation for time series data mining with shape based similarity. Springer Data Mining and Knowledge Discovery. 2006;13(1):11–40.

[50] Aggarwal CC, Han J, Wang J, Yu PS. A framework for clustering evolving data streams. In: Proceedings of the International Conference on Very Large Databases (VLDB). Berlin, Germany; 2003. pp. 81–92.

[51] Delp E, Saenz M, Salama P. Block truncation coding. In: Bovik A, editor. The Handbook of Image and Video Processing. Academic Press; 2005. pp. 661–672.

[52] Freris N, Vlachos M, Turaga D. Cluster-aware compression with provable K-means preservation. In: Proceedings of the SIAM Conference on Data Mining (SDM). Anaheim, CA; 2012. pp. 82–93.

[53] van der Schaar M, Andreopoulos Y. Rate-distortion-complexity modeling for network and receiver aware adaptation. IEEE Transactions on Multimedia (TMM). 2005;7(3): 471–479.

[54] Kamilov U, Goyal VK, Rangan S. Optimal quantization for compressive sensing under message passing reconstruction. In: Proceedings of the IEEE International Symposium on Information Theory (ISIT). Saint Petersburg, Russia; 2011. pp. 459–463.

[55] Boufounos P. Universal rate-efficient scalar quantization. IEEE Transactions on Information Theory. 2012;58(3):1861–1872.

[56] Lughofer E. Extensions of vector quantization for incremental clustering. Pattern Recognition. 2008;41(3):995–1011.

[57] Vapnik V. Statistical Learning Theory. John Wiley & Sons, Inc.; 1998.

[58] Kira K, Rendell L. A practical approach to feature selection. In: Proceedings of the International Conference on Machine Learning (ICML). Aberdeen, Scotland; 1992. pp. 249–256.

[59] Witten IH, Frank E, Hall MA, editors. Data Mining: Practical Machine Learning Tools and Techniques. 3rd edn. Morgan Kauffman; 2011.

[60] Kohavi R, John GH. Wrappers for feature subset selection. Artificial Intelligence. 1997;97(1):273–324.

[61] Russel S, Norvig P. Artificial Intelligence: A Modern Approach. Prentice Hall; 2010.

[62] Gacek A, Pedrycz W. ECG Signal Processing, Classification and Interpretation: A Comprehensive Framework of Computational Intelligence. Springer; 2012.

[63] Mallat S. A Wavelet Tour of Signal Processing, The Sparse Way. Academic Press; 2009.

[64] Abdi H, Williams L. Principal component analysis. Wiley Interdisciplinary Reviews: Computational Statistics. 2010;2(4):433–459.

[65] Roweis S. EM algorithms for PCA and SPCA. In: Proceedings of the Annual Conference on Neural Information Processing Systems (NIPS). Denver, CO; 1997. pp. 626–632.

[66] Lee J, Verleysen M. Nonlinear Dimensionality Reduction. Springer; 2007.

[67] Hoffmann H. Kernel PCA for novelty detection. Pattern Recognition. 2007;40(3): 863–874.

[68] Papadimitriou S, Sun J, Faloutsos C. Streaming pattern discovery in multiple time-series. In: Proceedings of the International Conference on Very Large Databases (VLDB). Trondheim, Norway; 2005. pp. 697–708.

[69] Masaeli M, Fung G, Dy J. From transformation-based dimensionality reduction to feature selection. In: Proceedings of the International Conference on Machine Learning (ICML). Haifa, Israel; 2010. pp. 751–758.

[70] Zhao Z, Liu H. Spectral feature selection for supervised and unsupervised learning. In: Proceedings of the International Conference on Machine Learning (ICML). Corvalis, OR; 2007. pp. 1151–1157.

[71] Song L, Smola A, Gretton A, Borgwardt K, Bedo J. Supervised feature selection via dependence estimation. In: Proceedings of the International Conference on Machine Learning (ICML). Corvalis, OR; 2007. pp. 823–830.

[72] Gashler M, Ventura D, Martinez T. Iterative non-linear dimensionality reduction with manifold sculpting. In: Proceedings of the Annual Conference on Neural Information Processing Systems (NIPS). Vancouver, Canada; 2008. pp. 513–520.

[73] Rosman G, Bronstein M, Bronstein A, Kimmel R. Nonlinear dimensionality reduction by topologically constrained isometric embedding. International Journal of Computer Vision. 2010;89(1):56–68.

[74] Lin J, Vlachos M, Keogh E, Gunopulos D. Iterative incremental clustering of data streams. In: Proceedings of the International Conference on Extending Database Technology (EDBT). Heraklion, Greece; 2004. pp. 106–122.

[75] Zhu Y, Shasha D. StatStream: statistical monitoring of thousands of data streams in real-time. In: Proceedings of the International Conference on Very Large Databases (VLDB). Hong Kong, China; 2002. pp. 358–369.

[76] Guha S, Gunopulos D, Koudas N. Correlating synchronous and asynchronous data streams. In: Proceedings of the ACM International Conference on Knowledge Discovery and Data Mining (KDD). Washington DC; 2003. pp. 529–534.

[77] Yi BK, Sidiropoulos N, Johnson T, Jagadish HV, Faloutsos C, Biliris A. Online data mining for co-evolving time sequences. In: Proceedings of the IEEE International Conference on Data Engineering (ICDE). San Diego, CA; 2000. pp. 13–22.

[78] Sugiyama M, Kawanabe M, Chui PL. Dimensionality reduction for density ratio estimation in high-dimensional spaces. Neural Networks. 2010;23(1):44–59.

[79] Fang J, Li H. Optimal/near-optimal dimensionality reduction for distributed estimation in homogeneous and certain inhomogeneous scenarios. IEEE Transactions on Signal Processing (TSP). 2010;58(8):4339–4353.

[80] Marquardt D. An algorithm for least-squares estimation of nonlinear parameters. Journal of the Society for Industrial and Applied Mathematics. 1963;11(2):431–441.

[81] Wall L. Programming Perl. 3rd edn. O'Reilly Media; 2000.

[82] Aho A. The AWK Programming Language. Addison Wesley; 1988.

[83] Tidwell D. XSLT. 2nd edn. O'Reilly Media; 2000.

[84] Taubman D, Marcellin M. JPEG2000 Image Compression, Fundamentals, Standard and Practice. Kluwer Academic Publishers; 2002.

[85] Garofalakis M, Gehrke J, Rastogi R. Querying and mining data streams: you only get one look (tutorial). In: Proceedings of the ACM International Conference on Management of Data (SIGMOD). Madison, WI; 2002. p. 635.

11 Stream analytics: modeling and evaluation

11.1 Overview

In this chapter we focus on the last two stages of the data mining process and examine techniques for modeling and evaluation. In many ways, these steps form the core of a mining task where automatic or semi-automatic analysis of streaming data is used to extract insights and actionable models. This process employs algorithms specially designed for different purposes, such as the identification of similar groups of data, of unusual groups of data, or of related data, whose associations were previously unknown.

This chapter starts with a description of a methodology for offline *modeling*, where the model for a dataset is initially learned, and online *evaluation*, where this model is used to analyze the new data being processed by an application (Section 11.2). Despite the use of algorithms where a model is learned offline from previously stored training data, this methodology is frequently used in SPAs because it can leverage many of the existing data mining algorithms devised for analyzing datasets stored in databases and data warehouses.

Offline modeling is often sufficient for the analytical goals of many SPAs. Nevertheless, the use of online modeling and evaluation techniques allows a SPA to function autonomically and to evolve as a result of changes in the workload and in the data. Needless to say this is the goal envisioned by proponents of stream processing. Thus, in the rest of the chapter, we examine in detail online techniques for modeling or mining data streams.

The mining process is classically divided into five different tasks. We use the same organization to provide a comprehensive overview of them in a streaming context. The first task, *classification*, focuses on assigning one of multiple categories or labels to each data item (tuple) from a dataset, based on the values its attributes have (Section 11.3). The second task, *clustering*, consists of assigning tuples to groups, or clusters, so that the tuples in the same cluster are more similar to each other than to those in other clusters (Section 11.4). The third task, *regression*, is used to estimate the relationship between one or more *independent variables*, such as the attributes of a tuple, and a dependent variable whose value must be predicted (Section 11.5). The fourth task, *frequent pattern* and *association rule* mining, focuses on discovering recurrently occurring patterns and structures in various types of data, and using this information to identify rules that capture dependencies and relationships between the occurrences of items or attributes (Section 11.6). Finally, the last task, *anomaly detection*, aims at detecting patterns in a given data set that do not conform to an established normal behavior (Section 11.7).

As in Chapter 10, we formally define each of these problems, provide a summary of the well-known algorithms and of the extensions required to apply them in a stream processing scenario. We also discuss the implementation of one specific illustrative technique for each class and briefly review the existing research in the area.

11.2 Offline modeling and online evaluation

As previously described, the last two stages of the mining process involve *modeling* and *evaluation*. Modeling, or *model learning*, is the process of applying knowledge discovery algorithms to the data to extract patterns and trends of interest, resulting in one or more models. Evaluation, or *model scoring*, involves applying the derived models to new data to generate forward-looking predictions, and to understand the model performance.

Several commercial and open-source data mining tools implement the modeling and evaluation steps for each of the different classes of algorithms. These tools include MATLAB [1], R [2], SPSS [3], SAS [4], and Weka [5], among others.

In many real-world scenarios, it is common to use a large repository of previously collected data for model learning, and then use these models for scoring as new data flows into an application. For instance, consider an application whose goal is to spot fraudulent credit card transactions. In this case, a large collection of transactions, along with information about previously detected fraudulent transactions, is first mined to build a model that captures the patterns that permit separating regular transactions from abnormal ones. Each pattern is based on transaction attributes such as its monetary value, the location where the transaction originated, and the frequency of certain types of purchase for different types of customers.

This model is then *scored* against newly arriving credit card transactions to predict if individual transactions, or even groups of transactions, match the patterns that indicate fraud, as expressed by the model. In this way, a credit card company can identify fraudulent transactions as soon as they occur and, sometimes, even before additional transactions take place, limiting the amount of financial loss that it might incur.

Clearly, scoring new transactions is a stream processing problem. First, it is associated with streaming data. Second, there is a time constraint on providing a response, in this case either accepting or declining an ongoing transaction.

On the other hand, the modeling process, which in this case involves examining large repositories of previously stored transactions, is not necessarily a stream processing problem since it can be performed offline using traditional data mining modeling tools.

Figure 11.1 illustrates how offline model learning and online scoring can be performed. In this figure, we show historical data being processed offline by traditional data mining tools producing the data mining model. Subsequently the model is imported into a SPA to score the new, incoming data. In this diagram, we also see that (some of the) newly arrived data is stored, thereby allowing the periodic recreation of the model by re-running the modeling step.

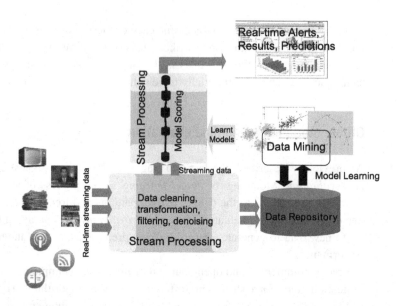

Figure 11.1 Offline model building and online scoring for continuous data analysis.

Standard data mining model representation

The interaction between the continuous online model scoring and the offline model building processes poses a requirement on standardizing the representation of models. The Data Mining Group (DMG) [6], a vendor-led standardization consortium, has developed the Predictive Model Markup Language (PMML) [7], an XML-based language, to provide a representation for models related to predictive analytics and to data mining tasks. Models of different types and classes such as classification, regression, and clustering can be represented using the PMML notation.

Once represented in this format, a model can then be shared between different PMML-compliant applications and tools. For instance, it is possible to create a model using a particular vendor's tool and use other PMML-compliant applications to visualize, analyze, or score it against new data.

The PMML schema includes the following sections:

(a) A *header* with information about the model itself, including, for example, its version number.
(b) A *data dictionary* that lists the possible attributes used by a model, including their specific types, and value ranges.
(c) A set of *data transformations* that lists transformation steps used to convert new data into the format that the model expects.
(d) A *model* that expresses the actual model representation (e.g., a neural network is indicated by the PMML XML element `NeuralNet`).
(e) A *mining schema* that, among other things, lists the actual tuple attributes used by the model, and their relative importance, as well as information on how to deal with missing values.

(f) A set of *targets* that expresses post-processing steps to be applied to the result obtained by using the model (e.g., converting continuous values into discrete ones).

(g) A set of *outputs* that lists the attributes output by the model including the confidence associated with a prediction as well as any other scores associated with using the model.

The PMML XML-schema is standardized and developers can write their own custom code to read or write PMML. Indeed, PMML-compliant tools generally include built-in support to export and import PMML models.

PMML and Streams

Streams can directly use and manipulate PMML models via its *mining toolkit* [8]. It consists of four SPL primitive operators, Classification, Clustering, Regression and Associations, that implement a subset of data mining functions and models (Table 11.1).

Given a classification model, the Classification operator processes incoming tuples to predict a categorical label for them. In the credit card transaction example we just discussed, such labels could indicate that a transaction is either a "fraud" or "normal." The classification model may be binary, where a scoring operation results in one of two categorical classes, or M-ary, where a scoring operation results in one of M categorical classes. The names of these categorical classes are specified in the PMML model specification file.

Figure 11.2 demonstrates how the Classification operator is used in general. As can be seen, the operator takes the name of the PMML file holding the classification model as input using its model parameter. The model is loaded from the file system when the operator is instantiated and can be any of the classification models shown in Table 11.1.

The operator instantiation also requires parameters that specify how the attributes in the incoming stream are to be mapped to the attributes required by the mining model. Such configuration is necessary because the model may have been built on training data where the original attributes were named differently from the names of the attributes seen in the incoming tuples.

Table 11.1 PMML models supported in Streams' mining toolkit.

Operator	Model	PMML version
Classifier	Decision tree	2.0–3.0
	Naïve Bayes	2.0–3.2
	Logistic regression	2.0–3.2
Clustering	Demographic clustering	2.0–3.0
	Kohonen clustering	2.0–3.0
Regression	Polynomial regression	2.0–3.0
	Linear regression	2.0–3.0
	Transform regression	2.0–3.0
Associations	Association rules	2.0–3.2

```
1   stream <InType, tuple<rstring predictedClass, float64 confidence>> Out
2     = Classification(In)
3   {
4     param
5       model: "modelFileName";
6       inputStreamAttrName_1: "modelAttrName_1";
7       ...
8       inputStreamAttrName_k: "modelAttrName_k";
9   }
```

Figure 11.2 Using the `Classification` operator.

```
1    <MiningSchema>
2      <MiningField usageType="active" name="AGE" importance="0.249"/>
3      <MiningField usageType="active" name="NBR_YEARS_CLI" importance="0.065"/>
4      <MiningField usageType="active" name="AVERAGE_BALANCE" importance="0.18"/>
5      <MiningField usageType="active" name="MARITAL_STATUS" importance="0.024"/>
6      <MiningField usageType="active" name="PROFESSION" importance="0.023"/>
7      <MiningField usageType="active" name="BANKCARD" importance="0.449"/>
8      <MiningField usageType="supplementary" name="GENDER"/>
9      <MiningField usageType="predicted" name="ONLINE_ACCESS"/>
10   </MiningSchema>
```

Figure 11.3 An excerpt of a PMML model file.

For instance, consider the `MiningSchema` excerpt from the sample PMML file shown in Figure 11.3. In this case, all of the model attributes whose `usageType` is tagged as `active` must be present as attributes in the incoming stream. Their names and corresponding mapping to incoming tuple attribute names must be indicated as parameters to the operator instantiation.[1] Hence, for this particular PMML excerpt, the invocation of the `Classifier` operator is as shown in Figure 11.4.

In this example,[2] it is also worth noting that this `Classifier` operator instance appends two new attributes to the outgoing tuples: `predictedClass`, the predicted class computed by the model expressed as a `rstring` value, and `confidence`, the confidence score for the prediction expressed as a `float64` value.

The other operators in the mining toolkit are used in a very similar fashion. Specifically, given a set of predefined clusters expressed as a PMML model, the `Clustering` operator processes incoming tuples to determine which cluster each one is closest to. Given a regression model expressed as a PMML model, the `Regression` operator processes incoming tuples to predict a numeric real-valued output from the attributes in the tuple. Similarly, given a set of association rules expressed as a PMML model, the `Associations` operator applies them to the data to determine which rules are satisfied. Generic examples of how these operators may be invoked are depicted by Figure 11.5.

As seen, the instantiation parameters for all three operators are similar to that for the `Classifier` operator. Naturally, however, each operator produces a slightly different

---

[1] Attributes whose `usageType` are tagged as `supplementary` are not required by the model. Hence, they do not need to be mapped, even if they are present in the incoming stream. An attribute whose `usageType` is tagged as `predicted` corresponds to the name of the output generated from the training data.

[2] This operator along with others in the mining toolkit are documented in its reference guide [8].

```
1   type InType = tuple<int32 age, int32 yearsCli, float64 avgBalance,
2     rstring maritalStatus, rstring profession, bool hasBankcard>;
3
4   stream<InType, tuple<rstring predictedClass, float64 confidence>> Out
5     = Classification(In)
6   {
7     param
8       model: "bankonlineaccess.pmml";
9       age: "AGE";
10      yearsCli: "NBR_YEARS_CLI";
11      avgBalance: "AVERAGE_BALANCE";
12      maritalStatus: "MARITAL_STATUS";
13      profession: "PROFESSION";
14      hasBankcard: "BANKCARD";
15  }
```

Figure 11.4 Instantiating the Classification operator.

```
1   stream<InType, tuple<int64 clusterIndex, float64 clusteringScore>> Out
2     = Clustering(In)
3   {
4     param
5       model: "modelFileName";
6       inputStreamAttrName_1: "modelAttrName_1";
7       ...
8       inputStreamAttrName_k: "modelAttrName_k";
9   }
10
11  stream<InType, tuple<float64 predictedVal, float64 predictedStdDev>> Out
12    = Regression(In)
13  {
14    param
15      model: "modelFileName";
16      inputStreamAttrName_1: "modelAttrName_1";
17      ...
18      inputStreamAttrName_k: "modelAttrName_k";
19  }
20
21  stream<InType, tuple<rstring head, float64 support, float64 confidence>> Out
22    = Associations(In)
23  {
24    param
25      model: "modelFileName";
26      inputStreamAttrName_1: "modelAttrName_1";
27      ...
28      inputStreamAttrName_k: "modelAttrName_k";
29  }
```

Figure 11.5 Instantiating the Clustering, Regression, and Associations operators.

result based on the nature of the mining problem. The Clustering operator returns the index of the cluster its internal model indicates that an incoming tuple belongs to, along with a confidence score. The Regression operator returns the predicted value along with the standard deviation associated with this prediction. And, finally, the Associations operator returns the *head* of the rule that matches the incoming tuple, as well as the associated *support* and *confidence* values (Section 10.2). When multiple rules match the input, one tuple is emitted per rule that matched.

Summary

In general a combination of offline modeling and online evaluation can be used to address the needs of several stream mining applications. While conforming to a standard

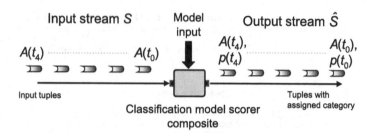

Figure 11.6 The classification model scoring process.

such as PMML allows easy reuse of models, it is also possible to implement customized logic for scoring when an application has more sophisticated or non-standard requirements. For instance, scoring using a decision tree (Section 11.3) may be as simple as a cascaded set of *if-else* conditions. This logic can be implemented easily and efficiently.

The fundamental advantage of combining offline modeling with online evaluation is that we can exploit the readily available and mature tools for offline modeling, without sacrificing the real-time performance needed for scoring streaming data.

Nevertheless, when the data characteristics change, the learned models must also evolve. Otherwise, the evaluation performance degrades over time. In the next few sections we will see how we can measure mismatches between a model and the data as well as how to use this information to adapt it, online. We will also look into how to apply online learning techniques to streaming data.

11.3   Data stream classification

As described in Section 10.2, the stream classification problem consists of assigning one of multiple categories[3] to a tuple based on the values its attributes have. In a classification problem, the tuple's attributes may or may not be numeric and can take arbitrary values. The classification categories are pre-defined and known a priori.

The *scoring* or *evaluation* part of the classification process is where a classification model scorer (*classifier*, for short) assigns a category to a tuple, as shown in Figure 11.6. More formally, the classifier assigns a category $p(t)$ (from a set L) to each incoming tuple $A(t)$, possibly appending this information to the original tuple.

Consider again the application aimed at spotting potential fraud in a stream of credit card transactions. In this scenario, each incoming tuple corresponds to an individual transaction whose attributes include a customer's identifier, the point of sale, and its amount. The two classification categories are $L = \{fraudulent, legitimate\}$.

The classifier's task consists of assigning one of the two categories to each incoming tuple. Note, however, that the assigned category may not necessarily be correct as it simply corresponds to a prediction originated from the classifier model.

---

[3] The terms *label*, *class*, and *outcome* are also used in data mining and statistics literature to refer to a category.

Table 11.2 The fraud detection confusion matrix.

	Predicted *fraudulent*	Predicted *legitimate*
True *fraudulent*	N_{TP}	N_{FN}
True *legitimate*	N_{FP}	N_{TN}

The *modeling* part of the classification process consists of a supervised learning step where a function that maps the values of the tuple attributes to a category is derived. The learning step employs a training data set where each tuple has been previously associated with one of the $|L|$ categories[4] from the set L, by a domain expert. The resulting function is the classification model.

Classifier accuracy

The accuracy of a classifier depends both on the characteristics of the data as well as on the classifier itself. Hence, it is important to be able to quantify the accuracy of a particular classifier for a specific problem.

The accuracy of a classifier with $|L|$ classes can be computed from an $|L| \times |L|$ *confusion matrix* $\mathbf{C}$. In this matrix, each column represents the number of instances in a predicted class, while each row represents the number of instances in an actual class.

In Table 11.2 we show a simple two-class confusion matrix for the fraud detection problem. In this case, C_{ij} represents the count of samples from the test set that are labeled as belonging to class j (*fraudulent*), but whose true class is actually i (*legitimate*).

The entries in the confusion matrix represent the counts of the number of tuples with each combination of predicted class and true class. Specifically in this two-class example, these counts are referred to as the number of true positives (N_{TP}), the number of true negatives (N_{TN}), the number of false negatives (N_{FN}), and the number of false positives (N_{FP}). N_{TP} and N_{TN} count the number of instances where the classifier made a correct prediction. Conversely, N_{FP} and N_{FN} represent classification errors.

There are other well-known measures that can be derived from the confusion matrix, for instance, the precision p and the recall r. In the fraud detection problem, precision is the fraction of the transactions that are correctly labeled *fraudulent* over the total number of transactions labeled *fraudulent* (correctly or not). And recall is the fraction of transactions that are correctly labeled *fraudulent* over the total number of transactions correctly labeled *fraudulent* plus the number of *fraudulent* transactions incorrectly labeled *legitimate*. These metrics can be computed as follows:

$$p = \frac{N_{TP}}{N_{TP} + N_{FP}}$$

and

$$r = \frac{N_{TP}}{N_{TP} + N_{FN}}$$

---

[4] $|L|$, the cardinality of the L set, represents the number of classification categories.

Similarly, the probability of detection p_D and the probability of false alarm p_F are also often used. These metrics may be computed as follows:

$$p_D = \frac{N_{TP}}{N_{TP} + N_{FN}}$$

and

$$p_F = \frac{N_{FP}}{N_{FP} + N_{TN}}$$

p_D and p_F can be plotted together on a chart referred to as the Receiver Operating Characteristic (ROC) chart. It can be used to characterize the performance of a classifier at different operating points as the classifier is evaluated empirically (Section 12.3.4).

Incremental learning

While the modeling step may be performed offline (Section 11.2), our focus in the rest of this section will be on online modeling. The learning step in this case is incremental and the resulting model is updated with the arrival of more incoming tuples.

When learning is performed incrementally, it is also necessary to know the *true* category $l(t)$ associated with a tuple $A(t)$. Clearly, $l(t)$ and the prediction $p(t)$ might not be the same, because of imperfections in the classifier. Usually $l(t)$ will become available within a certain amount of delay after the prediction $p(t)$ is made.[5]

Revisiting our earlier example, the prediction indicating whether a credit card transaction is *fraudulent* (or not) must be made early enough such that different actions can be taken to minimize monetary losses. For instance, if a transaction is deemed *fraudulent*, it can be investigated more closely, additional confirmation can be sought from the customer, or, the transaction can be preemptively denied. Naturally, any of these actions must be taken before it is known whether the transaction is truly *fraudulent*.

Once the true category for a tuple is known, the information can be used to update the classification model and improve its accuracy. For instance, making correct predictions can be rewarded (Figure 11.7), whereas a mistake might result in a penalty being applied.

The diagram in Figure 11.7 includes a stream that provides the true category associated with (some of) the original tuples for which classification predictions were made. Note that the output of the *scorer* itself is routed to the *learner* such that it can also use that information to gauge the accuracy of the model as more information about the true categories becomes available.

The output of the *learner* operator consists of model updates that are routed to the *scorer* via an output interface, which can be implemented using different inter-process communication mechanisms, including by streaming. Indeed, the separation of roles between *learner* and *scorer* is simply an application design decision and other configurations are certainly possible (Chapter 9).

While the model can be updated after every new true category tuple is processed by the *learner* operator, sharing with the *scorer* might occur less frequently. Clearly, the ability to tune the appropriate model update interval exposes a tradeoff between the

---

[5] In the most general case, the true category ($l(t) \in L$) may not be known for *all* previously classified tuples, even after a delay, but it often will be for at least some of the tuples.

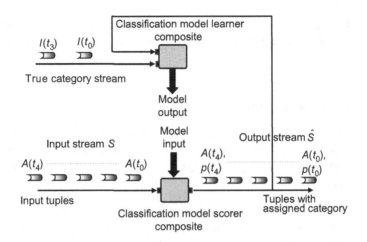

Figure 11.7 The classification model learning process.

communication overhead incurred by sharing the model and the reduction in accuracy, as the *scorer* might not have the most up-to-date model.

Techniques

There are several well-known classification techniques used in traditional data mining. A first group includes linear algorithms [9] that compute a linear combination of features that separates two or more classes of objects. These methods include Linear Discriminant Analysis (LDA) and the associated Fisher's Linear Discriminant, logistic regression, Naïve Bayes (NB) classifiers, and the perceptron, a type of neural network. They are well suited for online scoring as well as for incremental learning.

Other popular classification techniques [9] rely on models of different types: Gaussian Mixture Model (GMM), Support Vector Machine (SVM), decision tree, neural network, Bayesian network, k-Nearest Neighbors (KNN), and Hidden Markov Model (HMM). These techniques were designed for traditional offline data mining. While they are all suitable for online scoring, they are not necessarily adequate for incremental learning in the form they were originally proposed. Nevertheless, with the interest in stream mining, incremental variations for some of these algorithms have already been proposed [10].

In addition to these techniques, there are also meta-learning techniques that build on top of these algorithms to improve their classification performance. These techniques include *boosting* [11] and ensemble methods [12], which leverage multiple classification models to obtain better predictive performance than could be attained from any of the individual models by themselves.

The choice of the appropriate classification algorithm for a SPA is obviously driven by its analytical and computational requirements. Fundamentally, a designer must understand the characteristics of the application's workload as well as of the specific candidate algorithms, such as its accuracy as well as memory and computational demands (perhaps by experimentation) before a decision can be made (Section 12.3).

Finally, a designer must also determine the best parameterization for the selected classification algorithm, a task that is sometimes challenging, even for data mining experts.

11.3.1   Illustrative technique: VFDT

The Very Fast Decision Tree (VFDT) classification technique [13] supports incremental scoring and learning, making it very well suited for use in SPAs. Before presenting the VFDT algorithm, we formalize the classification problem when expressed as a decision tree.

The classification problem
Given a set of Q training tuples $A(t_i)$, each with N attributes $a_j(t_i)$, and their associated *true* categories $l(t_i)$, with $0 \leq i < Q$ and $0 \leq j < N$, the classification model learning problem consists of determining a function Γ to map a tuple's attribute values to a corresponding category. This function can then be used to make predictions and to classify new tuples.

The learning step
The learning of a decision tree involves determining the tree structure, the attribute to test at each node, and the appropriate test to be performed. The decisions to split a node, which are made during the construction of the tree, are carried out by optimizing a local measure using a split evaluation function G. Typical functions rely on the information gain [14] metric, used in the C4.5 algorithm [15], or on the Gini index [14, 16], used in the Classification And Regression Tree (CART) algorithm [17].

The basic intuition behind splitting a node is to carry it out in such a way that the resulting children nodes are as homogeneous as possible with respect to the category associated with the training tuples matching these nodes. To understand this idea, let's examine a simple example where G is the information gain metric.

Using techniques from information theory, the information gain or mutual information between two random variables A and B is computed as follows:

$$IG(A; B) = H(A) - H(A|B)$$

where $H(\cdot)$ represents the *entropy*[6] of a random variable. $H(A)$ and $H(A|B)$ can be computed as follows:

$$H(A) = -\sum_{a} P(A = a) \cdot \log_2(P(A = a))$$

---

[6] The *entropy* is a measure of uncertainty in a random variable, expressed by the average number of bits needed to represent one symbol in a message. The entropy also quantifies the uncertainty involved in predicting the value of a random variable. For example, specifying the outcome of a fair coin flip, with two possible outcomes, provides less information than specifying the outcome from a roll of a die, with six possible outcomes.

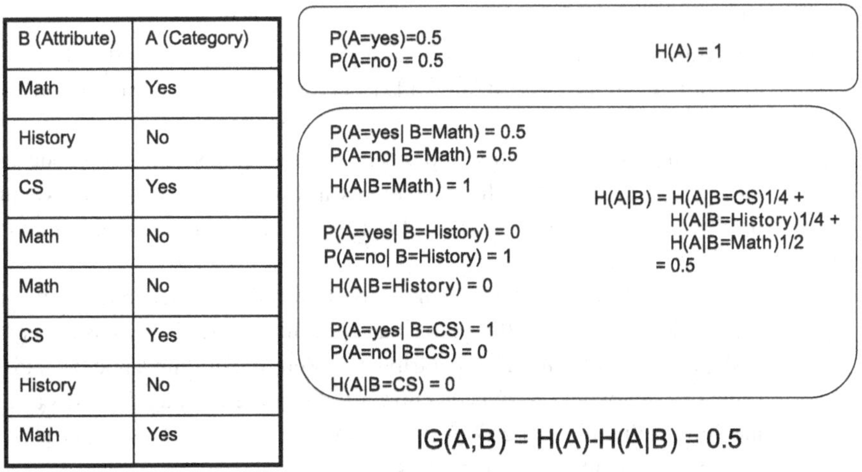

B (Attribute)	A (Category)
Math	Yes
History	No
CS	Yes
Math	No
Math	No
CS	Yes
History	No
Math	Yes

P(A=yes)=0.5
P(A=no) = 0.5   H(A) = 1

P(A=yes| B=Math) = 0.5
P(A=no| B=Math) = 0.5
H(A|B=Math) = 1   H(A|B) = H(A|B=CS)1/4 +
                                          H(A|B=History)1/4 +
P(A=yes| B=History) = 0   H(A|B=Math)1/2
P(A=no| B=History) = 1   = 0.5
H(A|B=History) = 0

P(A=yes| B=CS) = 1
P(A=no| B=CS) = 0
H(A|B=CS) = 0

IG(A;B) = H(A)-H(A|B) = 0.5

Figure 11.8 Computing the information gain.

and

$$H(A|B) = \sum_b P(B = b) \cdot H(A|B = b)$$

Intuitively, the $IG(A; B)$ information gain represents the reduction of uncertainty in random variable A, given what is known about random variable B. The larger this reduction, the more predictive power there is, as illustrated by the example in Figure 11.8.

This example [18] considers a tuple attribute representing a student's major, whose value is either "Math," "CS," or "History." And the classification categories indicate whether a student enjoys the movie "Gladiator" or not ($L = \{Yes, No\}$).

In this case, the goal is to evaluate how predictive a student's major is concerning his or her enjoyment of the movie. The entropy $H(A)$ for the category attribute is computed as follows:

$$H(A) = -\big(P(A = Yes) \cdot \log_2(P(A = Yes)) + P(A = No) \cdot \log_2(P(A = No))\big)$$

Taking a closer look at Figure 11.8, it is easy to see that knowing that a student majors in "History" or "CS" is a good predictor to whether the student likes the movie. On the other hand, knowing that the student is a "Math" major is not very predictive. Overall, knowing the major results in an uncertainty reduction of around 0.5 bits.

This example illustrates the main insight behind the structure of a decision tree. When it is being built, the fundamental operation consists of replacing the leaves by *test* nodes, starting at the root. The attribute to test at each node is chosen by comparing all of the available attributes and choosing the most predictive, according to their G score.

The learning step

An offline decision tree *learner* assumes that all of the training data is available either on disk or in memory. Instead, the VFDT classification model learner updates the decision tree incrementally, as training samples arrive one by one.

Specifically, the first few tuples are used to choose the root node and its associated test. Subsequently, the next tuples are passed down to the existing leaves and used to choose the appropriate split attributes at that level, proceeding down the tree, recursively.

Using the Hoeffding bound,[7] it can be shown that the accuracy of the online learning process is not significantly worse than for offline learning [13]. Specifically, to ensure, with probability $1 - \delta$, that the attribute chosen for splitting a node at a given time is the same one that would have been chosen if the entire training data was available at once, the split evaluation score G of the best attribute to split must be at least ϵ larger than that of the second best attribute to split.

Based on statistical guarantees associated with the Hoeffding bound, for certain split evaluation functions[8] G, ϵ can be computed as follows:

$$\epsilon = \sqrt{\frac{R^2 \cdot \ln(1/\delta)}{2 \cdot q}}$$

where q is the number of tuples seen so far and R is the range of the score as computed by the split evaluation function G. Note that the value of ϵ gets smaller as more tuples are processed.

The algorithm

Algorithm 10 describes the VFDT model learning steps, which are incremental and performed as new incoming tuples, labeled with the true category, arrive. Its performance has been shown to be comparable to several state-of-the-art non-incremental decision tree algorithms.

Algorithm 10 makes two considerations. First, each attribute i of a tuple takes one of multiple discrete values from the set Z_i ($a_i(t) \in Z_i$). Second, the total number of categories L is fixed and known a priori.

The algorithm starts with a root node and incrementally splits each leaf node. The decision on whether and how to split a node is made to satisfy the Hoeffding bound, ensuring that its probabilistic guarantees hold.

The incremental counts n_{ijkm} are used to simplify the computation of the score produced by the split evaluation function G. They hold the number of tuples whose i-th attribute value has index j in Z_i, that belong to the k-th category in L, and that map to the tree node v_m.

The attribute to split on is decided by finding the attribute with the largest split evaluation score among all attributes in the tuple. Algorithm 11 demonstrates how one such score, based on information gain, can be computed.

[7] The Hoeffding inequality [19] provides an upper bound on the probability that the sum of random variables deviates from its expected value.

[8] The score generated by such a function must be computed as an average over the score values of individual tuples at a node.

Algorithm 10: VFDT model learner.

Param : N, number of attributes
Param : Z_i for $0 \leq i < N$, list of values for attribute at index i
Param : L, list of categories
Param : G, split evaluation function
Param : δ, one minus the probability of selecting the right attribute to split
$DT \leftarrow$ root node v_0 ▷ Decision tree with the root node — not split yet
$\mathbf{Z} \leftarrow \left(\bigcup_i Z_i \right) \cup Z_\emptyset$ ▷ All attributes plus null attribute (represents not splitting)
$\mathbf{Z}_0 \leftarrow \mathbf{Z}$ ▷ All possible attributes to split on for root node
▷ keep a counter n_{ijkm} for each attribute (i), value (j), category (k), node (m)
▷ n_{ijkm} counts the # of tuples with a given attribute value (at index j in Z_i) and category (at index k in L) at a given node (v_m)
$n_{ijk0} \leftarrow 0, \forall 0 \leq i < N, 0 \leq j < |Z_i|, 0 \leq k < |L|$ ▷ Init counters for the root node ($m=0$)
while *not end of the stream* **do**
    read tuple $A(t)$ and its category $l(t)$ ▷ Next tuple in the stream
    propagate $A(t)$ and $l(t)$ to leaf node v_m ▷ Use the node tests in the tree
    for $Z_i \in \mathbf{Z}_m$ **do**   ▷ For each possible split attribute
        $j \leftarrow$ index of element $a_i(t)$ in Z_i ▷ For the attribute value at hand
        $k \leftarrow$ index of category $l(t)$ in category set L ▷ For the category at hand
        $n_{ijkm} \leftarrow n_{ijkm} + 1$ ▷ Increment the counter maintained at the leaf node
    $p_m \leftarrow \mathrm{argmax}_k \sum_j n_{0jkm}$ ▷ Find the most frequent category at node v_m
    ▷ p_m can be used for quick prediction at leaf node v_m (not used for learning)
    if $|\{k : n_{0jkm} > 0\}| > 1$ **then**   ▷ There is more than 1 category at v_m
        continue   ▷ No more splitting can happen here
    $\bar{G}_{im} \leftarrow G(\{m_{jk} = n_{ijkm}\}), \forall Z_i \neq Z_\emptyset \in \mathbf{Z}_m$ ▷ Evaluate all attribute splits
    $\bar{G}_{\emptyset m} \leftarrow G(\{m_{\emptyset k} = \sum_j n_{0jkm}\}) \in \mathbf{Z}_m$ ▷ Evaluate no split (null attribute)
    $Z_u \leftarrow \mathrm{argmax}_{Z_i \in \mathbf{Z_m}} \bar{G}_{im}$ ▷ Attribute domain with the highest score
    $Z_v \leftarrow \mathrm{argmax}_{Z_i \in \mathbf{Z_m} \setminus Z_u} \bar{G}_{im}$ ▷ Attribute domain with the 2nd highest score
    $q \leftarrow \sum_{j,k} n_{0jkm}$ ▷ Number of tuples at the tree node v_m
    $\epsilon \leftarrow \sqrt{\dfrac{R^2 \cdot \ln(1/\delta)}{2 \cdot q}}$ ▷ Here R is the range of the evaluation function
    if $\bar{G}_{um} - \bar{G}_{vm} > \epsilon$ *and* $Z_u \neq Z_\emptyset$ **then**   ▷ Gain difference is sufficient
        replace v_m in DT by a node that splits on attribute Z_u
        for *each* $x \in Z_u$ **do**   ▷ For each branch of the split
            add node $v_{m'}$ to DT as a child of v_m, $m' = |DT|$ ▷ Add a new node
            ▷ The node $v_{m'}$ will get tuples from v_m that have $a_u(t) = x$
            $\mathbf{Z}_{m'} \leftarrow \mathbf{Z}_m - Z_u$ ▷ Cannot split on the same attribute again
            $n_{ijkm'} \leftarrow 0, \forall i \text{ s.t. } Z_i \in \mathbf{Z}_{m'}, 0 \leq j < |Z_i|, 0 \leq k < |L|$ ▷ Init counters for $v_{m'}$

Note that the tree construction algorithm also includes a null attribute $Z_\emptyset$ that corresponds to not splitting a node at all. With the aid of this attribute two conditions must hold before the algorithm decides to split a node. First, the split evaluation score for any attribute of the tuple must be greater than the score for the null attribute. Second, the split evaluation score for the candidate attribute must be greater than the second best score by more than the threshold value ϵ. When using the information gain metric, ϵ is defined as follows:

$$\epsilon = \sqrt{\frac{\log_2(|L|)^2 \ln(1/\delta)}{2 \cdot q}}$$

Algorithm 11: Split evaluation function based on information gain. The R value for this function is $\log_2(|L|)$, where $|L|$ is the number of categories.

Param : $\{m_{jk}\}$, the number of tuples with j-th attribute value and k-th category

$s_j \leftarrow \sum_k m_{jk}, \forall j$ ▷ The number tuples with j-th attribute value

$s \leftarrow \sum_j s_j$ ▷ Total number of tuples

$p_{jk} \leftarrow m_{jk}/s_j, \forall j, k$ ▷ Frequency of k-th category for j-th attribute value

$H_j \leftarrow -\sum_k p_{jk} \cdot \log_2(p_{jk}), \forall j$ ▷ Information entropy for j-th attribute value

$p_j \leftarrow s_j/s, \forall j$ ▷ Frequency of j-th attribute value

$H' \leftarrow \sum_j p_j \cdot H_j$ ▷ Information entropy for the *split* set of tuples

$p_k \leftarrow \sum_j m_{jk}/s, \forall k$ ▷ Frequency of the k-th category

$H \leftarrow -\sum_k p_k \cdot \log_2(p_k)$ ▷ Information entropy for the entire set of tuples

return $H - H'$ ▷ The information gain from splitting

Hence, the estimate of the information gain G after seeing q tuples lies within ϵ of the true value with probability $1 - \delta$. Consequently, by performing a split only when the difference in information gain between the top two attributes is greater than ϵ, the algorithm ensures that, with probability $1 - \delta$, the same attribute would have been selected for the split had an offline decision tree algorithm, with all of the training data available, been used.

Once the split is performed, the new node does not include the attribute on which the split was just made ($\mathbf{Z}_{m'}$), so that other attributes can be tested at the new leaf node.

The scoring step

As we have seen, the VFDT classification model, as well as the model generated by other decision tree-based methods, is a tree where each node represents a test on an attribute of the tuples to be classified, and each branch from a node corresponds to a possible outcome of the test. At the lowest level of the tree, each leaf contains a class prediction (Figure 11.9).

In this example, at the root node, the attribute $a_k(t)$ for an incoming tuple $A(t)$ is tested first. Based on which branch value it matches, the scoring algorithm proceeds down that branch, until a leaf node is reached. At a leaf node, a prediction is made by looking at the most frequent category for this leaf as determined during the learning step.

For instance, it can be seen that the root node v_0 splits on the k-th attribute. Assuming that this attribute has three possible values: $[x_{k0}, x_{k1}, x_{k2}]$ and $A(t)$'s $a_k(t)$ attribute has value x_{k0}, the search progresses to node v_1. Node v_1 splits on the k'th attribute, which can assume two values: $[x_{k'0}, x_{k'1}]$. $A(t)$'s $a_{k'}(t)$ attribute has value $x_{k'0}$, hence the search progress to node v_4, a leaf node. The most frequently occurring category in v_4 is p_4, so it becomes the category assigned to $A(t)$. In this example, only equality tests were used. In practice, the tests may include different conditions as well as numeric and non-numeric values.

11.3.2   Advanced reading

There are several memory and disk optimizations proposed as extensions to the basic VFDT algorithm we described [13]. The implementation of the basic algorithm is

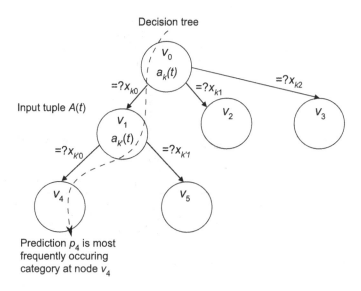

Decision tree

Input tuple A(t)

Prediction p_4 is most
frequently occuring
category at node v_4

Figure 11.9 Scoring using a decision tree.

available as an open-source library [20] and can be easily wrapped by an SPL primitive
operator.

The VFDT method has been extended to deal with dynamically changing streams.
This extension is referred to as Concept-adapting Very Fast Decision Tree learner
(CVFDT) [21] and employs the regular VFDT algorithm over sliding windows to peri-
odically update the classifier. Jin and Agrawal [22] have also improved the original
VFDT algorithm, boosting its efficiency and decreasing its space requirements for a
given level of accuracy.

Finally, several stream classification algorithms have been proposed to effectively
take temporal locality into account. Some of these algorithms were designed to be
purely single pass adaptations of conventional classification algorithms [13], whereas
others are more effective in accounting for the evolution of the underlying data
stream [21, 23]. Another recent innovation in on-demand stream classification methods
is the use of micro-clustering (Section 11.4) for classification purposes.

11.4 Data stream clustering

Clustering is a well studied problem in the data mining literature [24]. As seen in Section 10.2, the stream clustering problem consists of assigning a set of tuples to groups,
or clusters, so that the tuples in the same cluster are more similar to each other than
to those in other clusters. Hence, a cluster is a group of tuples with a low distance
between one another, forming a dense area of the data space, or of a particular statistical
distribution.

To employ clustering, the tuple attributes are required to be numeric or be converted to
a numeric representation ahead of time [25]. The numeric values are necessary because

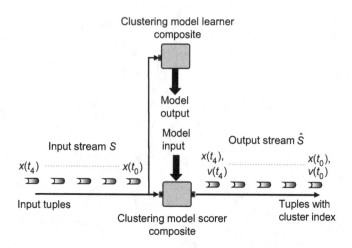

Figure 11.10 The clustering model learning and scoring process.

a cluster is determined by computing the distance between each tuple as a function of a *similarity metric*. The actual number of clusters used to represent a dataset may or may not be known a priori and may also need to be determined during the model learning step.

The *scoring* or *evaluation* part of the clustering process involves using a clustering model scorer, the *clusterer*, and a set of pre-determined clusters, computed during a learning step, to determine which cluster an incoming tuple is most similar to. As a result, the *clusterer* assigns a cluster index $v(t)$ to each incoming tuple $A(t)$, possibly appending this information to the original tuple. The overall modeling as well as the scoring and evaluation part of the clustering process are shown in Figure 11.10.

An application scenario

Consider a hypothetical application used by a telecommunications provider aimed at identifying users with similar service usage profiles, such that promotions can be designed for them. In this case, clustering model learning may be used to group together users into similar service profiles.

Specifically, service profiles of new users or of users currently browsing the provider's website can be sent to a clustering model scorer (as tuples) to determine which previously identified cluster the current profile matches best. This information, coupled with promotion plan information, may then be used to place more relevant ads or to send promotions to the user in real time.

Modeling

The *modeling* part of the clustering process consists of an unsupervised learning step where the similarity between tuples, based on their attribute values, is computed, and the set of clusters that include tuples similar to each other is identified.

In some cases, the number of clusters is also determined based on the properties of the tuples. Note that this is very different from the classification modeling process (Section 11.3), where the ground truth with each tuple's "true label" is available at training time. Conversely, clustering does not rely on the presence of any such information and depends solely on the properties of the incoming data. In Figure 11.10, it can be seen that the learner does not take any input besides the incoming stream.

There are four broad classes of clustering algorithms, each of which employs a different notion of what a cluster is and, correspondingly, makes use of different similarity metrics.

Centroid-based clustering

In centroid-based clustering methods, each cluster is represented using a *centroid* vector and a new tuple is assigned to the cluster with the *nearest* centroid vector, which is identified by computing the distance between this tuple and the centroids for the different clusters.

Specifically, given a set of numeric tuples $\mathbf{x}(t_i)$ with $0 \leq i < Q$, the modeling problem consists of finding the best k clusters C_j and their corresponding centroid vectors μ_j $(0 \leq j < k)$ such that the sum of intra-cluster distances is as small as possible. The corresponding optimization problem can be stated as follows:

$$\mathrm{argmin}_{\{C_j, \mu_j\}} \sum_{j=0}^{k-1} \sum_{\mathbf{x}(t) \in C_j} \left| \mathbf{x}(t) - \mu_j \right|^2$$

where $| \cdot |^2$ represents the squared norm of the resulting vector.

Note that in this formulation, the Euclidean distance between a tuple's corresponding vector and the centroid vector for the cluster is used. Also, in this case, the number of clusters k is known a priori. As mentioned, in practice, k may be unknown, and there are extensions to this problem that also try to determine its value. In general, this optimization is known to be in the Non-deterministic Polynomial time (NP)-hard class of algorithmic problems, and approximate solutions are used instead.

Among the most popular of these solutions is Lloyd's algorithm [26], a method also used in quantization, more commonly referred to as the *k-means algorithm*. In this method the centroid vectors μ_j are set to be the means of each cluster, i.e. $\mu_j = \frac{1}{|C_j|} \sum_{\mathbf{x}(t) \in C_j} \mathbf{x}(t)$. Hence, the mean vector itself does not necessarily correspond to an actual tuple from the training set.

The *k-means algorithm* is a gradient descent-based approach guaranteed to provide a local optimum for the objective function defined earlier. It starts by picking k randomly chosen mean vectors[9] μ_j, followed by a Nearest Neighbors (NN) assignment of the tuples in the training dataset to these vectors. In other words, a tuple $\mathbf{x}$ is assigned to cluster j if $\forall i \neq j, |\mathbf{x} - \mu_j| \leq |\mathbf{x} - \mu_i|$, where $0 \leq i, j < k$.

From then on, the mean of the cluster is iteratively recomputed as the sample average of all the tuples assigned to the cluster, followed again by an NN reassignment of tuples

---

[9] There has also been research focusing on identifying the set of initial vectors [27].

to clusters. This process is repeated until it converges. Convergence is achieved when the mean vectors do not change more than a previously specified threshold.

There are also other algorithmically similar approaches: the *k-medoid algorithm* and the *k-median algorithm*. In the *k*-medoid clustering, the centroid vector must correspond to an actual tuple from its cluster, while in the *k*-median clustering the centroid vector is picked to be the median of the tuples in the cluster.

The scoring of a centroid-based clustering model is straightforward and similar to NN classification [28]. Simply put, a new tuple is assigned to the cluster whose centroid vector is closest to it. Clearly, scoring is computationally cheap and parsimonious in its use of memory, requiring only the calculation of *k* distances per incoming tuple, relying on a model representation that consists of only the clusters' centroid vectors.

Distribution-based clustering

Distribution-based clustering algorithms are closely related to methods in statistics that try to identify the distribution that best fits a set of samples. In these algorithms, a cluster groups tuples most likely to have been drawn from the same distribution. The collection of all tuples is represented using a weighted mixture of Gaussian distributions, a GMM, where each distribution has a different mean and variance. A mixture model with *k* Gaussians may be represented as:

$$\sum_{j=0}^{k-1} w_j \cdot \mathcal{N}(\mu_j, \Sigma_j)$$

where $\mathcal{N}(\mu_j, \Sigma_j)$ is the multivariate Normal distribution's Probability Density Function (PDF):

$$\mathcal{N}(\mu, \Sigma) = \frac{1}{\sqrt{2 \cdot \pi \cdot \det(\Sigma)}} \cdot e^{-\frac{1}{2}(\mathbf{x}-\mu)^T \times \Sigma^{-1} \times (\mathbf{x}-\mu)}$$

where operation $\det(\cdot)$ represents the determinant of a matrix.

In this case, the modeling step involves estimating the best weights w_j, the means μ_j, as well as the covariance matrix Σ_j for each Gaussian in the model, such that they capture the distribution of the tuples as accurately as possible.

During this step the Expectation Maximization (EM) [29] algorithm is used to model the tuples using *k* Gaussian distributions. These distributions are initialized randomly and their means and variances are iteratively optimized to better fit the data set. On convergence, tuples that are more likely to have been drawn from the same Gaussian distribution are said to belong to the same cluster.

As in centroid-based methods, scoring is straightforward. It requires storing the GMM and determining which Gaussian distribution a new tuple $\mathbf{x}$ is most likely to belong to, by computing a *likelihood score*. The likelihood score $\mathcal{L}$ is computed by evaluating the Gaussian function at the value of the tuple, for each Gaussian in the mixture, and selecting the one for which the likelihood is the largest. Hence for Gaussian *j* in the mixture, we have:

$$\mathcal{L}(j|\mathbf{x}) = w_j \cdot \mathcal{N}(\mu_j, \Sigma_j)$$

We then select the Gaussian i that has the largest likelihood score, i.e., $i = \text{argmax}_j \mathcal{L}(j|\mathbf{x})$ Intuitively, if we plot the Gaussian mixtures, the tuple is assigned to the Gaussian that has the largest height at its value.

Hierarchical clustering

In hierarchical clustering, also known as connectivity-based clustering, a hierarchy of clusters is built on the principle that tuples that are close to each other, measured by a distance function, must belong to the same cluster [30]. Common distance measures include the Euclidean distance,[10] the Mahalanobis distance [31], and the Manhattan distance [32].[11]

While it is straightforward to define the distance between two tuples, this method also requires a measure of the distance between two clusters. Different alternatives exist, each one leading to a different clustering algorithm and include: (a) *single linkage measure*, which captures the smallest distance between a tuple from the first cluster and any tuple from the second cluster; (b) *complete linkage measure*, which computes the maximum distance between a tuple from the first cluster and any tuple from the second cluster; and (c) *average linkage measure*, which computes the average distance between a tuple from the first cluster and any tuple from the second cluster.

To construct the cluster hierarchy, two different approaches can be used: *agglomerative* or *divisive*. The agglomerative approach is bottom-up, starting with each tuple corresponding to its own cluster and proceeding by merging clusters based on their distance from each other. In other words, clusters that have the smallest distance between them are merged. In contrast, the divisive approach is top-down. All tuples are initially assigned to one cluster, which is subsequently split, recursively. In both cases, the splits and merges are often made in a greedy manner.

The resulting structure of hierarchical clustering is often visualized using a *dendrogram*. While this method is simple to grasp, the results are not always easy to use as no unique partitioning of the data set exists. Instead the result is a hierarchy of clusters that needs to be interpreted appropriately. Hence, scoring such models can be complex and very dependent on an application's specific analytical task.

Connectivity-based clustering methods have also motivated the development of density-based clustering methods, which we describe next.

Density-based clustering

In density-based clustering, a cluster is defined as an area with a high density of tuples. Tuples that end up placed in sparsely populated areas are usually considered to be noise, and tuples placed in border areas are not part of any of the clusters.

One of the popular density-based clustering methods is Density-Based Spatial Clustering of Applications with Noise (DBSCAN) [33], which groups tuples or subsets

---

[10] The two-dimensional Euclidean distance between two points is the length of the line segment connecting them. The same idea can be generalized to a higher number of dimensions.

[11] The Manhattan distance, or taxicab distance, is the distance a car would drive in a city laid out in square blocks, if there are no one-way streets. The same idea can be generalized to a higher number of dimensions, rather than just the two dimensions of a city grid.

of tuples using measures similar to the ones used in connectivity-based clustering. Thus, two tuples or two sets of tuples are grouped when their distance is smaller than a threshold. Note that this type of clustering can create arbitrarily shaped clusters.

Scoring a density-based cluster model requires identifying the cluster a new tuple is connected to, by measuring the appropriate distances while accounting for the threshold used. As a result, a tuple may be identified as belonging to no cluster if it lies in sparse areas.

Discussion

Among the clustering techniques, the k-means algorithm is frequently used because of its theoretical properties. For instance, the space partitions created by each cluster correspond to Voronoi regions [26], allowing the model properties to be understood analytically. Furthermore, its close relationship to NN classification makes it easy to implement and interpret. On the other hand, given k-means' reliance on gradient descent and its dependency on initialization conditions, the algorithm may lead to uncertainties in the clustering results, which must be carefully resolved.

Distribution-based clustering suffers from the drawback that the appropriate modeling function must be selected before the modeling process begins. Such selection relies on assumptions on the data distribution that may not always be valid in practice.

Hierarchical clustering provides a multi-resolution way of looking at groupings of tuples and can, sometimes, provide visually interpretable results. On the other hand, because tuples can be seen as belonging to different clusters depending on the level of the hierarchy that is used, the semantic meanings of each cluster may be hard to determine.

Initial settings and interpretation difficulties can also occur with density-based clustering methods. They are sensitive to the choice of the threshold and allow for some tuples to belong to none of the clusters. On the other hand, given a fixed threshold, density-based clustering is deterministic. This is in contrast to the k-means and the distribution-based methods, whose outcome is a function of the initialization conditions.

As with all modeling approaches, clustering algorithms require careful experimentation and tuning of their parameters, including the number of clusters k, the similarity and distance metrics, and their thresholds.

Finally, it is often difficult to adapt an offline clustering algorithm for use in a streaming context. The difficulty stems from the single pass over the data constraint normally faced by high-performance SPAs. While some work in this direction has been done [34], in practical terms it is usually more important, analytically, to be able to examine the set of clusters over user-specified time horizons. For instance, an analyst may want to observe the behavior of the clusters in the data stream over the past week, month, or year, to better understand the underlying properties of the data. Hence, typical streaming extensions of clustering methods focus on storing *intermediate* cluster statistics using incremental methods so clusters can be extracted on demand. This is described next.

11.4.1 Illustrative technique: CluStream microclustering

The CluStream clustering technique [35] is a stream processing algorithm designed to compute intermediate cluster statistics from streaming data, in terms of cluster feature vectors [36], to support centroid-based stream clustering.

The cluster feature vectors keep track of the first and second order moments of the incoming data, which satisfy the *additivity property* and can be used to compute a host of other parameters important for clustering, including a cluster's centroid and radius.

The additivity property implies that the first and second order moments can be maintained as a simple incremental addition of statistics over data points. Hence, they can be computed in a streaming manner from the incoming tuples. Furthermore, it is also possible to obtain the statistics over a particular time period by subtracting the statistics at the beginning of the period from the statistics at the end of the period.

The CluStream algorithm makes use of microclusters for this purpose. A stream microcluster $\mathcal{M}$, capturing properties of Q tuples $\mathbf{x}(t_i)$, $0 \le i < Q$ in a stream, can be defined by the following set of attributes.

The sum of the tuple attributes:

$$\mathbf{m1} = \sum_{i=0}^{Q-1} \mathbf{x}(t_i)$$

The sum of the squares of the tuple attributes:

$$\mathbf{m2} = \sum_{i=0}^{Q-1} \mathbf{x}(t_i) \bullet \mathbf{x}(t_i)$$

where the $\bullet$ operation is an attribute by attribute product of the two vectors, resulting in the squared value of each attribute in the vector.

The sum of the timestamps:

$$\tau 1 = \sum_{i=0}^{Q-1} t_i$$

The sum of the square of the timestamps:

$$\tau 2 = \sum_{i=0}^{Q-1} (t_i)^2$$

Finally, each microcluster is also associated with an *id* identifier. The *id* can be either a singleton or a list comprising all of the identifiers of microclusters that have been merged to form this particular microcluster. Hence, each stream microcluster $\mathcal{M}$ is defined by a 6-tuple with the following format:

$$\mathcal{M} = \{\mathbf{m1}, \mathbf{m2}, \tau 1, \tau 2, Q, id\}$$

A microcluster's attributes can be computed incrementally from the incoming tuples. As each microcluster gets updated, it includes information since the beginning of the computation up to the current time step. Hence, to identify the actual clusters for a specific period of time, *snapshots* of the microclusters are stored at different instants of time in a *pyramidal time frame* as follows.

A snapshot is associated with the *order* in which it was built, from 1 to $\log_\alpha(T)$, where T is the time elapsed since the start of the stream. Order i microcluster snapshots are stored whenever the clock value from the beginning of the stream is a multiple of α^i, where α is a previously selected constant. For each order, only the last $\alpha + 1$ snapshots are retained, resulting in a total of $(\alpha + 1) \cdot \log_\alpha(T)$ snapshots being stored.

A post-processing *macro*-clustering step is performed to reconstruct the actual clusters from the information stored in the microcluster snapshots. To ensure that the clusters yielded by the macroclustering step are comparable to offline clustering, a large number of microclusters is necessary. In the CluStream algorithm q microclusters, $\mathcal{M}_j$ ($0 \leq j < q$), are computed and maintained. The number q is chosen to be significantly larger than the expected number of natural clusters in the stream to ensure accuracy.

The algorithm is bootstrapped using a standard offline k-means algorithm where the first Q' tuples in the stream are used to create q clusters. Hence, the number Q' must be selected such that it is large enough to allow the extraction of q clusters from it.

Once the initial q clusters are created, the corresponding microclusters' attributes are computed and, as tuples arrive, they are continuously updated. Whenever a new tuple $\mathbf{x}(t_i)$ is received, three choices are possible: (i) the tuple belongs to an existing microcluster, (ii) the tuple is an outlier and doesn't belong to any of the microclusters, and (iii) the tuple represents an evolution in the stream characteristics and, hence, a new microcluster must be created.

To determine if a tuple belongs to an existing microcluster, the algorithm locates the microcluster $\mathcal{M}_p$ whose centroid is the closest to it by computing the Euclidean distance between them. Additionally, the algorithm computes the cluster extent as β (also called the maximal boundary factor) times the sum of the standard deviation in each dimension (corresponding to the values for each tuple attribute). The extent for a microcluster with a single tuple is set heuristically as γ times the extent of its nearest microcluster, where γ is another user-specified parameter.

Hence, the extent for microcluster $\mathcal{M}_p$ is computed as follows:

$$E_p = \frac{\beta}{Q_p} \cdot \left(\sqrt{Q_p \cdot \mathbf{m2}_p - \mathbf{m1}_p \bullet \mathbf{m1}_p} \right)^{\mathrm{T}} \times \mathbf{1}$$

where $\mathbf{1}$ is a vector with 1 in each of its N dimensions and $\sqrt{\cdot}$ denotes an element-wise square root on the vector.

If the new $\mathbf{x}(t_i)$ tuple lies within the extent of its closest microcluster, the corresponding microcluster attributes are updated using this tuple. Otherwise, a new microcluster must be created. To accommodate it, however, either an existing microcluster must be deleted or two existing microclusters must be merged.

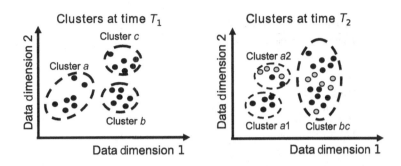

Figure 11.11 Evolution of microclusters over time as new tuples arrive.

To delete an existing microcluster an estimate of its recency is made using measures $\tau 1$ and $\tau 2$, and by computing the mean and variance of the timestamps in this microcluster, assuming a Gaussian distribution of timestamps with those parameters.

The recency is then estimated as the time past which the top $\frac{n}{2 \cdot Q}$-percentile of the points within this distribution lie. For microcluster $\mathcal{M}_p$ with timestamp mean $\mu_p = \frac{\tau 1_p}{Q_p}$ and timestamp standard deviation $\sigma_p = \frac{\sqrt{Q_p \cdot \tau 2_p - \tau 1_p^2}}{Q_p}$, the percentile area $\kappa(x)$ is computed as follows:

$$\kappa_p(x) = \int_x^{\infty} \frac{1}{\sqrt{2\pi \cdot \sigma_p^2}} \cdot e^{\frac{-(t-\mu_p)^2}{2 \cdot \sigma_p^2}} \, dt$$

Hence, the recency timestamp r_p, or *relevance stamp*, is estimated as the value of x such that $\kappa_p(x) = \frac{n}{2 \cdot Q_p}$, where n is also an algorithm parameter used to control how much importance is given to the most recent n points in the microcluster.

The microcluster with the smallest estimated relevance stamp is identified and, if this value is less than δ (a pre-defined relevance timestamp threshold), the microcluster is deleted. On the other hand, if none of the microclusters have a small relevance stamp, none of them can be deleted and the closest two microclusters, measured in terms of the distance between their centroids, are merged into a single one. Using this procedure, illustrated by Figure 11.11, exactly q microclusters are retained at all times.

In the diagram, at time T_1, there are $q = 3$ microclusters but, because of the arrival of new tuples at time T_2, clusters b and c are merged into a new, larger, microcluster bc, a new microcluster $a2$ is created, and microcluster a is updated, becoming microcluster $a1$.

Algorithm 12 formalizes all of these steps. In its description the *kmeans()* function implements the standard k-means algorithm for initialization where the initial tuples are used as we previously indicated.

After that, finding the closest microcluster to an incoming tuple requires comparing the distance between the tuple and the centroid of each of the q stored microclusters. The extent E_p for a microcluster is computed as discussed earlier. If an incoming tuple $\mathbf{x}(t)$ is found to lie within the extent of its closest microcluster $\mathcal{M}_p$, each of its attributes, $\mathbf{m}1, \mathbf{m}2, \tau 1, \tau 2$, are appropriately updated as is its Q_p count.

Algorithm 12: CluStream microclustering.

Param : α, snapshot parameter
Param : β, maximal boundary factor
Param : γ, extent of microcluster with a single tuple
Param : n, recency estimation interval
Param : Q', number of tuples to store during initialization
Param : q, number of microclusters
Param : δ, relevance timestamp threshold
Param : $\hat{T}$, controls the number of orders for saving snapshots

$\mathcal{M} \leftarrow kmeans(Q', q)$ $\triangleright$ Find the initial set of microclusters
$T \leftarrow Q'$ $\triangleright$ Initialize the timestamp

while *not end of the stream* **do**
  read tuple $\mathbf{x}(t)$ $\triangleright$ Get the current tuple
  $\mathcal{M}_p \leftarrow \operatorname{argmin}_{\mathcal{M}_i} |(1/Q_i) \cdot \mathbf{m1}_i - \mathbf{x}(t)|$ $\triangleright$ Find the closest microcluster
  if $Q_p > 1$ **then** $\triangleright$ We have a non-singleton cluster
     $E_p \leftarrow (\beta/Q_p) \cdot \left(\sqrt{Q_p \cdot \mathbf{m2}_p - \mathbf{m1}_p \bullet \mathbf{m1}_p}\right)^T \times \mathbf{1}$ $\triangleright$ Compute the extent
  else $\triangleright$ We have a singleton cluster
     $\triangleright$ Find the nearest non-singleton (size > 1) cluster
     $\mathcal{M}_r \leftarrow \operatorname{argmin}_{\mathcal{M}_i \text{s.t.} Q_i > 1} |(1/Q_i) \cdot \mathbf{m1}_i - (1/Q_p) \cdot \mathbf{m1}_p|$
     $E_p \leftarrow E_r \cdot \gamma$ $\triangleright$ Scale extent of $\mathcal{M}_r$ to compute $\mathcal{M}_p$'s extent
  if $|\mathbf{x}(t) - (1/Q_p) \cdot \mathbf{m1}_p| < E_p$ **then** $\triangleright$ The new point is inside of $\mathcal{M}_p$
     $\mathbf{m1}_p \leftarrow \mathbf{m1}_p + \mathbf{x}(t)$ $\triangleright$ Update $\mathcal{M}_p$ — $\mathbf{m1}_p$
     $\mathbf{m2}_p \leftarrow \mathbf{m2}_p + \mathbf{x}(t) \bullet \mathbf{x}(t)$ $\triangleright$ Update $\mathcal{M}_p$ — $\mathbf{m2}_p$
     $\tau1_p \leftarrow \tau1_p + t, \tau2_p \leftarrow \tau2_p + t^2$ $\triangleright$ Update $\mathcal{M}_p$ — $\tau1_p$ and $\tau2_p$
     $Q_p \leftarrow Q_p + 1$ $\triangleright$ Update $\mathcal{M}_p$ — Q_p
  else $\triangleright$ The new point is outside of $\mathcal{M}_p$
     $y \leftarrow \infty, j \leftarrow 0$ $\triangleright$ The lowest relevance value and micro cluster index
     for $p \leftarrow 0$ *to* $q - 1$ **do** $\triangleright$ For each existing microcluster
       $\mu_p \leftarrow \tau1_p/Q_p, \sigma_p \leftarrow \sqrt{Q_p \cdot \tau2_p - \tau1_p^2}/Q_p$ $\triangleright$ Timestamp mean, stddev.
       $r_p \leftarrow r$ s.t. $\kappa_p(r; \mu_p, \sigma_p) = n/(2 \cdot Q_p)$ $\triangleright$ Find cluster's time relevance
       if $y > r_p$ **then** $\triangleright$ A less relevant cluster is seen
         $y \leftarrow r_p, j \leftarrow p$ $\triangleright$ Update the lowest relevance and its index
     if $y < \delta$ **then** $\triangleright$ Least relevant cluster is not sufficiently relevant
       delete $\mathcal{M}_j$ $\triangleright$ Remove the irrelevant microcluster
     else $\triangleright$ Least relevant cluster is sufficiently relevant (cannot delete)
       $\mathcal{M}_a, \mathcal{M}_b \leftarrow \operatorname{argmin}_{a,b} |(1/Q_a) \cdot \mathbf{m1}_a - (1/Q_b) \cdot \mathbf{m1}_b|$ $\triangleright$ Closest pair
       $\mathcal{M}_{p'} \leftarrow CreateMicroCluster()$ $\triangleright$ Create a merged cluster
       $\mathbf{m1}_{p'} \leftarrow \mathbf{m1}_a + \mathbf{m1}_b$ $\triangleright$ Update $\mathcal{M}_{p'}$ — $\mathbf{m1}_{p'}$
       $\mathbf{m2}_{p'} \leftarrow \mathbf{m2}_a + \mathbf{m2}_b$ $\triangleright$ Update $\mathcal{M}_{p'}$ — $\mathbf{m2}_{p'}$
       $\tau1_{p'} \leftarrow \tau1_a + \tau1_b, \tau2_{p'} \leftarrow \tau2_a + \tau2_b$ $\triangleright$ Update $\mathcal{M}_{p'}$ — $\tau1_{p'}, \tau2_{p'}$
       $Q_{p'} \leftarrow Q_a + Q_b$ $\triangleright$ Update $\mathcal{M}_{p'}$ — $Q_{p'}$
       delete $\mathcal{M}_a, \mathcal{M}_b$ $\triangleright$ Remove the two, now merged, microclusters
     $\mathcal{M}_{p''} \leftarrow CreateMicroCluster()$ $\triangleright$ Create a new cluster for $x(t)$
     $\mathbf{m1}_{p''} \leftarrow \mathbf{x}(t)$ $\triangleright$ Set $\mathcal{M}_{p''}$ — $\mathbf{m1}_{p''}$
     $\mathbf{m2}_{p''} \leftarrow \mathbf{x}(t) \bullet \mathbf{x}(t)$ $\triangleright$ Set $\mathcal{M}_{p''}$ — $\mathbf{m2}_{p''}$
     $\tau1_{p''} \leftarrow t, \tau2_{p''} \leftarrow t^2$ $\triangleright$ Set $\mathcal{M}_{p''}$ — $\tau1_{p''}, \tau2_{p''}$
     $Q_{p''} \leftarrow 1$ $\triangleright$ $\mathcal{M}_{p'}$ is a singleton
  for $i \leftarrow 1$ *to* $\lfloor \log_\alpha(\hat{T}) \rfloor$ **do** $\triangleright$ For each order
    if $T \mod \alpha^i = 0$ **then** $\triangleright$ Time to create a snapshot for order
     store snapshot of microclusters
  $T \leftarrow T + 1$ $\triangleright$ Increment the timestamp

Alternatively, if the tuple is found to lie outside the extent of its closest microcluster, the microcluster with the smallest relevance timestamp is computed. If its relevance timestamp is smaller than δ, it is discarded.

If none of the current microclusters have a small relevance timestamp, the closest two microclusters are found and merged. The merge process for microclusters $\mathcal{M}_a$ and $\mathcal{M}_b$ to create the new microcluster $\mathcal{M}_{p'}$ requires setting its attributes, including its tuple count $Q_{p'}$ and its identifier $id_{p'}$, which will be a superset of the contents of both the id_a and id_b lists. In this way, the algorithm keeps track of the microclusters that were merged over time.

Once either a delete or a merge operation is performed, a new microcluster with the current tuple $\mathbf{x}(t)$ is created. Finally, based on the pyramidal time frame, a snapshot of the current microclusters is saved.

To generate the actual clusters from the stored microclusters, a macroclustering step is required. This step can make use of any centroid-based clustering algorithm, including the k-means algorithm, which is used to merge the microclusters appropriately over a specified time horizon h and for the desired number of clusters k. In rough terms, the microclusters that are selected, given h and k, are treated as pseudopoints, which are then reclustered, yielding the desired k clusters [35].

11.4.2   Advanced reading

The microclustering technique has been shown to be much more effective and versatile than the k-means-based stream clustering technique proposed earlier [34]. CluStream has also been extended to handle a variety of other problems [10].

For instance, when clustering high-dimensional data [37, 38], the same microcluster statistics can be used for maintaining the characteristics of the clusters, although it also becomes necessary to maintain additional information to keep track of the projected dimensions in each cluster. The projected dimensions can be used in conjunction with the cluster statistics to compute the projected distances that are required for intermediate computations.

Similarly, when clustering uncertain data, such as in sensor networks, where the data is generally noisy, it may be desirable to incorporate a degree of uncertainty into the clustering process. To do so, the microcluster statistics are augmented with the information about the underlying uncertainty, adding a degree of robustness to the clustering process [39].

Finally, when clustering text and categorical data, microclustering has also been shown to be usable by tweaking the statistics that are stored by the algorithm [40]. Specifically, the counts of the frequencies of the discrete attributes in each cluster are maintained and the inter-attribute correlation counts are also computed.

Several other clustering techniques designed for specific use cases and data types exist. For instance, *biclustering* or *coclustering* techniques [41] have been introduced to handle problems in bioinformatics and gene expression research. These techniques, also known as two-mode clustering [42], jointly consider clustering columns as well as rows of a combined data matrix whereas traditional clustering algorithms may be viewed as clustering columns of a matrix where each one stores individual data items. This approach translates into the ability to simultaneously cluster both genes and conditions allowing knowledge discovery on gene expression data.

There are also several algorithms for graph clustering including spectral clustering [43] and clique detection [44]. These problems are central to the analysis of graphs representing computer and telecommunication network topologies and social networks.

Finally, while we have focused on *hard* clustering algorithms where a tuple belongs to a single cluster, there has been research into *soft* or *fuzzy* clustering [45] algorithms, where a tuple is assigned a degree or likelihood of belonging to a cluster, but is not explicitly associated to it.

11.5   Data stream regression

Regression [46, 47] is a technique used to estimate the relationship between one or more *independent variables* such as the attributes of a tuple and a dependent variable, whose value must be predicted.

Similar to classification, regression can be used for forecasting. But, in contrast to classification, where the predictions usually result in a category from a discrete and finite set, regression is applied to numeric, and not necessarily discrete, values and the resulting prediction is also numeric.

Consider a simplified scenario where the incoming tuples to a finance application contain the current stock prices for a set of companies from a particular business sector. Suppose this application is tasked with predicting the stock price for a particular company within a certain window of time in the future to manage short-term trading decisions. A regression model can be designed to capture the relationship between the past values of the stock prices to generate the necessary short-term predictions.

The estimation of the regression function corresponds to the modeling (or learning) part of the regression analysis, while the use of the function for prediction corresponds to the evaluation or scoring part of the analysis.

These steps are depicted by Figure 11.12 where a Stream Processing Application (SPA) employs a *learner* operator, which continuously refines the regression function, and a *scorer* operator, which makes a prediction $y(t)$ for each incoming tuple $x(t)$.

As is the case with online classification (Section 11.3), when the true values $v(t)$ corresponding to these predictions become available (if at all), they can be used by the learner, along with the data on prior predictions, to update the model and improve future predictions.

Revisiting the stock price prediction scenario, we could have $x(t)$ as a vector of *current* stock prices for the companies in the technology sector and $y(t)$ as the predicted price for a particular stock two minutes in the future. Once two minutes have elapsed, $v(t)$ would be the actual true price for that stock.

In this scenario, if predictions are generally reliable, there would be a two-minute window in which a trading algorithm could identify opportunities for purchasing or selling a particular stock and, hence, make a profit. Once the true value $v(t)$ becomes

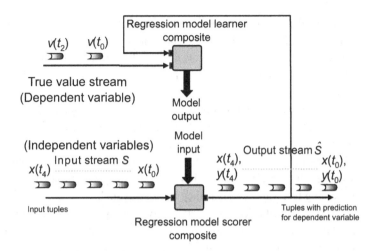

Figure 11.12 The regression model learning and evaluation process.

available, this application's learner operator can use this information as well as its own prior predictions to further improve the model.

Regression classes

There are two main classes of regression models: *parametric* and *non-parametric*. A parametric regression model [46] represents the relationship between the independent and dependent variables using a regression function with a finite number of (unknown) parameters. These parameters are computed during the learning step of the regression analysis.

In contrast, a non-parametric regression model [47] makes no assumptions about the function that represents the relationship between the independent and the dependent variables. Hence, the learning step of a non-parametric regression analysis involves estimating the regression function as well as its parameters.

Parametric regression

The parametric regression class includes several well-known techniques such as the simple linear regression method as well as the linear and non-linear least squares regression methods.

A linear regression model generates a prediction $y(t)$ by computing the following value:

$$y(t) = \sum_{i=0}^{N-1} w_i \cdot x_i(t)$$

where $x_i(t)$ represents the value of the i-th attribute of an N-attribute tuple and w_i corresponds to a weight associated with that attribute in the linear combination.

In matrix/vector form, this equation can be re-written as follows:

$$y(t) = \mathbf{w}^{\mathrm{T}} \times \mathbf{x}(t)$$

where $\mathbf{w}$ is the $N \times 1$ weight vector and $\mathbf{w}^{\mathrm{T}}$ is its transpose.

As seen, a linear regression method assumes a *linear* relationship between the independent variables and the dependent variable. Its learning step consists of a *data fitting* process to estimate the weights w_i for the linear function describing this relationship, by minimizing the distance between a prediction $y(t)$ and its corresponding true value $v(t)$.

When parametric regression analysis is used in an offline learning and online scoring configuration, the optimal weight vector **w** is determined by training on an existing set of tuples. The fixed weight vector is then used to score new tuples.

In contrast, when parametric regression analysis is used in an online learning and scoring configuration, the weight vector must be estimated incrementally, from the streaming true values $v(t)$, as they become available.

In this case, the learning step involves computing the weight vector such that the prediction $y(t)$ is as close as possible to the true value $v(t)$. Different measures, referred to as *loss functions*, may be used to quantify the distance between the prediction and the true value. Common loss functions include the squared loss, the hinge loss, the absolute loss, and the ϵ insensitive loss [29].

In general, in an online regression configuration different weight vectors are used for scoring as they are potentially updated every time new true values become available. The fresher weight vector is computed by combining the new true values with the information captured by the prior iteration of the vector. In most cases, only the latest version of the weight vector is stored and used for scoring.

Non-parametric regression
The non-parametric regression class also includes several well-known techniques such as kernel regression, Non-Parametric Multiplicative Regression (NPMR), as well as regression trees.

Kernel regression is used to estimate the conditional expectation of a random variable to find a non-linear relationship between independent and dependent variables. Its learning phase employs a weighting function, referred to as a kernel function, to direct the data fitting in the learning phase of the method. It has been used in image processing and reconstruction as well as for video upscaling.

NPMR can also model non-linear interactions among independent variables such as the behavior of organisms sharing a habitat in ecological modeling problems [48]. Its distinctive feature is that its learning step consists of optimizing the data fitting process *without* a reference to a specific global model. Rather, it makes use of a local model as well as of a kernel function. The local model describes the shape of the function used to fit the values defined by the independent variables *near* a target point, and the kernel function describes the weight that defines what *near* means for a particular local model.

Finally, a regression tree can also be used to learn both linear as well as non-linear relationships between independent and dependent variables. Since the dependent variable is not categorical (as is the case for a decision tree), a regression model is fitted using each of the independent variables. Specifically, for each independent variable, the data is split at several points, eventually creating a decision tree. At each split point, the error between the predicted value and the actual one is computed and compared across the independent variables. The combination of independent variable and split point yielding the lowest prediction error is chosen as the actual split point, a process that is repeated recursively. In cases where the relationship between independent and dependent variables is thought to be linear, a standard classification tree algorithm such as CART [17] can be used. On the other hand, if the relationship is thought to be non-linear, algorithms such as C4.5 [15] are used instead.

Discussion

Generally, parametric regression is computationally simpler than non-parametric regression. Yet a parametric regression model may not be able to capture some of the subtleties associated with real-world problems [48] and, hence, might produce unreliable predictions.

Once again, a clear understanding of the regression problem to be addressed, as well as of the computational environment where regression is going to be used, is necessary to decide which type of analysis to employ. In practice, several parametric standard linear and least squares regression algorithms have been used effectively as part of SPAs. One of these techniques, linear regression with Stochastic Gradient Descent (SGD), is described in detail next.

11.5.1   Illustrative technique: linear regression with SGD

SGD [49] is an optimization method used to minimize an objective function expressed as the sum of multiple differentiable functions.

Consider a linear regression problem where the regression model is to be learned over a set of Q tuples. This problem can be expressed as an optimization problem seeking to minimize an objective function, in this case, a loss function, capturing the squared error between the predicted and true values with the following form:

$$\Phi(\mathbf{w}) = \sum_{i=0}^{Q-1} \left(v(t_i) - \mathbf{w}^T \times \mathbf{x}(t_i) \right)^2$$

Note that in this loss function the squared error between the true value $v(t)$ and the prediction $\mathbf{w}^T \times \mathbf{x}(t)$ is summed up across multiple tuples, or time steps. This function is then optimized by computing an appropriate weight vector so its minimum value can be obtained.

This process can be implemented offline using a standard gradient descent technique, where all tuples, grouped in a batch, are considered together. Or it can be done iteratively using the SGD method, making it suitable for use by SPAs.

In this case, the SGD optimization method ensures that, for each incoming tuple with a true value (a time step), the weight vector, $\mathbf{w}_{(k)}$ at time t_k, is updated by modifying the last computed weight vector $\mathbf{w}_{(k-1)}$ as follows:

$$\mathbf{w}_{(k)} = \mathbf{w}_{(k-1)} - \alpha \cdot \nabla \Phi_k|_{\mathbf{w}=\mathbf{w}_{(k-1)}}$$

where $\nabla \Phi_k$ represents the gradient of the objective function considering only the tuple $x(t_k)$, $\nabla \Phi_k|_{\mathbf{w}=\mathbf{w}_{(k-1)}}$ is the value of the gradient evaluated using the weight vector from the prior time step t_{k-1}, and α is the step size or the *learning rate*.

Intuitively, at each iteration, the update factor moves the weight vector in the direction of the gradient of the loss function such that, over time, the local minimum for this function can be achieved.[12]

Considering each time step's squared error-based loss function:

$$\Phi_k(\mathbf{w}) = \left(v(t_k) - \mathbf{w}^T \times \mathbf{x}(t_k)\right)^2$$

Its gradient may be computed as follows:

$$\nabla \Phi_k = -2 \cdot (v(t_k) - \mathbf{w}^T \times \mathbf{x}(t_k)) \cdot \mathbf{x}(t_k)$$

And, hence, each update step takes the following form:

$$\nabla \Phi_k|_{\mathbf{w}=\mathbf{w}_{(k-1)}} = -2 \cdot (v(t_k) - \mathbf{w}^T_{(k-1)} \times \mathbf{x}(t_k)) \cdot \mathbf{x}(t_k)$$

To avoid convergence problems, the learning rate α can be reduced over time to ensure that the weight vector is not perturbed significantly at each step as the method gets closer to the solution. For instance, if i is the current tuple count, the learning rate can be set to $\alpha_0/\sqrt{i}$, where α_0 is the initial learning rate.

The SGD linear regression algorithm

The model learning steps outlined above can be formalized algorithmically (Algorithm 13). This algorithm starts by initializing $\mathbf{w}$ with a uniform weight vector. The ith step, carried out when a true value tuple is received, consists of computing an update vector using the gradient value followed by a recomputation of the weight vector. At the same time, the learning rate is also updated by decreasing it by a $\sqrt{i}$ factor.

Algorithm 13: SGD linear regression model learner.

Param : N, number of attributes in each tuple
Param : α_0, learning rate parameter
$\mathbf{w} \leftarrow \frac{1}{N} \cdot \mathbf{1}_{N \times 1}$ ▷ Initialize the weights
$i \leftarrow 0$ ▷ Tuple counter
$\alpha \leftarrow \alpha_0$ ▷ Initialize the learning rate
while *not end of the stream* **do**
 read tuple $\mathbf{x}(t)$ and its true value $v(t)$ ▷ Read the next tuple
 $y(t) \leftarrow \mathbf{w}^T \times \mathbf{x}(t)$ ▷ Compute the prediction
 $\mathbf{u} \leftarrow -2 \cdot (v(t) - y(t)) \cdot \mathbf{x}(t)$ ▷ Compute the update vector
 $\mathbf{w} \leftarrow \mathbf{w} - \alpha \times \mathbf{u}$ ▷ Update the weights
 $i \leftarrow i + 1$ ▷ Update the tuple counter
 $\alpha \leftarrow \alpha_0/\sqrt{i}$ ▷ Reduce the learning rate

---

[12] As with the standard gradient descent method, convergence to a global minimum cannot be guaranteed.

Because the update vector is continously modulated by a reducing step size α, for practical purposes, the weight vector is likely to stabilize to a local minimum, particularly if there is a steady state relationship between the incoming tuples and the current state of the model. If, however, this relationship changes over time due to changes in the incoming tuple flow, the weight vector must be updated to reflect this change.

As discussed earlier, evaluating a regression model is very simple. The prediction $y(t)$ for any given incoming tuple $\mathbf{x}(t)$ is computed as a function of the the *current* value of the weight vector:

$$y(t) = \mathbf{w}^T \times \mathbf{x}(t)$$

The convergence of the SGD-based linear regression method has been analyzed extensively using the theory of convex minimization and of stochastic approximation. When the learning rate α decreases at an appropriate rate, it, in most cases, converges to a global minimum [49]. The larger the value of α_0 the faster the initial learning rate, but the greater the chance of overshoot or undershoot.

Also, as described in Algorithm 13, over time α will get progressively closer to zero and the weights will stop evolving. There are different ways to overcome this effect based on the desired convergence property. One solution includes resetting the value of α when it is believed that the characteristics of the stream have changed (as measured using statistical tests). This adjustment must be made carefully to avoid excessive weight fluctuation.

11.5.2   Advanced reading

The implementation of the SGD-based linear regression algorithm can be extended in several ways to increase its performance as well as to address problems related to the constantly evolving data characteristics [50, 51, 52] arising in SPAs, where online learning must be used for forecasting.

Performance-related improvements include modifing the update vector to avoid a residual sign change after each update step by clippling $\mathbf{w}_{(k+1)}$ such that sgn $(v(t_k) - \mathbf{w}_{(k)}^T \times \mathbf{x}(t_k))$ is the same as sgn $(v(t_k) - \mathbf{w}_{(k+1)}^T \times \mathbf{x}(t_k))$. In this case, if the sign of the residual changes after the update, the weight $w_{(k+1)}$ is recomputed such that the residual goes to zero (this can be solved for easily). This change ensures that the perturbation to the weights at each step is controlled, allowing the use of larger values for the learning step α, and, ultimately, generating better predictions.

Along the same lines, another improvement aimed at using a more aggressive learning rate consists of modifying the update step as follows:

$$\mathbf{w}_{(k+1)} \leftarrow \mathbf{w}_{(k)} - \frac{\left(1 - e^{-\alpha}\right)}{\mathbf{x}(t_k)^T \times \mathbf{x}(t_k)} \cdot (y(t_k) - v(t_k)) \cdot \mathbf{x}(t_k)$$

In this case the extra normalization terms ensure that larger values can be used for α, yielding better predictions, but also reducing α's impact over time, keeping the weight vector stable.

Improvements related to coping with the dynamic nature of streaming data range from modifying the algorithm, so it responds to different importance levels associated with the incoming true value tuples, to the use of adaptive gradients.

In the first case, the algorithm is simply tweaked to take into consideration the importance of a particular true value tuple, before updating the weight vector. This change is useful in scenarios where certain tuples are more *reliable* than others or when forecasting for certain types of tuples is more important than others.

In the second case, the algorithm is changed to allow for the management of different learning rates for different $\mathbf{x}(t)$ attributes. In this case, α will no longer be a scalar value, but a vector holding attribute-specific step sizes. This change is useful when some tuple attributes are more predictive than others.

Finally, in certain cases, when it becomes necessary to model non-linearity, the SGD-based linear regression algorithm can still be used by making use of artificial attributes that are added to a tuple to capture a specific type of non-linearity. For instance, if an attribute of the form $x_k(t)^2$ is added to a tuple $\mathbf{x}(t)$, the algorithm will compute a linear weight on the squared value of the original $x_k(t)$ attribute as well.

11.6   Data stream frequent pattern mining

Frequent pattern mining aims at discovering frequently occurring patterns and structures in various types of data. Depending on the mining task and input data, a pattern may be an itemset, a collection of items that occur together; a sequence, defining a particular ordering of events; or a subgraph or a subtree, representing a common structure found in a larger one [53].

Frequent patterns provide an efficient summarization of the data from both static datasets as well as from continuous sources, enabling the extraction of knowledge by applying additional mining algorithms. Hence, this technique is often used by other mining tasks such as association rule mining, classification, clustering, and change detection.

In this section, we focus on the use of frequent patterns in association rule mining. The original motivation for addressing this problem was driven by the need to analyze grocery store transaction data to determine customer purchasing behavior, so that decisions on product pricing, promotions, and store layout could be made more effectively. Equivalent problems have been formulated for many other analytical tasks, making association rule mining a very popular data analysis technique.

In a continuous processing environment such as the one depicted by Figure 11.13, two streams are typically used as input, one for modeling and another for association rule scoring. In practice, these streams may be the same, but the second is a delayed version of the first and contains additional data, including the ground truth.

In Figure 11.13, the *frequent pattern mining operator* extracts the relevant frequent patterns and stores them in memory. The *association rule mining operator* performs its modeling by retrieving these frequent patterns and processing them to determine the association rules. These rules are then provided to the *scorer operator* as the current model. The *scorer* takes in incoming tuples and determines which

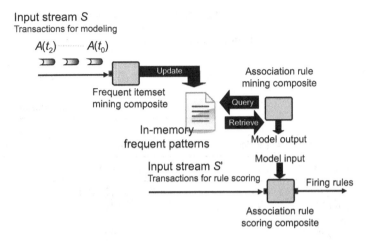

Figure 11.13 The association rule modeling and scoring process.

(if any) association rules are triggered based on the contents of the tuple being processed.

Frequent itemset problem formulation

A *k-itemset* X is defined as a set containing k *items*, out of a collection of all possible items $\mathcal{I}$, in short, $X \subset \mathcal{I}$. For instance, given the items sold by a grocery store, $X = \{\text{milk}, \text{chips}\}$ is a possible 2-itemset.

A *transaction* $\mathcal{T}$ over the set of all items $\mathcal{I}$ can be represented by a 2-tuple (tid, I), where *tid* is a transaction identifier and I is an itemset. Hence, in the grocery store example, a transaction may be the purchase made by a shopper, with *tid* representing a combination of the shopper identifier and the time of the purchase (Table 11.3). For instance, if a shopper purchases milk, eggs, cheese, and chips, $\mathcal{T} = (\langle shopperId, ts \rangle, \{\text{milk}, \text{eggs}, \text{cheese}, \text{chips}\})$. The transaction is said to *support* an itemset X if $X \subseteq I$. Clearly, this sample transaction $\mathcal{T}$ supports the itemset X defined earlier.

A *transaction set* $\mathcal{D}$ is a set of transactions over the set of items $\mathcal{I}$. For instance, it can represent the collection of all transactions that took place in a particular grocery store over a certain period of time.

The *cover* set of an itemset X with respect to transaction set $\mathcal{D}$ is the set of transaction identifiers for all of the transactions in $\mathcal{D}$ that support itemset X, defined as follows:

$$cover(X, \mathcal{D}) = \{\mathcal{T}.tid \mid \mathcal{T} \in \mathcal{D}, X \subseteq \mathcal{T}.I\}$$

Given this definition, the *support* and *relative support* of an itemset X in $\mathcal{D}$ can be trivially computed as follows:

$$support(X, \mathcal{D}) = |cover(X, \mathcal{D})|$$

$$relative\ support(X, \mathcal{D}) = \frac{support(X, \mathcal{D})}{|\mathcal{D}|}$$

where $|\cdot|$ denotes a set's cardinality, i.e., the number of elements this set holds.

Table 11.3 A grocery store transaction set.

Transaction id	I: itemset
1	{milk, chips, beer}
2	{chips, beer}
3	{pizza, milk}
4	{chips, pizza}

Table 11.4 The itemsets and their support extracted from the transaction set in Table 11.3.

Itemset	Cover	Support
{}	{1, 2, 3, 4}	4
{chips}	{1, 2, 4}	3
{beer}	{1, 2}	2
{pizza}	{3, 4}	2
{milk}	{1, 3}	2
{beer, chips}	{1, 1}	2
{beer, milk}	{1}	1
{chips, pizza}	{4}	1
{chips, milk}	{1}	1
{pizza, milk}	{3}	1
{beer, chips, milk}	{1}	1

An itemset X is *frequent* in $\mathcal{D}$ if its $support(X, \mathcal{D}) \geq \sigma$, with $0 \leq \sigma \leq |\mathcal{D}|$ defined as a *support threshold*. For instance, in the grocery store example, the different itemsets, their cover set, and their support with respect to the transaction set $\mathcal{D}$ from Table 11.3 are summarized in Table 11.4.

Given these definitions and the support threshold σ, the goal of frequent itemset mining is to find the set of all frequent itemsets $\mathcal{F}(\mathcal{D}, \sigma)$ in a transaction set $\mathcal{D}$.

Association rule problem formulation

An *association rule* is an expression on two itemsets, X and Y, of the form $X \Rightarrow Y$, where $X \cap Y = \{\}$. X is the *body* or *antecedent* of the rule and Y is the *head* or *consequent* of the rule. One such rule indicates that, if a transaction supports itemset X, it also supports itemset Y. In plain terms, both X and Y co-occur in the set of transactions.

A rule is *confident*, if its confidence exceeds a pre-established confidence threshold γ, with $0 \leq \gamma \leq 1$, where a rule's *confidence* is computed as follows:

$$confidence\,(X \Rightarrow Y) = \frac{support\,(X \cup Y, \mathcal{D})}{support\,(X, \mathcal{D})}$$

Therefore, given a set of frequent itemsets, the goal of association rule mining is to identify all $X, Y \subseteq \mathcal{I}$ such that the rules using these itemsets are confident:

$$\mathcal{R}(\mathcal{D}, \sigma, \gamma) = \{X \Rightarrow Y \mid X \cap Y = \emptyset, X \cup Y \in \mathcal{F}(\mathcal{D}, \sigma), confidence(X \Rightarrow Y, \mathcal{D}) \geq \gamma\}$$

Table 11.5 The association rules, their support, and confidence extracted from the set of frequent itemsets in Table 11.4.

Association rule	Support	Confidence
{beer} $\Rightarrow$ {chips}	2	100%
{beer} $\Rightarrow$ {milk}	1	50%
{chips} $\Rightarrow$ {beer}	2	66%
{pizza} $\Rightarrow$ {chips}	1	50%
{pizza} $\Rightarrow$ {milk}	1	50%
{milk} $\Rightarrow$ {beer}	1	50%
{milk} $\Rightarrow$ {chips}	1	50%
{milk} $\Rightarrow$ {pizza}	1	50%
{beer, chips} $\Rightarrow$ {milk}	1	50%
{beer, milk} $\Rightarrow$ {chips}	1	100%
{chips, milk} $\Rightarrow$ {beer}	1	100%
{beer} $\Rightarrow$ {chips, milk}	1	50%
{milk} $\Rightarrow$ {beer, chips}	1	50%

Revisiting the grocery store example, the corresponding association rules, their support, and confidence are summarized in Table 11.5.

Modeling algorithms

Neither frequent itemset nor association rule mining can be solved exhaustively. There are potentially $2^{|\mathcal{I}|}$ possible different itemsets. It is computationally infeasible to even enumerate them all when $\mathcal{I}$ is a large set. In practice, several algorithms have been designed to exploit certain properties of itemsets so they can function efficiently in practice. For instance, many of these algorithms leverage the monotonicity of the support measure, which decreases as an itemset increases in size.

In fact, frequent pattern mining has been extensively analyzed for the offline case of disk-resident data sets. The Agrawal, Imielinski, Swami (AIS) algorithm [54] was the first such algorithm. Other algorithms, such as the Apriori algorithm [55], the Eclat algorithm [56], the FP-growth algorithm [57] have since been developed [53].

When it comes to streaming data, multiple formulations of the problem are possible [58, 59]. The first one is the *entire stream approach*, where frequent patterns are mined over the entire length of a stream. The main difference between this approach and offline pattern mining algorithms is that the frequent patterns must be mined in a single pass over the incoming data. Yet, most of the offline algorithms require multiple passes and, hence, are not usable in a streaming context.

On the other hand, frequent pattern *counting* can be implemented using sketch-based algorithms to identify frequent patterns. As seen in Section 10.6, sketches are often used to determine heavy hitter values in a stream, therefore, using a similar approach in this case is straightforward. Manku and Motwani [60] proposed the single pass *lossy counting* algorithm, which was the first one to employ this idea (Section 11.6.1).

The second formulation is the *sliding window approach*. As a stream changes characteristics over time, it is often desirable to determine the frequent patterns over a particular sliding window. The Moment algorithm embodies this idea [61]. Of course, even when restricted to the data in a window, there is an implicit assumption that the number of frequent patterns is manageable and can be held in main memory.

The third formulation is the *damped window approach*. In this case, a *decay factor* is introduced such that different weights can be assigned to data from different periods of time. In this way, each tuple's weight is multiplied by a factor f (where $f < 1$) every time a new tuple arrives. The resulting effect corresponds to an exponential decay applied to older tuples. Such a model is quite effective for streams that evolve over time as recent tuples are deemed as more important than older ones. Chang and Lee [62] proposed one such algorithm, which maintains a lattice to record frequent itemsets and their counts.

Scoring

Evaluating incoming tuples against an association rule model is quite simple. Every tuple is treated as a transaction and tested against the rules in the model. The scoring step involves identifying the matching association rules where the matching of a rule $X \Rightarrow Y$ is established by assessing whether the transaction in a tuple supports a rule's body itemset (X).

Once a matching rule is found, other itemsets (Y) that often co-occurred in past transactions can be determined, resulting in a prediction that can be used to guide an action to be taken by a SPA. For instance, if a customer purchases {milk} and this transaction matches a previously identified rule {milk} $\Rightarrow$ {coffee}, it is likely that this customer may also want to purchase {coffee}. Hence, a grocery store could instantaneously offer a discount to improve the likelihood of an additional purchase happening in the near future.

Note that an association rule scorer operator, `Associations`, is available in SPL's mining toolkit [8] (Section 11.2).

Discussion

The use of frequent pattern and association rule mining over streaming data must be considered carefully, particularly when it comes to setting the parameters that guide how a specific algorithm operates, as well as its use of memory and computing cycles.

On the one hand, counting, a very simple operation, forms the essence of most of these algorithms, on the other hand, the number of such counts grows in line with the number of itemsets and, as a consequence, can rise rapidly depending on the problem setting. As mentioned before, experimentation and empirical evaluation with snippets of the data that is expected to be processed in production by a particular SPA is warranted.

11.6.1 Illustrative technique: lossy counting

The lossy counting algorithm [60] was designed to identify frequent itemsets over an entire stream. It was among the first proposed algorithms for frequent itemset mining on streams and has been shown to have excellent performance on real-world as well as benchmark datasets.

The algorithm, in its base form, identifies frequent *items*, that is, 1-itemsets, over a stream of incoming transactions, using a single pass strategy. In the rest of this section we will discuss this simpler case, but the algorithm, as described in its original publication, has been extended to identify k-itemsets as well.

In general, the 1-itemset result can be computed exactly, if space is not a constraint. In practice, a grocery store may stock thousands of items, a problem that also occurs in other domains, making the use of an exact algorithm infeasible. In these cases, when the number of 1-itemsets is large, employing the lossy counting algorithm is especially beneficial as it trades off accuracy for a bounded use of resources, particularly memory, providing also precise error guarantees.

Algorithm parameters
Lossy counting is an approximate algorithm controlled by two user-defined parameters, a relative support threshold σ_{rel} and an error threshold ϵ. The first parameter σ_{rel} is relatively easy to set as it is driven by application requirements. In contrast, the error threshold is harder to stipulate. A general rule of thumb by the algorithm's authors is that it should have a value that is an order of magnitude smaller than the relative support threshold: $\epsilon \sim \frac{1}{10}\sigma_{rel}$.

Considering these settings and that Q incoming transactions have been processed, the algorithm makes three guarantees: (1) all 1-itemsets whose true support exceed $\sigma_{rel} \cdot Q$ are identified correctly; (2) no 1-itemset whose true support is less than $(\sigma_{rel} - \epsilon) \cdot Q$ will be output; and (3) the estimated support for any 1-itemset is at most $\epsilon \cdot Q$ less than its true support.

The algorithmic steps
The algorithm partitions the incoming stream into W-transaction windows, where W is set as $W = \left\lceil \frac{1}{\epsilon} \right\rceil$. Each window is labeled with an integer identifier w, starting from 1. For each stored 1-itemset I, the algorithm maintains a record with three attributes in the form $c = \{I, f, \Delta\}$, where f represents the estimated count of the itemset and Δ is the largest possible error in this estimate.

When a transaction in the form of a new tuple $A(t_Q)$ arrives, each 1-itemset in the transaction triggers an update of the data structure c. If this 1-itemset I is already in C, its frequency count is updated, otherwise a new record $c = \{I, 1, w-1\}$ is created where $w = \left\lceil \frac{Q}{W} \right\rceil$ becomes the current window identifier.

Subsequently, at each window boundary, some records might be deleted to ensure that memory is used efficiently. Specifically, when the condition $Q \mod W = 0$ is true, each c record is screened and, if $f + \Delta \leq w$ holds, this record is deleted. The pseudo-code in Algorithm 14 formalizes both of these steps.

Algorithm 14: Lossy counting for 1-itemsets.

Param : σ_{rel}, relative support threshold
Param : ϵ, tolerable error threshold
$C \leftarrow \{\}$ ▷ Set of all data structures
$Q \leftarrow 1$ ▷ Number of tuples so far
$W = \left\lceil \frac{1}{\epsilon} \right\rceil$ ▷ Window size
while *not end of the stream* **do**
    read tuple $A(t)$ ▷ Next tuple/transaction in the stream
    $w \leftarrow \left\lceil \frac{Q}{W} \right\rceil$ ▷ The window id
    for $I \in A(t)$ **do**   ▷ For each item in the tuple
        if $c_I \in C$ **then**   ▷ If item is already in memory
            $c_I.f \leftarrow c_I.f + 1$ ▷ Increment its frequency
        else   ▷ If the item is not in memory
            $c_I \leftarrow (I, 1, w - 1)$ ▷ New entry with $f = 1, \Delta = w - 1$
            $C \leftarrow C \cup c_I$ ▷ Add the item into memory
    if $Q \bmod W = 0$ **then**   ▷ If at a window boundary
        for $c_I \in C$ **do**   ▷ For each item in memory
            if $c_I.f + c_I.\Delta \leq w$ **then**   ▷ Frequency is at most $\epsilon \cdot Q = w$
               $C \leftarrow C - c_I$ ▷ Remove the item from memory
    if *user requests set of frequent items* **then**
        for $c_I \in C$ **do**   ▷ For each item in memory
            if $c_I.f \geq (\sigma_{rel} - \epsilon) \cdot Q$ **then**   ▷ Frequency is sufficiently large
               output c_I ▷ Report item as frequent (relative freq. is $\frac{f+\Delta}{Q}$)
    $Q \leftarrow Q + 1$ ▷ Increment the number of tuples

In essence, when an itemset's relative support is determined to be less than, or equal to ϵ, it is deleted from memory. When a new item is seen, either for the first time or after it has been previously deleted, it is stored in memory and its past support, which is no longer known, is optimistically assumed to be ϵ and saved in the Δ attribute of its corresponding c record. As a result, the algorithm maintains any item whose relative support is greater than ϵ and its current frequency f cannot undershoot the actual value by more than $\epsilon \cdot Q$. A graphical example showing this algorithm at work is depicted by Figure 11.14.

More interestingly, the lossy counting algorithm guarantees that the number of items kept in memory is at most $\frac{1}{\epsilon} \cdot \log(\epsilon \cdot Q)$. To exemplify how substantial the drop in memory utilization can be, consider a scenario where 1 billion items might exist. If $\sigma = 0.1$ ($\epsilon = 0.01$), at most 3712 records are kept.

The last step in the continuous operation of the algorithm is to check for user scoring requests seeking the current frequent 1-itemsets. When such a request occurs after Q transactions have been processed, all 1-itemsets with $f \geq (\sigma_{rel} - \epsilon) \cdot Q$ are output. Revisiting the diagram in Figure 11.14, the algorithm's instantaneous results can be derived from the current tally at the bottom of the visual diagram.

11.6.2   Advanced reading

Algorithms that output itemsets which do not meet the relative support threshold are termed *false positive* algorithms. For instance, the lossy counting algorithm may yield

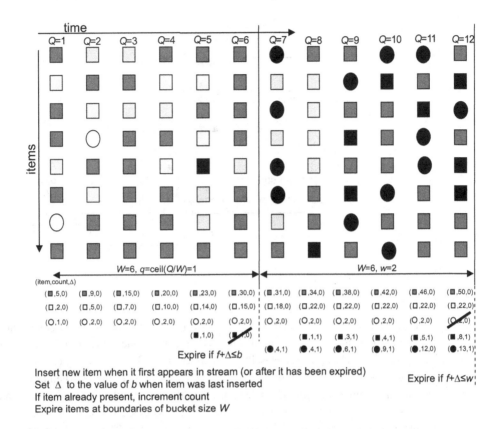

Figure 11.14 Lossy counting algorithm. Each shape represents an item and each column is a transaction.

false positive itemsets with relative support in the range $[(\sigma_{rel} - \epsilon) \cdot Q, \sigma_{rel} \cdot Q]$. Along the same lines, algorithms that miss itemsets that meet the desired relative support threshold are called *false negative* algorithms.

Some of the other algorithms for frequent itemset mining on streams that have been proposed fit these two categories. And, in generall, all of them [63] can be grouped according to certain characteristics based on their window management model; their update method, either tuple or batch and whether they produce an exact or approximate result. The availability of these variations allows a developer to better pick an algorithm that more appropriately matches the analytical and performance needs of a specific SPA.

11.7   Anomaly detection

The anomaly detection problem consists of identifying patterns in data that do not conform to *expected* behavior [64]. Depending on the application domain, an unexpected pattern is sometimes referred by different names such as an anomaly, an outlier, an exception, an aberration, a surprise, a peculiarity, or a contaminant. Anomaly

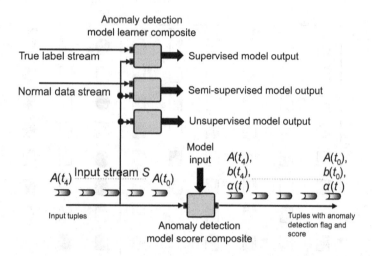

Figure 11.15 The anomaly detection scorer and modelers.

detection can be performed on tuples with numeric as well as non-numeric attributes and implemented as a continuous process such as the one depicted in Figure 11.15.

As with techniques for classification (Section 11.3), the *scoring* or *evaluation* part of the anomaly detection process is where the scorer determines whether one or more incoming tuples are anomalous or not. Note that the problem is analogous to a binary classification problem. In this case, the scorer assigns a boolean value *true* to a flag attribute b indicating that an anomaly has been detected. The scorer may also compute a numeric score α to indicate how anomalous this occurrence is so multiple anomalies can be ranked.

Anomaly detection algorithms have a wide applicability with documented uses in analyzing network traffic, monitoring complex computing systems such as data centers, as well as in analyzing credit card and retail transactions, manufacturing processes, and in surveillance and public safety applications.

Consider an example where a SPA is inspecting the TCP/IP traffic of a corporate network on a continuous basis. In this hypothetical application, each incoming tuple $A(t)$ might contain a summary of the network traffic behavior for a certain host in a computer network. In this case, the application can assign a high score to a tuple to indicate an anomaly when that host is, for example, seen establishing a connection to a server previously flagged as part of an illegal botnet.

Types of anomalies

There are three different kinds of anomalies that usually are of interest to a data analysis application.

A *point anomaly* occurs when an individual tuple is deemed anomalous with respect to the rest of the tuples seen so far. In this case, never-before-seen values and abnormally large or small values can indicate an anomaly.

A *contextual anomaly* occurs when a tuple is deemed anomalous with respect to a particular context, but is not necessarily abnormal across all tuples seen so far. Referring back to the network traffic application, if there is a sudden increase in network activity at night when usually there is little activity, this event could be labeled anomalous, even if the traffic amount is smaller than what is typically seen during the day.

The context for anomaly detection may be defined in different ways. For instance, it can take into account temporal proximity (such as in the network traffic scenario), spatial proximity, as well as other types of measures that are relevant to a particular application.

A *collective anomaly* occurs when a collection of related tuples is deemed anomalous with respect to the entire set of tuple seen so far. While individual tuples may not be anomalous, the collection of related tuples is different enough from the rest of the tuples seen so far.

Algorithms

Several anomaly detection algorithms have been designed to spot point anomalies. Also by transforming the data (Section 10.9), for instance, by aggregating it, it becomes possible to identify contextual and collective anomalies with these same algorithms. Nevertheless, devising and calibrating a transform for this purpose might not be simple.

Learning a model for anomaly detection can be done in different ways (Figure 11.15). Supervised learning approaches may be used where ground truth on whether a tuple is anomalous or not is provided, either at an initial training period or, continuously, once it becomes known. Naturally, obtaining the ground truth for *every* tuple can be impractical. Thus, in a semi-supervised approach, ground truth for a subset of tuples, typically the ones deemed normal, is provided to the algorithm. Finally, in an unsupervised approach, no ground truth information is provided at all. Instead an algorithm relies solely on the observed tuples to determine which ones are normal and which ones aren't. Whatever the learning method, anomaly detection algorithms typically assume that only a small fraction of the entire stream of tuples is anomalous.

Anomaly detection algorithms can be segmented into a few different classes: *classification*-based methods, *clustering*-based methods, NN-based methods, *statistical* methods, *information theoretic* methods, and, finally, *spectral* methods.

Classification-based algorithms

A classification-based algorithm employs either a supervised or a semi-supervised learning approach. When supervised learning is used, examples with both normal and abnormal tuples must be part of the training dataset. In this case, any classification algorithm (Section 11.3) can be used.

In contrast, when semi-supervised learning is used, only data labeled normal is available in the training dataset, and a *one-class* classification algorithm [65] can be used to distinguish the normal tuples from the abnormal ones. Several well-known learning algorithms for building models such as SVMs, neural networks, Bayesian networks, and decision trees can provide one-class classification [64].

When using a classification-based algorithm and the data labeled normal can be broken down into subclasses, for instance, normal weekday network traffic and normal weekend network traffic in the scenario we used earlier, multiple one-class classifiers can be trained for each normal subclass. In this case, if none of the classifiers label the data as normal, it is then considered anomalous.

Clustering-based algorithms

Several well-known clustering algorithms (Section 11.4) such as k-means, Self-Organizing Maps (SOMs), and Expectation Maximization (EM) can be used for anomaly detection. Depending on the algorithm, either unsupervised or semi-supervised learning can be employed. In the first case, cluster modeling and scoring are performed online as tuples flow in. In the latter case, modeling is first performed on the training dataset and, subsequently, scoring is used to distinguish normal tuples from anomalous ones.

Irrespective of the algorithm, the basic idea consists of assessing whether an incoming tuple belongs to one of the clusters in the model (Section 11.7.1). For instance, a tuple is labeled anomalous if it is located far from any of the clusters' centroids. Similarly, when relying on a density-based clustering technique, the labeling can also take into consideration each cluster's density and, in this case, make its call based on the fact that a normal tuple must belong to a large and dense cluster, while an anomalous one must belong to a small and sparse cluster.

Nearest Neighbors (NN)-based algorithms

NN-based anomaly detection techniques are similar to clustering-based methods in that an incoming tuple that is far from its nearest neighbor, or that lies in a region that is not dense, is labeled as an anomaly. Another similarity is that both unsupervised and semi-supervised learning can be used. In contrast, however, NN methods consider the distance from an incoming tuple to its nearest neighbor as opposed to a cluster's centroid.

Several alternatives exist. For instance, a KNN-based anomaly detection algorithm computes the distance of a new tuple to its k closest neighbors and a combined measure such as the maximum, the average, or the sum of these distances is then used for labeling. Another variation uses a measure of density in the labeling process. For instance, given a fixed radius r, a count is computed to assess how many neighbors for an incoming tuple exist in that area. If this number is small, the tuple is labeled anomalous.

Statistical algorithms

Statistical methods for anomaly detection fit a statistical model, which can be derived by unsupervised or semi-supervised learning, to a training dataset with normal samples and then use a statistical *inference test* to determine whether a new tuple is anomalous or not. Specifically, if a tuple has a low probability of being generated from the pre-computed model, it is assumed to be an anomaly.

Many approaches for statistical anomaly detection use parametric statistical models where the form of the probability density function (Section 10.4) is assumed to be known (e.g., a Gaussian distribution), and only its parameters are estimated from the training data.

Several statistical tests can then be used to determine whether a new tuple is normal or anomalous. Non-parametric techniques commonly use the histogram computed from the data to approximate the underlying density and make no assumptions on the form of the density function. In this case, a common test consists of computing the distance from a tuple to the mean of the distribution. If a maximum threshold is violated, the tuple is deemed anomalous. For instance, given a distribution with mean μ and standard deviation σ, a tuple $\mathbf{x}(t_i)$ is labeled anomalous if $|\mathbf{x}(t_i) - \mu| > 3 \cdot \sigma$, where $3 \cdot \sigma$ is a preset threshold. Other commonly used statistical tests include Student's t-test, Hotelling t^2 test, and Grubb's test [66, 67].

Information-theoretic-based algorithms
Information-theoretic methods take a tuple dataset D and identify the anomalous subset I such that $C(D) - C(D - I)$ is as large as possible. The measure C used to make the anomaly (or not) determination can be the entropy, the Kolmogorov complexity, or other information-theoretic measures (Section 10.4).

These methods tend to use unsupervised learning and are better applicable to data that has an inherent structure, such as graphs. Also they are usually better suited to offline analysis, first because they work best when the complete tuple dataset D is available and, second, because determining I requires computationally expensive optimization and search strategies.

Spectral-based algorithms
Spectral anomaly detection methods operate under the assumption that data, when projected onto a subspace, can be separated into normal and anomalous classes. Several of these techniques, which typically use unsupervised learning, employ Principal Component Analysis (PCA) (Section 10.8) for the projection step, and a tuple that is close to the mean is assumed to be normal.

Discussion
As seen, anomaly detection techniques borrow heavily from existing data mining techniques, which are often used unchanged. The choice of the appropriate algorithm depends on whether supervised, semi-supervised, or unsupervised learning can be employed, as well as on the characteristics of the data to be analyzed. For instance, information-theoretic techniques are well suited for analyzing graph data, but not for analyzing streaming data in a SPA.

More importantly, each application has its own definition of what an anomaly is as well as of the *cost* associated with detecting it. For instance, when monitoring a patient in an Intensive Care Unit (ICU), identifying and acting on an anomalous event is critically important, as it might lead to a life threatening outcome. Clearly, the accuracy and the costs of mislabeling must be translated into specific requirements and constraints on the anomaly detection method a SPA might choose to use.

Finally, as we discussed before, the choice of a particular method must also take into account its computational complexity. For instance, statistical methods can be computationally simpler than other methods and, hence, should be used when processing resources are particularly scarce.

11.7.1 Illustrative technique: micro-clustering-based anomaly detection

The CluStream microclustering algorithm we examined in Section 11.4.1 can also be used for anomaly detection, using either unsupervised or semi-supervised learning.

Recalling its processing steps, this algorithm computes intermediate cluster statistics referred to as microclusters. Whenever a new tuple $\mathbf{x}(t_i)$ is received, it can either (i) be found to belong to an existing microcluster, (ii) be labeled as an outlier and thus not be associated with any of the microclusters, or (iii) represent an evolution in the stream characteristics, requiring a new microcluster to be created. In the first case, a tuple is deemed normal. In cases (ii) and (iii), because it does not belong to any existing microcluster, it might be an anomaly.

Note that the CluStream algorithm creates a new microcluster in both of these cases. However, if the tuple turns out to be an isolated anomaly, it is expected that the new microcluster will expire after some time as no other tuples will be associated with it. Alternatively, if the tuple represents a shift in the properties of the stream, the corresponding microcluster is likely to endure, even after a certain period of time.

Thus, assessing whether a new tuple belongs to case (ii) or (iii) requires waiting for a while. Yet, for practical purposes, this distinction may actually be irrelevant. Even if a new tuple represents a shift in the stream, the initial tuples associated with the shift consist of previously unseen behavior and, hence, must be labeled anomalous.

Therefore, the only change to the algorithm, when used in anomaly detection, is the signaling of an anomaly associated with a tuple whenever a new microcluster is created. In this case, the distance between the tuple and its closest microcluster, normalized by the cluster extent, can be used as the anomaly score. Considering the implementation steps formalized in Algorithm 12, the score α can be computed as follows:

$$\alpha = \frac{1}{E_p} \cdot \left| (1/Q_p) \cdot \mathbf{ml}_p - \mathbf{x}(t_i) \right|$$

After computing this score for a tuple, if it is far from a small microcluster, i.e. one with a small extent, this tuple more likely represents a larger anomaly than if it is far from a large microcluster.

11.7.2 Advanced reading

Anomaly detection is still a very active area of research. For instance, there has been work on developing contextual and collective anomaly detection methods that do not require transforming the data, so a point anomaly detection algorithm can be used. Examples of these techniques include a framework for autonomously learning how to separate unusual bursty events from normal activity [68], as well as an algorithm for finding time series discords, which are subsequences that are different from all the previous ones and, hence, unusual [69].

On another front, novel techniques have been investigated to better support distributed anomaly detection applications [70]. And, finally, privacy-preserving anomaly detection has also been an active area of research [71, 72].

11.8 Concluding remarks

In this chapter we examined several stream mining algorithms and techniques for the modeling and evaluation stages of the mining process. It included online techniques for classification, clustering, regression, frequent pattern and association rule mining, and anomaly detection.

In many cases the algorithms discussed here can be directly implemented in the context of a new SPA, but it is also possible that adaptations and improvements will become necessary to address application-specific needs. We expect that the interested readers will be able to build on these techniques as they engineer new and more sophisticated analytics. Indeed, in the next chapter, we describe three application prototypes and demonstrate, in practice, how a complex SPA can make use of sophisticated analytics, and also address real-world scalability and performance issues, using many of the principles and algorithms discussed in these last three chapters.

References

[1] Mathworks MATLAB; retrieved in April 2012. http://www.mathworks.com/.

[2] The R Project for Statistical Computing; retrieved in April 2012. http://www.r-project.org/.

[3] IBM SPSS Modeler; retrieved in March 2011. http://www.spss.com/software/modeler/.

[4] SAS Analytics; retrieved in June 2012. http://www.sas.com/technologies/analytics/.

[5] Weka Data Mining in Java; retrieved in December 2010. http://www.cs.waikato.ac.nz/ml/weka/.

[6] The Data Mining Group; retrieved in September 2010. http://www.dmg.org/.

[7] Guazzelli A, Zeller M, Chen W, Williams G. PMML: an open standard for sharing models. The R Journal. 2009;1(1):60–65.

[8] Streams Mining Toolkit; retrieved in July 2012. http://publib.boulder.ibm.com/infocenter/streams/v2r0/topic/com.ibm.swg.im.infosphere.streams.product.doc/doc/IBMInfoSphereStreams-MiningToolkit.pdf.

[9] Witten IH, Frank E, Hall MA, editors. Data Mining: Practical Machine Learning Tools and Techniques 3rd edn. Morgan Kauffman; 2011.

[10] Aggarwal C, editor. Data Streams: Models and Algorithms. Springer; 2007.

[11] Schapire RE, Singer Y. Improved boosting algorithms using confidence-rated predictors. Machine Learning. 1999;37(3):297–336.

[12] Polikar R. Ensemble based systems in decision making. IEEE Circuits and Systems Magazine. 2006;6(3):21–45.

[13] Domingos P, Hulten G. Mining high-speed data streams. In: Proceedings of the ACM International Conference on Knowledge Discovery and Data Mining (KDD). Boston, MA; 2000. pp. 71–80.

[14] Cover TM, Thomas JA. Elements of information theory. New York, NY: John Wiley & Sons, Inc and Interscience; 2006.

[15] Quinlan JR, editor. C4.5: Programs for Machine Learning. Morgan Kaufmann; 1993.

[16] Gini C. Variabilità e mutabilità: contributo allo studio delle distribuzioni e delle relazioni statistiche. Studi economico-giuridici; 1912. pp. 211–382.

[17] Breiman L, Friedman J, Olshen R, Stone C, editors. Classification and Regression Trees. Chapman and Hall; 1993.

[18] Decision Tree Course; retrieved in September 2012. http://www.cs.cmu.edu/~awm/tutorials.

[19] Hoeffding W. Probability inequalities for sums of bounded random variables. Journal of American Statistical Association (JASA). 1963;58(301):13–30.

[20] Very Fast Machine Learning Library; retrieved in September 2012. http://www.cs.washington.edu/dm/vfml/.

[21] Hulten G, Spencer L, Domingos P. Mining time changing data streams. In: Proceedings of the ACM International Conference on Knowledge Discovery and Data Mining (KDD). San Francisco, CA; 2001. pp. 97–106.

[22] Jin R, Agrawal G. Efficient decision tree construction on streaming data. In: Proceedings of the ACM International Conference on Knowledge Discovery and Data Mining (KDD). Washington, DC; 2003. pp. 571–576.

[23] Aggarwal CC, Han J, Wang J, Yu PS. On demand classification of data streams. In: Proceedings of the ACM International Conference on Knowledge Discovery and Data Mining (KDD). Seattle, WA; 2004. pp. 503–508.

[24] Xu R, Wunsch D. Clustering. Series on Computational Intelligence. John Wiley & Sons, Inc and IEEE; 2009.

[25] Joachims T. Learning to Classify Text Using Support Vector Machines: Methods, Theory and Algorithms. Kluwer Academic Publishers; 2002.

[26] Gersho A, Gray RM. Vector Quantization and Signal Compression. Kluwer Academic Publishers; 1991.

[27] Arthur D, Vassilvitskii S. k-means++: the advantages of careful seeding. In: Proceedings of the ACM/SIAM Symposium on Discrete Algorithms (SODA). New Orleans, LA; 2007. pp. 179–196.

[28] Cover T, Hart P. Nearest neighbor pattern classification. IEEE Transactions on Information Theory. 1967;13(1):21–27.

[29] Bishop C, editor. Pattern Recognition and Machine Learning. Springer; 2006. Chapters 3 and 7.

[30] Hastie T, Tibshirani R, Friedman J. The Elements of Statistical Learning: Data Mining, Inference, and Prediction. Springer; 2009.

[31] Mahalanobis PC. On the generalized distance in statistics. Proceedings of the Indian National Science Academy. 1936;2(1):49–55.

[32] Deza M, Deza E. Encyclopedia of Distances. Springer; 2009.

[33] Ester M, Kriegel HP, Sander J, Xu X. A density-based algorithm for discovering clusters in large spatial databases with noise. In: Proceedings of the ACM International Conference on Knowledge Discovery and Data Mining (KDD). Portland, OR; 1996. pp. 226–231.

[34] Guha S, Mishra N, Motwani R, O'Callaghan L. Clustering data streams. In: Proceedings of the IEEE Symposium on Foundations of Computer Science (FOCS). Redondo Beach, CA; 2000. pp. 359–366.

[35] Aggarwal CC, Han J, Wang J, Yu PS. A framework for clustering evolving data streams. In: Proceedings of the International Conference on Very Large Databases (VLDB). Berlin, Germany; 2003. pp. 81–92.

[36] Zhang T, Ramakrishnan R, Livny M. BIRCH: an efficient data clustering method for very large databases. In: Proceedings of the ACM International Conference on Management of Data (SIGMOD). Montreal, Canada; 1996. pp. 103–114.

[37] Aggarwal CC, Wolf JL, Yu PS, Procopiuc C, Park JS. Fast algorithms for projected clustering. In: Proceedings of the ACM International Conference on Management of Data (SIGMOD). Philadelphia, PA; 1999. pp. 61–72.

[38] Aggarwal CC, Han J, Wang J, Yu PS. A framework for high dimensional projected clustering of data streams. In: Proceedings of the International Conference on Very Large Databases (VLDB). Toronto, Canada; 2004. pp. 852–863.

[39] Aggarwal CC, Yu PS. A framework for clustering uncertain data streams. In: Proceedings of the IEEE International Conference on Data Engineering (ICDE). Atlanta, GA; 2008. pp. 150–159.

[40] Aggarwal CC, Yu PS. A framework for clustering massive text and categorical data streams. In: Proceedings of the SIAM Conference on Data Mining (SDM). Bethesda, MD; 2006. pp. 479–483.

[41] Cheng Y, Church G. Biclustering of expression data. In: Proceedings of the International Conference on Intelligent Systems for Molecular Biology. Portland, OR; 2000. pp. 93–103.

[42] Mechelen IV, Bock H, Boeck PD. Two-mode clustering methods: a structured overview. Statistical Methods in Medical Research. 2004;13(5):363–394.

[43] Shi J, Malik J. Normalized cuts and image segmentation. IEEE Transactions on Pattern Analysis and Machine Intelligence (TPAMI). 2000;22(8):888–905.

[44] Bomze I, Budinich M, Pardalos P, Pelillo M. The maximum clique problem. Handbook of Combinatorial Optimization; 1999.

[45] Nock R, Nielsen F. On weighting clustering. IEEE Transactions on Pattern Analysis and Machine Intelligence (TPAMI). 2006;28(8):1223–1235.

[46] Fox J, editor. Applied Regression Analysis, Linear Models, and Related Methods. SAGE Publications; 1997.

[47] Takezawa K, editor. Introduction to Nonparametric Regression. John Wiley & Sons, Inc.; 2005.

[48] McCune B. Non-parametric habitat models with automatic interactions. Journal of Vegetation Science. 2006;17(6):819–830.

[49] Gardner WA. Learning characteristics of stochastic-gradient-descent algorithms: a general study, analysis, and critique. Signal Processing. 1984;6(2):113–133.

[50] Duchi J, Hazan E, Singer Y. An improved data stream summary: the count-min sketch and its applications. Journal of Machine Learning Research. 2010;12:2121–2159.

[51] McMahan B, Streeter M. Adaptive bound optimization for online convex optimization. In:

Proceedings of the International Conference on Learning Theory (COLT). Haifa, Israel; 2010. pp. 244–256.

[52] Karampatziakis N, Langford J. Online importance weight aware updates. In: Proceedings of the Conference on Uncertainty in Artificial Intelligence (UAI). Barcelona, Spain; 2011. pp. 392–399.

[53] Goethals B. Survey on Frequent Pattern Mining. Helsinki Institute for Information Technology Basic Research Unit; 2003.

[54] Agrawal R, Imielinski T, Swami AN. Mining association rules between sets of items in large databases. In: Proceedings of the ACM International Conference on Management of Data (SIGMOD). Washington, DC; 1993. pp. 207–216.

[55] Agrawal R, Srikant R. Fast algorithms for mining association rules in large databases. In: Proceedings of the International Conference on Very Large Databases (VLDB). Santiago, Chile; 1994. pp. 487–499.

[56] Zaki MJ. Scalable algorithms for association mining. IEEE Transactions on Data and Knowledge Engineering (TKDE). 2000;12(3):372–390.

[57] Han J, Pei J, Yin Y, Mao R. Mining frequent patterns without candidate generation. In: Proceedings of the ACM International Conference on Management of Data (SIGMOD). Dallas, TX; 2000. pp. 1–12.

[58] Jin R, Agrawal G. An algorithm for in-core frequent itemset mining on streaming data. In: Proceedings of the IEEE International Conference on Data Mining (ICDM). Houston, TX, USA; 2005. pp. 201–217.

[59] Giannella C, Han J, Pei J, Yan X, Yu P. Mining frequent patterns in data streams at multiple time granularities. In: Kargupta H, Joshi A, Sivakumar K, Yesha Y, editors. Data Mining: Next Generation Challenges and Future Directions. MIT Press; 2002. pp. 105–124.

[60] Manku GS, Motwani R. Approximate frequency counts over data streams. In: Proceedings of the International Conference on Very Large Databases (VLDB). Hong Kong, China; 2002. pp. 346–357.

[61] Chi Y, Wang H, Yu PS, Muntz RR. Moment: maintaining closed frequent itemsets over a stream sliding window. In: Proceedings of the IEEE International Conference on Data Mining (ICDM). Brighton, UK; 2004. pp. 59–66.

[62] Chang JH, Lee WS. Finding recent frequent itemsets adaptively over online data streams. In: Proceedings of the ACM International Conference on Knowledge Discovery and Data Mining (KDD). Washington, DC; 2003. pp. 487–492.

[63] Cheng J, Ke Y, Ng W. A survey on algorithms for mining frequent itemsets over data streams. Knowledge and Information Systems. 2008;16(1):1–27.

[64] Chandola V, Banerjee A, Kumar V. Anomaly detection: a survey. ACM Computing Surveys. 2009;41(3).

[65] Tax D. One-class classification: concept-learning in the absence of counter-examples [Ph.D. Thesis]. Delft University of Technology; 2001.

[66] Devore J. Probability and Statistics for Engineering and the Sciences. Brooks/Cole Publishing Company; 1995.

[67] Asadoorian M, Kantarelis D. Essentials of Inferential Statistics. 5th edn. University Press of America; 2008.

[68] Ihler AT, Hutchins J, Smyth P. Adaptive event detection with time-varying Poisson processes. In: Proceedings of the ACM International Conference on Knowledge Discovery and Data Mining (KDD). Philadelphia, PA; 2006. pp. 207–216.

[69] Keogh EJ, Lin J, Lee SH, Van Herle H. Finding the most unusual time series subsequence:

algorithms and applications. Knowledge and Information Systems. 2007;11(1):1–27.

[70] Zimmermann J, Mohay GM. Distributed intrusion detection in clusters based on non-interference. In: Proceedings of the 2006 Australasian Workshop on Grid Computing and E-Research. Hobart, Tasmania, Australia; 2006. pp. 89–95.

[71] Vaidya J, Clifton C. Privacy-preserving outlier detection. In: Proceedings of the IEEE International Conference on Data Mining (ICDM). Brighton, UK; 2004. pp. 233–240.

[72] Bao HT. A distributed solution for privacy preserving outlier detection. In: Proceedings of the International Conference on Knowledge and Systems Engineering (KSE). Hanoi, Vietnam; 2011. pp. 26–31.

Part V

Case studies

12 Applications

12.1 Overview

In previous chapters we covered different aspects of the stream processing paradigm, including its software engineering and systems aspects, as well as its analytical foundations. In this chapter, we make use of this knowledge by taking a closer look into the design of real-world applications.

We present three applications from different business domains: a large-scale systems operations monitoring application used to manage a portion of an IT software environment, a patient monitoring application used in the context of clinical healthcare, and an application for semiconductor fabrication process control used in a manufacturing environment.

Each of these applications is described from the perspective of a system analyst who has to design the appropriate software solution. Hence, we cover the entire software engineering process, starting from gathering the requirements, designing and implementing the application components, and finalizing with the stages of optimization and refinement.

The first application, which we label Operations Monitoring (OM), is focused on analyzing log messages and events capturing the resource usage of various operational processes in a large-scale networked system shared by multiple users and processes (Section 12.2). Its aim is to continuously monitor the state of that environment, enforcing appropriate resource usage policies, as well as detecting (and potentially predicting) system outages.

This application is a front-end to different analytic applications. Its primary challenge consists of dealing with the scalability and performance requirements associated with handling large volumes of streaming data. We thus highlight design and implementation choices that emphasize how these challenges were met.

The second application, which we label Patient Monitoring (PM), is focused on analyzing streaming physiological data flows from hospitalized patients (Section 12.3). Its aim is to *predict* health complications before they occur.

This application faces two intertwined challenges: providing good analytics, while making efficient use of the available computational resources. We thus explore and highlight tradeoffs between the use of computational resources and the quality of the application's predictive results, evaluating different combinations of offline and online learning.

Our design of the PM application also explores a secondary software engineering facet, which is the integration of custom-built code with off-the-shelf analytical libraries, in this case the Weka [1] machine learning toolkit.

The third application, which we label Semiconductor Process Control (SPC), is focused on analyzing data from a chip fabrication facility to identify potential faults, as well as to optimize the chip testing process (Section 12.4).

Similar to the PM application, the challenge faced by the SPC application is related to the design of its custom streaming analytics. In this case, the application must cope with incremental updates and provide automatic adaptation in response to the time-varying nature of the data it analyzes. Our study of the SPC application also explores, briefly, the design of its GUI, which allows users to continuously evaluate its analytical results.

12.2   The Operations Monitoring application

12.2.1   Motivation

Consider an environment where a large number of streams of log messages, originating from different software components belonging to a corporate Information Technology (IT) infrastructure, must be monitored and managed. One such environment is present in network service providers where voice, SMS, and data services are made available directly to users, or to other applications over a cell phone network.

In such environments, the software components that support these services can generate log messages that capture instantaneous snapshots of their internal operation, including information about their resource utilization.

Maintaining this multi-component software infrastructure in good working order is critically important to a network service provider. The use of stream processing techniques can help with the detection of existing problems (e.g., the violation of pre-defined resource utilization thresholds), as well as with the prediction of future problems (e.g., potential outages), which are pieces of information that can be used to improve the overall quality of service delivered to customers.

We describe a flexible *mediation application*, built on a SPS, that continuously collects the streaming log messages and performs appropriate filtering and transformation, such that different analytics can be incrementally deployed to perform specific monitoring and detection tasks. The design goals for this mediation application are as follows:

(a) Isolate the analytic applications from implementation details, including the specific protocols and data formats used to ingest the logs produced by the software components to be monitored.
(b) Provide sufficient scalability to handle a large number of log messages and be flexible enough to use more hosts, should they become available.
(c) Enable different analytic applications to subscribe to the content they are individually interested in, minimizing the transmission of irrelevant information.

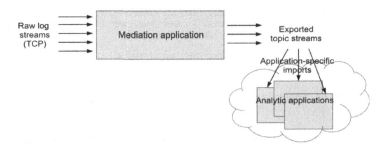

Figure 12.1 The operational monitoring environment.

Figure 12.1 shows a high-level overview of the proposed design. On the left, *raw* log streams, delivered as TCP connections, are ingested by the mediation application, and the corresponding *structured* topic streams are exported for downstream consumption by different analytic applications.

12.2.2   Requirements

The key requirements of the mediation application include the capability to manage its expected *workload* both in terms of the volume of input data it must process and the results it streams out; to provide *scalability* and *load balancing* as a function of how well the allocated computational resources are used; to minimize the propagation of errors and disturbances to other parts of the IT environment; and, finally, to provide a degree of *fault tolerance*, delivering the required quality of service.

Workload
The mediation application must ingest data from one or more *sources* of log messages. We use *nSources* to represent the number of such sources. Each one is tapped via a TCP connection and carries single-line textual log messages. Each source log stream may contain one or more topics of interest, each corresponding to a different message format. We use *nTopics* to represent the number of topics of interest.

The mediation application must create *nTopics* streams, where each topic stream contains tuples originating from one of the *nSources* log streams. Typical settings for *nSources* and *nTopics* are 1000 and 8000, respectively.

The resulting topic streams are produced after parsing the original textual log messages, providing a common structure to this data. In this way, analytic applications can specify topic-based subscriptions and receive only the matching tuples, without any additional parsing.

Figure 12.2 shows examples from a raw log source. These two lines belong to two different topics. In particular, the EVENT attribute determines the topic of the log. For the first log line, the topic is Receiving, and for the second line, the topic is Sending. It is important to note that the two topics have different raw log formats, which translate to different schemas for the topic streams generated by the mediation application. In this example, both log lines belong to the same source, identified by SRC_A1. In addition

```
<2011-03-11 16:41:11.676 EST><SRC_A1><DEBUG><LAYER = P2P, EVENT = Receiving, PIN
  = 32453847, TYPE = 1, METHOD = onReceive, DST_PIN = 20748804, SEQ = 1, VERIFIED
  = true, TIMEOUT = 259200, SIZE = 124>

<2011-03-11 16:41:11.676 EST><SRC_A1><DEBUG><LAYER = P2P, EVENT = Sending, PIN =
  20742309, SRC_PIN = 32453948, TYPE = 1, METHOD = send, SIZE = 124>
```

Figure 12.2 Sample log messages processed by the OM application.

to the various log attributes and their source information, a timestamp is also associated with each log message.

Scalability and load balancing

The mediation application may seem very simple, yet the scale of the mediation task creates significant challenges. To see this, consider a simple design where each source log stream is processed by a source operator, split *nTopics*-way and parsed into tuples. The resulting *nSources* × *nTopics* streams can then be merged into the final *nTopics* streams based on the topic formats and, subsequently, exported.

Despite this apparent simplicity, this design has several major limitations. First, considering the expected values associated with *nSources* and *nTopics*, both in the thousands, several million stream connections will have to be established (*nSources* × *nTopics*). This outcome can severely hinder the application's scalability, particularly when this many physical socket connections (in the worst case, one socket connection per stream) may be used in a multi-host deployment of the application. Second, this design also requires several million operators, which substantially complicates job management and system administration tasks.

Finally, the use of *nSources* independent source operators makes it difficult to balance the load across hosts, as each original log stream might have a different outflow data rate.

Hence, a satisfactory design must include a better way of partitioning the workload, allowing the application to grow, as more computational resources are added, and to provide a balanced load distribution as it scales up.

Interaction with other applications

The mediation application will be part of a common distributed IT environment. As a result, another design consideration relates to the interaction among the multiple interconnected systems and applications, both upstream (e.g., the log producers) and downstream (e.g., the consumer analytics) from the mediation application.

Interactions, particularly deleterious ones, can arise in many different ways. For instance, when TCP connections are employed, there is always the possibility of creating back pressure (Section 6.3.4), where a sluggish data consumer may slow down its data producer. Furthermore, this type of degradation can propagate through multiple levels of consumer/producer relationships and affect a larger portion of the distributed environment.

In a candidate design, neither the mediation application, nor the future analytic applications it is intended to support, should exert back pressure on the external log sources.

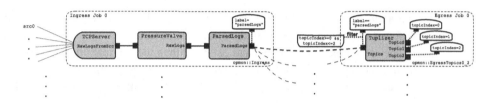

Figure 12.3 A condensed view of the OM's mediation application.

In other words, it is the mediation application's responsibility to shed load to prevent performance degeneration in external systems.

Fault tolerance
In the context where the OM application will be used, *short* outages where tuples from some of the log streams are not available is considered tolerable. Hence, to minimize operational disruptions, the ingress and egress sides of the application should be independently restartable. In this way, the stalled flow of data can readily resume after failed segments are restarted, potentially on different hosts.

12.2.3   Design

We will now examine a design for the mediation application[1] that addresses the requirements informally outlined in Section 12.2.2. An overview for this design is depicted by Figure 12.3.

Workload partitioning
The physical configuration we sketched earlier presents clear scalability and system management limitations, which all stem from a naïve design for partitioning and processing log message streams.

We now consider an alternate data partitioning organization. To handle the input log streams, the application can instead employ *nIngress ingress* partitions to perform the ingestion, parsing, and splitting of the incoming log streams.

On the exporting side, the application can make use of *nEgress egress* partitions that will perform merging, tuple creation, and exporting of the structured streams.

In this way, the application will require *nIngress* × *nEgress* number of connections between the ingress and egress partitions. Because of partitioning, the values of *nIngress* and *nEgress* can be made considerably smaller than *nSources* and *nTopics*, respectively.

Naturally, this design exposes a tradeoff: fewer streams at the expense of multiplexing multiple topics in the same stream and, eventually, additional processing to route and separate topics when necessary.

To multiplex logs from different topics into a single stream, we use a generic schema during ingress processing, delaying the creation of topic-specific schemas to egress processing. This generic schema contains the source and the topic for the log message,

---

[1] Additional material related to the OM application can be accessed at www.
    thestreamprocessingbook.info/apps/om.

and a binary blob storing the parsed log message. In this way, messages belonging to different topics can be transported by the same stream.

This design is more scalable and balanced than the original one. However, realizing these benefits is predicated on the actual implementation being able to extract computational benefits from the partitioning scheme. We will outline how this can be done, next.

First, *nIngress*, the number of ingress partitions, can be set to be the same as the number of hosts allocated to ingest the original log streams. Indeed, each ingress partition can be made multi-threaded, allowing it to concurrently handle multiple TCP connections and fully utilize the available CPU cores on each host.

Second, *nEgress*, the number of egress partitions, can be set to be the same as the number of hosts times the number of cores per host, assuming that the hosts are architecturally homogeneous. Unlike ingress partitions, the egress partitions are single-threaded.

Concretely, if the computational environment consists of eight 4-core hosts where four are used for data ingress and four for data egress, there will be four ingress partitions ($nIngress = 4$) and 16 egress partitions ($nEgress = 16$), resulting in $nIngress \times nEgress = 4 \times 16 = 64$ stream connections.

Note that, when an application is designed with workload partitioning in mind, it becomes topologically elastic. In other words, the same application can be distributed on a different number of hosts and automatically make use of different physical layouts.

In this way, an application can better match and use the computational resources allocated to it. In general, different settings for *nIngress* and *nEgress* can be used to realize different compute-multiplexing tradeoffs. For instance, in Figure 12.4 we show a full-scale view of a mediation application with $nSources = 9$ and $nTopics = 12$; where we set $nIngress = 2$ and $nEgress = 4$ for a deployment on four dual core hosts ($nHosts = 4$ and $nCores = 2$).

Application decomposition

The SPL *solution* corresponding to the mediation application comprises multiple jobs, including *ingress* and *egress* jobs (shown, respectively, on the left and on the right in Figure 12.4). The collection of ingress jobs is used to distribute the original TCP log streams for further processing, whereas the collection of egress jobs is used to export the resulting topic streams for downstream consumption.

Each ingress job is responsible for a single ingress partition, and all ingress partitions are implemented by a single SPL application. Note that different ingress jobs are simply different instantiations of the same application (with the appropriate partition configuration) on different hosts.

Likewise, each egress job is responsible for a single egress partition. In this case, however, each partition corresponds to a slightly different SPL application and, when deployed, a slightly different job. This variation is a consequence of each topic stream having a unique stream schema, which results in different SPL code for each partition. Nevertheless, from an implementation standpoint, this can be elegantly handled by SPL's built-in code generation support, as will be described later.

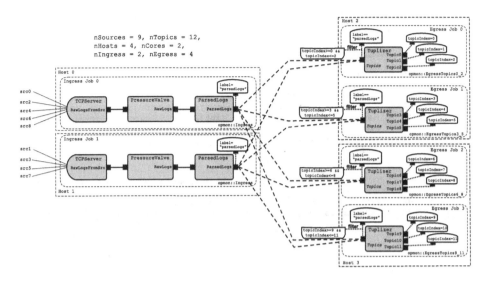

Figure 12.4 The partitioned mediation application placed on two hosts.

Ingress partition

The main composite of the `Ingress` application (Figure 12.5) employs a `partitionColocation` directive (line 32) as part of its `config` clause. This setting instructs the compiler to place all of the application's operators into a single partition named `ingress`. This partition is also marked `relocatable` (line 33) and `restartable` (line 34), addressing one of the fault tolerance requirements described earlier.

The malleability of the `Ingress` application comes from its several submission-time parameters:

(a) The `nSources` parameter (line 7) specifies the number of input log streams.
(b) The `nIngress` parameter (line 8) specifies the number of ingress partitions to be used by the solution.
(c) The `port` parameter (line 9) specifies the port on which the application listens for incoming TCP connections carrying log messages. Since each host will only hold a single ingress partition, this value can be the same for all partitions.
(d) The `nCores` parameter (line 10) specifies the number of cores to be used. This value can also be fixed,[2] in cases where the mediation application is deployed on a homogeneous cluster.

Note that there is no parameter to differentiate one ingress partition from another. As a result, any of the ingress partitions can handle any of the incoming TCP connections. This configuration aids with fault tolerance and also provides an extra degree

---

[2] It can, alternatively, be computed at runtime by interrogating the OS.

```
1    namespace opmon;
2    use com.ibm.streams.tcp::TCPServer;
3    use com.ibm.streams.ext.util::PressureValve;
4    composite Ingress
5    {
6      param
7        expression<int32> $nSources: (int32) getSubmissionTimeValue("nSources");
8        expression<int32> $nIngress: (int32) getSubmissionTimeValue("nIngress");
9        expression<int32> $port: (int32) getSubmissionTimeValue("port");
10       expression<int32> $nCores: (int32) getSubmissionTimeValue("nCores");
11     graph
12       stream<rstring line> RawLogsFromSrc = TCPServer() {
13         param
14           port: $port;
15           numThreads: $nCores;
16           connectionCap: (int32) ceil((float32)$nSources / (float32)$nIngress);
17       }
18
19       stream<rstring line> RawLogs = PressureValve(RawLogsFromSrc) {
20         param
21           bufferSize: 10000;
22       }
23
24       stream<int32 srcIndex, int32 topicIndex, blob parsedBinTuple>
25         ParsedLogs = Parser(RawLogs) {}
26
27       () as Exporter = Export(ParsedLogs) {
28         param
29           properties: { label = "parsedLogs" };
30       }
31     config
32       placement: partitionColocation("ingress");
33       relocatable: true;
34       restartable: true;
35   }
```

Figure 12.5 Source code for the ingress partition.

of flexibility. Specifically, a round-robin DNS-based load balancing scheme [2] can be employed by the external sources to distribute the original log stream connections to ingress partition instances.

The ingress partition comprises four operators arranged in a pipeline: a TCPServer (line 12), a PressureValve (line 19), a Parser (line 25), and an Export (line 27) operator. The TCPServer operator,[3] a multi-threaded TCP server, accepts and services multiple incoming connections from external clients, in this case, the external log generators.

The number of threads and the specific port a TCPServer operator employs can be configured using the port (line 14) and numThreads (line 15) parameters. In the Ingress application, the number of threads is set based on the number of cores, nCores, available in the host.

In the Ingress application, the TCPServer expects that the TCP connections will be transporting the incoming log streams as lines of plain text. It will then create a tuple from each line and place the line's content into a tuple attribute named line.

---

[3] The TCPServer operator is not part of SPL's standard toolkit, but available as a third-party extension from the Streams Exchange [3], a community-owned repository of SPL operators, toolkits, and applications.

The maximum number of simultaneous connections the `TCPServer` operator accepts is bounded by the `connectionCap` (line 16) parameter. This threshold is computed as the result of $\frac{nSources}{nIngress}$ ratio, ensuring that each ingress partition handles an equal share of the external loggers.

The `TCPServer` operator is followed by a `PressureValve` operator.[4] It is used to manage temporary overloading and to prevent back pressure. As a result, if the downstream operators are not able to ingest the incoming data at the rate it is produced, the `PressureValve` operator temporarily buffers the incoming data, but eventually starts dropping tuples, if the overload condition persists and its internal buffer fills up.

The `PressureValve` operator is followed by a `Parser` operator whose application-specific logic converts a log line into a tuple. This tuple's schema (line 24) contains attributes with the index of the source log stream from which the tuple was received (`srcIndex`), the index of the topic the tuple's content is associated with (`topicIndex`), and the parsed tuple as a `blob` (`parsedBinTuple`). This `blob`, an opaque collection of bytes, encodes a valid SPL `tuple`, which makes it computationally cheap and also straightforward to re-create later in the flow, when it will be needed.

The `Parser` operator's implementation can follow one of two approaches. Either *interpretation*, where the operator reads the description of all topic formats during initialization and, based on this information, interprets the incoming lines of text to create the corresponding tuple representation. Or *code generation*, where the parsing routines used by the operator are generated a priori, again based on the description of the topic formats. In this case, the parsing code is custom-built at compile-time and is as efficient as possible. Our design for the mediation application employs the latter approach.

While considerably more efficient at runtime, the downside of the code generation method is that adding new log formats incurs the one-time cost of regenerating the `Parser`'s internal logic, a step that can be avoided with interpretation. Yet, since adding new formats requires editing the egress partition code and rebuilding it, this extra compile-time cost is relatively inexpensive.

The last operator in the pipeline is an `Export` operator, which outputs the `Parser` operator's results, labeling it with a stream property, `label="parsedLogs"` (line 29). As mentioned earlier, the topic streams are distributed across the egress partitions, which subscribe to the streams generated by the `Export` operator. While each egress partition receives streams from all of the ingress partitions, its interest is solely on the tuples belonging to its range of topics of interest. This refinement is carried out by making use of content filters (Section 4.2.2).

Egress partition
As in the `Ingress` application, the main composite of the `Egress` application employs a `partitionColocation` directive (line 39) as part of its `config` clause.

---

[4] The `PressureValve` operator is similar to the `InputBackPressure` operator seen in Section 9.2.1.2 and also available from the Streams Exchange.

```
1    namespace opmon;
2    composite EgressWithTopics0_2
3    {
4      param
5        expression<int32> $startTopicIndex: 0;
6        expression<int32> $endTopicIndex: 2;
7      graph
8        stream<int32 srcIndex, int32 topicIndex, blob parsedBinTuple>
9            ParsedLogs = Import() {
10          param
11            subscription: label == "parsedLogs";
12            filter: topicIndex >= $startTopicIndex &&
13                    topicIndex <= $endTopicIndex;
14        }
15
16       (stream<int32 srcIndex, timestamp tm, int32 trafficSrc, int32 trafficDst,
17             int32 appId, int32 appVersion, float32 flowAmount> TopicStream0;
18        stream<int32 srcIndex, timestamp tm, int32 duration, int32 problemCode,
19             int32 problemIP, float32 severity> TopicStream1;
20        stream<int32 srcIndex, int32 outageCode, int32 serviceId,
21             list<tuple<timestamp start, timestamp end>> times> TopicStream2
22       ) as TopicsStreams = Tuplizer(ParsedLogs) {}
23
24       () as Exporter0 = Export(TopicStream0) {
25         param
26           properties: { topicIndex = 0 };
27       }
28
29       () as Exporter1 = Export(TopicStream1) {
30         param
31           properties: { topicIndex = 1 };
32       }
33
34       () as Exporter2 = Export(TopicStream2) {
35         param
36           properties: { topicIndex = 2 };
37       }
38     config
39       placement: partitionColocation("egress");
40       relocatable: true;
41       restartable: true;
42   }
```

Figure 12.6 Source code for egress partition handling topics 0 to 2.

As a result, all of its operators are placed into a single partition named `egress`. This partition is also marked as `relocatable` (line 40) and `restartable` (line 41).

In Figure 12.6, the source code for an instance of this application is shown. The `EgressWithTopics0_2` composite operator handles $\frac{nTopics}{nEgress} = \frac{12}{4} = 3$ topics, in this case ranging from 0 to 2 (lines 5 and 6), respectively. Other topics are handled by additional instances of this composite employed by the mediation application. Each instance is slightly different due to the different incoming streams schemas. The specific topic range handled by each egress partition is identified by the parameters `$startTopicIndex` and `$endTopicIndex` defined in the composite operator's `param` clause (lines 5 and 6).

Each egress partition's composite operator includes an `Import` operator with the `label=="parsedLogs"` expression (line 11), indicating its interest in the streams exported by the `Export` operators from the ingress applications.

Note that the `Import` operator also specifies a content filter (line 12), in this case, defined by the expression `topicIndex >= $startTopicIndex && topicIndex <= $endTopicIndex`. This filtering is internally executed by the Streams runtime for each ingress partition, minimizing the tuple traffic flowing from ingress to egress components. In this way, each egress partition only receives the tuples it must handle.

The incoming tuples received by the `Import` operator are passed to a user-defined, multi-output port capable, `Tuplizer` operator (line 22), which converts the binary attribute `parsedBinTuple` into an actual tuple.[5]

Finally, each of the `Tuplizer`'s outgoing streams are exported (lines 24, 29, and 34). The `Export` operators performing this task use a property named `topicIndex` to indicate the topic being exported. These streams are the final output generated by the mediation application. Hence, an external application interested in subscribing to specific topic streams from one or more sources can use a combination of a stream subscription and a content filter expression to have the data routed to it.

Finally, we revisit the issue of the egress application design. The SPL language provides built-in scripting support using Perl. Scripting is used in what we refer to as *mixed-mode* source code. The mixed code is only relevant at compile-time and it requires a pre-processing step, which emits pure SPL code.

Mixed-mode source code is stored in a file with extension `.splmm`, in contrast to pure SPL code, which is stored in files with extension `.spl`. As seen in Figure 12.7, a Perl code segement is delimited by `<%` and `%>` (line 2).

The mixed-mode script reads the number of egress partitions and the list of topic formats as pre-processing arguments (lines 3–4). The script iterates over the number of partitions and computes the topic range for each partition (lines 5–15). It then uses this information to emit SPL code with several specializations.

First, the script uses the range to create a unique name for the main composite (line 17). Second, it defines the start and end topic indices to be handled by this composite (lines 19–20). Third, it defines the `Tuplizer` operator's outgoing streams and their schemas (lines 28–32). And, finally, it defines the list of `Export` operators (lines 33–38), one per topic this composite is responsible for.

Now, if we take a second look at Figure 12.6, it is clear that that code is simply the result of one expansion of the script from Figure 12.7.

12.2.4 Analytics

The mediation application can provide support to multiple external consumers, forming the larger OM application. We now briefly examine the `UsageMonitor` application (Figure 12.8), one of the OM's simplest consumers, as an example of the interaction between the mediation application and its consumers.

---

[5] Recall that the data in the `blob` is already encoded as a tuple and, hence, this conversion does not require any additional code.

```
1   namespace opmon;
2   <%
3   my $nEgress = shift(@ARGV);
4   my @topicFormats = @ARGV;
5   my $nTopicsPerPartition = int(scalar(@topicFormats) / $nEgress);
6   my $nExcessTopics = scalar(@topicFormats) % $nEgress;
7   for (my $i=0; $i<$nEgress; ++$i) {
8     my ($startTopicIndex, $endTopicIndex);
9     if ($i<$nExcessTopics) {
10      $startTopicIndex = $i * ($nTopicsPerPartition + 1);
11      $endTopicIndex = $startTopicIndex + $nTopicsPerPartition;
12    } else {
13      $startTopicIndex = $i * $nTopicsPerPartition + $nExcessTopics;
14      $endTopicIndex = $startTopicIndex + $nTopicsPerPartition - 1;
15    }
16  %>
17    composite EgressWithTopics<%=$startTopicIndex%>_<%=$endTopicIndex%> {
18      param
19        expression<int32> $startTopicIndex: <%=$startTopicIndex%>;
20        expression<int32> $endTopicIndex: <%=$endTopicIndex%>;
21      graph
22        stream<int32 srcIndex, int32 topicIndex, blob parsedBinTuple>
23          ParsedLogs = Import() {
24          param
25            subscription: label == "parsedLogs";
26            filter: topicIndex>=$startTopicIndex && topicIndex<=$endTopicIndex;
27          }
28        (<%for(my $index=$startTopicIndex; $index<=$endTopicIndex; ++$index) {
29            my $format = $topicFormats[$index-$startTopicIndex];
30            print "stream<int32 srcIndex, $format> Topic$index";
31            print ";" unless $index==$endTopicIndex;
32        }%>) as Topics = Tuplizer(ParsedLogs) {}
33        <%for(my $index=$startTopicIndex; $index<=$endTopicIndex; ++$index) { %>
34          () as Exporter<%=$index%> = Export(Topic<%=$index%>) {
35            param
36              properties: { topicIndex = <%=$index%> };
37          }
38        <%}%>
39      config
40        placement: partitionColocation("egress");
41        restartable: true;
42        relocatable: true;
43    }
44  <%}%>
```

Figure 12.7 Mixed-mode source code for the egress partitions.

The `UsageMonitor` application employs an `Import` operator (line 10) to sub-scribe to tuples associated with the topic containing data usage information from third-party applications. As seen, in the `Import` operator's `subscription` expression, this topic's index is 0 (line 12). The specific data usage for each third-party application hosted by the network service provider can be tallied up using the information contained in the `flowAmount` and `appId` attributes.

For this purpose, the `UsageMonitor` application employs an `Aggregate` opera-tor (line 16) to generate minute-by-minute data usage summaries partitioned by `appId`. A `Filter` operator is then used to identify the applications whose data usage amounts violate preset thresholds, presumably consuming more resources than established in their service contracts.

The `UsageMonitor` application uses only a single topic stream from the medi-ation application. Other consumers may have sophisticated subscriptions to perform considerably more detailed analysis, involving cross-correlation of data extracted from

```
1   namespace opmon;
2   composite UsageMonitor
3   {
4     param
5       expression<float32> $usageThreshold:
6             (float32) getSubmissionTimeValue("usageThreshold");
7     graph
8       stream<int32 srcIndex, timestamp tm, int32 trafficSrc,
9             int32 trafficDst, int32 appId, int32 appVersion,
10            float32 flowAmount> DataUsage = Import() {
11        param
12          subscription: topicIndex == 0;
13      }
14
15      stream<int32 appId, float32 flowAmount, timestamp tm>
16          UsageSummary = Aggregate(DataUsage) {
17        window
18          DataUsage: tumbling, delta(tm, 60.0), partitioned;
19        param
20          partitionBy: appId;
21        output
22          UsageSummary: flowAmount = Sum(flowAmount);
23      }
24
25      stream<UsageSummary> OverusingApps = Filter(UsageSummary) {
26        param
27          filter: flowAmount > $usageThreshold;
28      }
29  }
```

Figure 12.8 Usage monitoring.

different topic streams. Clearly, the design of the mediation application affords a great deal of flexibility to its data consumers.

12.2.5 Fault tolerance

The mediation application is stateless and is tolerant to limited amounts of data loss. When a host is down, the log streams that were handled by the set of partitions running on that host will be temporarily unavailable, until the flow can be re-established, possibly by switching these partitions to a new host.

Recovering from partial failures on the ingress side is straightforward. As seen, the design of the ingress application is flexible with respect to fault tolerance, since a partition can transparently handle any one of the incoming TCP connections.

As a result, multiple options, with different administrative and capital costs, are available to address failures in any of the ingress hosts. First, the TCP connections to the failed host can be easily re-routed to healthy ones. Second, a new ingress partition can be instantiated on a spare host and the TCP traffic routed to it. Indeed, standby ingress partitions can be provisioned a priori, in cases where more stringent Quality of Service (QoS) must be delivered even when facing failures or sudden increases in the workload.

Similar recovery options are also available in the presence of failures in any of the egress hosts. However, the use of hot standby partitions is not possible. Unlike ingress partitions, egress ones are not identical and, hence, additional recovery steps are necessary.

Downstream from the mediation application, most of the analytics that are part of the OM application can also survive minor disruptions in the data flow. For instance, the `UsageMonitor` application might not temporarily flag customers who are violating their service agreement when the data usage stream is offline. Nonetheless, as soon as the flow resumes, its detection logic also comes back without any further intervention.

12.3   The Patient Monitoring application

The description of the OM application in the previous section was focused on scalable and resilient processing of voluminous logging data. We deliberately minimized the discussion of the analytics deployed in the context of that application in favor of focusing more on its systems-oriented aspects.

In this section we shift our focus to demonstrate how powerful analytical capabilities can be leveraged when data is continuously processed, mined, and analyzed as part of a unified data analysis process. We consider a clinical setting with patients in a hospital who are continuously monitored for different health conditions, but focus primarily on the analysis of an Electrocardiogram (ECG) signal. While some of the techniques we describe are specific to this application, many of its design principles and analytics are generic, and can be adapted for use by other classes of applications.

12.3.1   Motivation

Many clinical information systems acquire physiological sensor readings from patients as continuous streams, where multiple signals are collected at periodic intervals. They contain relevant information about the current state of a patient. When this data is correlated with information extracted from other types of medical records, including laboratory test results, and a patient's prior history from Patient Data Management Systems (PDMSs) and Clinical Information Systems (CISs), physicians can be given data-backed facts to aid them in monitoring a patient's progress and devising an appropriate course of treatment.

Another important aspect of continuous data analysis is the potential for the detection of early onset conditions associated with critical illnesses. Signal processing and knowledge discovery techniques can be used to isolate subtle, yet clinically meaningful correlations that are often *hidden* in the multimodal physiological data streams, and in the history captured by the patient's medical records. Clearly, finding, isolating, and extracting this information can lead to earlier interventions and/or prophylactic strategies.

Early research in this area has been promising. For instance, at the University of Virginia, Griffin and Moorman [4] have studied the correlation of heart rate variability with the onset of sepsis. Vespa *et al.* [5] have studied the correlation of features extracted from Electroencephalograms (EEGs) with the onset of delayed ischemia from cerebral vasospasm in Neurological Intensive Care Units (ICUs). And IBM researchers have engaged with the Hospital for Sick Children in Toronto and the University of Ontario Institute of Technology to devise methods aimed at detecting sepsis [6].

12.3.2 Requirements

Applications in a clinical setting must be able to ingest and analyze physiological signals of several kinds. One of the most important is the ECG, a transthoracic interpretation of the electrical activity of the heart over a period of time, which is detected by electrodes attached to the outer surface of the skin across the chest. The ECG is typically sampled in the 200 to 500 Hz frequency range, with 12 electrodes.

The continuous analysis of the ECG signal directly yields individual heartbeats. The extraction of this signal involves detecting the QRS complex[6] in the signal. After that, measures like the average heart rate of the patient and its variability can be computed directly. This basic information has been shown to be predictive of the onset of many complications for patients treated in an ICU.

In general, an ECG signal has five deflections [7], named P to T waves. The Q, R, and S waves occur in rapid succession and are the result of a single event, thereby the name QRS complex. A Q wave is a downward deflection after the P wave. An R wave follows it as an upward deflection. An S wave is any downward deflection following an R wave. And, finally, the T wave follows the S wave (Figure 12.9). The interval between successive R wave peaks is used to capture the inter-beat interval and, hence, to compute the heart rate. Several algorithms can be employed for this task [8, 9].

A difficulty in computing the heart's inter-beat interval arises because of *ectopic heartbeats*, which are also depicted by Figure 12.9. Ectopic heartbeats are small variations in otherwise normal heartbeats, leading to extra or skipped heartbeats, and are usually harmless. Although they may occur without an obvious cause, they are typically correlated with electrolyte abnormalities in the blood, as well as with ischemia, a decrease in blood supply to the heart.

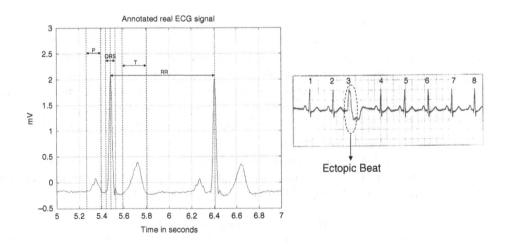

Figure 12.9 ECG heartbeat and ectopic heartbeat.

---

[6] The QRS complex is the name given to the combination of three of the graphical deflections in an ECG tracing. It is indicative of the depolarization of the right and left ventricles of the human heart.

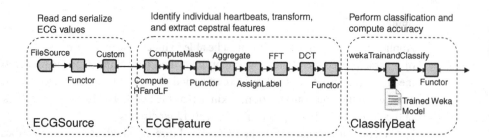

Figure 12.10 The PM application.

When ectopic heartbeats are present in the ECG signal they cause the computation of the heart rate and of its variability to degrade. Thus, there has been significant effort in filtering them out. This is the problem we tackle in the rest of this section.

12.3.3　Design

The implementation of ectopic heartbeat detection processing logic as a SPA[7] includes multiple steps (Figure 12.10). First, the ECG data must be ingested, potentially from multiple patients. Second, the ECG data must be converted into a suitable representation, a step that consists of processing the raw incoming signal to extract specific features and the heartbeat profile. Third, the data must be classified to decide whether an individual heartbeat is *ectopic* or *normal*. Finally, the incoming signal can be cleaned out by filtering out ectopic heartbeats. These three analytical steps lead to a natural decomposition of the analytical application.

Data ingest
A typical ICU has hundreds of different types of sensors, each with their own hardware as well as data encoding formats and interfaces. Managing this diversity is no simple matter. Indeed, multiple vendors are in the business of providing the wiring and device drivers necessary to communicate with these sensors, outputting an aggregated set of data streams, with different sensor signals from multiple patients.[8] Generally, a SPA's data ingestion component connects to one of these data aggregators to receive and decode the raw sensor data.

To maintain our focus on the analytic aspect of the PM application, rather than employing real data feeds,[9] we will make use of annotated data from the Massachusetts Institute of Technology (MIT)'s Physiobank [12], a database of physiological signals for

---

[7]　Additional material related to the PM application can be accessed at www. thestreamprocessingbook.info/apps/pm. The website includes the implementation details for the operators not shown here.

[8]　For instance, the DataCaptor terminal device [10] can connect to eight different patient sensors and convert the devices' RS232 [11] output into a TCP or UDP stream. This terminal device can be placed near the data sensors to forward the signals to a server capable of collecting simultaneous incoming data streams from multiple patients.

[9]　The utilization of *canned* data to design and test a new SPA is a common approach to refine an initial prototype.

biomedical research. Specifically, we will use data from the Arrhythmia Database containing half-hour excerpts of two-channel ambulatory ECG recordings, obtained from 47 subjects studied by the Beth Israel Hospital (BIH) Arrhythmia Laboratory in Boston, between 1975 and 1979.

In this data, 23 recordings were chosen at random from a set of 4000 24-hour ambulatory ECG recordings collected from a mixed population of inpatients (about 60%) and outpatients (about 40%). The other 24 recordings were selected from the same set to include less common, but clinically significant, arrhythmias[10] that would not be well represented in a small random sample.

The recordings were digitized at 360 samples per second per channel with an 11-bit resolution over a 10 mV range. Two or more cardiologists independently annotated each record and disagreements were subsequently resolved, yielding a total of approximately 110, 000 computer-readable reference annotations.

To read these recordings, we designed the ECGSource operator (Figures 12.10 and 12.11). The original CSV data contains multiple fields (see schema in line 1):

```
1   type ECGRawTupleT = tuple<float64 ts, list<float64> tss,
2     list<float64> ECGValues1, list<float64> ECG2Values2,
3     list<rstring> annot>;
4   type ECGTupleT = tuple<ECGRawTupleT, tuple<rstring patientID>>;
5   type ECGBeatT = tuple<float64 ts, list<float64> ECGValues1, rstring patientID,
6     rstring label>;
7   type ECGValueT = tuple<float64 ts, float64 ECGValue, rstring annot,
8     rstring patientID>;
9
10  composite ECGSource(output stream<ECGValueT> serializedECGData) {
11    param
12      expression<rstring> $inputFileName: getSubmissionTimeValue("inputFile");
13
14    graph
15      stream<ECGRawTupleT> ECGRawData = FileSource() {
16        param
17          file: $inputFileName;
18          format: csv;
19          hasDelayField: true;
20      }
21
22      stream<ECGTupleT> ECGData = Functor(ECGRawData) {
23        output
24          ECGData: patientID = $inputFileName,
25                   ECGValues1 = ECGValues1 * 1000f;
26      }
27
28      stream<ECGValueT> serializedECGData = Custom(ECGData) {
29        logic
30          onTuple ECGData: {
31            mutable int32 count = 0;
32            while(count < size(ECGValues1)) {
33              mutable ECGValueT t = { ts = tss[count], patientID = patientID,
34                  ECGValue = ECGValues1[count], annot = annot[count] };
35              count++;
36              submit(t, serializedECGData);
37            }
38          }
39      }
40  }
```

Figure 12.11 The ECGSource operator designed to process data from the MIT Physiobank.

---

[10] An arrhythmia is an alteration in the rate or rhythm of the heartbeat.

a timestamp `ts`, used to emulate the delay between two samples; a list of ECG values (from two channels generated by different electrodes on the patient body, `ECGCValues1` and `ECGValues2`); and, associated with each of these values, a timestamp stored in the list `tss` and an annotation ("N" or "A") stored in the list `annot`.[11] Note that, while two channels of data are available, the downstream operators will only make use of the `ECGValues1` attribute.

The layout of the input data read from a file using a `FileSource` operator (line 15) requires some amount of pre-processing to make it adequate for the downstream tasks. First, the `ECGSource` composite operator employs a `Functor` operator (line 22) to annotate each outgoing tuple with the name of the file from which the original data was read. This file name is used in lieu of a patient identifier so readings from multiple patients can be processed simultaneously.

Second, since the data corresponding to a single heartbeat may span multiple rows, it is first expanded into individual samples (from the list of values in `ECGValues1`), before these samples can be regrouped into a heartbeat representation. A `Custom` operator (line 28) performs this serialization by outputting tuples with the `ECGValueT` schema (line 7).

Data representation

The next task consists of extracting the features needed by the QRS complex detection algorithm we employ [8], so the individual heartbeats can be reconstructed. This task is implemented by the `ECGFeature` composite operator (Figures 12.10 and 12.12–12.13). It carries out a multi-step process, which starts with a high pass and low pass filter implemented by the `ComputeHFandLF` composite operator (line 11). This operator filters the ECG sample data and generates higher order moments as required by the detection algorithm.

After filtering, the `ComputeMask` composite operator (line 14) applies different thresholds to the moments to identify heartbeat boundaries, computing a mask, stored in the `qrsMask` attribute, which will be used to extract the heartbeat.

Each heartbeat is extracted by a `Punctor`[12] operator (line 17), which injects punctuation marks into the stream to delineate heartbeat boundaries as specified by the `qrsMask` attribute.

Once the heartbeat boundaries are streamed out, an `Aggregate` operator (line 24) is used to collect the samples for each heartbeat into lists. Following this, the collected samples are processed by the `AssignLabel` composite operator (line 36), which combines the annotations for all the samples into a `label` for this heartbeat. Note that this label corresponds to the "ground truth," stating whether the heartbeat is considered

---

[11] As mentioned, the ECG signals in the database are pre-annotated by expert physicians as *normal* or *abnormal*. This annotation represents the ground truth, which can be used for evaluating the accuracy of an ectopic heartbeat detection algorithm.

[12] A `Punctor` operator is used to transform incoming tuples in much the same way as the `Functor` operator does. Yet, its main function is to output window punctuations under specific conditions. It must be configured with a `punctuate` parameter that specifies this condition as well as a `position` parameter that specifies whether the punctuation is sent `before` or `after` the next tuple.

```
1    use com.ibm.streams.timeseries::FFT;
2    use com.ibm.streams.timeseries::DCT;
3
4    composite ECGFeature(input stream<ECGValueT> serializedECG;
5                         output stream<ECGBeatT> reducedCepstralBeat) {
6      param
7        expression<int32> $nFeatures;
8      graph
9        /* Apply low and high pass filters in a cascaded manner to estimate higher
                order moments in the signal. This requires applying specified weights
                to delayed versions of the signal */
10       stream<ECGValueT, tuple<float64 filteredval>>
11          ecgFeatures = ComputeHFandLF(serializedECG) {}
12
13       /* Compute a mask to identify heartbeat based on predefined thresholds on
                the higher order moments */
14       stream<ECGValueT, tuple<float64 qrsMask>> mask = ComputeMask(ecgFeatures)
                {}
15
16       /* Aggregate the samples over the mask to extract a heartbeat */
17       stream<Mask> PunctMask = Punctor(mask) {
18         param
19           punctuate: (qrsMask > mask[1].qrsMask);
20           position: before;
21       }
22
23       stream<ECGTupleT, tuple<list<float64> qrsMasks>>
24          maskedAggregate = Aggregate(PunctMask) {
25         window
26           PunctMask: tumbling, punct();
27         output
28           maskedAggregate: ts = Min(ts), tss = Collect(ts),
29             patientID = First(patientID), ECGValues1 = Collect(ECGValue),
30             annot = Collect(annot), qrsMasks = Collect(qrsMask);
31       }
32
33       // to be continued ...
```

Figure 12.12 Feature extraction for MIT Physiobank data.

ectopic or normal, based on annotations provided by the experts. This label will be used later for assessing the accuracy of the classifiers.

After extracting the heartbeat, two transforms, Fast Fourier Transform (FFT) and Discrete Cosine Transform (DCT), are applied to the data to extract its frequency domain representation. These transforms filter out some of the noise and produce a compact representation of the original signal (Section 10.9).

This process is implemented using three operators. The first transform is carried out by the FFT operator (line 40), which is part of SPL's time series toolkit. It computes the power spectrum (line 46)[13] on the incoming tuple and appends it to the outgoing tuple.

Subsequently, the power spectrum is run through the DCT operator (line 49), which is also part of SPL's time series toolkit. It computes the cepstral coefficients required by the rest of the algorithm by applying the DCTTransform built-in function. The size of the transform and the input time series are specified in similar fashion to the FFT operator, by setting their corresponding parameters.

---

[13] The power spectrum of a signal corresponds to the signal's power at each frequency and is computed as the sum of the squares of the real and imaginary parts of the FFT-transformed signal. The FFT operator computes it using the output function Power().

```
34    // ... continued
35    /* Annotate the extracted heartbeat with a single label */
36    stream<ECGBeatT> labeledBeat = AssignLabel(maskedAggregate) {}
37
38    /* Transform data to extract cepstral features */
39    stream<ECGBeatT, tuple<list<float64> PSD> >
40        fftBeat = FFT(labeledBeat) {
41      param
42        inputTimeSeries: "ECGValues1";
43        resolution: 512u;
44        algorithm: complexFFT;
45      output
46        fftBeat: PSD = Power();
47    }
48
49    stream<ECGBeatT> cepstralBeat = DCT(fftBeat) {
50      param
51        inputTimeSeries: "PSD";
52        algorithm: DCT;
53        size: 512u;
54      output
55        cepstralBeat: ECGValues1 = DCTTransform();
56    }
57
58    stream<ECGBeatT> reducedCepstralBeat = Functor(cepstralBeat) {
59      logic
60        state:
61          list<uint32> idx = makeSequence(1, $nFeatures);
62        output
63          reducedCepstralBeat: ECGValues1 = at(ECGValues1,idx);
64      }
65    }
```

Figure 12.13 Feature extraction for MIT Physiobank data (continued).

Finally, the tuples resulting from the DCT operator are processed by a Functor (line 58) to select the first $nFeatures of the cepstral coefficients for further processing [8, 9].

In essence, our design consists of assembling the raw data into a list of samples, which is then used to compute multiple temporal properties from the signal. Subsequently, the samples are stitched together into a heartbeat and a compact representation is extracted by applying transforms to that data.

Looking more closely at the use of the operators from the time series toolkit, the resolution (line 43) and size (line 53) parameters, employed by the FFT and DCT operators, respectively, indicate the largest temporal extent of the signal being transformed.

In this case, the heartbeat is assumed to have less than 512 samples. The compaction properties of these transforms ensure that the energy of the result is concentrated in a small number of non-zero low-frequency coefficients (typically 16–32). The compressed data, whose size is 10 to 20 times smaller than the original data, still retains a large amount of fidelity with respect to the original signal. As the data reduction (Section 9.2.2) comes at the expense of additional computation, it is important to minimize the communication costs by intelligently fusing the intermediate operators (Section 9.3.3).

Classification

The ectopic heartbeat detection problem can be seen as a classification problem (Section 11.3), yielding one of two possible classes: ectopic or normal. In the experimental study described in this section, we will consider two possible algorithmic alternatives [13]: a decision tree classifier as well as a Naïve Bayes (NB) one.

These two alternatives are evaluated with two different configurations: first, offline learning with online scoring and, second, a hybrid configuration where the models are incrementally updated online, based on the arrival of new data.

As mentioned earlier, a decision tree classifier employs a tree structure where its leaves represent class labels and branches represent conjunctions of features that are evaluated against a tuple, when scoring is performed.

The decision tree is computed using the C4.5 algorithm [13],[14] a non-incremental learning algorithm, unlike VFDT (Section 11.3.1), that builds the tree from offline training data. C4.5 is recursive and, for each node added to the decision tree, it greedily selects the tuple attribute that most *effectively* splits its set of samples into class subsets, which are then subjected to the same process. The effectiveness of the split is computed as a function of the normalized information gain, or the difference in entropy, that results from choosing a particular attribute. Scoring with this algorithm is carried out by traversing the tree and applying the conditions at each node to a tuple to be classified. For a batch of m training n-attribute tuples, the complexity of the learning algorithm is $O(mn^2)$, whereas the complexity for scoring a tuple is $O(n)$.

The NB classifier relies on applying Bayes' theorem with strong independence assumptions. In other words, the classifier assumes that the presence or absence of a particular feature associated with a class is unrelated to the presence or absence of any other feature, given the class variable.

The NB model has very efficient learning and scoring models. The Maximum-Likelihood Estimation (MLE) method[15] is often used for the estimation of parameters of the NB model during learning. In this case, the MLE method is used to compute the relative frequencies of different attribute values in the training set, for a given label. In some cases, for instance, when attributes values are drawn from a continuous domain, a parametric model (e.g., a Gaussian or Gaussian Mixture Model (GMM)) is assumed for the features from the training set to reduce complexity. For a batch of m training n-attribute tuples, the NB classifier learning complexity is $O(mn)$, whereas the complexity for scoring is $O(n)$.

Using Weka

Both of these classifiers have multiple open-source and commercial implementations. In the PM application, we employed the Weka framework [1]. Weka's algorithms are implemented as Java classes that that can be used by custom applications to ingest (an *instance* class), to pre-process, and to score (a *filter* class) the data, as well as to learn (a *classifier/clusterer* class) and to evaluate the performance of a data mining model (an *evaluation* class).

---

[14] As will be seen later, we employed the Weka implementation of this algorithm and, hence, we will use its Weka name, J48, later in this section.

[15] The MLE is a method for estimating the parameters of a statistical model. In broad strokes, given a training dataset and its presumed statistical model, this method selects values for the model parameters, which yield a distribution that gives the observed data the highest probability, maximizing its likelihood function.

```
1    type AccuracyT = tuple<float64 ts, rstring patientID, rstring label,
2      rstring prediction, int64 Ntruepos, int64 Nfalsepos, int64 Ntrueneg,
3      int64 Nfalseneg>;
4
5    composite ClassifyBeat(input stream<ECGBeatT> reducedCepstralBeat;
6                           output stream<AccuracyT> accuracy) {
7      param
8        expression<rstring> $learnAlgo;
9        expression<int64> $nFeatures;
10       expression<int64> $learnInterval;
11       expression<rstring> $modelName;
12     graph
13       stream<ECGBeatT, tuple<rstring prediction>>
14         classifiedECGBeat = wekaTrainAndClassify(reducedCepstralbeat) {
15       param
16         modelType: $learnAlgo;
17         modelName: $modelName;
18         numberOfFeatures: $nFeatures;
19         numberOfTrainingSamples: $learnInterval;
20       }
21
22       stream<AccuracyT> accuracy = Functor(classifiedECGBeat) {
23         logic state: {
24           mutable int64 Ntp = 0;
25           mutable int64 Ntn = 0;
26           mutable int64 Nfp = 0;
27           mutable int64 Nfn = 0;
28         }
29         onTuple classifiedECGBeat: {
30         if(prediction == label) {
31           if(label=="0")
32             Ntn++;
33           else
34             Ntp++;
35         } else {
36           if(label=="0")
37             Nfp++;
38           else
39             Nfn++;
40         }
41       }
42       output
43         accuracy: Ntruepos=Ntp, Ntrueneg=Ntn, Nfalsepos=Nfp, Nfalseneg=Nfn;
44     }
45   }
```

Figure 12.14 Classification for ectopic heartbeat detection.

By customizing these classes, the services provided by Weka can be used by any Java application. In our implementation (Figure 12.14), where the classifier and additional helper logic are encapsulated in the ClassifyBeat composite operator, we employed SPL's Java-wrapping mechanisms to develop the wekaTrainAndClassify operator (line 14), which envelopes the Weka algorithms. This operator can be used for online scoring, with either a static or a periodically updated model.

This operator must be configured with the actual classification algorithm to be applied to the data by providing an appropriate value for the modelType parameter. The value $learnAlgo, which is a parameter to the ClassifyBeat composite, controls which specific algorithm to use. The acceptable options encompass all of the classification algorithms supported by Weka, for example: "J48" (a decision tree), "NB" (a NB classifier), "MLP" (a Multi-Layer Perceptron neural network), "RBF" (a Radial Basis Function neural network), "RDF" (a Random Decision Forest ensemble classifier), or "SLR" (a Single Logistic Regression model).

The model update period is controlled by the $learnInterval parameter (line 10). When set to 0, the classifier uses a static model, loaded at startup from the location specified by the $modelName parameter (line 11), and scores the incoming data, appending a prediction attribute (line 13) to the original tuple.

When the $learnInterval is greater than 0, a brand new model is recreated every time $learnInterval tuples have been accumulated. The process of learning a new model is carried out on a different thread, and the prior model is used for scoring until a new model is ready.

This new model is created using the accumulated tuples and their associated expert annotations, emulating a (somewhat artificial) scenario where the true classes for these tuples have already been identified. In practice, however, there might be a substantial delay until this information becomes available for use, if it ever does.

The ClassifyBeat composite operator also employs a Functor operator (line 22) to measure the accuracy of the classification process by computing a confusion matrix (Section 11.3). The counts corresponding to the entries in this matrix are computed by comparing the label and the prediction attributes (lines 30–40).

12.3.4   Evaluation

As discussed earlier (Section 11.3), the accuracy of a classifier must be experimentally studied to understand its performance when subjected to real data. In the rest of this section, we discuss how one such empirical study can be carried out, how to exploit tradeoffs between different model learning scenarios, as well as how different parameter settings affect the PM application.

The empirical study employed data corresponding to 20 patients from the MIT Physiobank database. In the study, each patient has a dedicated instance of the ECGSource and ECGFeature operators. Each patient's data stream is exported to a single instance of the ClassifyBeat operator where the classification results are produced. The study comprised the set of experiments summarized in Tables 12.1 and 12.2 and focused on two performance metrics: the probability of detection (p_D) and the probability of false alarm (p_F).

As seen in the tables, the study includes three main scenarios: offline learning combined with offline scoring (offline data mining), offline learning with online scoring, and online learning and scoring.

In the case of online learning and scoring, different parameter settings for $nFeatures (i.e., the number of retained cepstral coefficients defined in line 9 in

Table 12.1 Empirical study configurations: offline learning.

Experiment	Learning/scoring	Algorithm	Learning strategy
1	offline / offline	J48	10-fold cross validation
2	offline / offline	NB	10-fold cross validation
3	offline / online	J48	learn on the first half of the data
4	offline / online	NB	learn on the first half of the data

Table 12.2 Empirical study configurations: online learning and scoring.

Experiment	Algorithm	`$nFeatures` / `$learnInterval`
5	J48	16 / 500
6	J48	16 / 1000
7	J48	16 / 2000
8	J48	32 / 500
9	J48	32 / 1000
10	J48	32 / 2000
11	NB	16 / 500
12	NB	16 / 1000
13	NB	16 / 2000
14	NB	32 / 500
15	NB	32 / 1000
16	NB	32 / 2000

Figure 12.14) as well as `$learnInterval` (i.e., the number of accumulated tuples, i.e. heartbeats, before a new learning cycle, defined in line 10) were used to modify the behavior of the algorithms. In general, the use of smaller values for `$nFeatures` should lead to decreased computational complexity, but also to lower accuracy. Conversely, the use of smaller values for `$learnInterval` implies more frequent model rebuilding and, hence, higher computational cost, which should hopefully lead to higher classification accuracy.

Classification accuracy
The primary set of results, in the form of accuracy metrics, are depicted by Figure 12.15, where the numbers correspond to the experiments listed in Tables 12.1 and 12.2. In this chart, better configurations will simultaneously be more precise, correctly labeling a heartbeat (a high p_D), and generate a low rate of false alarms (a low p_F). Thus, the closer the point for an experiment is to the upper left side of the chart, the better its performance. From the results, several observations can be made.

The first one is that the J48 algorithm results are consistently better than those of the NB algorithm. In Figure 12.15, it is easy to see that points 1, 3, and 5–10, which correspond to different experiments using the J48 algorithm, cluster on the upper left corner of the chart.

Switching the focus to the experiments where offline learning is employed (Table 12.1), we considered two different training strategies: (a) ten-fold cross validation, where the training data is randomly partitioned into ten subsets and multiple learning rounds are performed; and (b) fixed training dataset, where the first half of the data is used for learning and the second half is used for testing the algorithms' accuracy.

Thus, considering these training alternatives, the second observation is that the ten-fold cross validation is usually a better learning strategy as experiments 1 and 2 outperform 3 and 4, respectively.

Considering now the online learning experiments (Table 12.1), we tested different settings for the `$nFeatures` as well as for the `$learnInterval` parameter.

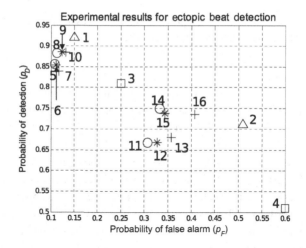

Figure 12.15 Accuracy results for the ectopic heartbeat detection problem.

We can see that increasing the value for $nFeatures improves the classification performance as, unsurprisingly, more fidelity in the signal representation translates into increased accuracy. Hence, for the J48 algorithm, experiments 8–10 display higher performance than experiments 5–7. Similarly, for the NB algorithm, experiments 14–16 display higher performance than experiments 11–13.

We can also see that reducing the value for $learnInterval leads to improved performance in a more subtle way. While the detection probability is nearly the same, it comes with a reduced false alarm probability. Clearly, learning more frequently allows the classifier to adapt to changes in data characteristics much more quickly. Note that, in these experiments, the data characteristics factor in differences across patients as well as changes in a single patient's profile as he/she goes through different physiological states during hospitalization.

Finally, we contrast the results of an analysis conducted in batch mode, where the data mining task is carried out offline, versus the online strategy.

We can see that neither approach seems to have consistently better performance. Focusing on the J48 algorithm, experiment 1 (offline) has higher p_D than experiments 5–10 (online). However, experiment 1 also has a correspondingly higher p_F. Indeed, the (approximately) best online result, obtained with the configuration for experiment 8, is roughly comparable to the result obtained with the configuration for experiment 1. Hence, selecting the best alternative depends on which combination of p_D and p_F should be ranked higher, a decision that has to be made in line with the application requirements set forth by the designer.

Interestingly, in the case of the generally inferior NB algorithm, the online data mining experiments (11–16) actually outperform all of the offline learning approaches. In this case, experiment 14 appears to be the one yielding the better overall performance.

We must emphasize that employing offline data mining in the context of the PM application is not realistic. In fact, many clinical decisions resulting from the information captured by the ECG signal must be made in near real-time to adapt to the time-varying nature of individual patients' heartbeats.

Resource utilization

In the prior analysis, we implicitly assumed that the classifier has sufficient computational resources and the algorithms' learning and scoring steps can keep up with the input data rates. Yet, in a production environment, the PM application can face resource shortages, but must remain functional. Thus, we also have to examine the impact of resource constraints on the classification algorithms.

To perform this analysis, we artificially created a resource bottleneck by replaying the data faster than real-time, while maintaining the computational resources allocated to the task fixed. In this way, the classifier will eventually not be able to keep up with the data ingestion rate. Furthermore, to prevent the classifier from exerting back pressure on the data sources, a load shedding operator was added to the application flow graph. Its fixed-size queue and discard policy cause data to be dropped when the application's analytical segment can no longer keep up with the incoming data.

Naturally, casting some of the input data aside can affect the classification as certain heartbeats will be missed altogether and the model will lose accuracy with less data to train the algorithm. In our study, up to 60% of the data was lost. These results, profiling the J48 algorithm under severe resource constraints, are shown in Figure 12.16.

It is clear that the lack of computational resources can make a significant difference in the performance of the algorithm. In the plot on the left, looking at the amount of dropped data for some of the experiments, it can be seen that increasing the $learnInterval leads to higher discard rates. For instance, by comparing the results for experiments 5–7, it is easy to see that experiment 5 is associated with the lowest number of dropped tuples. In this case, as the $learnInterval is increased, the complexity of each learning run grows significantly larger, as each time an enlarged collection of tuples is used in that process. This increase in complexity outweighs any reduction in performance due to less frequent learning. The same trend holds when we

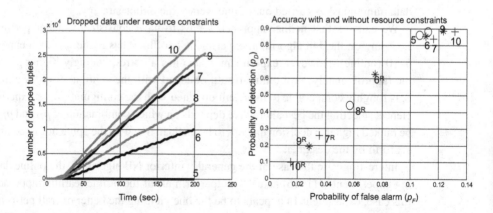

Figure 12.16 Performance results for the J48 algorithm under resource constraints.

examine the results for experiments 8–10, as well as the results for the NB algorithm, which we don't show here.

In the plot on the right, a data point is shown for each experiment indicating its performance with and without resource constraints (an R is used to mark the resource-constrained data points). As can be seen, generally, discarding data impacts the classification performance adversely and, not surprisingly, a greater amount of data discarded results in greater accuracy degradation.

Interestingly, this observation leads to a reversal of the conclusions we drew under no resource constraints, when analyzing the impact of setting the $nFeatures param-eter. Since increasing the value for $nFeatures leads to a higher computational cost when running the classification algorithm and, consequently, greater data loss in constrained scenarios, the performance degradation due to drops outweighs any perfor-mance improvements that result from having higher signal fidelity, where more features of the signal are used. Hence, unlike the case with no resource constraints, increasing $nFeatures leads to worse performance under resource constraints.

Nonetheless, the impact of the $learnInterval parameter is consistent across the two cases, as a small $learnInterval allows the algorithm to adapt quickly to new data characteristics[16] and, hence, to perform better.

The important lesson to draw from these experiments is that to understand the per-formance implications of different parameter settings in a data mining algorithm, it is necessary to consider the system context under which the analytic will be used. In this way, the proper adjustments can be made to maximize its accuracy and performance, before the application is deployed in a production environment.

12.4 The Semiconductor Process Control application

In this section we demonstrate how continuous analytics can be used in semiconductor manufacturing and test process control in the context of SPC, an application that was designed and developed in collaboration with test engineers from the IBM manufactur-ing division. In the rest of this section, we examine one part of the overall application, but omit certain details that are proprietary and confidential.

As was the case for the PM application, we focus primarily on the analytics and the effect of its design choices on the application's performance. In this case, however, rather than relying on an external data mining library, we examine how custom, con-tinuous, and incremental data analysis, alongside an integrated GUI application, can be created from scratch.

12.4.1 Motivation

Great advances in semiconductor technology and circuit design techniques have been observed over the last decades. Yet, chip manufacturing is a very challenging

---

[16] Note that it is possible to have a setting in which a smaller value for the $learnInterval parameter leads to worse performance. This situation may arise, for instance, when obtaining or loading a new model becomes more onerous as a new model is loaded more frequently.

undertaking, primarily because of the complexity and sensitivity of its production process. The SPC application was designed to address some of the challenges faced in this multi-stage process.

The fabrication process

The semiconductor manufacturing process requires a sequence of operations on each wafer to create the multi-layered physical and electrical structures supporting the Very Large Scale Integration (VLSI) circuitry. This process includes multiple iterative steps: the *application of photo resist, photolithography, etching, deposition, polish,* and *oxidation.*

Each of these processing steps is complex, requiring a careful interplay between many electrical, physical, chemical, and mechanical components. A typical manufacturing process pipeline employs several tools and each of them performs a processing step, with semiconductor wafers being routed from one tool to the next dynamically, to maximize utilization.

Process instrumentation

During manufacturing, tool operation parameters, wafer defects, and test results are collected and analyzed in real-time to provide information for process improvement and to drive the process control. The information includes: (a) *trace and process data* from the individual tools, capturing their process attributes; (b) *Fault Detection and Classification (FDC) data* from multi-variate analysis of the trace data, capturing variations in the process steps; (c) *measurement data* from the wafers, including different physical properties such as the refractive index and the metal deposition thickness; (d) *inline test data* from different electrical tests performed on the chips during manufacturing; and (e) *final test data* from different electrical, functional, and logic tests performed on the individual chips after the process is complete.

These different types of data are streamed out and available at different sampling rates and times (Figure 12.17). Furthermore, certain types of data, for example, measurement and inline test data, are often available as aggregated statistics, computed on a sampled subset of wafers and chips to reduce manufacturing costs.

Defect analysis

Defects in the fabrication process may occur due to several operational, mechanical, or chemical control errors, but also due to environmental uncertainty. Detecting these defects is typically carried out via automated defect analysis, consisting of a set of *inline* defect analysis tests and a set of comprehensive *final* analysis tests.

Inline tests are performed during manufacturing. The data and results for these tests are available in near real-time. Final tests, on the other hand, are performed after the wafer is fully produced and include comprehensive electrical, logic, and characterization tests that determine the *functional envelope* in which the chips can operate.

Final tests also include several detailed measurements of multiple electrical parameters, using different test configurations, for each pin on each chip, for each wafer,

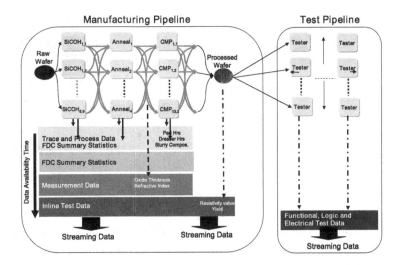

Figure 12.17 The semiconductor manufacturing and testing pipeline.

resulting in hundreds of thousands of measurements for a single wafer. Additionally, subsets of these tests are repeated on each wafer, under different conditions, to simulate different chip usage patterns. Clearly, the test process consumes a significant amount of time and generates several high volume data streams.

A streaming problem
Process control requires the online monitoring of this complex sequence of manufacturing steps. The impact of misconfigurations and process drifts within individual tools on the final chip *yield* is often subtle and hard to identify. In fact, there are complex interactions across the multiple manufacturing stages that, in aggregate, can result in reduced chip yields and performance.

As a result, significant resources have been invested in building systems and analytic solutions for semiconductor manufacturing process control, including real-time monitoring systems and algorithms that can facilitate the online processing, feature extraction, correlation, and statistical analysis of process data.

12.4.2   Requirements

The SPC application must allow manufacturing and test engineers to effectively monitor the process and predict end-to-end yields, such that they can adaptively control the chip manufacturing and test process.

The application's unique monitoring and forecasting challenges require the design and implementation of parametric and non-parametric models that can be used in real-time for forecasting. Better prediction capabilities can be instrumental in minimizing the Mean Time to Detect Failures (MTDF), and, thus, can improve the end-to-end process control.

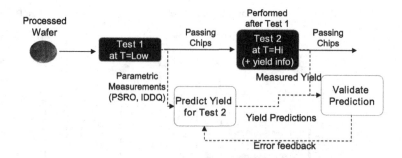

Figure 12.18 Two-stage chip testing.

Analytics: a test process example

Consider a specific test process to determine the Performance Limited Yield (PLY) metric of each chip on a wafer. This test process looks at several different parametric measures on each chip and, using a set of statistical binary classifiers, labels each chip as *good* or *bad*. PLY is a number between 0 and 100 and a threshold pre-set by a process engineer is used to separate individual chips into good or bad.

The test (Figure 12.18) is repeated at two different temperature levels, Low and High, to ensure that the chip can operate under different environmental conditions. Only chips that pass the first (Low temperature) test are tested at the High temperature. In Figure 12.18, the dark boxes correspond to actual steps in the test process performed by Automated Test Equipments (ATEs), with solid lines indicating a flow of chips from one step to another. The light boxes correspond to analytic steps to be performed by the SPC application, with the dashed lines corresponding to stream flows between the ATEs and the application, as well as internal SPC flows.

Two types of parametric measurements of interest are the average for the Performance Sort Ring Oscillator (PSRO) readings and the sum of the Direct Drain Quiescent Current (IDDQ) readings across each chip. The PSRO[17] frequency reading is a measure of the chip performance in terms of hardware speed. The IDDQ[18] reading is a proxy for the presence of short and/or open circuits on the chip.

In practice, several hundreds of such measurements are taken across the chip, but to simplify the analysis test engineers use, respectively, their average and sum, stored in a two-dimensional feature vector. In this feature space, of which Figure 12.19 is an example, each point represents the log(IDDQ) and PSRO metrics for one chip, with the different shades of gray indicating the yield as measured by the High temperature test. White points (left) have higher yield than black points (right). Note that the actual values on each axis are obscured to prevent the releases of proprietary information.

The goal of our analysis is to predict the yield at High temperature using the parametric measurements obtained from the Low temperature test. As described later, this prediction requires a dynamic segmentation of the two-dimensional space, based on the

---

[17] PSROs are embedded in a chip during its production and their frequency measurements are a quick indicator of the overall fabrication process quality.

[18] The IDDQ reading is obtained by measuring the supply current, *Idd*, in the circuit's quiescent state, hence the name *Iddq*.

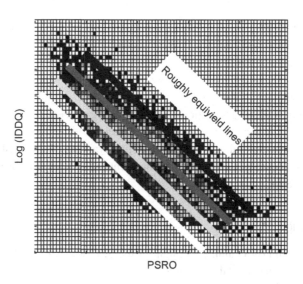

Figure 12.19 Using a linear fit predictor.

parametric measurements and on the delayed feedback about the true yield. If a high confidence prediction that a chip is likely to be *bad* can be made, re-testing can be avoided, reducing costs, saving time, and improving the MTDF.

In addition to the interesting analytical requirements faced by the SPC application, other requirements also make stream processing a particularly suitable paradigm for it, as discussed next.

Continuous data sources

The raw data, consisting of test results on individual chips, is generated continuously as discrete events from the ATEs, with explicit timestamps associated with each item. Clearly, an application designed to process this data must also be able to ingest it continuously.

Distributed processing

Test results are generated by different ATEs, potentially distributed across multiple test sites. Hence, an application analyzing this data must also be distributed and implement mechanisms to cope with operational and latency requirements that are specific to such an environment. These requirements can be naturally addressed in a well-designed SPA.

Continuous analytics

Since the test data is normally streamed out, it is only natural that the analyses and their predictions are also made in a continuous fashion, as soon as new data becomes available.

A streaming design not only allows the application to keep up with the incoming data, minimizing the needs for storage, but also lowers the reaction time between detecting an actionable event and making decisions to adjust the manufacturing or test process. In our case, as we have described, if a decision can be reached quickly as part of a continuous analytical task, an unnecessary second test on a chip might be avoided based on the outcome of the first test.

Continuous adaptation
Subtle random shifts in the manufacturing process impact the quality of the chips as well as their corresponding test measurements over time. Hence, the forecasting analytics must continuously adapt to the time-varying characteristics of the data to maintain their accuracy. Clearly, this requirement can be addressed as part of the regular structure of a SPA.

12.4.3   Design

We will now examine a design for the SPC application[19] that addresses the forecasting requirements for the test process scenario outlined in Section 12.4.2.

We focus primarily on stream analytic algorithms for the forecasting portion of the SPC application. As part of our design process, we evaluate two techniques: parametric segmentation using Principal Component Analysis (PCA) [14] and non-parametric local segmentation using a Nearest Neighbors (NN) classifier [15]. We also consider the weighted combination of these classifiers. And, finally, we empirically evaluate the incremental construction and tuning of these models based on ground truth available as delayed feedback, specifically when the actual results for the High temperature tests become available later on.

Data acquisition and pre-processing
Data acquisition is carried out using lightweight TCP clients on each ATE, which stream test results to a `TCPSource`, configured as a TCP server. In this way, data from the ATEs is delivered reliably to the SPC application and new sources (or new ATEs) can be dynamically added to the environment.

The ATE data arrives in the Transducer Electrical Data Sheet (TEDS)[20] binary format, which must be decoded to extract the IDDQ and PSRO measurements for each chip.

A chip is associated with about 200 of each of these measurements, which are grouped into a single tuple identified by a combination of the wafer and chip identifiers. The tuple also includes a measured yield for the chip as a result of the Low temperature test.

The data from the High temperature test, with the ground truth, used in the incremental learning process is also obtained using the same infrastructure. Naturally, the ground truth is only available after a delay, and only for certain chips. Hence, the

---

[19] Additional material related to the the SPC application can be accessed at www.thestreamprocessingbook.info/apps/spc.

[20] TEDS [16] is a set of standards defined by Institute of Electrical and Electronics Engineers (IEEE) to allow access to transducer data.

segment of the application responsible for updating the online classification models also includes an additional pre-processing step whereby the computed feature vector from the Low temperature test is buffered and later joined with the yield results from the High tempertaure test.

Data transformation

A chip's 200-dimensional feature vector must be summarized using a data reduction operation (Section 9.2.2.1). In this way, it can be visualized and analyzed more conveniently. As described earlier, a simple reduction operation, at the expense of some impact on the accuracy, can be employed: the PSRO readings are averaged and the IDDQ readings across the entire chip are added up, resulting in a two-dimensional feature vector per chip.

Designing a forecasting model

As we have seen in the PM application, it is necessary to have a good understanding of the problem and of the mining methods before selecting a particular technique. Preliminary offline analysis of the data and experimentation are both fundamental in this process.

In the test data scenario, the distribution of the feature and label space captured by Figure 12.19 can be very informative, but only after the data is appropriately pre-processed. In this diagram, it is easy to pick out multiple regions with nearly linear equi-yield regions, indicating that these two features, pre-processed in the way they were, can be effectively used for forecasting.

Thus, one important modeling step is to devise *how* to pre-process the raw test data. For this problem, it turns out that the original data captured in the feature space shows interesting properties when binned using a rectangular grid, where each grid cell is painted with a color based on the *mean* yield of all chips that lie inside its region.

A linear fit model

In Figure 12.19, the color of the grid changes from white to black and the shade indicates the yield, with the darkest shade representing the lowest yield. As can be seen, the lines we drew, which are nearly parallel to the principal axis, tend to have grid cells of roughly the same shade along them, suggesting the possibility of using a geometry-based model for prediction. These lines signify a strong *linear* trend for areas of similar yield. Indeed, while several standard data mining techniques could have been directly applied to this problem, it is always important to understand the feature and label space to determine if simple, intuitive models can be used. With these preliminary observations in mind, a Line Fit (LF) model using PCA to determine the principal axis, followed by binning the parameter space, seems like an adequate approach.

In this model, the expected yield in each bin is computed as the mean yield of all points that lie within it. The number of bins is selected based on the desired granularity of the partitioning, and may be dynamically changed based on the error feedback.

The LF model then consists of the parallel lines and the yield per bin. Therefore, a prediction operation using this model can project new feature vectors onto the line orthogonal to the parallel lines and use the appropriate yield bin as the expected value for the yield. Note also that the LF model builder can be designed to allow for its lines to be rotated around the principal axis. Hence, different search strategies can be used within $\pm\theta$ degrees of the original axis orientation, to determine the best line-fit model.

Clearly, this approximate model is computationally simple to compute and store, as well as to use during online classification. Historically, similar methods have been used in the manufacturing environment.

When used in the SPC application, the streaming version of the LF modeling process can be completely automated, eliminating the need for manual data collection.

A k-Nearest Neighbors model

Despite the LF model's simplicity, lack of robustness and over-fitting can occur when using it. Indeed, as part of designing an application forecasting mechanism, it is important to consider and compare alternative techniques. In this case, we also make use of a non-parametric classifier, using the k-Nearest Neighbors (KNN) algorithm.

When qualitatively comparing a KNN to an LF model, advantages of the former include the fact that it can discover arbitrary, non-linear, class separation boundaries. Indeed, the KNN classification does not impose any pre-defined models on the data. When it comes to accuracy, the precision of the KNN algorithm tends to increase with the amount of historical data points that have been accumulated [15]. Nevertheless, because SPC is a continuous application, the accumulation of data has to be restricted to a certain fixed-size sliding window, to bound the amount of memory used by the algorithm.

Our algorithm simply utilizes the yield of the k nearest neighbors of an x_{Low} testing point, the feature vector for a Low temperature test, and predicts its yield based on the median yield of its k *closest* neighbors.[21]

Model fusion

Considering that both the LF and the KNN models are available, another possibility is to combine them, an approach that can potentially be more accurate than either model by itself.

Hence, we also experiment with a weighted combination of the prediction results from the two classifiers using a linear regression model (Section 11.4), where the weights associated with the two classifiers are estimated by employing the Stochastic Gradient Descent (SGD) method [17].

Model initialization

The construction of these models requires setting different parameters, for instance, making a choice for the value of k to be used by the KNN algorithm. This step, as was

---

[21] Note that other ways of combining the yield of the neighbors, e.g., using the average, can also be used.

the case with selecting the techniques themselves, is carried out by experimenting with the training data offline. In this case, we employed a historical dataset consisting of data for 3000 chips used for training, and 3000 other chips used for validation.

In the case of the LF classifier, PCA was used over the training set to estimate the principal axis of projection. Subsequently, an exhaustive search within a $\pm\theta$ angle of the principal axis was carried out to determine the best classifier. In the case of the KNN classifier, the bootstrap method [18] was used to determine the optimum k on the training and validation data sets. With the parameters for each classifier in place, the combined classifier could then be calibrated by determining each model's individual weights in the mix.

Incremental model update

The LF and KNN predictive models used by the SPC classifiers must be updated frequently to catch up with changes introduced as a result of random perturbations in the manufacturing process. These perturbations create complex interactions between the test parameters and chip yields.

Additionally, the fabrication process drifts over time, and updating the models becomes necessary to incorporate information about chip properties over more recent time horizons. To determine the appropriate time for a model update, additional analytics that capture and detect model drift [19, 20, 21] can be used. In practice and through experimentation, the simpler approach of implementing periodic updates has proven effective. The update period was determined using analysis on the training and validation data.

For the LF model, it is possible to use incremental PCA techniques [14], which call for certain approximations to be made. For our specific problem, a simpler and exact method can be used instead. PCA requires the computation of the sample mean and covariance matrix from the data [22]. Both of them can be incrementally computed from the temporal samples. Additionally, the eigenvalues and eigenvectors of a 2×2 matrix can be computed analytically, as solutions to a quadratic equation. Hence, PCA can be conveniently implemented as an incremental computation in our simple application. Note that instead of searching exhaustively for the best angle around the principal axis, we restricted the search to the more recent data samples, stored in a sliding window.

The KNN model can also be incrementally maintained as long as the closest neighbors can be searched in a window, and this is the approach we took.

The choice of the window size and slide factor enable different compute-accuracy tradeoffs for our models. Intuitively, the slide of the window controls the periodicity of the model update and the size of the window controls how much of the recent history is used to update it. The larger the slide, the less frequent the model is updated, which reduces computational complexity and may, correspondingly, reduce its accuracy when the underlying data characteristics change rapidly. Conversely, the larger the window size, the more history is used to update the model, incurring additional memory and computation costs, but with the potential benefit of having a more accurate model.

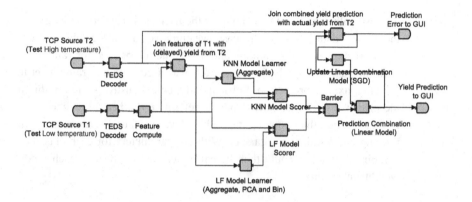

Figure 12.20 The flow graph for yield prediction segment of the SPC application.

The choice of the appropriate window size and slide must be made after examining the appropriate data characteristics and the desired tradeoffs between accuracy and complexity. In our implementation, we use the training and validation datasets to determine these tradeoffs, and fix the window size to 1000 chip yield measurements, and the slide to 200 chip yield measurements. These parameters were determined empirically, and may be different under different manufacturing and environmental conditions.

Finally, we use the ensemble model (consisting of the LF and KNN models) to provide robustness to this sliding window computation. As described in Section 11.5.1, the SGD method is a purely incremental algorithm, especially with progressive validation, so the weights of the combined method can be adjusted incrementally.

Implementation

The SPC application flow graph depicted by Figure 12.20 corresponds to the source code shown in Figures 12.21–12.23. This SPA includes the elements necessary to ingest the test data, learn and score with the different prediction methods, and output the results shown in real-time by a GUI-based application.

T1RawStream and T2RawStream, the streams in lines 23 and 30, respectively, consist of binary data, encoded using the TEDS format. These streams are received from two different types of ATEs: one for test T1 (the Low temperature test) and the other for test T2 (the High temperature test). In both cases TCPSource operators configured as TCP servers that can read binary data are used so multiple ATEs can send their data to the application.

Once the data is received, it must be decoded, a process implemented by the TEDSDecoder operator (lines 37 and 42). This operator can be parameterized to decode both the per chip parametric measurements and the resulting yield. The parametric measurements are aggregated to populate a feature vector, by averaging the PSRO readings and computing the logarithm of the summed up IDDQ readings, which are used later for forecasting. The creation of the feature vectors is implemented by the Custom operator in line 47.

```
1   type ATERawTupleT = tuple<list<byte> tedsPayload>;
2   type T1DecodedFeatureT = tuple<timestamp ts, list<float64> psro, list<float64> iddq,
3     rstring chipID, rstring waferID>;
4   type FeatureVectorT = tuple<timestamp ts, float64 PSROavg, float64 log_IDDQsum,
5     rstring chipID, rstring waferID>;
6   type TrainingFeatureT = tuple<FeatureVectorT, tuple<float64 T2yield>>;
7   type YieldPredictionT = tuple<timestamp ts, float64 yieldPrediction, rstring chipID,
8     rstring waferID>;
9   type PredictionErrorT = tuple<YieldPredictionT, tuple<float64 T2yield>>;
10  type T2DecodedYieldT = tuple<timestamp ts, float64 T2yield, rstring chipID,
11    rstring waferID>;
12  type LFModelT = tuple<list<float64> principalAxis, list<float64> binThresholds,
13    list<float64> T2yieldPerBin>;
14  type KNNModelT = tuple<list<list<float64> samples, list<float64> T2yieldPerSample>;
15  type CombinedModelT = tuple<float64 LFweight, float64 KNNweight>;
16
17  composite YieldMonitor {
18    param
19      expression<uint32> $inputPortT1: getSubmissionTimeValue("inputPortT1");
20      expression<uint32> $inputPortT2: getSubmissionTimeValue("inputPortT2");
21      expression<float64> $timeBetweenTests: getSubmissionTimeValue("timeBetweenTests");
22    graph
23      stream<ATERawTupleT> T1RawStream = TCPSource() {
24        param
25          role: server;
26          port: $inputPortT1;
27          format: bin;
28      }
29
30      stream<ATERawTupleT> T2RawStream = TCPSource() {
31        param
32          role: server;
33          port: $inputPortT2;
34          format: bin;
35      }
36
37      stream<T1DecodedFeatureT> T1Features = TEDSDecoder(T1RawStream) {
38        param
39          decodeType: "feature";
40      }
41
42      stream<T2DecodedFeatureT> T2Features = TEDSDecoder(T2RawStream) {
43        param
44          decodeType: "yield";
45      }
46
47      stream<FeatureVectorT> featureVectors = Custom(T1Features) {
48        logic
49          onTuple T1Features: {
50            mutable FeatureVectorT t = { ts = ts, chipID = chipID, waferID = waferID,
51              PSROavg = avg(psro), log_IDDQsum = log(sum(IDDQ)) };
52            submit(t, featureVectors);
53          }
54      }
55
56  // to be continued ...
```

Figure 12.21 The yield prediction segment of the SPC application.

The feature vectors are then transmitted to both predictors, the LFScorer and the KNNScorer operators in lines 71 and 73, respectively. These operators are implemented as primitive SPL operators. Each of these scorers also requires a forecasting model and both can load an initial baseline model at startup. Subsequently, they also receive model updates from their respective online learners seen in lines 67 and 69.

```
57      // ... continued
58      stream <TrainingFeatureT> trainingFeatures = Join(featureVectors; T2Features) {
59        window
60          featureVectors: sliding, time($timeBetweenTests);
61          T2Features: sliding, count(1);
62        param
63          match: featureVectors.chipID == T2Features.chipID &&
64                 featureVectors.waferID == T2Features.waferID;
65      }
66
67      stream<LFModelT> LFModels = LFLearner(trainingFeatures) {}
68
69      stream<KNNModelT> KNNModels = KNNLearner(trainingFeatures) {}
70
71      stream<YieldPredictionT> LFPredictions = LFScorer(featureVectors; LFModels) {}
72
73      stream<YieldPredictionT> KNNPredictions = KNNScorer(featureVectors; KNNModels) {}
74
75      // to be continued ...
```

Figure 12.22 The yield prediction segment of the SPC application (continued).

The results from the two predictors are synchronized and then put together using a Barrier operator (line 98), which passes them to a Custom operator (line 78) that performs a linear combination, generating a prediction based on the *combined* models. Note that the logic in this Custom operator can also update the default weights (0.5, at startup) assigned to the LF and KNN predictions on request.

In addition to the scoring operators, the application includes the processing logic to continuously update the three models. These updates are performed by the LFLearner, KNNLearner, and LinearRegression operators.

To incrementally update the models, the *true* yields that are received after a chip has gone through test T2 must be synchronized with the feature vectors received after test T1. Hence, a Join operator (line 58) is used to correlate the featureVectors and the T2Features streams. Since the T2 results arrive after the T1 feature vectors, this Join operator must be configured with a time-based window with size $timeBetweenTests to account for this delay.

The LFLearner operator (line 67) incrementally computes the covariance matrix and mean, before analytically determining the principal axis. It then performs a search inside a sliding window to determine the best angle around the principal axis, as well as the threshold associated with the binning along the direction perpendicular to the principal axis. As seen earlier, a chip that lies in a specific bin will be forecast to have the average yield of the chips in that bin.

The KNNLearner composite (line 69) primarily aggregates the feature vectors and the yield associated with them, such that the KNN scorer can determine the appropriate neighboring samples to make its prediction. Again, a fixed-size sliding window of recent history is maintained by this composite.[22]

---

[22] Our implementation can maintain separate windows for the LF and the KNN models. However, in our experiments we used the same window size and slide for the model update.

```
76      // ... continued
77
78      stream<YieldPredictionT> combinedPredictions = Custom(mergedPredictions; combinedModels) {
79        logic
80          state: {
81            mutable float64 LFweight = 0.5;
82            mutable float64 KNNweight = 0.5;
83          }
84          onTuple combinedModels: {
85            LFweight = LFweight;
86            KNNweight = KNNweight;
87          }
88          onTuple mergedPredictions: {
89            mutable YieldPredictionT t = { ts = ts, chipID = chipID, waferID = waferID,
90                yieldPrediction = LFweight*LFpred + KNNweight*KNNpred };
91            submit(t, combinedPredictions);
92          }
93      }
94
95      stream<CombinedModelT> combinedModels = LinearRegression(predictionErrors) {}
96
97      stream<timestamp ts, float64 LFpred, float64 KNNpred, rstring chipID, rstring waferID>
98          mergedPredictions = Barrier(LFPredictions; KNNPredictions) {
99        output
100         mergedPredictions: KNNpred = KNNPredictions.yieldPrediction,
101                            LFpred = LFPredictions.yieldPrediction;
102     }
103
104     stream<PredictionErrorT> predictionErrors = Join(combinedPredictions; T2Features) {
105       window
106         combinedPredictions: sliding, time($timeBetweenTests);
107         T2Features: sliding, count(1);
108       param
109         match: combinedPredictions.chipID == T2Features.chipID &&
110                combinedPredictions.waferID == T2Features.waferID;
111     }
112
113     () as Sink1 = GUISink(combinedPredictions) {}
114
115     () as Sink2 = GUISink(predictionErrors) {}
116   }
```

Figure 12.23 The yield prediction segment of the SPC application (continued).

Finally, the LinearRegression operator (line 95) updates its model using the SGD method, adjusting the linear weights used by the *combined* model. To update the linear regression model (line 95), it is necessary to estimate the prediction error. This is carried out by a Join operator (line 104) that synchronizes the combinedPredictions stream with the T2Features stream.

The results, comprising the predictions, as well as the computed prediction errors, are sent to a GUI-based application (Section 12.4.5) using two specially written GUISink operators (lines 113 and 115). These sinks carry out the appropriate format XML conversions, as well as establish the external connections and perform the real-time data transfer operations.

12.4.4   Evaluation

Considering the implementation described in the prior section, we studied the accuracy of five different predictors in forecasting the expected yield of the High temperature test based on feature vectors constructed from the results of the Low temperature test.

Table 12.3 Observed error rates for different prediction methods.

	week 1	week 2	week 3	week 4	week 5
LF-fixed	26%	35%	43%	44%	48%
LF-incremental	24%	31%	28%	25%	21%
KNN-fixed	22%	27%	32%	37%	38%
KNN-incremental	20%	24%	22%	19%	17%
combined	19%	21%	19%	14%	12%

These five predictors made use of the two individual models, LF and KNN, trained in two different ways. In the first configuration, the predictors were trained once, with the 3000 sample training set and we refer to them as LF-fixed and KNN-fixed. In the second configuration, the predictors make use of incremental model updates, after being bootstrapped with the same training data, and we refer to them as LF-incremental and KNN-incremental. The last configuration consists of the combined model, using incremental updates for each of the two individual models.

In Table 12.3, we summarize the error rates of each prediction method observed during a 5-week period. The error rate is computed as a relative error in yield prediction, that is, the average of the relative error for each chip yield. This value is computed as $\frac{|y - \hat{y}|}{y} \cdot 100$, where y is the *true* yield and $\hat{y}$ is the prediction from one of the models.

As seen in the table, the incremental model updates consistently help reduce the error rates when compared to one of the configurations that uses a fixed model trained only on historical data. This observation is valid for the LF as well as for the KNN model.

We can also see that the KNN model reliably outperforms the LF model across all of the time periods, whether making use of incremental updates or not. This fact indicates that the line-fit based model is, at best, a coarse approximation of the data-space.

Finally, the *combined* model displays the best forecasting performance across all predictors, as it optimizes the linear combination based on the previously seen prediction error.

Discussion

This experimental evaluation is only the first step towards creating a production-grade application. In fact, there are several issues a designer must still consider. We highlight two of them next.

First, when a model is incrementally updated as was the case for some of our predictors, the High temperature test must also be performed in some chips that are predicted to have a low yield. This is necessary to ensure that the model update can account for potential drift in the predictions associated with low-yield chips. Furthermore, the application could also make use of machine learning techniques to de-bias for this type of skew [13].

Second, our evaluation was the result of a single experiment. In general, a more comprehensive study is necessary to ensure that an application's analytical and performance requirements are adequately met by the prototyped design.

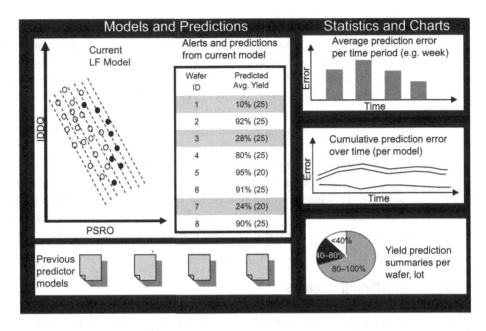

Figure 12.24 *Yield Monitoring* application's GUI.

12.4.5   User interface

We have not explored the user interface design for any of the application examples discussed so far, having focused primarily on their system-oriented and data mining aspects. Yet, the User Experience (UX) facet of a SPA is very important, as an effective design can be one more analytical asset, allowing the integration of human expertise into such applications.

While UX considerations are not the focus of this book, we briefly describe one of SPC's GUI applications, focusing on the analytical segment we discussed earlier. In this case, live results from the application are streamed to the user, processed, and displayed by the *Yield Monitoring* browser-based application. A rendering of this GUI is shown in Figure 12.24.

The actual interface was implemented using Adobe Flex [23] (now known as Apache Flex [24]), a toolkit used in the development of cross-platform GUI-based applications using Adobe's Flash platform. The data between the analytic application and the interface is exchanged using an XML document, transmitted over a TCP connection.

The main elements of the *Yield Monitoring* application include visualizations that provide an analyst with a real-time glimpse into the state of the fabrication process. The GUI is able to depict the current linear model, so an assessment of the model fit can be easily done by an analyst. This up-to-date information can also be contrasted and compared with prior models, since the application also stores them and their prediction error, allowing a visual inspection of the learning process over time.

The GUI also depicts yield predictions for recently examined chips, before the High temperature test has taken place. These predictions are color-coded based on the severity of the forecast. For instance, chips displayed in dark color are expected to have *low* yield and, hence, are of greater interest. These results are aggregated per wafer (as an average yield), and presented as alerts in a table, such that process and test engineers can closely examine this data in more detail. Note that the values in parenthesis indicate the number of chips that each prediction is based on, providing a measure of its associated confidence.

The *Yield Monitoring* application also implements a dashboard providing an overview of the forecasting process. Various statistics are computed on the fly and updated continuously, allowing the inspection of the actual yield and the forecasting accuracy. Graphical plots include historical curves that show the prediction error over time, the prediction error per week, as well as the prediction error per wafer lot.

12.5   Concluding remarks

In this chapter, we examined three SPA prototypes designed to deal with real-world problems. Each one of these prototypes had a different focus and emphasized a different set of system-oriented or analytical techniques, discussed elsewhere in this book.

The first application, OM, dealt primarily with scalability, flexible data delivery, and fault tolerance in the context of operational monitoring of a multi-component corporate IT infrastructure. We illustrated how a large number of high-throughput and multi-format log message streams can be parsed, routed, and re-organized to provide structured data to external consumer applications via a flexible, scalable, and fault-tolerant subscription mechanism.

The second application, PM, focused primarily on analytics. We saw how ectopic heartbeat detection over real-time ECG streams can be performed. While we only partially explored how such an application could be implemented in a clinical environment, we provided an in-depth look at the the application's feature extraction as well as classification tasks. We also analyzed and empirically evaluated different stream mining strategies, comparing their effectiveness in terms of accuracy and performance in a resource-constrained environment.

Finally, the third application, SPC, focused on adaptive feedback and prediction analytics. We saw how this SPA can be used to forecast chip yield and how it can continuously update and improve its prediction model based on delayed feedback.

References

[1] Weka Data Mining in Java; retrieved in December 2010. http://www.cs.waikato.ac.nz/ml/weka/.

[2] Kopparapu C. Load Balancing Servers, Firewalls, and Caches. John Wiley & Sons, Inc.; 2002.

[3] Streams Exchange; retrieved in May 2012. https://www.ibm.com/developer works/mydeveloperworks/groups/service/html/communityview? communityUuid=d4e7dc8d-0efb-44ff-9a82-897202a3021e.

[4] Griffin MP, Moorman JR. Toward the early diagnosis of neonatal sepsis and sepsis-like illness using novel heart rate analysis. Pediatrics. 2001;107(1):97–104.

[5] Vespa P, Nuwer M, Juhasz C, Alexander M, Nenov V, Martin N, et al. Early detection of vasospasm after acute subarachnoid hemorrhage using continuous EEG ICU monitoring. Electroencephalography and Clinical Neurophysiology. 1997;103(6):607–615.

[6] Blount M, Ebling M, Eklund M, James A, McGregor C, Percival N, et al. Real-time analysis for intensive care: development and deployment of the Artemis analytic system. IEEE Engineering in Medicine and Biology Magazine. 2010;29(2):110–118.

[7] Bonow R, Mann D, Zipes D, Libby P. Braunwald's Heart Disease: A Textbook of Cardiovascular Medicine. Elsevier Saunders Co.; 2012.

[8] Pan J, Tompkins WJ. A real-time QRS detection algorithm. IEEE Transactions on Biomedical Engineering (TBME). 1985;32(3):230–236.

[9] Hamilton PS, Tompkins WJ. Quantitative investigation of QRS detection rules using the MIT/BIH arrhythmia database. IEEE Transactions on Biomedical Engineering (TBME). 1986;33(12):1157–1165.

[10] CapsuleTech; retrieved in December 2010. http://www.capsuletech.com/.

[11] Electronic Industries Association – Engineering Department. Interface Between Data Terminal Equipment and Data Communication Equipment Employing Serial Binary Data Interchange. EIA; 1969.

[12] MIT PhysioBank: Arrhythmia Database; retrieved in December 2010. http://www.physionet.org/physiobank/database/mitdb/.

[13] Hastie T, Tibshirani R, Friedman J. The Elements of Statistical Learning: Data Mining, Inference, and Prediction. Springer; 2009.

[14] Vieira Neto H, Nehmzow U. Incremental PCA: an alternative approach for novelty detection. In: Proceedings of Autonomous Robotic Systems Conference (TAROS). London, UK; 2005. pp. 227–233.

[15] Duda RO, Hart PE. Pattern Classification and Scene Analysis. John Wiley & Sons, Inc.; 1973.

[16] IEEE. Standard for a smart transducer interface for sensors and actuators. IEEE Standards Association; 2007. IEEE 1451.5-2007.

[17] Gardner WA. Learning characteristics of stochastic-gradient-descent algorithms: A general study, analysis, and critique. Signal Processing. 1984;6(2):113–133.

[18] Hall P, Park BU, Samworth RJ. Choice of neighbor order in nearest-neighbor classification. Annals of Statistics. 2008;36(5):2135–2152.

[19] Aggarwal CC. A framework for diagnosing changes in evolving data streams. In: Proceedings of the ACM International Conference on Management of Data (SIGMOD). San Diego, CA; 2003. pp. 575–586.

[20] Dasu T, Krishnan S, Venkatasubramaniam S, Yi K. An information-theoretic approach to detecting changes in multi-dimensional data streams. In: Proceedings of the Symposium on the Interface of Statistics, Computing Science, and Applications (Interface). Pasadena, CA; 2006.

[21] Kifer D, Ben-David S, Gehrke J. Detecting change in data streams. In: Proceedings of the International Conference on Very Large Databases (VLDB). Toronto, Canada; 2004. pp. 180–191.

[22] Dongarra J, Duff I, Sorensen D, van der Vorst H, editors. Numerical Linear Algebra for High Performance Computers. Society for Industrial and Applied Mathematics; 1998.

[23] Adobe. Using Adobe Flex 4.6. Adobe Systems; 2011.

[24] Apache Flex; retrieved in September 2012. `http://http://incubator.apache.org/flex/`.

Part VI

Closing notes

13 Conclusion

Stream processing has emerged from the confluence of advances in data management, parallel and distributed computing, signal processing, statistics, data mining, and optimization theory.

Stream processing is an *intuitive* computing paradigm where data is consumed as it is generated, computation is performed at wire speed, and results are immediately produced, all within a continuous cycle. The rise of this computing paradigm was the result of the need to support a new class of applications. These analytic-centric applications are focused on extracting intelligence from large quantities of continuously generated data, to provide faster, online, and real-time results. These applications span multiple domains, including environment and infrastructure monitoring, manufacturing, finance, healthcare, telecommunications, physical and cyber security, and, finally, large-scale scientific and experimental research.

In this book, we have discussed the emergence of stream processing and the three pillars that sustain it: the *programming paradigm*, the *software infrastructure*, and the *analytics*, which together enable the development of large-scale high-performance SPAs.

In this chapter, we start with a quick recap of the book (Section 13.1), then look at the existing challenges and open problems in stream processing (Section 13.2), and end with a discussion on how this technology may evolve in the coming years (Section 13.3).

13.1 Book summary

In the two introductory chapters (Chapters 1 and 2) of the book, we traced the origins of stream processing as well as provided an overview of its technical fundamentals, and a description of the technological landscape in the area of continuous data processing.

In Chapters 3–6 we focused on the development of applications and the techniques employed in flow-driven programming, as well as on the more conventional software engineering topics of modularity, extensibility, and distributed processing. We provided a detailed introduction to Streams' SPL language. We also included a discussion of visualization and debugging techniques associated with SPAs.

Subsequently, we described the internal structure of a SPS. In Chapters 7 and 8, we discussed the middleware layers and SPS-based services that provide the necessary support to SPAs, shielding them from the complexities of a distributed runtime environment. We then illustrated these services concretely, with an in-depth look into the InfoSphere Streams SPS, its services, and application programming interfaces.

After covering the application development and system infrastructure areas, we switched to application design principles (Chapter 9) and analytics (Chapters 10 and 11). We first described specific implementation principles that can make a SPA reliable, efficient, and scalable. We then focused on the unique analytical requirements of stream processing. As discussed, SPAs must employ lightweight data analysis techniques and algorithms suitable for continuous data exploration, with the additional constraint that multiple passes over the incoming data may be infeasible and, in extreme cases, some of the data might have to be skipped or dropped altogether.

Finally, the last part of the book was dedicated to examining application case studies. The focus of Chapter 12 was on providing an overview of the design, software engineering, and analytic challenges facing a developer whose goal is (to either) build a new application, or to use elements of stream processing as part of a larger data processing task.

In essence, this book provided a comprehensive overview of the stream processing field, ranging from its origins and its technological and analytical underpinnings, to the practical engineering aspects of implementing production-grade applications.

13.2   Challenges and open problems

Stream processing has three technical foundations: *programming language, middleware*, and *analytics*. All of these areas have seen significant advances that made it possible to develop stream processing systems and applications. However, additional research is needed to support the rapidly increasing data volumes and workloads as well as the higher degree of analytic and engineering complexity associated with emerging business and scientific problems.

These requirements impose additional challenges in the areas of software engineering, integration, scalability, and distributed computing, as well as in continuous and adaptive analytics, as we will discuss next.

13.2.1   Software engineering

Corporate data processing environments are complex. SPAs are expected to function and to inter-operate with other applications and components in such environments. One of the first questions faced by a designer contemplating the use of stream processing relates to the use of existing or legacy analytical and data processing assets that might already be in place. Once this issue is addressed and a new application has been developed, its production life cycle and evolution must be carefully managed. SPAs, due to their continuous operation, introduce new twists to three classical software engineering problems: how to ascertain good quality, how to allow components to be modified and upgraded, as well as how to manage maintenance cycles.

Thus, we discuss four of these challenges in more detail: the problems of incorporating existing software assets into newly built SPAs, *testing*, *versioning*, and *maintenance*.

Blending with existing software assets

Business and scientific organizations have invested heavily in analytical toolkits written using languages such as C++ and Java, as well as in scripting languages, including Python, Perl, and many others. Additionally, languages specifically designed for data analysis such as MATLAB [1, 2], Mathematica [3, 4], and R [5, 6] are all extensively used.

These analytical assets are in many cases used for periodic and ad-hoc *offline* data analysis tasks. Their primary users are generally expert data engineers and scientists, who sometimes do not make an effort to ensure that their code is reusable. Furthermore, in some cases, these assets are not suitable for *online* data analysis and, hence, cannot be directly embedded in a SPA. Yet, leveraging existing analytics as part of continuous data processing is frequently the goal behind the development of new SPAs.

Integrating an institution's legacy intellectual property and code into a stream processing solution entails decisions that range from a complete rewrite of the necessary code, which is expensive, to leveraging the existing code using software bridges, which is usually architecturally complex and, sometimes, performance taxing.

There is no simple solution when it comes to the integration of legacy code, particularly when it comprises sophisticated data analysis operations. In practice, the integration between existing code and a new application relies on hand-crafted software bridges as well as on wrapping technologies (e.g., Swig [7, 8]) and inter-operation protocols (e.g., Java Native Interface (JNI) [9]).

We believe that tooling to aid and guide the porting of legacy libraries, as well as general-purpose wrapping layers, will become increasingly more important as the need for integration increases. Likewise, the need for profiling and performance analysis tooling, so that the existing code can be evaluated in the context of a new SPA, will also increase in importance.

Finally, there is pressing need for the development of public open-source toolkits and cookbooks comprising common stream processing operators and applications (as has happened in other domains [10]) to help application designers. Efforts like this are only beginning in the community around Streams [11].

Testing

An integral component of the software development cycle with direct impact on software quality is functional testing. This step helps to ensure that an application behaves as specified. It includes testing individual application components, the overall application, as well as the middleware and application combined.

For SPAs, it is necessary to scrutinize the application for performance and stability degradation that might occur as an application is updated or deployed in a different physical configuration, leading to violation of QoS requirements set forth during its design phase.

While many of these problems are well-understood and testing frameworks and tools exist, testing SPAs, particularly distributed ones, is challenging. Some of the difficulties are similar to the ones faced by other distributed computing models, and arise from the inherent non-determinism faced by a distributed application. Indeed, in many cases, it is not feasible to exactly reproduce the set of events that might expose a bug.

Other difficulties stem from the same characteristics that make these applications powerful analytical tools, in particular, their ability to evolve dynamically and adapt in the presence of unknown variations in the workload. Clearly, it is not possible to emulate and induce all of the conditions that might arise in production.

As a result, while existing testing technologies can help, addressing non-determinism as well as emulating complex runtime conditions to ensure reasonable test coverage remains difficult. Hence, the field of SPA testing and, possibly, the field of developing test-friendly SPAs are both fertile ground for new research and engineering contributions.

Versioning

In the complex IT environment present in large organizations, the software ecosystem can be substantially uneven when it comes to versions of OSs, packages, and individual libraries. This situation is further complicated by the disparate collection of additional hardware, and the drivers that make these computers work. Concretely, different versions of application programming interfaces, communication protocols, and external libraries might coexist at any given point in time.

SPS middleware and the applications they support must contend with this situation as they often require the interconnection of different parts of the corporate software ecosystem. Built-in architectural support for *versioning*, *dependency tracking*, as well as for *provisioning* are mostly absent from stream processing middleware.

Nevertheless, each of these problems has engineering solutions, for instance, the ability to associate versions to APIs found in WSDL [12], and the dependency tracking capabilities and the software asset management provided by technologies such as the Apache Maven [13] build automation tool.

Bringing in capabilities of this type into a SPS and into SPAs requires active research and development.

Maintenance cycle management

In several domains the need for continuous processing requires applications to function in a non-stop or *quasi*-non-stop mode. Nevertheless, these applications and some of their components must be *refreshed* from time to time, either to fix bugs or to upgrade one or more components with additional functionality. These tasks can only be effectively accomplished if provisions for these types of maintenance operations are made when an application is designed.

Unfortunately, well-established software engineering techniques and system support for implementing and managing periodic maintenance cycles for continuous applications do not exist yet. In practice, organizations develop their own ad-hoc policies and mechanisms to ensure that an application, or some of its components can be upgraded periodically or on demand. Similarly, home-grown mechanisms are used for selectively turning on and off new features after a new round of improvements.

In essence, two problem areas remain to be addressed. First, the techniques and corresponding mechanisms for managing partial downtime, as well as for hot swapping application components, must be devised to ensure that a critical application can continue to run, even as the application or its underlying computational resources face an upgrade cycle. Second, the techniques and corresponding mechanisms for dynamically activating and deactivating changes must be developed. In this way, recently installed changes can be reverted quickly in the event of a post-update problem.

13.2.2 Integration

As mentioned in Section 1.2, a SPA might be only one component of a larger Information Technology (IT) *solution*, comprising other conventional applications and technologies. Clearly, the *integration* between a SPA and the non-stream processing components must be effectively addressed as part of the design process. Decisions made at this stage range from the relatively simple, when the integration points are located at the edges of the stream processing component, to very complex, when the point of contact between streaming and non-streaming components happens at the core of either application.

In general, technologies designed for providing inter-operation often lag behind the continuous evolution in the software landscape. We highlight two challenges in this space: addressing the problem of diversity in *communication substrates* and the *software provisioning* problem.

Diversity in communication substrates
When we consider the communication mechanisms provided by most SPSs, the primary way of interconnecting operators and larger application components (e.g., composite operators) is by establishing one or more stream connections. While this form of integration relies on a communication abstraction that is natural to SPAs, it lacks the request–response or RPC [14] nature of many earlier inter-process communication frameworks.

In many situations, the use of RPCs is not only more adequate for certain types of inter-component interaction, but it is also more intuitive and less resource intensive than setting up two unidirectional stream connections.

Likewise, other styles of communication, provided, for example, by high-performance processing frameworks such as MPI [15] can, in many cases, better address the communication needs of particular subsystems, even when they are part of a SPA.

While an application developer can make use of RPC- or MPI-style interactions, web service technologies [16], and CORBA [17] when developing new operators and application components, the coexistence of these other forms of communication with the native stream connections is usually opaque to the SPS.

As discussed in Section 13.2.3, the fact that these interactions are not *managed* by the middleware creates several difficulties for debugging, visualization, and profiling tools.

In other words, because the middleware is not aware of other communication substrates, tools of that kind cannot make use of comprehensible constructs and abstractions to simplify the understanding of the dynamic behavior of an application.

Similarly, because the use of external communication primitives is usually buried inside low-level operators, they are harder to track and to understand by people other than the original application developers.

There are several architectural difficulties here. First, the overall system is prone to unintended interactions between its components. In fact sometimes these unintentional interactions occur at a lower level, such as at the TCP or UDP layer, where different substrates might interfere with each other. Second, applications may face subtle bugs that are difficult to reproduce and, therefore, difficult to fix. And, finally, there is an unnecessary increase in architectural complexity as an application and the underlying middleware must cope with different type systems, different threading models, and different memory management mechanisms introduced by the external communication substrate.

Additional research and substantial engineering effort is necessary to seamlessly (or natively) provide these other types of communication mechanisms with SPSs. Nevertheless, the benefits, including the use of a homogeneous type system, service dispatching structure, lock and memory management inside an operator, as well as inside other application components, are clear.

Building upon such integration, debugging, visualization, and performance analysis tooling must also be reengineered to *understand* these mixed modes of component interoperation, and to provide abstractions that enable developers to fully understand the dynamics of a complex, distributed application.

Provisioning

Another dimension to the integration process concerns the issue of provisioning software assets required by an application. In most SPSs, making available the external dynamic libraries, model bootstrapping files, and other assets is a task typically left to system administrators. In most cases, these assets must be pre-installed in locations accessible by applications that depend on them, prior to their launching.

With the use of large clusters and cloud computing, ad-hoc provisioning becomes error-prone and cumbersome for two main reasons. First, it might not be known a priori which hosts an application will be deployed on and, second, in some cases an asset the application depends on might be dependent on others. Ideally, OSs, language runtime environments, and SPSs should natively incorporate better support for provisioning software assets.

Hence, there is a need for a mechanism to explicitly declare application dependencies as well as a way to retrieve the chain of transitively dependent assets. Similarly, there is also a need for an application launcher, either as a part of the OS or as a part of the SPS. This launcher should fetch and deploy an application's assets as part of its bootstrapping process.

Both areas require additional research and engineering so that comprehensive platform- and language-independent provisioning mechanisms can be put in place. Such mechanisms can greatly simplify the task of deploying and managing large-scale SPAs, as well as other distributed applications that share the same computational environment.

13.2.3 Scaling up and distributed computing

In Chapter 9, we described ways to parallelize (Section 9.3.2) and to improve the performance of SPAs (Section 9.3.3). Despite these techniques, there are three areas where the current state-of-the-art is inadequate.

The first one is in *auto parallelization*, i.e., the set of language, compiler, and middleware techniques to support the automatic parallelization of applications. Intelligent parallelization is essential to exploit the changes in the computing landscape, where multi-core chips, Graphics Processing Units (GPUs), and cluster and cloud computing are now the norm. Despite these trends in the hardware side, the efficient use of these computing resources still hinges on the manual parallelization of applications.

The second challenge lies in the management of large-scale *in-memory* distributed state, employed by data mining and other algorithms (Section 10.1). The scaling up of applications and algorithms to handle ever increasing continuous workloads imposes larger demands for these *in-memory* models, where indexing, searching, and updating operations must be carried out while maintaining consistency, balanced distribution, and fault tolerance.

The third challenge originates from the emerging area of *big data* analytics and applications. The integration of disparate collections of data sets, along with the desire to leverage sophisticated and complex data analysis, imposes additional demands on the computing infrastructure, not only in the form of software engineering design techniques, but also in integration with the rest of the computing infrastructure.

Auto parallelization

In Section 9.3.2, we mentioned that an application's data flow graph might exhibit sites with untapped pipeline, data, and task parallelism. While we studied several design principles and techniques to exploit these parallelization opportunities (Section 9.3.2.2), there are many challenges that remain to be addressed, primarily on automating this process.

First, there is the issue of transparently and automatically *finding* the sites in an application that are amenable to the use of parallelization techniques. Second, the specific parallelization techniques targeting applications must be *elastic* and *adaptive*. In other words, these techniques must be able to reshape the application to respond to changes in the workload as well as to changes in the availability of resources. Some progress has been made in this area [18], but substantial additional research and engineering is necessary before a comprehensive suite of techniques is available for use.

Third, assuming that a set of candidate sites for parallelization has been found, automatically identifying the most *profitable* configuration and the specific set of parallelization techniques that maximizes throughput, reduces latency, and preserves the semantics and correctness of an application can be a formidable task.

Different parallelization alternatives are affected by the rates of the data streams consumed and generated by an application, the individual costs of streaming operators used in the application, as well as the capacity and availability of computational resources. As a result, identifying the *best* parallelization strategy requires compile-time analysis, run-time profiling, cost models, and decision-making techniques. Moreover, because of the dynamics of the application and of the environment, optimization decisions must be revisited periodically. Preliminary progress has been made in this research area [19, 20], but substantial challenges in compiler technology, tooling, and infrastructure support remain unaddressed.

Distributed state management

Large-scale applications often create and maintain a sizable amount of state as they process the incoming data. Examples include communication graphs between IP addresses in network analysis SPAs, and options prices for different instruments in finance engineering SPAs. This large amount of state also tends to be distributed across multiple hosts.

In general, distributed state management is not provided by SPSs. It is either hand-crafted from scratch by an application developer, or implemented by leveraging external mechanisms, such as Distributed Hash Tables (DHTs) [21] and other specialized techniques [22, 23].

In essence, shared access to an application's state is often provided via ad-hoc, and usually cumbersome, third-party mechanisms. These mechanisms are generally not *managed* by a SPS and, therefore, not integrated with any stream processing language. These shortcomings hinder the ability to debug, visualize, and profile state management operations and, as a result, the entire application.

Clearly, deeper integration of these types of mechanisms via application programming interfaces and language extensions, making them a *managed resource*, can be very beneficial. First, it provides an intuitive method for streaming algorithms to make use of distributed data structures, freeing the algorithm from explicitly dealing with out-of-core state. Second, it can simplify the integration between the producers of the distributed state and its consumers, whether they are part of the same application or not.

Furthermore, *managed* distributed shared state can supply semantics that are more flexible [23, 24, 25] than the *atomicity, consistency, isolation,* and *durability* (or ACID semantics) [26] typically provided by relational databases. Indeed, specializations can be made along each of these four axes, based on a better understanding of the application use cases.

Hence, we expect that additional research will identify language extensions and data management techniques that can simplify the creation and management of large amounts of distributed state, by better integrating them with SPSs and SPAs.

Big data

The term *big data* is used to denote a dynamic or static data set whose scale makes it hard to use conventional databases for storage, processing, and data analysis.

In many ways, stream processing and its ability to process data *online* supports the needs of continuous *big data* analysis. Furthermore, stream processing analytics and software infrastructure have also been shown to work well when used to process vast quantities of *offline* data [27, 28, 29].

Nevertheless, in many cases, massive data warehouses must still be maintained, either because we want to analyze and query older data in conjunction with more recent additions, or because we want to make use of newer analytical techniques that were not available when some of the older data was collected. The research challenges in this space are many, particularly at the intersection between offline and online data analysis.

These challenges include support for a more efficient integration between SPSs and *big data* software platforms such as Apache Hadoop [30], and repositories hosted by distributed file systems including the Apache Hadoop Distributed File System (HDFS). Additional challenges are driven by the need to develop techniques for enabling *efficient* and *scalable* processing of continuously incoming data in conjunction with previously stored data.

In general, we expect that new research will lead to techniques that are more appropriate for stream processing in distributed file systems, indexing, data partitioning and declustering, parallel I/O, and query planning and scheduling.

13.2.4   Analytics

In Chapters 10 and 11 we described the stream data mining process, including the steps of data acquisition, pre-processing and transformation, as well as modeling and evaluation.

Among the challenges that remain to be tackled in the data pre-processing and transformation steps, we highlight the design and implementation of algorithms that expose accuracy–complexity tradeoffs so they can be effectively exploited by applications. In fact, improving the ability to take advantage of these tradeoffs is one of the ways in which the growing scalability and diversity needs of applications can be addressed.

Similarly, the modeling and evaluation steps can benefit from further improvements in offline learning and online scoring techniques that make use of concept drift detection and ensemble learning [31], allowing these techniques to be effectively used by developers to improve their data analysis tasks.

Certain methodological advancements could also be extremely fruitful as a means to improve the use of sophisticated analytics in stream processing. For example, when designing a stream data mining task for a SPA, it is not immediately clear which combination of offline and online analysis will work best, in general, nor will work best it clear that the *best* combination can be automatically determined for every problem.

When it comes to online learning (Chapter 11), improvements in existing techniques as well as the development of additional single pass algorithms and incremental strategies for model building are essential to enable accurate, adaptive, faster, and computationally cheaper analytics.

Beyond algorithmic issues, there are also several software engineering aspects that must be considered. While stream programming languages like SPL support the extensions required for implementing different types of analytic algorithms from scratch, reusing and adapting existing legacy analytics, as well as making use of functionality provided by external analytic platforms and applications are both desirable capabilities.

For example, a developer might want to leverage algorithms and functionality available in software tools including MATLAB [1], R [5], SPSS [32], SAS [33], and Weka [34]. As we mentioned earlier in this chapter, integrating a SPA with these software packages and platforms imposes additional requirements on versioning, provisioning, maintenance, dependency checking, and debugging.

Furthermore, the ability to dynamically exchange models across different platforms and applications can also be useful in weaving together stream processing and non-stream processing data analysis applications.

Finally, with the emergence of *big data*, it becomes necessary to switch from pure data analysis to an *augmented* analytical process that also carries out data exploration. In this case, the goal is to *automatically* discover information of interest from streams and large repositories containing both structured and unstructured data.

The challenge in this case consists of the development of techniques that enable *automated* data exploration or, in other words, that provide the capability to locate and extract novel and relevant pieces of information, along with the contextualization of such information. Also, as is the case with the applications we have studied, these newer techniques must also be designed with dynamic adaptation in mind so they can keep up with the changes in the data, as the external world evolves.

The development of these techniques will fundamentally shift the focus from an intervening human analyst, to an intelligent system that can autonomically steer the data analysis. In essence, rather than requiring human analysts to carefully ask the *right* questions and, subsequently, attempt to cope with the resulting flood of information, the aim is to switch to algorithms and applications that automatically find, investigate, and act on *anything* that looks *interesting* and/or *surprising*.

13.3   Where do we go from here?

Stream processing has seen a substantial amount of progress in its theoretical foundations as well as in the engineering of language and distributed SPSs to support continuous data processing. Yet, many of the ways in which this technology can be used are still unexplored and there are several different avenues of research and engineering that can be pursued to evolve and expand it.

The widespread use of public and private computing clouds, the instrumentation of larger swaths of the computing infrastructure and of the many physical systems that must be monitored and analyzed only exacerbates the need for continuous processing.

Moreover, in many corporate environments, a multitude of existing systems must still be transformed from *store and process* applications to continuous processing systems.

This transformation will unlock more value from the data they accumulate and process. As a result, both the automated and the human-driven decision-making process can be improved, reliably shortening the cycle between *sensing* and *responding*.

We believe that the stream processing paradigm will also find its way into *personal* applications. The ubiquity of networked personal devices such as smart phones, tablets, as well as of many household devices, coupled with the abundance of sources that continuously generate data of personal interest, including social network sites, geo-location services, and private subscriptions to news alerts is likely to lead to applications that cater to each one of us, individually.

In many ways, the existing technological framework already provides all of the basic building blocks necessary to implement continuous and complex data analysis at wire speeds. In the near future, the central challenges in making effective use of these applications are twofold. First, there is a mismatch between what these continuous application can extract and output, and what humans can absorb and act on. Second, there is the issue of fine tuning the degree of autonomy that these applications can have, to automatically adjust their analytical parts in response to environment, system, or resource variations.

In both cases, techniques to *intelligently* mine, represent, summarize and rank the continuously extracted knowledge, and to identify new and unexpected information will become increasingly important.

In essence, we expect fundamental breakthroughs in two areas. First, in user-experience, where techniques will be devised to prioritize, index, and convey the most relevant insights, based on the context and goals of the decision maker.

Second, they will occur in analytics, where techniques will be devised to automatically *suggest* and *employ* innovative ways of analyzing the incoming data to extract new insights from relationships that might not be initially apparent to human analysts.

An exciting future lies ahead!

References

[1] Mathworks MATLAB; retrieved in April 2012. http://www.mathworks.com/.
[2] Palm W. Introduction to MATLAB for Engineers. McGraw Hill; 2010.
[3] Wolfram Research – Mathematica; retrieved in April 2012. http://www.wolfram.com/.
[4] Wolfram S. The Mathematica Book – Version 4. Cambridge University Press; 1999.
[5] The R Project for Statistical Computing; retrieved in April 2012. http://www.r-project.org/.
[6] Everitt B. A Handbook of Statistical Analyses Using R. 2nd edn. Chapman & Hall and CRC Press; 2009.
[7] Simplified Wrapper and Interface Generator; retrieved in April 2012. http://www.swig.org/.
[8] Cohn R, Russel J. SWIG. VSD; 2012.
[9] Gordon R. Essential JNI: Java Native Interface. Prentice Hall; 1998.

[10] Press WH, Flannery BP, Teukolsky SA, Vetterling WT. Numerical Recipes: The Art of Scientific Computing. Cambridge University Press; 1992.

[11] Streams Exchange; retrieved in May 2012. `https://www.ibm.com/developer works/mydeveloperworks/groups/service/html/communityview? communityUuid=d4e7dc8d-0efb-44ff-9a82-897202a3021e`.

[12] Christensen E, Curbera F, Meredith G, Weerawarana S. Web Services Description Language (WSDL) 1.1. World Wide Web Consortium (W3C); 2001. `http://www.w3.org/TR/wsdl`.

[13] Apache Maven; retrieved in April 2012. `http://maven.apache.org/`.

[14] Sun Microsystems. RPC: Remote Procedure Call Protocol Specification Version 2. The Internet Engineering Task Force (IETF); 1988. RFC 1050.

[15] Gropp W, Lusk E, Skjellum A. Using MPI: Portable Parallel Programming with Message-Passing Interface. MIT Press; 1999.

[16] Booth D, Haas H, McCabe F, Newcomer E, Champion M, Ferris C, *et al.* Web Services Architecture – W3C Working Group Note. World Wide Web Consortium (W3C); 2004. `http://www.w3.org/TR/ws-arch/`.

[17] The Object Management Group (OMG), Corba; retrieved in September 2010. `http://www.corba.org/`.

[18] Schneider S, Andrade H, Gedik B, Biem A, Wu KL. Elastic scaling of data parallel operators in stream processing. In: Proceedings of the IEEE International Conference on Parallel and Distributed Processing Systems (IPDPS); 2009. pp. 1–12.

[19] Wolf J, Bansal N, Hildrum K, Parekh S, Rajan D, Wagle R, et al. SODA: An optimizing scheduler for large-scale stream-based distributed computer systems. In: Proceedings of the ACM/IFIP/USENIX International Middleware Conference (Middleware). Leuven, Belgium; 2008. p. 306–325.

[20] Wolf J, Khandekar R, Hildrum K, Parekh S, Rajan D, Wu KL, et al. COLA: Optimizing stream processing applications via graph partitioning. In: Proceedings of the ACM/IFIP/USENIX International Middleware Conference (Middleware). Urbana, IL; 2009. pp. 308–327.

[21] Stoica I, Morris R, Karger D, Kaashoek F, Hari. Chord: A scalable peer-to-peer lookup protocol for internet applications. In: Proceedings of the ACM International Conference on the Applications, Technologies, Architectures, and Protocols for Computer Communication (SIGCOMM). San Diego, CA; 2001. p. 149–160.

[22] memcached – A Distributed Memory Object Caching System; retrieved in April 2012. `http://memcached.org/`.

[23] Losa G, Kumar V, Andrade H, Gedik B, Hirzel M, Soulé R, et al. Language and system support for efficient state sharing in distributed stream processing systems. In: Proceedings of the ACM International Conference on Distributed Event Based Systems (DEBS). Berlin, Germany; 2012.

[24] Fox A, Gribble SD, Chawathe Y, Brewer EA, Gauthier P. Cluster-based scalable network services. In: Proceedings of Symposium on Operating System Principles (SOSP). Saint Malo, France; 1997. pp. 78–91.

[25] Shen K, Yang T, Chu L. Clustering support and replication management for scalable network services. IEEE Transactions on Parallel and Distributed Systems (TPDS). 2003;14(11):1168–1179.

[26] Gray J. Transaction Processing: Concepts and Techniques. Morgan Kaufmann; 1992.

[27] Beynon M, Ferreira R, Kurc T, Sussman A, Saltz J. DataCutter: middleware for filtering very large scientific datasets on archival storage systems. In: Proceedings of the IEEE Symposium on Mass Storage Systems (MSS). College Park, MD; 2000. pp. 119–134.

[28] Ferreira R, Moon B, Humphries J, Sussman A, Miller R, DeMarzo A. The virtual microscope. In: Proceedings of the AMIA Annual Fall Symposium. Nashville, TN; 1997. pp. 449–453.

[29] Kumar V, Andrade H, Gedik B, Wu KL. DEDUCE: At the intersection of MapReduce and stream processing. In: Proceedings of the International Conference on Extending Database Technology (EDBT). Lausanne, Switzerland; 2010. pp. 657–662.

[30] Apache Hadoop; retrieved in March 2011. http://hadoop.apache.org/.

[31] Polikar R. Ensemble based systems in decision making. IEEE Circuits and Systems Magazine. 2006;6(3):21–45.

[32] IBM SPSS Modeler; retrieved in March 2011. http://www.spss.com/software/modeler/.

[33] SAS Analytics; retrieved in June 2012. http://www.sas.com/technologies/analytics/.

[34] Weka Data Mining in Java; retrieved in December 2010. http://www.cs.waikato.ac.nz/ml/weka/.

Keywords and identifiers index

Index

Printed in the United States
By Bookmasters